Outstanding Contributions to Logic

Volume 31

Outstanding Contributions to Logic puts focus on important advances in modern logical research. Each volume is devoted to a major contribution by an eminent logician. The series will cover contributions to logic broadly conceived, including philosophical and mathematical logic, logic in computer science, and the application of logic in linguistics, economics, psychology, and other specialized areas of study.

A typical volume of Outstanding Contributions to Logic contains:

- A short scientific autobiography by the logician to whom the volume is devoted
- The volume editor's introduction. This is a survey that puts the logician's contributions in context, discusses its importance and shows how it connects with related work by other scholars
- The main part of the book will consist of a series of chapters by different scholars that analyze, develop or constructively criticize the logician's work
- Response to the comments, by the logician to whom the volume is devoted
- A bibliography of the logician's publications Outstanding Contributions to Logic is published by Springer as part of the Studia Logica Library.

This book series, is also a sister series to Trends in Logic and Logic in Asia: Studia Logica Library. All books are published simultaneously in print and online. This book series is indexed in SCOPUS.

Proposals for new volumes are welcome. They should be sent to the editor-in-chief sven-ove.hansson@abe.kth.se

Sophia Arbeiter · Juliette Kennedy
Editors

The Philosophy of Penelope Maddy

Springer

Editors
Sophia Arbeiter
Department of Philosophy
University of Pittsburgh
Pittsburgh, PA, USA

Juliette Kennedy
Department of Mathematics and Statistics
University of Helsinki
Helsinki, Finland

ISSN 2211-2758 ISSN 2211-2766 (electronic)
Outstanding Contributions to Logic
ISBN 978-3-031-58424-4 ISBN 978-3-031-58425-1 (eBook)
https://doi.org/10.1007/978-3-031-58425-1

This Springer imprint is published by the registered company Springer Nature Switzerland AG
The registered company address is: Gewerbestrasse 11, 6330 Cham, Switzerland

Introduction

Penelope Maddy was trained as a Mathematician, a Logician, a Set Theorist, as well as a Philosopher. So how are we to place her work in the philosophical landscape? *Second Philosophy* comes immediately to mind. A Second Philosopher is not a first philosopher, someone who places a priori methodology at the center of her inquiry. Descartes, for example, sought a general theory of the world, of existence, knowledge, and truth on the basis of a priori considerations—which science is then built upon. A Second Philosopher reverses this direction of inquiry, granting a priori considerations a place only *in light of* and *compatible with* empirical inquiry. Is Maddy therewith an inheritor of the Logical Empiricists? A Quinean Naturalist? An Ordinary Language Philosopher? Here is her most recent answer to these questions:

> Within the limits of contemporary taxonomies, Second Philosophy would have to count as a version of methodological naturalism, usually glossed as the assumption that the only reliable way of finding things out is to employ the method of science. The trouble with this, it seems to me, is that 'scientific method' is a moving target; any attempt to pin it down would artificially limit the crucial, ongoing processes of expansion and correction. (Maddy, 2022, p. 1)

Second philosophical inquiry, then, begins with science—but what counts as science? Maddy doesn't provide us with an exact criterion, in fact she thinks of the various candidate definitions of the scientific method on the table, e.g. "the empirical method of acquiring knowledge" or "observation, hypothesis, experiment, analysis", as insufficient. This rejection says a lot about Maddy, who reminds us over and over again in her writing that the history of science cannot be summarized according to any such seemingly broad candidate definition, and that it would be wrong to expect future science to be any different. In consequence, Maddy's work is rich in historical examples and contextual analyses of actual scientific research. At the same time, Maddy also avoids giving a contextualist account of what science is, to wit: whatever counts as scientific method at a certain time constitutes the only reliable way of finding things out at that time.

Maddy introduces a main character—a *heroine*—in her work:

> I introduce a highly idealized inquirer called the Second Philosopher and characterize Second Philosophy as the answers she gives to traditionally philosophical questions. (2022, p. 2)

The Second Philosopher has various idealized properties: infinite time and curiosity, and a constitution that allows her to do nothing but pursue her questions. The Second Philosopher is fascinated by the world she finds herself in, and seeks to improve her beliefs through experimentation and via theory-construction—what we might call "the scientific method". Second philosophy encompasses the methods of inquiry used by Maddy's idealized heroine, as well as what she believes and asserts based on those. Consequently, Second Philosophy is not presented as another theory, a fixed set of beliefs, or a set of propositions to be affirmed. What the Second Philosopher believes is up to the actual inquiry, and cannot be given a priori.

If only we were like the second philosopher! Yet we aren't. Maddy is aware from the history of science that the best researchers have been biased in their interpretations, limited by their resources and restricted by their milieu. Thus, in a certain sense, Second Philosophy is unavailable to us. What the Second Philosopher believes about the world is what we can only *hope* to believe, if we were more perfect. What is the point, then, of introducing such an idealized character into philosophy? Why does Maddy prefer a characterization of Second Philosophy as based on this heroine, instead of a more complicated characterization of the scientific method? The idealized Second Philosopher is after all a *person* and has only human properties, just in a perfected form. Clearly, the foundation of any scientific practice for human inquirers plays a fundamental role for Maddy. *We* are engaged in inquiry, and our abilities and constraints shape what the scientific practice ends up being. *We* can improve, expand and correct our inquiry.

The second philosopher is above all *curious*: observing her world with fascination and awe, questioning why things are a certain way, and seeking to understand to the best of her abilities why they are that way. Curiosity is a human trait that is never expanded upon in Maddy's project, it's a given of Second Philosophy. Thus Maddy builds her philosophy around an inquirer who is like us—insofar as she is curious, experiments, and constructs theories—and unlike us in her perfection. Hence, Maddy's general point of view has a strongly humanistic foundation and is unique in the philosophical landscape.

As Maddy observes herself, Second Philosophy does not fit contemporary taxonomies well. Nevertheless we can briefly note three connections to 20th century philosophy in the background of Maddy's thought. First, the perspective of Wittgenstein's *Tractatus*, which finds resonance in Maddy's work. As does Maddy, Wittgenstein writes from the perspective of a unique and ideal inquirer, who studies how languages reaches "up to the world" and what the logical structure of thought is. He never characterizes this inquirer explicitly, but Wittgenstein's inquirer shares features with the Second Philosopher: in both cases philosophy is identified with the thoughts of this unique and ideal inquirer and cannot be packaged into theses. They also share an interest in the limits of philosophy, expressed in the *Tractatus* by the claim that the correct method in philosophy would be to only utter propositions of natural science (§6.53), and by Maddy as a commitment to pursuing her questions guided by the methods and results of science. The Logical Empiricists embraced this commitment, rejecting a unique faculty of philosophical insight and having little tolerance for certain (metaphysical) questions of philosophy. Maddy agrees. Hence a debt to Logical Empiricism, as espoused, for example, by Schlick, Carnap and Neurath, is undeniable. Finally, Maddy's heroine can also be viewed as a certain elaborate version of Austin's "Plain Man"—the person who perceives tables and chairs like the ordinary man on the street, for example, and judges accordingly that they exist. Just as for Austin, for Maddy common sense and ordinary judgments are to be embraced, at least at the beginning of inquiry. Indeed, the Second Philosopher ought to put any theory that conflicts with ordinary judgment under extraordinary scrutiny. In her rejection of many philosophical questions, Maddy inherits much from Ordinary Language Philosophy.

Maddy, then, is squarely engaged with the questions and framework of 20th century philosophy—from early Wittgenstein, to the Logical Empiricists, to Ordinary Language Philosophy, and to Quine. Nevertheless, it would be a mistake to see Maddy as addressing only 20th century concerns. While Maddy engages strongly with the questions and theories of that period, her unique method of *Second Philosophy* has the potential to transcend these limitations. In her view, she isn't the first to advocate for such an approach. 17th- and 18th-century thinkers—Descartes, Boyle, Locke, Newton, Berkeley and Reid—were, like her, *practitioners*, i.e. both scientists and philosophers. For Maddy, Second Philosophy is their rightful heir:

> The early modern notions and concerns live on in what we now think of as interdisciplinary work—in the foundations of mathematics, physics, biology, vision theory, consciousness studies, and more. (2022, p. 4)

In the early modern period the division between the disciplines of philosophy and science wasn't what it is now. From the perspective of Second Philosophy our present situation constitutes a misstep, in that while for instance Locke's philosophical work was informed by the best science of his time, the work of many contemporary philosophers of mind, for example, are not so informed. In contrast, the Second Philosopher has studied all that we know about the human brain from a scientific perspective, and only makes philosophical claims based on that knowledge. Thus for Maddy, Second Philosophy is *the* inheritor of this Early Modern tradition, based as it is on a unified understanding of science and philosophy. Insofar as there is a line through the history of philosophy that not just appreciates our best scientific understanding of the world, but is engaged primarily with science, it is Second Philosophy.

No area of Maddy's work manifests this commitment to science better than her work in the foundations of mathematics and set theory. Maddy inherited a concern for the foundations of mathematics from Frege and Hilbert, Carnap and Quine, among others. She then turned her fire on the foundations of set theory—set theory in its post-forcing phase, that is—and indeed was one of the first philosophers, if not the very first, to do so. The 20th century debate on the status of mathematical and set-theoretic truths is of course a vigorous one. Initially a realist on mathematical truth, Maddy the naturalist embraced the *difficult* case: an unquestionably scientific domain, namely set theory, which is seemingly untethered from the empirical domain—biting the bullet, as it were, given the empiricist strain in her work. For set theory, the Second Philosopher should strive towards

> a version of realism that genuinely accounts for the nature of set-theoretic language and practice, that respects the actual structure of set-theoretic justifications. (2011, p. 61)

Thus, the Second Philosopher starts by observing what set theorists are doing, what they achieved, and what they assume (the science), and only then considers what kind of philosophical justification is required. The central aim here is nothing less than the formulation of a naturalist justification of the higher infinite.

For the naturalist, this is new territory. The Quinean Naturalist rejects higher set theory, again citing its lack—as of this writing—of empirical applications. Maddy takes a more prudent view of the matter. For her, it would be imprudent to dismiss higher set theory:

> Now it might be tempting to suggest that this unfettered contemporary mathematics be sequestered in some way (...). The trouble is that bits of unfettered pure mathematics have turned out to be profoundly applicable—and there's no way to predict ahead of time which these will be (...). I think no honest observer, even one primarily interested in mathematics only for the sake of application, would think it prudent to rein it in. (2011, p. 95)

But how to naturalize higher set theory? The Second Philosopher is keen to sink her teeth into applications, but as things stand now large cardinals, forcing axioms and the like are only beginning to seep into core mathematics—albeit at an increasing pace. As for the possibility of a *bivalence* in set theory, there is a clear plurality of models of ZFC, witnessing any number of important but mutually inconsistent propositions, and seemingly little in the way of grounds for deciding among them—again, as of this writing. Maddy's essay for this volume "On Multiverism" lays out a nuanced position here: sovereign with respect to the various multiversist/universist views lately on the table, "unencumbered by metaphysical delicacies". There would be a great deal to say about Maddy's meticulous argument in a longer format, but what is also especially notable in her answer to this question here is the forefronting of mathematical depth in her monograph *Defending the Axioms*—an important concept, relative to mathematical practice, for which the philosopher of set theory has had, until now, no language. Once again Maddy embraces the difficult move, toward depth.

Second Philosophy is not mainstream in contemporary philosophy departments. Many understand philosophy in terms of an argumentative defense of a philosophical *thesis*, a stance Maddy rejects. Others may work on philosophical problems that Maddy considers non-problems, or that she refuses to provide an answer to. Still others think that genuine *philosophical* knowledge is possible, a type of knowledge that is separate from scientific knowledge—but this Maddy also rejects. In consequence, the Second Philosopher is silent on the many bread and butter issues of philosophy. For some, her silence is frustrating. For them, to say, "My view is that most compatible with science" is to give in to philosophical quiescence. Despite these differences Maddy never gives up on her peers: her papers and books always start off with mainstream questions, for instance, how to justify the axioms of set theory, or whether there are mathematical objects, or whether we have access to an external world. Then at some stage, her heroine, the Second Philosopher, enters the stage, and approaches the question from the empirical standpoint. In this sense, any philosopher, even those most skeptical of Maddy's method, has an easy way into her work. She can always start with her own (mainstream) questions and see how Maddy's response unfolds.

Besides such critics, there are plenty of philosophers who share her outlook on the connection between science and philosophy, even if they don't assume the label "Second Philosophy". For them, Maddy's principled understanding of philosophy *second* to science is the correct view. Some of their voices are collected in this book.

Penelope Maddy's philosophical work spans an unparalleled breadth, from the foundations of mathematics and set theory, to the philosophy of logic, to the history and philosophy of science, to historical work on Descartes, Kant, Wittgenstein, and Quine. Thus, this collection must by necessity encompass papers touching on only a small part of her work. The format of the volume is paper followed by Maddy's response. The selection of papers here is personal, reflecting the interests of the editors. (Indeed other editors could have drawn on a completely different group of contributors!) Rather than present or interpret each contribution in this introduction, we leave this pleasure to the reader. We make one exception in commenting on one of the contributions coming directly out of mathematical practice, written by the model- and set theorist Maryanthe Malliaris with Assaf Peretz. The authentic voice of the working mathematician is meant to honor the perspective of Maddy's work, rooted as it is in philosophical critique, but also rooted in the deepest regions of mathematics itself. We are also pleased to include three contemporary conversations on the state of set theory (taking place in 2023). The first is between Penelope Maddy and Menachem Magidor; the second is between David Asperó and Corey Switzer; the third is the transcription of a late-night group conversation between Hugh Woodin and a large group of participants at the Arctic Set Theory Workshop VI in Kilpisjärvi, Finland in February 2023. We are extremely grateful to Jeffrey R. Schatz, Corey Switzer, and Beau Mount, respectively, for their heroic work in creating transcriptions of these conversations. We placed Maddy's essay "On Multiversism" in this section, to be considered as firmly part of the conversation. Going forward, we very much look forward to Maddy's further reactions to the issues raised in these conversations, as set theory's magnificent future unfolds.

(Selected) Bibliography

Maddy, P, 1988a. "Believing the Axioms. I", *The Journal of Symbolic Logic*, Vol. 53, No. 2, pp. 481–511.

- 1988b. "Believing the Axioms. II", *The Journal of Symbolic Logic*, Vol. 53, No. 3, pp. 736–764.
- 1990. *Realism in Mathematics*, Oxford: OUP.
- 1997. *Naturalism in Mathematics*, Oxford: OUP.
- 2007. *Second Philosophy: A naturalistic Method*, Oxford: OUP.

– 2011. *Defending the Axioms: On the philosophical Foundations of Set-Theory*, Oxford: OUP.
– 2014. *The Logical Must: Wittgenstein on Logic*, Oxford: OUP.
– 2016. *What do Philosophers do? Skepticism and the Practice of Philosophy*, Oxford: OUP.
– 2020. (with Toby Meadows) "A reconstruction of steel's multiverse project", *Bulletin of Symbolic Logic* 26(2): 118–169.
– 2022. *A Plea for Natural Philosophy, And Other Essays*, Oxford: OUP.
– 2023. (with Jouko Väänänen) *Philosophical uses of Categoricity Arguments, Elements in the Philosophy of Mathematics*, Cambridge: CUP.

Sophia Arbeiter
Juliette Kennedy

Contents

Contributors

Carolin Antos Konstanz, Germany

John T. Baldwin Department of Mathematics, Statistics and Computer Science, University of Illinois Chicago, Chicago, IL, USA

Patricia Blanchette Department of Philosophy, University of Notre Dame, Notre Dame, IN, USA

John P. Burgess Department of Philosophy, Princeton University, Princeton, NJ, USA

Annalisa Coliva University of California, Irvine, CA, USA

Curtis Franks University of Notre Dame, South Bend, IN, USA

Deborah Kant Hamburg, Germany

Juliette Kennedy Department of Mathematics and Statistics, University of Helsinki, Helsinki, Finland

Mary Leng Department of Philosophy, University of York, Heslington, UK

Øystein Linnebo University of Oslo, Oslo, Norway

Penelope Maddy Department of Logic and Philosophy of Science, University of California, Irvine, CA, USA

Maryanthe Malliaris University of Chicago, Chicago, USA

Donald A. Martin University of California, Los Angeles, CA, USA

Beau Madison Mount Faculty of Philosophy, Oxford University, Oxford, UK

Alexander Paseau University of Oxford, Oxford, UK

Assaf Peretz New York City, NY, USA

Jeffrey R. Schatz Yale University, New Haven, CT, USA

Gila Sher University of California, San Diego, CA, USA

Zeynep Soysal University of Rochester, Rochester, NY, USA

John R. Steel University of California, Berkeley, USA

Corey Bacal Switzer Institut für Mathematik, Kurt Gödel Research Center, Universität Wien, Wien, Austria

Mark Wilson University of Pittsburgh, Pittsburgh, USA

Chapter 1
Intellectual Autobiography

Penelope Maddy

When I was introduced to the joys of set theory as a high schooler and learned of the Continuum Hypothesis, I was captivated: 'surely such an obvious and natural question must have an answer! But even the fundamental assumptions of mathematics aren't enough to settle it one way or the other. What in the world would a solution even look like?' My curiosity about these matters led me to the Berkeley math department for undergraduate study, a perfect time and place for anyone interested in the foundations of set theory. Tarski was running the Group in Logic and Methodology of Science, new large cardinal axioms were being posted on office doors, and Dana Scott had recently proved that the existence of a measurable cardinal was enough to guarantee that V isn't equal to L. This was a welcome result—no one thought V = L should be true—but I found myself asking: 'why this assumption? Why should we think there are measurable cardinals?' Beyond this—though large cardinalswere known not to settle CH—one could hope that another type of new axiom might.

The mathematical challenges of this program were obvious, but I was struck by another seemingly intractable question: what counts as a justification or a criticism of an axiom candidate? Mathematics proceeds by proof, but the starting points for proofs can't be proved. What could count as evidence one way or the other? What form could this type of evidence possibly take?

It's clear in retrospect that I was already well down the slippery slope from mathematics to philosophy, but at the time, it was a shock when Princeton responded to my application for graduate study in mathematics by admitting me to the program in history and philosophy of science. I assumed that HPS must be something like L&M—we worked with much less information in those days—so off I went.[1] As it

[1] Not quite immediately, though. First I spent some time in Bristol on a Marshall Scholarship, studying with Stephan Körner, one of the attendees at Wittgenstein's lectures in the 1930s.

P. Maddy (✉)
Department of Logic and Philosophy of Science, University of California, Irvine, CA, USA
e-mail: pjmaddy@uci.edu

S. Arbeiter and J. Kennedy (eds.), *The Philosophy of Penelope Maddy*,
Outstanding Contributions to Logic 31, https://doi.org/10.1007/978-3-031-58425-1_1

turned out, HPS wasn't at all like L&M; being in HPS was nearly indistinguishable from being in the graduate program in philosophy. I was in for a few rocky years of acclimatizing and re-tooling, but one stark lesson stood out: the philosophers I met tended to think that CH was a non-question; it isn't provable from ZFC and it isn't disprovable from ZFC, and that's all there is to know. The whole of my experience in the math department at Berkeley had taught me that this couldn't be right. Eventually, I took it as my mission to devise a firm philosophical ground for the legitimacy of CH so that my target question could be posed: how are axioms properly defended?

Gödel's well-known platonism was expressly designed to do this job—to establish CH as a legitimate question—but it was widely ridiculed in that philosophical climate, infused as it was with Benacerraf's famous epistemological challenge.[2] Proponents of the then-popular causal theories of knowledge and reference on which Benacerraf's argument is based often remarked that no one causally interacts with a set. My advisor, John Burgess, observed in response that no one causally interacts with a physical object, either, only with a time slice of its front surface. This got me to wondering about what makes an interaction with a time slice of a front surface into a perception of a physical object and whether something like this process, whatever it was, might be co-opted to argue for perception of sets of physical objects. And that question sent me off to the science library to see what the developmentalists and neuroscientists had to say.

This approach came so automatically to me that I didn't know what to make of it when one of my dissertation readers remarked uneasily on my mixing of metaphysical and epistemological questions with empirical psychology. Shrugging this off in the ignorance of youth, I took myself to have replaced Gödel's mysterious 'intuition' of sets with a scientifically respectable account of our ordinary perception of sets of ordinary physical objects. Then, as Gödel had described, the more theoretical parts of set theory were to be confirmed in relation to these intuitive parts, just as theoretical scientific claims are confirmed in relation to sense experience. The challenge, then, was to exploit this analogy to connect the confirmation of axioms with confirmation in the physical sciences.

After graduate school, I spent a few years at Notre Dame and a few more at the University of Illinois at Chicago (then 'Chicago Circle'), working on two parallel projects. The first was the next step on the central question: if axiom candidates are confirmed and disconfirmed by a process that runs parallel to theoretical science, then the rationality of arguments offered for and against should be assessed on this scientific standard. So I set out on collecting and cataloging those arguments, building on material in the dissertation and eventually spending time with the Cabal group of set theorists at UCLA and its environs. It was a remarkable group, as anyone familiar with the recent history of the subject can attest.[3] The result was 'Believing the axioms' (1988).

---

[2] Benacerraf (1973). It's always worth noting that Benacerraf didn't take himself to have refuted Platonism, only to have presented a challenge.

[3] Tony Martin was especially generous with his time and insight, for which I'm forever grateful.

The other project of those years returned to defending the legitimacy of CH.[4] I took myself to have responded to one line of anti-realist thought, rooted in Benacerraf, but remained troubled by Wittgenstein's complaints about mathematical realism as 'hocus-pocus', 'mathematical alchemy', and 'a sickness of our time'.[5] In the mid-80s, I tried hard to puzzle out where this hostility of Wittgenstein's was coming from and how to respond. This grew into a book-length manuscript, but there was a problem: early on, I sketched the set-theoretic realism from my dissertation as a foil for Wittgenstein's attack, intending to explore how this particular version could respond to his critique, but readers kept asking for more details of this realism until, as one astute advisor put it, 'the manuscript seems to be growing a second head'. In the end, by which time I'd reached my permanent home at UC Irvine, that extra head became *Realism in Mathematics* (1990). The residual material on Wittgenstein was relegated to a couple of articles (1984, 1986)—just as well, given that the philosopher at issue, I now realize, wasn't Wittgenstein at all, but Kripkenstein, an interpretive cul de sac I managed to escape a few years later.[6]

At this point, a neo-Gödelian realism had been defended at length, and the arguments offered in the literature and around the water cooler for and against new axiom candidates had been cataloged, so it was time to make good on the promise of defending the rationality of our axiom choices by showing that they're based on confirmation and disconfirmation in a sense analogous to confirmation and disconfirmation in natural science. Unfortunately, as I plugged diligently along on this quest, doubts began to creep in.

For one, some of the arguments I'd catalogued—arguments that my days among the mathematicians led me to regard as good arguments—some of those apparently good arguments didn't seem, if I was honest, to fit the mold of scientific evidence. Some could be construed to fit—like Martin's examples of 'prediction and confirmation' (Martin 1998)—but it also seemed to me perfectly reasonable, for example, to count the fact that large cardinal axioms provide a valuable measure of consistency strength as a point in favor of adding as many of them as are likely to be consistent. For that matter, why not include Foundation among our axioms, given that it allows the universe to be conveniently separated into ranks and doesn't appear to rule out anything we want? Why don't its calming effects on the behavior of transfinite cardinals count in favor of Choice, or its generation of an ordinal for every well-ordering in favor of Replacement?

But—and here's the problem—if the world of sets is objective in some sense parallel to the world of physical science, wouldn't adding an axiom because it has mathematically attractive features be as much wishful thinking as rejecting quantum mechanics because it's a mess? My set-theoretic realism—along with many other views now grouped together as forms of what I've come to call Robust Realism —adds what looks to be an inappropriate extra-mathematical condition: we might

[4] There was also the theory of proper classes (1983, 2000b) discussed in my exchange with Linnebo in this volume.

[5] Wittgenstein (1933/44), V.16, II.22, and II.23, respectively.

[6] See (1993) and subsequent writings on Wittgenstein.

have overwhelming mathematical reasons for liking a given axiom candidate, but we won't be justified in adding it to our list of fundamental assumptions unless we can also show that it satisfies the terms of my neo-Gödelian epistemology. This objection is obviously epistemic, but it's distinct from and prior to Benacerraf's concern: he observes that it's hard to see how we could come to know that an axiom is true; I suggest that it's inappropriate even to ask for more than the good mathematical reasons.

That was one worry. Another was that the case for my realism rested in part on Quinean indispensability arguments. This hadn't bothered me at the beginning, when I saw myself as trying to convince skeptical philosophers that CH remains a legitimate question and those philosophers all thought indispensability considerations to be decisive for ontology; I thought of this as a kind of philosophical jujitsu, using their own beliefs as leverage against them. But now that I was defending my own realism as a guide to the proper justification of axioms, I had to warrant those considerations myself, and the more I thought about it, the more uncomfortable I became. In my complacency, I'd been prepared to support my belief in the determinateness of CH by tracing a line in the air and declaring this to be a continuum, in which CH was either true or false. But is that spatial line really a continuum?

This question led me to realize that the use of continuum mathematics in science most often occurs in the context of idealizations, simplifications, and other useful falsehoods. The best candidate for a literal continuum seemed to be spacetime, so I turned to the physics—and found continuity to be an open question at the heart of the problem of quantum gravity. For that matter, the attitude of physicists toward their mathematics, and even toward the physical assumptions that go along with their use of that mathematics (like the continuity of spacetime), didn't seem tuned to any ontological ramifications: they didn't hesitate to add whatever mathematical machinery seemed useful and didn't take the success of the resulting theory to confirm either the existence of the abstracta involved or the truth of the attendant physical assumptions. In short, indispensability as a guide to ontology didn't hold up under scrutiny.

The third and final straw was a slow-dawning meta-philosophical realization. There I was, the realism I'd devised to legitimate the continuum problem in shambles. Could it be that the line I'd confidently drawn in space wasn't continuous after all? Could it be that continuous spacetime was no more than an effective mathematical model of a physical reality whose underlying structure we have yet to grasp (a view that a little digging revealed to be entertained by both Dedekind and Einstein)? Could it be that my decades-long confidence that CH has a determinate truth value was misplaced? With realism discredited, where could I turn for a new metaphysical perspective to shed light on these matters?

In this anxious moment, I was either saved or led fatally astray, depending on your viewpoint, by a pair of striking passages from Wittgenstein's *Philosophical Investigations* (Wittgenstein 1953). One comes in the course of his musings on the philosophy of psychology:

> The first step is the one that altogether escapes notice. We talk of processes and states, and leave their nature undecided. … But that's just what commits us to a particular way of looking at the matter. … the decisive move in the conjuring trick has been made, and it was the very one that seemed to us quite innocent. (*PI* 308)

Could it be, I thought, that I was making such a mistake when I assumed that what I needed was a replacement metaphysics? This brought to mind another passage, this one explicitly meta-philosophical. Wittgenstein points out that various philosophical debates arise from reactions that don't occur to us in ordinary life, that our ordinary practices don't in fact suffer from the difficulties philosophers imagine. What's gone wrong?

> If I'm inclined to suppose that a mouse comes into being by spontaneous generation out of grey rags and dust, it's a good idea to examine those rags very closely … But if I am convinced that a mouse cannot come into being from these things, then this investigation will perhaps be superfluous. But what it is in philosophy that resists such an examination of details, we have yet to come to understand. (*PI* 52)

Some unnoticed presupposition keeps the philosopher from examining the ordinary considerations and coming to appreciate their force. And there I was, having just collected a wealth of ordinary considerations, straight from the practice itself, but somehow imagining that I needed something else, a metaphysics, to support the search for new axioms!

Naturalistic philosophers have long cautioned that philosophy can't properly critique the effective methods of science on extra-scientific grounds, and yet there I was, thinking that philosophy was needed to support the methods of set theory. But if philosophy can't criticize from outside, it also can't support! This forehead-slapping realization was the final blow to my old realism and to the methodological relevance of metaphysics more generally.

The result was *Naturalism in Mathematics* (1997), where I argue that matters of truth and abstract ontology are irrelevant to the proper evaluation of mathematical methods and turn instead to careful examination of the gray rags and dust of the practice itself. My thought is that set theory has various mathematical goals and that opting for (or against) an axiom candidate on the basis of its effectiveness (or not) toward those goals is just ordinary means/ends reasoning. One such goal is to provide a 'foundation' (in purely mathematical terms[7]) for classical mathematics, a goal that councils us to develop a theory as generous as possible, so as not to curtail pure mathematics (this is Maximize), and as unified as possible, so as to provide univocal answers to questions of proof and existence (this is Unify). With these tools, $V = L$ can be shown to be restrictive, contradicting the admonition to Maximize, so its wide-spread rejection as an axiom is perfectly rational on purely mathematical grounds.

That $V = L$ would be a bad axiom isn't a surprising result, just confirmation of what most people already believed, but I regarded this as proof-of-concept: a substantive case against an axiom candidate could be mounted strictly in terms of mathematical

[7] See (1997), §II.2, or more recently (2017b) or (2019a).

goals and effective ways of achieving them. A mouse *could* be generated from the gray rags and dust; our presumption otherwise was a prejudice. This is nowhere near a solution for the CH, but it does go some way toward answering the prior question of what a solution might look like. If a new axiom or method had broad mathematical advantages and no disadvantages (with the exception, perhaps, of a tolerable risk of inconsistency that comes with increased consistency strength), and if this new axiom or method were to prove or disprove CH, could this count as a solution?[8]

Since I'd set philosophical matters of truth and ontology aside for methodological purposes in *Naturalism*, some readers concluded that I regarded these traditional issues in the philosophy of mathematics as pseudo-questions of some kind. This was never my position[9]: a naturalist is out to understand all aspects of the world and our place in it, so the nature of human mathematical activity is one of the many questions she faces. To address these questions, I realized that I had to back up to (at least) two broader issues before zeroing back in on the status of set-theoretic discourse. One was the nature of logic. The other was 'naturalism' itself, a term so elastic these days as to have outlived its usefulness. I'd been relying on Quinean naturalism as a point of departure for my mathematical naturalism, but as my disagreements with Quine mounted (about holism, about mathematics, about logic, about skepticism), it became clear that I had to be more explicit about my understanding of what the term entails. To emphasize that I was defining my own version, it seemed best to start fresh with a new coinage: Second Philosophy.

Familiar taxonomies would classify Second Philosophy as broadly methodological: the best, or perhaps only, way of finding out about the world and our place in it is empirical investigation, application of the scientific method. The trouble with this characterization is that the open-ended, ever-improving character of science makes it impossible to formulate a contentful criterion for what counts as 'scientific' (the famously hopeless demarcation problem). I figured the best way around this was to introduce a character, the Second Philosopher,[10] an idealized inquirer who begins with ordinary observation, progresses to generalizations, then controlled experimentation, theory formulation and confirmation, always circling back to correct her beliefs and improve her methods as she goes. I offer no argument that her way is the best or only way; I simply describe her approach to various traditional philosophical questions and some of her conclusions (with the implicit challenge to the resisting philosopher, 'what's wrong with that?'). Because she has no fixed distinction between science and non-science, her reaction to challenges like those of Descartes or Kant isn't to declare them 'unscientific', but to evaluate their various claims with open mind. (Unsurprisingly, they rarely pass muster from her perspective.) Along the way, she provides a straightforwardly realistic account of rudimentary logic as contingent

[8] This isn't what Hamkins (2015) would call a 'dream solution', because he requires that the new axiom be intrinsically justified.

[9] As is explicit in the book itself (see 1997, pp. 200–203).

[10] Called the 'Plain Inquirer' in (2017a), on analogy with Austin's Plain Man.

on very general structural features of the world.[11,12] Though it didn't occur to me at the time, in retrospect, this account of logic can be viewed, roughly, as what remains of the account of the perception of sets of physical objects from *Realism* when the empirical psychology is updated and the flawed support of indispensability considerations is removed.

This laid the groundwork for a return to the traditional metaphysical and epistemological questions about mathematics in *Defending the Axioms* (2011). The discussion there begins with the Robust Realism mentioned above, a general category that includes Gödel's realism and my own earlier set-theoretic realism, as well as a wide range of other views (e.g., Shapiro's *ante rem* structuralism (Shapiro 1997) and Hellman's modal-structuralism (Hellman 1989). A Robust Realist posits an objective subject matter for set theory that requires a non-trivial epistemology. The epistemological feature is more obvious in Gödel's intuition and my naturalized substitute, but Shapiro requires that a structural description be 'coherent' and Hellman that it be 'modally possible', where both these are effectively second-order (for example, only one of ZFC + CH and ZFC + ~ CH can qualify). Often this non-trivial epistemology is modeled on an analogy between confirmation in mathematics and confirmation in natural science, just the sort of thing I lost faith in after *Realism* when contemplating actual examples like those cited above. All versions of Robust Realism draw the same challenge as my own did: why aren't the purely mathematical considerations enough?

The mistake, I argue, is that Robust Realism puts the metaphysics first: such-and-such is the subject matter of set theory, the objective content we're out to track. The methods we actually use in practice are then assessed by their ability to do this and come up short: why isn't our preference for axioms with welcome mathematical consequences just wishful thinking? The moral of *Naturalism* is that the actual methods come first; the metaphysics, if there is one, had better match. So, the question becomes: what must sets be like that we can find out about them in these ways? The trivial answer is that sets just are the kind of thing that these methods track. This answer is given by the Thin Realist, who takes sets to be objective and abstract (nothing our methods tell us suggests that they depend on us or have spatiotemporal location) and offers only the trivial epistemology embodied in its metaphysics (sets just are ...).

So, because of what sets are, the Thin Realist can successfully investigate them using intra-mathematical means/ends analyses like those uncovered in *Naturalism*. It might seem, then, that sets are mere reflections of our set-theoretic methods, perhaps constituted by those methods, but I argue that, in fact, this isn't so, that they are entrees into strains of important or fruitful or objectively good mathematics—for which I use the term 'deep'. It follows that our methods manage to track these sets,

[11] Full classical logic then results from a pair of idealizations: the law of excluded middle and the material conditional. In (2014a), this account of logic ((2007), Part III, (2014c, 2018b)) is compared and contrasted with Wittgenstein's positions on logic, early and late, finding more common ground than might be expected.

[12] Other aspects of the second-philosophical stance are explored in (2007, 2017a, 2022h), and elsewhere.

when they do, not because sets are constituted by those methods, but because the ends in our means/ends arguments, the goals we pursue, are themselves deep, leading to deep mathematics. The notion of mathematical depth comes up in some of my replies, so for now I just note that it's what gives Thin Realism its anchor, its objective content.

But then again maybe it's pointless to insist on any metaphysics at all, even of the Thin variety. Maybe set theory isn't in the business of discovering truths; maybe it's just developing a theory of sets to do various mathematical jobs. This I call Arealism. The key insight is that the Thin Realist and the Arealist produce exactly the same set theory: their practice is guided by exactly the same mathematical considerations; both are responsible to the same facts of depth. This means that there's no practical difference between them, but I go further, arguing that there's no difference at all, no fact of the matter about which of them is right. Where one says 'these are good reasons to adopt this axiom', the other says 'these are good reasons to think this axiom is true'; where one says 'we may never be find compelling mathematical support for an axiom that settles CH' the other says 'CH is either true or false, but we may never be able to figure out which'—and the choice between these formulations, I claim, is a matter for free decision. Nothing in our ordinary uses of the words 'true' or 'exist' dictates one way or the other; we're free to adopt either idiom, whichever feels more natural, or even to go back and forth, depending on the context.

This line of thought has an important corollary. We've seen that we needn't worry, for example, over whether a mathematically welcome large cardinal *really* exists or the corresponding axiom is *really* true as the Robust Realist would have it. Their mathematical credentials are enough; our assessment should focus there and there alone. Another way to put this is that extrinsic considerations are what does the justifying (whether as 'confirmation of truth' for the Thin Realist or as 'compelling reasons to adopt' for the Arealist). Obviously, this reverses the traditional pecking order, which sees intrinsic justifications—a given claim is part of or dictated by the iterative conception of set—as the gold standard and extrinsic justifications as suspect or at best second-rate. From a second-philosophical perspective, the attraction of intrinsic thinking is that it gives us an intuitive picture to guide and stimulate our search for extrinsically beneficial avenues,[13] but an intuitive picture isn't an ontology or a semantics or a subject matter—it's just a psychological heuristic. Anything generated by such a heuristic has to earn its keep (Arealism) or its justification (Thin Realism) on extrinsic grounds, alone. If the mathematical benefits are strong enough, set-theorists will replace one such picture with another, as happened when the mathematical benefits of Choice helped motivate a switch from sets as extensions of concepts, or determined by definitions or rules, to the fully combinatorial iterative conception.

So *Defending the Axioms* completes a coarse-grained, second-philosophical account of the practice of set theory, its proper methods, its subject matter, and

[13] This generates a new perspective on Reinhardt's arguments (see the exchange with Linnebo, this volume).

the connection between them. Altogether, this amounts to a tentative methodological, ontological, and epistemological answer to that plaintive question from high school: what in the world would a solution to CH even look like? Suppose someone were to come up with a new axiom or method, perhaps inspired by new intrinsic considerations, say a new take on the iterative picture, or a new intuitive picture to take its place. Suppose further that this new axiom or method turned out to have many mathematical advantages and no known disadvantages, and suppose finally that it settles CH, one way or the other. My claim is that this would constitute a solution to problem of CH (subject to the usual caveats of the fallibilist), that it would be inappropriate to require in addition that the axiom be shown to be true or the method to be truth-tracking by some external measure. If this is right, then, it seems to me that the case of CH is largely a matter of waiting to see how the mathematics turns out!

Since that book, I've worked on a number of related topics: what a foundation does and why set theory still pulls its foundational weight despite challenges from category theory (2017b, 2019a); the significance of multiverse thinking in set theory (Maddy & Meadows 2020), 'On multiversism', this volume; a revival of an enhanced version of old-style, highly ridiculed if-thenism (really another approach to Arealism, 2022b). More broadly, my attempt to lay out a version of naturalism in *Second Philosophy* spurred an interest in other philosophical approaches—common sense philosophy, ordinary language philosophy, conceptual analysis, therapeutic philosophy—and in external world skepticism, which I'd been using as a diagnostic tool (culminating in 2017a). This in turn deepened my engagement with the history of philosophy, starting with Kant (2007, §§I.4, III.2); then Wittgenstein (2014a, 2017a, 2022e), Moore (2017a, 2022a), Austin (2017a, 2022c), Carnap (2007, §I.5), and Quine (2007, §6) in the analytic tradition; and more recently the early moderns, especially Descartes, Locke, Berkeley, and Reid (2017a, 2022c, 2022d). Looking back, though, I can see that it's all just been methodology—what's the proper way to do this?, first mathematics, then philosophy—and it all grew out of that teenage fascination with the continuum problem.[14]

References

Benacerraf, P. (1973). *Mathematical truth.* Reprinted in [1983] (pp. 403–420).

Benacerraf, P., & Putnam, H. (Eds.). (1983). *Philosophy of mathematics* (2nd Edn.). Cambridge: Cambridge University Press.

Hamkins, J. (2015). Is the dream solution to the continuum hypothesis attainable? *Notre Dame Journal of Formal Logic 56*, 135–145.

Hellman, G. (1989). *Mathematics without numbers.* Oxford: Oxford University Press.

Lycan, W. (2000b). A theory of sets and classes. In G. Sher & R. Tieszen (eds.), *Logic and intuition.* (pp. 299–316). Cambridge: Cambridge University Press.

[14] I'm most grateful to Juliette and Sophia for editing and to the authors of the various chapters for contributing to this volume (the opportunity to engage with them while drafting the replies was a special treat). Many thanks to all!

Maddy, P. (1983). Proper classes. *Journal of Symbolic Logic, 48*, 113–139.
Maddy, P. (1984). How the causal theorist follows a rule. In H. Wettstein et al, (Eds.), *Causation and causal theories, midwest studies in philosophy* (Vol. 9, pp. 457–477). Minneapolis: MN, University of Minnesota Press.
Maddy, P. (1986). Mathematical alchemy. *British Journal for the Philosophy of Science, 37*, 279–314.
Maddy, P. (1988). Believing the axioms. *Journal of Symbolic Logic, 53*(481–511), 736–764.
Maddy, P. (1990). *Realism in mathematics*. Oxford: Oxford University Press.
Maddy, P. (1993). Wittgenstein's anti-philosophy of mathematics. In J. Czermak, & K. Paul (Eds.), *Wittgenstein's philosophy of mathematics*. In *Proceedings of the 15th international Wittgenstein symposium* (pp. 52–72). Vienna: Verlag Hölder-Pichler-Tempsky.
Maddy, P. (1997). *Naturalism in mathematics*. Oxford: Oxford University Press.
Maddy, P. (2000a). Naturalism and the a prior. In P. Boghossian & C. Peacocke (Eds.), *New essays on the a priori* (pp. 92–116). Oxford: Oxford University Press.
Maddy, P. (2007). *Second Philosophy*. Oxford: Oxford University Press.
Maddy, P. (2011). *Defending the axioms*. Oxford: Oxford University Press.
Maddy, P. (2014c). A second philosophy of logic. In P. Rush (Ed.), *The metaphysics of logic* (pp. 93–108). Cambridge: Cambridge University Press. (Reprinted in [2022h], pp. 216–233).
Maddy, P. (2014a). *The logical must*. Oxford: Oxford University Press.
Maddy, P (2017a). *What do philosophers do?* New York: Oxford University Press.
Maddy, P. (2017b). Set-theoretic foundations. In A. Caicedo, J. Cummings, P. Koellner & P. Larson (Eds.), *Foundations of mathematics: Essays in honor of W. Hugh Woodin's 60th birthday* (pp. 289–322). Providence, RI: American Mathematical Society.
Maddy, P. (2018b). Psychology and the a priori sciences. In S. Bangu (Ed.), *Naturalizing logico-mathematical knowledge: Approaches from psychology and cognitive science* (pp. 15–29). New York: Routledge. (Reprinted in [2012h], pp. 237–252).
Maddy, P. (2019a). What do we want a foundation to do? In S. Centrone, D. Kant & D. Sarikaya (Eds.), *Reflections on the foundations of mathematics* (pp. 293–311). Cham Switzerland: Springer.
Maddy, P., & Meadows, T. (2020). A reconstruction of Steel's multiverse project. *Bulletin of Symbolic Logic, 26*, 118–169.
Maddy, P. (2022a). Moore's hands. In [2022h] (pp. 135–147).
Maddy, P. (2022b). Enhanced if-thenism. In [2022h] (pp. 262–293).
Maddy, P. (2022c). *A plea for natural philosophy*. In: [2022h] (pp. 13–48).
Maddy, P. (2022d). Hume and Reid. Reprinted in [2022h] (pp. 93–134). (First appeared in 2011 under the title 'Naturalism and common sense', *Analytic Philosophy, 52*, 481–504).
Maddy, P. (2022e). Wittgenstein on hinges. In [2022h] (pp. 148–170).
Maddy, P. (2022h). *A plea for natural philosophy and other essays*. Oxford: Oxford University Press.
Martin, D.A. (1998). Mathematical evidence. In H.G. Dales & G. Oliveri (Eds.), *Truth in mathematics* (pp. 215–231). Oxford: Oxford University Press.
Shapiro, S. (1997). *Philosophy of mathematics*. New York: Oxford University Press.
Wittgenstein, L. (1933/44). *Remarks on the foundations of mathematics*, revised edition G.H. von Wright, R. Rhees, & G.E.M. Anscombe, eds., G.E.M. Anscombe, trans. MIT Press, 1978.
Wittgenstein, L. (1953). *Philosophical investigations*, revised 4th edition. In G.E.M. Anscombe, P.M.S. Hacker & J. Schulte (Eds.), Malden: MA, Blackwell.

Part I
Philosophy of Set Theory and Mathematics

Chapter 2
Generically Invariant Set Theory

John R. Steel

Abstract $\mathcal{L}_{MV}$ is a sublanguage of the standard language of set theory in which the mathematics based on set-forcing-invariant principles can be carried out, and in which set-forcing-sensitive questions have no obvious formalization. We discuss some philosophical and technical issues related to the expressive power of this language.

Keywords Continuum Hypothesis · Multiverse · Generic multiverse · Quine · Projective

2.1 Introduction

Large cardinal hypotheses have been very successful at removing the incompleteness of ZFC in the realm of statements about concrete objects like natural numbers, real numbers, and simply definable sets of real numbers. On the other hand, nothing like them can decide the Continuum Hypothesis (CH), and many other natural questions about arbitrary sets of reals and objects of still higher type are provably beyond their reach. There is at present no generally agreed upon approach to deciding such questions.

This leads us to ask what it is about the large cardinal hypotheses, or ZFC itself for that matter, that justifies them, and whether there might be some other class of principles, justified similarly or perhaps on different grounds, that would have similar success in the realm of arbitrary sets of reals and beyond. And how would we recognize such principles if we had them in hand? What is it to be a solution to the Continuum Problem?

Penelope Maddy has written extensively on such questions. Indeed, she is one of the first to have written from a philosopher's point of view about the evidentiary standards employed by modern set theorists. Her approach is that of the Naturalist,

J. R. Steel (✉)
University of California, Berkeley, USA
e-mail: coremodel@berkeley.edu

S. Arbeiter and J. Kennedy (eds.), *The Philosophy of Penelope Maddy*,
Outstanding Contributions to Logic 31, https://doi.org/10.1007/978-3-031-58425-1_2

informed by a careful analysis of the evolution of existing theory, and reluctant to use some pre-determined theory of evidence as a reason to reject what set theorists[1] regard as a very successful theory. It is probably not surprising that, as a set theorist, I am sympathetic to her approach. We have exchanged views many times over the years.

It would not be possible to engage here with the whole of Maddy's work in this area, but fortunately she and Toby Meadows have recently published Maddy and Meadows (2020), whose starting point is the semi-philosophical paper Steel (2014) that I wrote in 2014. My plan here is to make some replies to Maddy and Meadows (2020), connecting these to more general aspects of Maddy's work where appropriate. I shall also update the more technical side of Steel (2014), in a way that I hope will be accessible to most readers.

The paper is organized as follows. Section 2.2 is devoted to general philosophical issues in the background. Section 2.3 discusses the twin heuristic principles *Maximize* and *Unify* underlying the search for new axioms in set theory. Section 2.4 reviews some definitions from Steel (2014), and Sect. 2.5 replies to some of the criticisms of its more philosophical parts made by Maddy and Meadows. Section 2.6 looks at the idea that the generic multiverse has a core, elaborating on a discussion in Meadows' paper Meadows (2021) and responding to a criticism made there. Sect. 2.7 describes some recent technical progress related to the the idea that the generic multiverse has a core that resembles HOD in a model of the Axiom of Determinacy, and Sect. 2.8 makes some final remarks.

2.2 Some Philosophy

Maddy has been heavily influenced by Quine, as was I. We have both come to disagree with him, in somewhat different ways. Some of those differences are relevant to the philosophical critique of Steel (2014) in Maddy and Meadows (2020).

2.2.1 *Naturalism and Holism*

Naturalist epistemology takes our existing best theory of everything as its starting point. It rejects the radical re-building projects of "First Philosophy", with its idealized knower who begins from nothing. Our theory and its language have been evolving for a very long time, and although individual humans may make important additions or revisions, no one can start from scratch. One beautiful statement of this attitude is Neurath's well-known boat metaphor:

> Wie Schiffer sind wir, die ihr Schiff auf offener See umbauen müssen, ohne es jemals in einem Dock zerlegen und aus besten Bestandteilen neu errichten zu können.

[1] "Practitioners".

Holism adds the view that this theory does not have a distinguished starting point. The language in which we express it gains meaning, and the theory within that language that we adopt is confirmed, as a whole. Very well-developed parts of it (such as set theory) have the great virtue that we have organized them as axiomatic systems, but even then, the choice between equivalent axiomatizations is somewhat arbitrary.

One should hasten to add that "our existing best theory of everything" is itself an idealization. In even the fairly well-developed area of theoretical physics, what we have now are two very successful theories (General Relativity and Quantum Field Theory) that seem on the surface to be inconsistent with each other. Fortunately, our focus in this paper is mathematics, where (we claim) the superficially inconsistent alternative theories can be unified.[2] The holistic point is simply that our mathematical language gains meaning, and our mathematical theory is confirmed, by the way that it interacts with the rest of what we know.

Holism and naturalism lead to Quine's indispensability argument. To put it briefly: mathematics is essential to the rest of science, so let's not dispense with it. It might be hard to find anyone who would disagree so far; for Quine, this argument was a reason to reject his former nominalism, so it was accompanied by the claim that any consistent nominalist would have to dispense with mathematics. I agree with that, and it seems that perhaps Maddy no longer does[3], but it does not matter so much here. Burgess' paper Burgess (2024) discusses the ontological issue thoroughly, and I agree pretty much completely with his point of view.

It's not so important here because our focus is the epistemology of mathematics, not its ontology. In this context we are dealing with mathematical theories, not mathematical objects; with set theories, not sets. Bringing sets into the discussion prematurely can lead one astray, and in particular into object-oriented formulations of *maximize* which, in my opinion, are mistaken.

Quine went on to say that we *should* dispense with the part of mathematics that is not essential to the rest of science. Quine drew his dividing line rather cavalierly, saying of the 'higher reaches of set theory' that

> We see them as meaningful because they are couched in the same grammar and vocabulary that generate the applied parts of mathematics. We are just sparing ourselves the unnatural gerrymandering of grammar that would be needed to exclude them.

The "we" here is apparently the subset of humanity sufficiently educated to have an opinion, and the passage implies that Quine is putting himself in that category. It is not clear whether he objects to putting up with nonsense in order to be spared the effort involved in avoiding it.[4] Others have tried to draw a dividing line with much

---

[2] It seems that category theory has been successfully interpreted in set theory; see Maddy (2017) and Maddy (2024) for a recent discussion of the relationship between the two. The unification of superficially inconsistent alternative extensions of ZFC is the main topic of Steel (2014).

[3] See Maddy (2011) Chapter IV, which seems to embrace a form of instrumentalism.

[4] See Quine (1990, p. 94). It is ironic that Quine is using the notion of meaning here to make a philosophical claim, as he does when he claims that the "gratuitous flights of higher set theory" are

more care[5], often motivated by constructivist scruples rather than nominalist ones. The line is always well below ZFC.

Needless to say, I disagree. Pure mathematics is inseparable from the rest of mathematics, and mathematics is inseparable from the rest of science. Gödel's 2nd incompleteness theorem shows that the highest reaches of set theory have consequences for the world we directly observe.[6] Those consequences do not affect the subject taught in Physics departments, but consequences like Projective Determinacy do have a bearing on the subject taught in Real Analysis courses. The subject taught in Real Analysis courses is important in Physics and many other areas. Higher set theory also interacts with the more abstract parts of Analysis, Algebra, and Topology.

The idea of dispensing with higher set theory, or "rounding it out"[7] in any old way, may seem to be hard-headed pragmatism, but it is not. Here as elsewhere, powerful unifying principles are preferable to "good enough for now", and it is both important and difficult to find the right ones. Long term pragmatism counsels that while there are basic unsolved problems in pure set theory, we should continue to develop it, regardless of applications. That can only make eventual applications more likely.

The more cogent objection to the work done on large cardinals and their consequences is that it is premature, that it deals with objects and notions so far removed from the rest of mathematics and science that long term pragmatism has no advice to give. The many mutually inconsistent ways of extending ZFC that we have discovered could be taken as evidence of this. What answers this objection is the "one road upward" phenomenon (cf. Steel (2000)): at the level of statements about the concrete (natural numbers, real numbers, and absolutely definable sets of real numbers) we have not found any divergences, but rather many different routes to the same theory.[8] The many different routes to it suggest that this theory will not wash away in the long term.

Maddy also rejects the dispensabilty sequel to the indispensability argument, but we may have some disagreement on its flaws. In Maddy (2007, p. 95) she objects to confirmational holism:

> What's gone wrong with the Quinean picture is confirmational holism. This case suggests that we cannot regard a scientific theory as a homogeneous whole, confirmed as a unit, that a consistent naturalist must recognize different types of evidence and various different roles hypotheses can play in our theorizing.

"on a par rather with uninterpreted systems". Quine (1984, p. 788.) An inability to interpret higher set theory would seem to be a good reason to withhold judgement.

[5] See for example Feferman (1992) and Feferman (1998).

[6] If there is an inaccessible cardinal, then in the next 25 years no contradiction in ZFC will be discovered. This is an experiment that will be performed. Of course, to decide how it turns out we need a proof-checking method; as always, what is confirmed is a theory as a whole.

[7] Quine (1984, p. 788).

[8] This "theory of the concrete" contains all $(\Sigma^2_1)^{\mathrm{Hom}_\infty}$ truths. (See Sect. 2.7.) There are various general theorems that capture aspects of this "one true hierarchy" phenomenon, for example Wadge's Lemma and the related prewellordering of jump operators (cf. Steel (1982)), and the Comparison Lemma of inner model theory. Montalban and Walsh have recently proved an interesting result in this direction at the level of Peano Arithmetic. See Montalban and Walsh (2019).

The case under discussion is the evidence for the existence of atoms and molecules produced by Einstein's explanation of Brownian motion and Perrin's experiments supporting it. I would read the main moral of that case differently, as being that confirmation accumulates in stages, and that having evidence from different directions is important.[9] If anything, this seems to support the holistic view, since Physics and Chemistry were being asked to face the tribunal together.

Maddy continues

> The Second Philosopher sees the evidential relationships of modern science in its many branches as complex and varied, to be studied and assessed in their particular contexts of inquiry, not obviously subject to characterization like 'observation and the hypothetico-deductive method'.

I would put it differently. Observation and the hypothetico-deductive method is fine as a general characterization, it just doesn't take you very far. A careful assessment in the particular context of inquiry, in our case set theory and the foundations of mathematics, is necessary if you hope to say something useful.

I have brought out these perhaps minor differences because Maddy sometimes writes as if pure mathematics in general, and set theory in particular, had somehow become a domain of inquiry separate from the rest of mathematics and science, subject to its own standards of evidence.[10] Pushed too far, this makes set theory into an art form, with the artists and critics being one and the same.

Such a fictionalist view misses crucial connections. The theory of natural numbers, real numbers, and definable sets of real numbers that we get from large cardinal hypotheses is a key part of the justification for believing them. Those concrete objects show up everywhere in mathematics and its applications. The theory of them we get from large cardinal hypotheses is the only rational route we have toward going beyond ZFC in this realm of the concrete. This is not because we have only investigated a few different ways of extending ZFC, it is because each of the many ways we have investigated is either limiting, or leads to this road.

In this holist view, our mathematical beliefs are ultimately justified by by the central role they play in the whole of our conceptual system. In a sense, all the evidence for them is extrinsic. This may seem paradoxical; do we really believe $2 + 2 = 4$ because of its consequences?

For the holist, this is an illformed question, because our belief that $2 + 2 = 4$ cannot be isolated from our other beliefs. One must ask why we believe, or should believe, some reasonably large set of connected propositions. One cannot believe $2 + 2 = 4$ except as part of some such complex.

---

[9] I would certainly agree that an epistemolgist should "recognize different types of evidence and various different roles hypotheses can play in our theorizing." In the epistemology of set theory, Maddy has gone well beyond Quine in that regard.

[10] For example: " my naturalist takes mathematics to be independent of both first philosophy and natural science (including naturalized philosophy that is continuous with science)–, in short, from any external standard. Maddy (1997, p. 184). Also "she [the second philosopher] sees no opening for the familiar tools of that [scientific] perspective to provide supports, correctives, or supplements to the actual justificatory practices of set theory." Maddy (2011, p. 55). I may well be reading more into these passages than is intended.

Moving to that level, the question is still ambiguous. Are we asking the causal question or the justificatory one? And who is the "we"? If the question is the causal one, and the "we" refers to individual human beings alive today, then all sorts of mathematical beliefs are forced upon us as being true, by our biology, our teachers, or random factors. If "we" refers to humanity as a whole, there is at least 70,000 years of linguistic evolution[11], preceded by the evolution of neural structures over many millions of years, that have produced our core mathematical beliefs. They are indeed inevitable, at least for the foreseeable future. At this level, the causes seem more closely connected to the justification. Humans evolved this conceptual system because it does certain jobs well. But if we try to justify some extension of that system, we must say what those jobs are, and why the expanded system would do them better. This seems to be the realm of extrinsic evidence.

In the context of evaluating new axioms for set theory, the interesting question is "why should we adopt them?", the "we" in question is humanity as a whole, and the important evidence given for or against will be extrinsic. The immediate appeal of some axiom can only suggest it as a candidate for more consideration.

The naturalist idea of looking at the history of mathematical belief acquisition as a guide to what standards may be appropriate now does have merit. There is quite a lot to look at there, most of it belonging to prehistory. Fortunately, the part most relevant to evaluating new axioms for set theory belongs to the last 100–200 years. Moreover, the theories we are considering now can be rigorously formalized, so that our discussion of them can be informed by real proofs of various claims. Metamathematics is possible, and it is essential to the epistemology of set theory. One might think of the process of strengthening the current foundations of set theory as linguistic evolution gone self-conscious.

2.2.2 Meaning Skepticism

One of the central criticisms of Steel (2014) by Maddy and Meadows has to do with its use of the notion of *meaning*. I shall make some general comments here, and more specific replies in Sect. 2.5.

Of course, philosophers have theorized about the meaning of mathematical statements a lot. The notion has been mis-used. In particular, the logical positivists' use of the claim that true mathematical statements are true solely in virtue of their meaning was very strongly criticized by Quine in his well known debate with Carnap, and in my opinion most of the criticism was just.

Nevertheless, it is pretty much impossible to discuss the epistemology of set theory without bringing the meaning of statements in the language of set theory into the picture. What we are evaluating are *interpreted* theories. How they are interpreted, and translated into one another, is important.

---

[11] See https://en.wikipedia.org/wiki/Blombos_Cave.

In the context of new axioms, it can be hard to distinguish sharpening the meaning of the language from discovering new truths. Perhaps there is some analog of the uncertainty principle that says this can never be fully done. But we do make translations, and they can be good or bad. We do it in daily life, in science, and in mathematics. We can define meaning operationally as that which is preserved by a good translation.

The reason that ZFC counts as a foundation for mathematics is that we can translate all mathematical language into the language of set theory, and prove the translated theorems from the axioms of ZFC. Taking over a term used by Enderton, Maddy calls this translation an "embedding", but "translation" or "interpretation" seem more accurate to me.[12]

Philosophers have written extensively on the notion of meaning, and used it in various contexts. It's in the job description of an *analytic* philosopher. Paraphrasing Hilbert, one might say that depriving an analytic philosopher of the notion of meaning is like depriving a boxer of the use of his fists. One can hope that something in this philosophical literature is relevant to the possibility of meaning indeterminacy in the language of set theory.[13]

But one need not have a theory of a notion in order to employ it. Steel (2014) is based on the idea that we do in real life identify and remove meaning indeterminacy, by either trimming back some piece of syntax, or fleshing out (further determining) the meaning of some existing syntax. There are episodes in the history of science that seem to fit this pattern.[14] The inclusion of the Axiom of Extensionality in ZFC can be seen as meaning clarification, in that it abandons the too-vague notion of property in favor of extensions. Several of the other axioms signal that one has abandoned the notion of set as division of all existing objects into members and non-members in favor of the iterative concept. Perhaps it will be useful to think of the resolution of CH as an analogous process. From an Olympian point of view, it is clear that the meaning humanity now assigns to the syntax of the language of set theory must have evolved over time, because at one time homo sapiens did not speak anything like the language of set theory, and its use of the language must have evolved in stages. We don't need a philosophical theory of meaning clarification to see it at work.

Let me contrast this view with the discussion in D. A. Martin's paper Martin (2018) of the possibility that what he calls *the concept of the sets* does not determine a truth value for CH. I think the main difference is that Martin's *concept of the sets* is not the same as what I am calling the meaning we currently assign to the syntax of $\mathcal{L}_{\in}$. Martin's *concept of the sets* does not seem to evolve over time, or depend on humans.[15] In Martin's view, understanding or "grasp" of the concept evolves, and perhaps that current understanding would correspond to what I am calling the

---

[12] See Maddy (2017, p. 290).

[13] Meaning holism seems closest to the notion of meaning that would be relevant as an underpinning for the philosophical parts of Steel (2014).

[14] For example, in Special Relativity, one abandons talk of an absolute time ordering, then fleshes out the meaning of "A happened before B" using the speed of light and the Minkowski metric.

[15] Martin says that he stands in between Gödel and Feferman on the nature of mathematical concepts, and that "it is more correct to say that they were discovered than that they were created by us".

meaning currently assigned to LST. Perhaps what I will describe as removing a meaning indeterminacy in $\mathcal{L}_\in$ would be described by Martin as improving our grasp on the concept of the sets by discovering a particularly basic new truth.

Martin describes his sense of *the concept of the sets* in Martin (2005), where it functions as a boundary condition on universes of sets[16] that instantiate it. As Martin puts it,

> When we work on the mathematical subject of set theory, we can think of what we are doing as finding out what is implied by the concept, what has to be true of any instance of the concept.

Martin argues that if neither CH nor its negation is implied by the concept of the sets, then this concept is not instantiated. Nevertheless, set theory can go on, because it is really about what the sets "would be like" if the concept were instantiated. As fas as I can see, the counterfactual doesn't add anything (Martin also leans to that view), so the conclusion should be that instantiation is irrelevant, and that set theory is about what is implied by the concept of the sets.

This makes it hard to find a distinction between ϕ and "ϕ is implied by the concept of the sets". If there is only one *concept of the sets*, then it seems to be functioning as an ideal meaning for the language of set theory. For the naturalist epistemologist, actual humanity and the way it actually uses language are a more promising place to start.

2.3 Maximize and Unify

In Maddy (1997) Maddy identifies two heuristic principles guiding the development of set theory, and calls them *Maximize* and *Unify*. Her statement of Maximize in Maddy (1997) is

> The set theoretic arena in which mathematics is modelled should be as generous as possible; the set theoretic axioms from which mathematical theorems are proved should be as powerful and fruitful as possible.

I agree with the sentiment, and could certainly endorse everything after the semicolon. The more metaphorical first clause leads in the wrong direction, in my opinion. The metaphor is elaborated in her statement of *Generous Arena*, which seems to have supplanted *Maximize* in the discussion of Maddy (2017):

> Set theory's universe, V, provides the **Generous Arena** in which all this takes place, and that's why the 'final court' condition takes the form it does: to be a full participant in mathematical interaction, a so-and-so must appear along-side of the full range of its fellows, with all the tools of construction and interaction fully available.

Martin (2018, p. 2). So he does not rule out Feferman's view that they are social constructs at this point.

[16] Alternatively, "concepts in the straightforward sense".

The more object-oriented parts of these formulations have led Maddy to attempt formalizations of *Maximize* based on the idea of a theory "providing an isomorphism type".[17] The attempted formalizations led to attempts to define *fair interpretation* formally: "ϕ is a fair interpretation of T in T' (where T extends ZFC) iff (i) T' shows ϕ is an inner model, and (ii) for all $\sigma \in T$, T' proves σ^{ϕ}." An inner model is taken to be a transitive class, perhaps cut off at some inaccessible cardinal. There are counterexamples to the attempts in Maddy (1997) and Maddy (1998) to define *Maximize* based on this notion of "fair interpretation".[18] In my opinion, the root of the problem is the idea that one should compare T with T' by considering the "sets they refer to", their "ontologies".

Talk of "the ontology of T" involves quantifying into a modal context. To compare "the ontology of T" to that of S, you need to translate the language of T into that of S. That's where the action is. Moreover, a good translation need not map referring expressions to referring expressions. In the set theory context, T might be interpreted by the user of S as the theory of what is forced in some partial order, for example.[19]

The idea of maximizing sets, rather than set theories, is sometimes given as a rationale for adopting forcing axioms like MM. I find it hard to see the power in such arguments. Why couldn't one argue that CH implies there is a wellorder of $\mathbb{R}$ of length ω_1, while MM implies there is no such thing, so CH implies there are "more sets of reals" than does MM? Moreover, if one takes the fact about the forcing relation that $M[G]$ has more sets than M as a guide to theory choice, isn't one led to "all sets are countable"? We end up with a theory (second order arithmetic) that is much weaker than ZFC in the sense that really matters, namely interpretative power.[20]

Woodin once argued for the axiom (*), now known to be a consequence of MM^{++}[21], on the basis that it maximizes the Π_2 theory of H_{ω_2} in a certain sense.[22] The maximization here is at the level of theories, not sets, but the order we are maximizing in is just inclusion. In other words, the translations have to map Π_2 sentences about H_{ω_2} to themselves. This is not the right notion of maximization, the one that is inspired by the role of set theory as a universal foundation.[23]

---

[17] Cf. (Maddy 1997, Chap. 6) and Maddy (1998).

[18] See Maddy (1997, p. 225). There are further counterexamples in Hamkins (2014).

[19] The Bennaceraf problems arise from the demand that a translation that is well defined on sentences be well defined on formulae.

[20] Hamkins' paper Hamkins (2014) seems to advocate "all sets are countable" on precisely these grounds. He then attempts to recover logical strength via the instrumentalist dodge (cf. Steel (2000)). One symptom of the problem is his claim (Hamkins, 2014, p.19) that the map $t(\psi) =$ "there is a transitive model of $\mathsf{ZFC} + \psi$" is a translation. This is not true; $t(\psi \wedge \psi)$ is not equivalent to $t(\psi) \wedge t(\phi)$. You need to pick a transitive model that is independent of ψ, and there is no way to do that. The fundamental problem is that the confused notion of "more sets" is being used in a context where "more logical strength" is what is appropriate.

[21] See Aspero and Schindler (2021).

[22] See for example Woodin (2001).

[23] Axiom (*) implies $\neg$CH, but a theory that maximizes the Σ^2_1 theory of the reals in the inclusion order must include CH.

The notion of fair interpretation that seems appropriate in a general formulation of *Maximize* is the the notion of a good translation. I doubt that this can be formally defined, but we do recognize good and bad translations in practice.

This leads to *maximize interpretative power* as the guiding heuristic for the discovery and evaluation of new axioms in set theory. Our foundational theory of sets should be such that all mathematical theories can be interpreted into it in a way that preserves their meaning. In this formulation, *Maximize* applies to set theories, rather than sets, and this seems more appropriate for an epistemological principle. *Maximize* is a principle of rationality, a broad rule of evidence, not a law of nature. It is parallel to *Unify* in that respect.[24]

To stay at the level of theories, rather than sets, is appropriate for a discussion of what theory one should believe, and of the extent to which believing one alternative conflicts with believing another. As a set theorist who believes some quite strong set existence principles, I am a realist. But set theory and the epistemology of set theory are two different subjects. The fundamental questions of epistemology have to do with meaning, evidence, and belief. In our case, the questions are "what would it be for humanity to add Axiom X to ZFC; how would we know that had happened?", and "what reasons could legitimately be given for or against doing that?". The first has to do with meaning: in an old slogan, the meaning is the use, and to adopt the axiom is to use it in a certain way. The second has to do with evidence. Of course, one can short-circuit a discussion of evidence by saying "adopt it if and only if it is true", but that goes nowhere. One hopes that it is possible to do better.

Maximizing interpretative power captures the central role that large cardinal hypotheses play in extending ZFC. They are our source of interpretative power. They are our most important tool for climbing the consistency strength hierarchy, or equivalently, generating new Π_1^0 truths. As we do that, we generate more complicated truths about the concrete in an orderly fashion. The apparent wellordering of natural consistency strengths is mirrored by the inclusion order on their canonical inner models, and natural models for all these natural theories can be built from the canonical inner models by forcing. This is how set theorists convince themselves that such theories are consistent.[25, 26]

Maddy states *Unify* in Maddy (1997) (pp. 208–209) as follows:

> One methodological consequence of adopting the foundational goal is immediate: if your goal is to provide a single system in which all objects and structures of mathematics can be modelled or instantiated, then you must aim for a single fundamental theory of sets. This methodological goal is just the flip side to one of the common objections to set theoretic foundations…

---

[24] Jeffrey Schatz's thesis Schatz (2019) discusses and compares the two versions of *Maximize*.

[25] There are some qualifiers omitted here. The most important is that this picture has as of now been only partly verified, and there are many fundamental open problems to do with filling it in a lower levels and extending to higher ones.

[26] The consistency strength hierarchy and its higher reaches do not seem to show up naturally in the language of category theory. That gives the language of set theory an advantage as an ultimate foundational language, a language into which all other mathematical language can be interpreted.

At this point she quotes MacLane

> ...Cohen's method of forcing ...leads to the construction of many alternative models of set theory. Another result is the introduction of a considerable variety of axioms meant to supplement ZF ...for these reasons 'set' turns out to have many meanings, so that the purported foundation of all Mathematics upon set theory totters. (MacLane, 1986, 358–9).

Maddy then continues

> We arrive at the methodological goal UNIFY by running this argument in reverse. If set theorists were not motivated by a maxim of this sort, there would be no pressure to settle CH, to decide the questions of descriptive set theory, or to choose between alternative new axiom candidates; it would be enough to consider a multitude of alternative set theories.

Unify is not directly stated here. In Maddy (2017) and Maddy (2024) it seems to be replaced by the combination of *Shared Standard* and *Generous Arena*. She says for example (Maddy, 2024, p. 13)

> We need to bear in mind that the cash value of 'these things exist in V' is just 'the existence of (surrogates for) these things can be proved from the axioms of set theory'—a straightforward manifestation of set theory's role as a Shared Standard of proof. To say that 'the universe of sets is the ontology of mathematics' amounts to claiming that the axioms of set theory imply the existence of (surrogates for) all the entities of classical mathematics— a simple affirmation of set theory's role as Generous Arena.

Maddy goes on to warn against figurative ontological talk, but in my opinion the paragraph immediately above has not gone far enough in that direction. The notion of an "entity of classical mathematics" brings in the objects in a discussion that should be devoted to theories.

What MacLane and others miss is how interrelated the many different extensions of ZFC that we have investigated are. The common theory of the concrete that is generated by climbing the consistency strength hierarchy is a dramatic manifestation of this interconnectedness. MacLane, and perhaps Maddy at points, miss this because they do not regard the interpretations given by forcing and inner models to count as unification. But they are. They enable set theorists to use each other's work. The multiverse language and theory are an attempt to make this underlying unity more visible, but set theorists are already well aware of it, and they make use of it.[27]

Hamkins misses this unity in a different way. If one strips away the Platonist imagery, his advice to "embrace as real" all universes in his multiverse is equivalent to the formalist's advice to investigate all ("interesting") consistent theories, without any overall framework relating the results of one investigation to those of another. One might say that MacLane sees a mess, and shies away, whereas Hamkins sees the unity of the generic multiverse, but then obscures it with a mess.[28]

---

[27] The paper Blue (2024) by Douglas Blue contains many examples of proofs that use the existence of generic extensions of V with some property to deduce conclusions about V itself.

[28] Concerning the ontological question, if we are using $\mathcal{L}_\in$ in the standard way, the answer to "how many universes of sets are there?" was provided by Cantor and Russell: none. Set theorists occasionally talk about proper classes, but as I understand that talk, one can translate it into $\mathcal{L}_\in$ with its standard interpretation.

2.4 Generically Invariant Set Theory

Large cardinal hypotheses are cofinal in the part of the interpretability hierarchy we know about. But, like ZFC itself, they are set-forcing-invariant, so they cannot decide CH and the many other statements that are not set-forcing-invariant.

This makes it natural to look for a sublanguage of $\mathcal{L}_\in$ in which the mathematics based on set-forcing-invariant principles can be carried out, and in which set-forcing-sensitive questions have no obvious formalization. This is done in Steel (2014), by introducing a *multiverse language* $\mathcal{L}_{MV}$ and an open-ended *multiverse theory* MV stated in that language. $\mathcal{L}_{MV}$ employs the usual syntax of $\mathcal{L}_\in$, but with two sorts, for the *worlds* and for the *sets*. Informally stated, the axioms of MV are:
Axioms of MV*:*

$(1)_\varphi$ φ^W, for every world W. (For each axiom φ of ZFC.)
(2) (a) Every world is a transitive proper class. An object is a set just in case it belongs to some world.
(b) If W is a world and $\mathbb{P} \in W$ is a poset, then there is a world of the form $W[G]$, where G is $\mathbb{P}$-generic over W.
(c) If U is a world, and $U = W[G]$, where G is $\mathbb{P}$-generic over W, then W is a world.
(d) (Amalgamation.) If U and W are worlds, then there are G, H set generic over them such that $W[G] = U[H]$.

A theorem of Laver shows that Axiom (2)(c) can be stated in $\mathcal{L}_{MV}$.[29]

The natural way to get a model of MV is as follows. Let M be a transitive model of ZFC, and let G be M-generic for $\mathrm{Col}(\omega, < \mathrm{OR}^M)$.[30] The worlds of the multiverse M^G are all those W such that

$$W[H] = M[G \restriction \alpha],$$

for some H set generic over W, and some $\alpha \in \mathrm{OR}^M$.

Laver's theorem implies that the full first order theory of M^G is independent of G, and present in M, uniformly over all M. That is, there is a recursive translation function t such that whenever M is a model of ZFC and G is $\mathrm{Col}(\omega, < \mathrm{OR}^M)$-generic over M, then

$$M^G \models \varphi \Leftrightarrow M \models t(\varphi),$$

for all sentences φ of the multiverse language. $t(\varphi)$ just says "φ is true in some (equivalently all) multiverse(s) obtained from me".

---

[29] See Laver (2007). The result was re-discovered by Woodin.

[30] One might assume that M is countable, so that generics actually exist. We shall only be interested in various first order facts about M and M^G, so Lowenheim-Skolem lets us eliminate this assumption in various well-known ways.

If $\mathcal{W}$ is a model of MV, then for any world $M \in \mathcal{W}$, there is a G such that $\mathcal{W} = M^G$. Thus assuming MV indicates that we are using the multiverse language as a sublanguage of the standard one, in such a way that t is a correct translation. Also, it is clear that if φ is any sentence in the multiverse language, then MV proves

$$\varphi \Leftrightarrow \text{ for all worlds } M, t(\varphi)^M \Leftrightarrow \text{ for some world } M, t(\varphi)^M.$$

Thus everything that can be said in the multiverse language can be said using just one world-quantifier.

One can add large cardinal hypotheses that are preserved by small forcing to the theory MV as follows: given such a large cardinal hypothesis φ, we add "φ^W, for all worlds W" to MV. By adding large cardinal hypotheses to MV this way, we get as theorems "for all worlds W, φ^W", for any φ in the theory of the concrete they generate.

The $\mathcal{L}_\in$ sentence CH is not generically invariant, and hence not provably-in-ZFC equivalent to any sentence in ran(t). There is no obvious way to translate CH into the multiverse language.

One might call MV and its extensions in $\mathcal{L}_{MV}$ *generically invariant set theory.* One can find more thorough explanations and proofs of the elementary facts about it in Maddy and Meadows (2020).

$\mathcal{L}_{MV}$ is motivated by the idea that one might remove a meaning-indeterminacy in $\mathcal{L}_\in$ by "trimming back", that is, restricting it to a sublanguage. $\mathcal{L}_{MV}$ and its interpretation in LST via t are a convenient way to do that. It is somewhat unfortunate that *multiverse* has the connotations it does. $\mathcal{L}_{MV}$ is meant to be an artificial language, with no meaning except that given by the translation t. It does not go beyond $\mathcal{L}_\in$ in expressive power, it drops back within it. Whether that is a proper drop is the main question about it; perhaps we trimmed too much. "More universes" does not mean a stronger language or theory, it means a potentially weaker one.

2.5 Some Replies to Maddy and Meadows

Let me address some specific questions and critiques raised in Maddy and Meadows (2020). The first concerns the motivation for MV and its language.

> Why wouldn't it be sufficient to isolate a set of axioms that captures this central idea well enough to generate a mathematically successful theory, even if it wasn't complete for some natural collection of toy models? Without a satisfactory answer to this question, we have no reason to adopt the axiomatizabilty requirement, and we're left without a principled argument for Amalgamation.

The answer here is that my goal was to capture an existing successful theory, generically invariant set theory, not invent a new one. The existing theory has been developed in $\mathcal{L}_\in$, but $\mathcal{L}_{MV}$ and its translation t into $\mathcal{L}_\in$ seem useful in isolating it. Amalgamation is true under this translation. It records the intention to be translatable via t into $\mathcal{L}_\in$. Amalgamation is not meant to be an independent insight into the nature of multiverses. It clarifies that you want to be understood as speaking a certain sublanguage of $\mathcal{L}_\in$.

The Maddy-Meadows passage seems to say that there could have been some way to assign meaning to $\mathcal{L}_{MV}$ beyond translating it into $\mathcal{L}_\in$. I don't think that is true. I doubt anyone is going to find mathematical concepts that cannot be expressed in $\mathcal{L}_\in$ any time soon. I think that any candidate for a new concept should be regarded as suspect until it can be formulated in $\mathcal{L}_\in$, especially if there is no language community with an established agreement on how the new concept is used. I myself do not feel capable of going beyond what can be expressed in $\mathcal{L}_\in$.

This point answers some related objections:

> It seems to us that the gloss 'settled/unsettled by the current meaning' is a problematic fit for the role of defective/virtuous just described. First, assuming $\mathcal{L}_{MV}$ sentences enjoy this virtue, it isn't obvious that a translation would preserve it.

A bad translation might not preserve the virtue, of course, but t is guaranteed to be a perfect translation, because it is the only way $\mathcal{L}_{MV}$ sentences have been given any meaning at all.

> Whatever being 'settled by the current meaning' comes to, it seems that there might well be a sentence of $\mathcal{L}_\in$ that translates to a synonymous sentence of $\mathcal{L}_N$ (the language of arthmetic), where the former is settled by the current meaning of $\mathcal{L}_\in$ but the latter is unsettled by the current meaning of $\mathcal{L}_N$ (e.g. a strong consistency statement).

Here the authors seem to be identifying "current meaning" with what is provable in some formal theory. I definitely do not want to do that. I would say that, as of now, there is no meaning-indeterminacy visible in $\mathcal{L}_N$.

> Why should the the meaning currently assigned to $\mathcal{L}_{MV}$ do any better at settling all sentences of $\mathcal{L}_{MV}$ than the meaning currently assigned to $\mathcal{L}_\in$ does at settling all sentences of $\mathcal{L}_\in$?

The answer here is that it cannot do any worse, and if there is meaning indeterminacy remaining in $\mathcal{L}_{MV}$, then we should deal with it.

The central question about the multiverse language is whether our trimming has gone too far. Do we lose expressive power if we stay within it? There are two possible answers: yes and no. In Steel (2014), "no" is elaborated as

Weak relativist thesis: (WRT) Every proposition that can be expressed in the standard language $\mathcal{L}_\in$ can be expressed in the multiverse language.

Thinking of $\mathcal{L}_\in$ as being equivalent to $\mathcal{L}_{MV}$ with a constant symbol $\dot{V}$ for a reference universe, the "yes" answer was expanded as

Strong absolutist thesis: (SAT) "$\dot{V}$" makes sense, and that sense is not expressible in the multiverse language.

These two theses regarding the semantic completeness of $\mathcal{L}_{MV}$ belong squarely to epistemology. The word "proposition" has a long history in philosophy, where it is used for "sentence meaning". Perhaps it comes attached to theories of meaning that are not suitable for this context.[31] Another way to state WRT is: "everything

[31] In particular, it seems plausible that in this context one should treat theories, rather than sentences, as the primary bearers of meaning.

that can be said in $\mathcal{L}_\in$ can be said in $\mathcal{L}_{MV}$". That is, $\mathcal{L}_{MV}$ has all the interpretative power present in $\mathcal{L}_\in$.

"Thesis" is a nod to the (perhaps illusory) parallel between the Weak Relativist Thesis and Church's Thesis. Neither is really capable of proof, and there are some similarities between their roles and the evidence for them. (Of course, the evidence for Church's Thesis is much stronger!)

WRT asserts that every sentence in $\mathcal{L}_\in$ that expresses a proposition is synonymous with one in $\text{ran}(t)$. Feferman would say "definite proposition". That may be better, although definiteness here is not a property of the proposition, but of its relationship to the sentence. Maddy and Meadows propose replacing synonymy with provable equivalence in ZFC, or perhaps some extension T of ZFC. I think that this is probably consistent with the intent of WRT, so long as T has an axiomatization consisting of sentences in $\text{ran}(t)$, and is a theory we believe.[32] If T has such an axiomatization, and $T \vdash \phi \leftrightarrow t(\psi)$, then letting θ be the conjunction of the axioms of T needed in the proof, $\theta \to \phi$ is logically equivalent to $\theta \to t(\psi)$, and the latter is logically equivalent to a sentence in $\text{ran}(t)$. If we have accepted T, then adding ϕ is equivalent to adding $\theta \to \phi$. So the Maddy/Meadows version of WRT implies the version above because logical equivalence implies synonymy.

WRT does allow that one might decide sentences like CH that are not logically equivalent to sentences in $\text{ran}(t)$ by asserting their equivalence with some ϕ in $\text{ran}(t)$. From the Weak Relativist point of view, this assertion involves a clarification of their meaning.

Maddy and Meadows object to the use of the notion of meaning in WRT. They write

> It's fairly easy to explicate this type of concern [that the meaning we currently assign to $\mathcal{L}_\in$ does not decide CH] in a metaphysical theory like Hamkins's or Woodin's - there is an abstract ontology of worlds in some of which CH is true and others of which it is false ...

In my opinion, talk about ontologies of worlds does not explain anything, it just muddies the whole question. We are discussing what theory to adopt, and the extent to which adopting one conflicts with adopting another. The subject matter in this discussion is interpreted theories; set theories, not sets or worlds.

Maddy and Meadows continue

> ...but we've seen that Steel's thinking is strictly linguistic. In that context, it's not obvious how to characterize the potential problem without providing a substantive theory of meaning (no hint of which appears in Steel (2014)).

Here I would say that it would certainly be good if there were a substantive theory of meaning that would underpin and elaborate on WRT. Maddy and Meadows are skeptical that any exists or can be developed. People have certainly written a lot about the meaning of mathematical statements, so at least some have been more hopeful in this regard.

However, I think WRT has content that does not rest on such a theory. Its role is to identify a potential meaning indeterminacy in $\mathcal{L}_\in$, not to say what meaning

[32] If not, then T-provable equivalence does not capture the intent of WRT.

indeterminacy in $\mathcal{L}_\in$ is. One way to do that in practice is to point to possible dis-ambiguations, and another is to trim away the allegedly meaningless part and claim that nothing has been lost. WRT makes a move in the latter direction. Moreover, when we trim back, the question as to whether the multiverse has a core immediately takes center stage, and this leads to what one might think of as a dis-ambiguation.

The strongest evidence for WRT is that the mathematical theory based on large cardinal hypotheses that we have produced to date can be naturally expressed in the multiverse sublanguage. Perhaps we lose something when we do that, some future mathematics built around an understanding of the symbol $\dot{V}$ that does not involve defining $\dot{V}$ in the multiverse language. But at the moment, it's hard to see what that is. WRT can be considered as a piece of advice: don't go looking for it. I don't think we can say yet with any confidence that this is good advice.

Maddy and Meadows have a different way of characterizing the virtue of generically invariant sentences of $\mathcal{L}_\in$, *impartiality*. Let me trace through their line of thought here. They begin

> We're assuming that our examination of the various candidates [for a foundational theory] shows them all to be on equal footing and that our best response is to trim the syntax of $\mathcal{L}_\in$.

To the extent this is meant to characterize my reason for isolating $\mathcal{L}_{MV}$, there is a subtle misunderstanding. It is based on something I wrote poorly:

> Before we try to decide whether some such theory is preferable to the others, let us try to find a neutral common ground on which to compare them. We seek a language in which all these theories can be unified, without bias toward any, in a way that exhibits their logical relationships, and shows clearly how they can be used together.

"Compare" here is misleading. "Unify" would be better. $\mathcal{L}_{MV}$ is not a device for comparing the merits of one of those theories with another, it is a way to make more visible their underlying unity.

Maddy and Meadows continue

> On those assumptions, consider the state of two imaginary set theorists, a universe theorist and a multiverse theorist. The universe theorist speaks $\mathcal{L}_\in$, embraces ZFC + LCs, and persists in trying to figure out the correct way to extend it; under our current assumptions, this universe theorist is just wrong, making a mistake. In contrast, our multiverse theorist is aware that no candidate is preferable, speaks $\mathcal{L}_{MV}$, and embraces MV. This multiverse theorist thinks, with considerable justification on our assumptions, that the universe theorist is missing the fact that all the candidate foundational theories represented by worlds in the multiverse have equal standing. To put this another way, we might say that from the multiverse theorist's perspective, the universe theorist's $\mathcal{L}_\in$ sentences may reflect an improper bias, restricting attention to one world, while all $\mathcal{L}_{MV}$ sentences are suitably impartial.

Of course, I find the part about universes and multiverses, and theories represented by worlds, vague and at the wrong level. The distinction I was trying to draw was between views on the semantic completeness of $\mathcal{L}_{MV}$. Weak relativism and Strong Absolutism are views on the semantics of $\mathcal{L}_\in$. They are not expressible in $\mathcal{L}_\in$.

Our "multiverse theorist" could adopt the syntax of $\mathcal{L}_\in$, and just be careful to stay in the range of the translation function. At the level I was going for, nothing has changed, she is just expressing herself differently. I am not sure whether she still counts as a multiverse theorist in the Maddy/Meadows sense.

The sentences in $\mathrm{ran}(t)$ are, up to logical equivalence, precisely those that are generically invariant. So they are impartial in that sense. But the main claim in WRT is that *it is indeed suitable* to remain within the generically invariant. That is, that there is no mathematical meaning beyond that which can be captured by generically invariant principles. Or, to retreat to a more reasonable claim, that there is none on the horizon now.

Let me summarize my disagreements with Maddy and Meadows by replying to their own summary.

> The substance of Steel's thought can be formulated more effectively in philosophically innocent mathematical terms. By these means, we steer away from the vagaries of mathematical meaning, truth, and existence and toward the methodologically central questions: how exactly do we select our theories and by what right?

I would agree that truth and existence are out of place in a discussion of what set theory to adopt and why. Not because they are vague, but because bringing them in leads to question-begging. Meaning is at the center of it, because we select our theories based on how they are interpreted. In this set theoretic context, our framework theory should be such that all others can be translated into it. Maddy's focus on methodology is not in conflict with this idea; after all, the logical positivists' slogan was that the meaning is the method of verification. If we understand "method of verification" in a suitably broad sense, this seems ok to me.

2.6 Does the Multiverse Have a Core?

Having retreated to $\mathcal{L}_{MV}$, one asks at once whether the absolutist's idea of a distinguished reference world is really gone. Perhaps there is an individual world that is definable in the multiverse language. An elementary forcing argument shows that if so, then there is a unique definable world, and this world is included in all the others.[33] In this case, we call this unique world included in all others the *core* of the multiverse. [34] This leads to what Steel (2014) calls the *Weak Absolutist Thesis*: The multiverse has a core.

It was a mistake to call this a thesis, and present it in parallel with the two semantic theses in the last section. "The multiverse has a core" is a statement in $\mathcal{L}_{\in}$, and in fact, in its sublanguage $\mathcal{L}_{MV}$. WRT and SAT are philosophical theses, and cannot be formulated in $\mathcal{L}_{\in}$. Whether the multiverse has a core is a question about sets, not set theories, and it can be formulated by the sentence $\exists U \forall W \forall x (x \in U \rightarrow x \in W)$ of $\mathcal{L}_{MV}$. Let us shorten this sentence to $\exists U (U = \mathfrak{C})$.

---

[33] This observation is due to Woodin.

[34] Fuchs, Hamkins, and Reitz, began the general study of such questions in what they call *set theoretic geology*. See Fuchs et al. (2015). Usuba's proof of their "downward directed grounds hypothesis" shows that W is the core iff W is what Fuchs et al. (2015) calls a *bedrock*.

Whatever one thinks of the semantic completeness of the multiverse language, it does highlight $\exists U(U = \mathfrak{C})$ as a fundamental question. Because the multiverse language is a sublanguage of the standard one, this is a question for everyone. If the multiverse has a core, then surely it is important, whether it is the denotation of the absolutist's $\dot{V}$ or not.

Fuchs, Hamkins, and Reitz have shown that neither MV nor its extensions by large cardinal hypotheses up to the level of supercompact cardinals decides whether there is a core to the multiverse, or the basic theory of this core if it exists. (See Fuchs et al. (2015).) But

Theorem 2.6.1 (Usuba 2017, 2019) *If there is an extendible cardinal, then* $t(\exists U(U = \mathfrak{C}))$.

The Fuchs-Hamkins-Reitz work shows that nothing follows from extendible cardinals concerning the basic theory of the core.

Usuba's theorem is certainly evidence that there is a core, but there is some reason to be hesitant. First, the large cardinal hypothesis is "global", that is, Σ_3 rather than Σ_2, and that is essential.[35] Second, strong evidence that there is a core should be evidence that there is a core with well-determined properties. The fact that the existence of extendible cardinals decides very little about the theory of the core weakens the evidence provided by Usuba's proof.

Before we turn to the possibility of a well-determined core, let us consider the semantic consequences of adopting "there is a core", in either syntax. There is a discussion of this question in Sect. 3 of Meadows' paper Meadows (2021), and I shall incorporate parts of it. I shall also reply to the "argument against the generic multiverse" made there.[36]

LST is equivalent to $\mathcal{L}^+_{MV}$, where $\mathcal{L}^+_{MV}$ is $\mathcal{L}_{MV}$ expanded by a constant symbol $\dot{V}$, in the following sense. If ϕ is a sentence of $\mathcal{L}_\in$, its translation in $\mathcal{L}^+_{MV}$ is just

$$u(\phi) = \phi^{\dot{V}}.$$

If ϕ is a sentence of $\mathcal{L}^+_{MV}$, then its translation in LST is

$$t^+(\phi) = \text{“}(V^G, \in, V) \models \phi\text{"},$$

where V^G is the (imaginary) "expanded" generic multiverse generated by V and V is treated as the interpretation of $\dot{V}$. The right hand side can be unpacked more fully using the definability of forcing. t^+ is well defined because the forcing is homogeneous. u actually acts on formulae, and is a standard relative interpretation.

---

35 The Σ_2 sentences are those that are, provably in ZFC, equivalent to sentences of the form $\exists\alpha(V_\alpha \models \varphi)$. This is the sense in which Σ_2 sentences are local. "There is a supercompact cardinal" is also Σ_3 rather than Σ_2. Extendibles are strictly stronger than supercompacts. Usuba observed that $t(\exists U(U = \mathfrak{C}))$ does not follow from the existence of supercompacts. Goldberg (2021) shows that nothing much weaker than the existence of extendibles suffice.

36 Meadows attributes the argument to Woodin.

Meadows (2021) shows using boolean ultrapowers that if V has a definable wellorder, then we can extend t^+ to a standard relative interpretation, but I don't see that this is important for our discussion. The extension does not preserve the meaning of formulae. Because the forcing is homogeneous, t^+ preserves $\wedge$ and $\neg$, moreover, letting MV^+ be the natural extension of MV to $\mathcal{L}^+_{MV}$,

(i) $\mathsf{ZFC} \vdash \phi \leftrightarrow t^+ \circ u(\phi)$, and
(ii) $\mathsf{MV}^+ \vdash \psi \leftrightarrow u \circ t^+(\psi)$,

for all sentences ϕ of $\mathcal{L}_\in$ and ψ of $\mathcal{L}^+_{MV}$. That is, ZFC proves that ϕ holds iff $\phi^{\dot{V}}$ holds in its expanded generic multiverse, and MV^+ proves that ψ holds iff $\dot{V} \models$ "ψ holds in my expanded generic multiverse". Adapting the terminology of Meadows (2021), let us say that ZFC and MV^+ are *weakly sententially equivalent via* (u, t^+).

Now let's consider possible weak sentential equivalences between theories in $\mathcal{L}_\in$ extending ZFC and theories in $\mathcal{L}_{MV}$ extending MV. Composing with (u, t^+) above, we are looking for such equivalences between extensions S of MV^+ and extensions T of MV. Because MV^+ has its symbol $\dot{V}$, it is natural to add "there is a core", that is $\exists U(U = \mathfrak{C})$, to both S and T.[37] As Meadows notes, the resulting theories are still not weakly sententially equivalent, the reason being that we have yet to settle anything about what is true of $\dot{V}$. The simplest way to do that is to add $\dot{V} = \mathfrak{C}$ to MV^+.

Let us write "$V = \mathfrak{C}$" for $t^+(\dot{V} = \mathfrak{C})$, that is, for the LST version of "V is the core of its generic multiverse".

Proposition 2.6.2 *The theories*

(i) $T_0 = \mathsf{MV} + \exists U(U = \mathfrak{C})$,
(ii) $T_1 = \mathsf{MV}^+ + \dot{V} = \mathfrak{C}$, *and*
(iii) $T_2 = \mathsf{ZFC} + V = \mathfrak{C}$

are weakly sententially equivalent.

T_0 and T_1 are clearly definitionally equivalent. The equivalence between T_1 and T_2 is given by (t^+, u). In other words, the $\mathcal{L}_\in$ sentence ϕ is translated to the $\mathcal{L}_{MV}$ sentence "ϕ is true in $\mathfrak{C}$", and the $\mathcal{L}_{MV}$ sentence ψ is translated to the $\mathcal{L}_\in$ sentence "ψ is true in the multiverse generated by me".

I believe the proper conclusion is that MV + $\exists U(U = \mathfrak{C})$ and ZFC + $V = \mathfrak{C}$ are equivalent as foundations. From the point of view of the Weak Relativist, passing from MV + $\exists U(U = \mathfrak{C})$ to $\mathsf{ZFC} + V = \mathfrak{C}$ amounts to making a definition. The real question was whether there is a core, and this is a question in the multiverse sublanguage.

It's worth noting that one could expand the list of weakly sententially equivalent theories given in the proposition. Let $\mathbb{P}$ be any homogeneous partial order in $\mathfrak{C}$ that is definable over $\mathfrak{C}$ (provably in ZFC); then the theory

$$T(\mathbb{P}) = \mathsf{MV}^+ + \text{“}\dot{V} \text{ is a } \mathbb{P}\text{-generic extension of } \mathfrak{C}\text{”}$$

[37] There may be interesting equivalences that do not involve doing that.

is also weakly sententially equivalent to T_0. In contrast to T_1, $T(\mathbb{P})$ is not a definitional extension of T_0. It seems to me that $T(\mathbb{P})$ is an odd way of presenting the same foundation as that in T_0, T_1, and T_2.

Meadows presents the equivalence of T_0 with T_2 as a practical argument "against the generic multiverse". It is not clear to me how one can argue against an object. More importantly, it is not clear to me how the equivalence between T_0 and T_2 could be an argument against one and in favor of the other.

I think the idea is probably that T_2 is a simpler presentation of T_0, with a more familiar syntax, so in the end it would be better to work with it. That is certainly true for elementary set theory, but anyone who is able to understand and work with the hypothesis $V = \mathfrak{C}$ will understand the simple equivalence between T_2 and T_0, and be able to shift between $\mathcal{L}_\in$ and $\mathcal{L}_{MV}$ easily. In any case, convenience is not what motivates T_0 and its syntax. $\mathcal{L}_{MV}$ is useful in isolating generically invariant set theory, and that helps us to understand what might justify T_2, and the extension $V = \text{Ult}-L$ of T_2 that we shall describe in the next section.

2.7 Absolute Ordinal Definability

Let us assume that the multiverse has a core. We still haven't gotten very far, because large cardinal hypotheses by themselves decide very little about the core. It seems reasonable to add that the sets in $\mathfrak{C}$ are all ordinal definable, but this still decides very little. The reason is that mild class forcings preserve the large cardinals, and can be used to construct models in which sets get into $\mathfrak{C}$ for reasons unrelated to the sets themselves, for example by being coded into the 2^α function at arbitrarily large α. The same construction shows that being ordinal definable, by itself, says very little about a set.[38] If we want to pin down $\mathfrak{C}$, we need an explanation for minimality and ordinal definability.

Inner model theory suggests a way to pin down $\mathfrak{C}$. The canonical inner model M_H for a large cardinal hypothesis H is its most concrete realization. Its construction yields a thorough *fine structure theory* for the model. We have constructed M_H for many H. Each M_H has a fine-grained hierarchy, and H is weaker than K if and only if M_H is a proper initial segment of the hierarchy of M_K. So the M_H we have constructed do fit together, in a hierarchy that seems to be the model-theoretic counterpart to the consistency strength hierarchy.[39]

---

[38] In Fuchs et al. (2015), Fuchs, Hamkins, and Reitz show this way that for any countable $M \models \mathsf{ZFC}$ and any ordinal α, M has a mild class forcing extension N such that $V_\alpha^M = V_\alpha^N$ and N satisfies "$\mathfrak{C}$ exists, and is equal to generic HOD". Here generic HOD consists of those sets that are ordinal definable in every set generic extension of V.

[39] Here I am assuming that M_H is pointwise definable, as it must be if it is truly minimal. If $\mathfrak{C}$ is an ultimate model along these lines, then the M_H would all be countable initial segments of it. See the introductory chapter of Steel (2024) for an overview of the inner model theory that is relevant in this section.

At the moment, we only have a theory of the M_H at limited large cardinal levels. If they have a general form, one that is independent of H and goes as far as the large cardinal hypotheses do, then this would be a candidate for the structure of $\mathfrak{C}$. Hugh Woodin has suggested that this is the case, and that this form is that of the HODs in certain models of the Axiom of Determinacy.

To me, this suggestion for pinning down $\mathfrak{C}$ looks like meaning clarification. It seems similar to what happened when we adopted the Axiom of Extensionality, or the Axiom of Regularity $V = WF$.

Let us give a brief explanation.

The sets in any M_H are ordinal definable in a certain generically absolute way.

Definition 2.7.1 Let $A \subseteq \omega^\omega$; then A is *homogeneously Suslin (*Hom_∞*)* iff for all κ, there is a system $\langle M_s, i_{s,t} \mid s, t \in \omega^{<\omega}\rangle$ such that

(1) $M_\emptyset = V$, and each M_s is closed under κ-sequences,
(2) for $s \subseteq t$, $i_{s,t} : M_s \to M_t$,
(3) if $s \subseteq t \subseteq u$, then $i_{s,u} = i_{t,u} \circ i_{s,t}$, and
(4) for all x, $x \in A$ iff $\lim_n M_{x \restriction n}$ is wellfounded.

Martin showed in 1968 that all Hom_∞ sets are determined, and that if there are arbitrarily large measurable cardinals, then all Π^1_1 sets are Hom_∞. Stronger large cardinal hypotheses imply that more complicated sets are Hom_∞. Work of Martin-Solovay Martin and Solovay (1969), Foreman-Magidor-Shelah (Foreman et al. (1998)), and various people in inner model theory led eventually to

Theorem 2.7.2 (Martin, S., Woodin 1985) *Assume there are arbitrarily large Woodin cardinals; then for any $A \in \mathrm{Hom}_\infty$, $L(A, \mathbb{R}) \models \mathsf{AD}^+$.*

AD^+ is a technical strengthening of the Axiom of Determinacy (AD), first isolated by Woodin.

Along with determinacy we get set-generic absoluteness.

Theorem 2.7.3 (Woodin 1987?) *If there are arbitarily large Woodin cardinals, then $(\Sigma^2_1)^{\mathrm{Hom}_\infty}$ truth is set-generically absolute.*

Recall that a set is *ordinal definable* (OD) iff it is definable over the universe of sets from ordinal parameters, and is *hereditarily ordinal definable* (HOD) just in case it and all members of its transitive closure are OD. The HOD of a determinacy model is close to the model itself; moreover these HOD's have Woodin cardinals.

Definition 2.7.4 *(Woodin)* $V = \mathrm{Ult} - L$ is the statement: There are arbitrarily large Woodin cardinals, and for any Σ_2 sentence φ of $\mathcal{L}_\in$: if if φ is true, then for some $A \in \mathrm{Hom}_\infty$, $\mathrm{HOD}^{L(A,\mathbb{R})} \models \varphi$.

This is read "V is ultimate L".

Theorem 2.7.5 (Woodin) *If $V = \mathrm{Ult} - L$, then*

(1) V is the core of its multiverse V^G.

(2) V is "generically absolute HOD".

See Woodin (2017).
One can state $V = \text{Ult} - L$ in the multiverse sublanguage.

The hope is that $V = \text{Ult} - L$ is consistent with all the large cardinal hypotheses, so that adopting it does not restrict interpretative power. Whether it is consistent with hypotheses significantly stronger than the existence of many Woodin cardinals is a crucial open problem.

At the same time, one hopes that $V = \text{Ult} - L$ will yield a detailed fine structure theory for V, removing the incompleteness that large cardinal hypotheses by themselves can never remove. It is known that $V = \text{Ult} - L$ implies the CH, and many instances of the GCH. Whether it implies the full GCH is a crucial open problem.

Both problems have to do with the theory of HOD in models of AD^+. They have been recognized as central questions in descriptive set theory since the early 1980s, and in inner model theory since the mid 1990s. Many people have worked on them and obtained various encouraging results.[40] In the period after Steel (2014) appeared I made some further progress in this direction. Let me state one of the main new theorems, assuming as I do some significant technical background.

Definition 2.7.6 *(AD^+)* A pointclass Γ is *long* iff there is an $A \in \Gamma$ such that A codes an (ω_1, ω_1) iteration strategy for a pure extender premouse with a long extender on its sequence. Otherwise Γ is *short.*

Theorem 2.7.7 (S. 2015–21, Steel (2024, 2024)) *Suppose there are arbitrarily large Woodin cardinals, and that there is a supercompact cardinal. Assume also that V is* uniquely iterable*; then*

(1) there is a short $\Gamma \subsetneq \text{Hom}_\infty$ such that $L(\Gamma, \mathbb{R}) \models \mathsf{AD}_\mathbb{R}$ and $HOD^{L(\Gamma,\mathbb{R})} \models$ "there is a subcompact cardinal", and
(2) there is a long Γ in Hom_∞.

Theorem 2.7.8 (S. 2015–21, Steel (2024, 2024)) *Assume $\mathsf{AD}^+ +$ "there is a long pointclass"; then for any short $\Gamma \subseteq P(\mathbb{R})$ such that $L(\Gamma, \mathbb{R}) \models \mathsf{AD}_\mathbb{R}$, $HOD^{L(\Gamma,\mathbb{R})}$ is a least branch premouse (so satisfies GCH, and has a fine structure).*

It should be possible to remove the iterability hypotheses in these theorems.[41] How to remove "V is uniquely iterable" from Theorem 2.7.7 is an instance of what has been the central problem in inner model theory since the mid-1980s. I suspect that removing the hypothesis "there is a long pointclass" from Theorem 2.7.8 is the most accessible of the big open problems to do with iterability.

Theorem 2.7.7 states that, granted its iterability hypothesis, $V = \text{Ult} - L$ is consistent with subcompacts. This is a significant step beyond the large cardinal hypotheses

[40] There is a survey of work in this area in Steel (2016).

[41] In Theorem 2.7.8, the iterability hypothesis is "there is a long pointclass". We suspect that conclusion (1) of 2.7.7 also follows from $\mathsf{AD}^+ +$ "there is a long pointclass", but the claim made in (Steel (2024), Theorem 1.5.1(2)) to have a proof of this was wrong.

that had been reached previously, and is therefore encouraging news. Subcompacts are near the upper limit of what can be modelled by mice with only short extenders. To reach larger cardinals like supercompacts, one would need a general comparison theorem for mice with long extenders. This is another long-open problem; there is a comparison theorem (modulo iterability) for mice in the lower reaches of the long extender realm, but no general one.[42]

If $V = \text{Ult}-L$ truly pins down the theory of the core, then it must decide GCH. To prove that $V = \text{Ult}-L$ implies a Π_2 sentence φ (like GCH), one must show that φ holds in $\text{HOD}^{L(A,\mathbb{R})}$ for *all* $A \in \text{Hom}_\infty$. One needs a general theory of these HOD's. Given what we know about "small" determinacy models, it seems likely that GCH is part of this theory. That leads to the following conjecture:

Conjecture 2.7.9 Assume $\mathsf{ZF} + \mathsf{AD}^+ + V = L(P(\mathbb{R}))$; then $\text{HOD} \models \mathsf{GCH}$.

In a sense this conjecture originates with the work of the Cabal in descriptive set theory in the late 1970s and early 1980s. Since then, we have seen that it belongs to inner model theory, and likely involves a general notion of mouse and iteration strategy, together with a general comparison lemma for these objects. Presumably these mice come equipped with a fine-grained hierarchy, the model-theoretic counterpart to the consistency strength hierarchy on the theories that hold in them or their generic extensions.

I think we are still pretty far from a proof of this conjecture. There are fragments of it that are more accessible, and would represent significant progress.

What we know from descriptive set theory, and from inner model theory where we have it, suggests that Conjecture 2.7.9 is true, and that its proof involves a general theory of mice and their iteration strategies. The mice must be capable of having long extenders, and probably supercompact cardinals and beyond. Although its less general approximations are much better targets at present, Conjecture 2.7.9 points to a long term future. The intricately structured world of inner model theory extends well beyond the part that we have discovered so far.[43]

2.8 Final Remarks

What would it be to adopt $V = \text{Ult}-L$ as a foundational axiom? Should we do that?

Adopting $V = \text{Ult}-L$ would not, and should not, mean ending the further development of theories like the forcing axioms. What can be forced is of permanent

[42] Those results are due to Woodin, Neeman, and the author. See Neeman and Steel (2016).

[43] Here let me record a disagreement with Bagaria et al. (2019). The fate of inner model theory does not depend on the truth of the Ultimate-L conjecture. The Ultimate-L conjecture has to do with one approach to constructing canonical inner models. There is a fair amount of evidence that this approach does not always work in the way that the Ultimate-L conjecture requires, but there are other approaches. In particular, it is a different approach that leads to the positive results on Conjecture 2.7.9 that we have now.

interest in set theory. As the title of Blue (2024) puts it, the generic multiverse is not going away. The goal here is to unify those theories in a framework that lets us use them together properly, not to eliminate them.

The key questions are whether the HODs of determinacy models are ultimate, and whether they are L-like. Gödel taught us not to dream of final theories, so *Ultimate* should be taken with a few grains of salt. What we would like to show is that we can interpret (at the sentential level) the natural theories like those we know in extensions of $\mathsf{ZFC} + V = \text{Ult} - L$. This seems to involve showing that the HODs of determinacy models can satisfy at least the local (Σ_2) forms of the very strong large cardinal hypotheses we know about. *L-like* means having a fine structure like that of L. At a minimum, we should be able to show that GCH holds. Experience suggests that if we do that, it will be the beginning of a thorough fine-structural analysis.

If all this works out, $V = \text{Ult} - L$ would be a clarificatory axiom, like the Axiom of Extensionality, or the Axiom of Regularity. It lets people know how you are using the language of set theory. There is as of now no reason to believe that we lose interpretative power by using the language of set theory this way, but it is too early to have a definite opinion.

References

Aspero, D., & Schindler, R. D. (2021). Martin's Maximum++ implies Woodin's axiom (*). *Annals of Mathematics, 193*, 793–835.

Blue, D., The generic multiverse is not going away. To appear.

Bagaria, J., Koellner, P., & Woodin, W. H. (2019). Large cardinals beyond choice. *Bulletin of Symbolic Logic, 25*, 283–318.

Burgess, J., Second philosophy: anti-nominalist reflections on Maddy's semi-nominalism, this volume.

Feferman, S. (1992). Why a little bit goes a long way. *Proceedings of the biennial meeting of the Philosophy of Science Association*, pp. 442–455.

Feferman, S. (1998). *In the light of logic*. New York: Oxford University Press.

Foreman, M., Magidor, M., & Shelah, S. (1998). Martin's maximum, saturated ideals, and nonregular ultrafilters. *Annals of Mathematics, 127*, 1–47.

Fuchs, G., Hamkins, J. D., & Reitz, J. (2015). Set-theoretic geology. *Annals of Pure and Applied Logic, 166*, 464–501.

Goldberg, G. (2021). Usuba's extendible cardinal. Preprint, arXiv:2108.06903 [math.LO].

Hamkins, J. D. (2014). A multiverse perspective on the Axiom of Constructibility. Lecture Notes Series, Institute for Mathematical Sciences, National University of Singapore. Infinity and Truth, pp. 25–45.

Laver, R. (2007). Certain very large cardinals are not created in small forcing extensions. *Annals of Pure and Applied Logic, 149*, 1–6.

MacLane, S. (1986). *Mathematics, form and function*. New York: Springer.

Maddy, P. (1997). *Naturalism in mathematics*. New York: Oxford University Press.

Maddy, P. (1998). $V = L$ and maximize. In *Logic Colloquium '95, Proceedings of the European Assn. of Symbolic Logic held in Haifa, Israel*, J. Makowsky et. al. eds, Lecture Notes in Logic vol. 11, pp. 134–152.

Maddy, P. (2007). *Second philosophy*. New York: Oxford University Press.

Maddy, P. (2011). *Defending the axioms*. New York: Oxford University Press.

Maddy, P. (2017). Set-theoretic foundations, In *Set-theoretic foundations, A. Caicedo et. al. eds., Foundations of mathematics: essays in honor of* W. Hugh Woodin's 60th birthday, Contemporary Mathematics, vol. 690, 289–320.

Maddy, P., What do we want a foundation to do? Reflections on Foundations: Univalent Foundations, Set Theory, and General Thoughts. S. Centrone et. al eds, 36 pp. To appear.

Maddy, P., & Meadows, T. (2020). A reconstruction of Steel's multiverse project. *Bulletin of Symbolic Logic, 26*, 119–169.

Martin, D. A. (2005). Gödel's conceptual realism. *Bulletin of Symbolic Logic, 11*, 207–224.

Martin, D. A. (2018). Completeness and incompleteness of basic mathematical concepts, Preliminary draft Feb. 2018, 36 pp., available at

Martin, D. A., & Solovay, R. M. (1969). A basis theorem for Σ^1_3 sets of reals. *Annals of Mathematics, 89*, 138–160.

Meadows, T. (2021). Two arguments aganst the generic multiverse. *Review of Symbolic Logic, 14*, 347–379.

Montalban, A., & Walsh, J. (2019). On the inevitability of the consistency operator. *Journal of Symbolic Logic, 84*, 205–225.

Neeman, I., & Steel, J. R. (2016). Equiconsistencies at subcompact cardinals. *Archive for Mathematical Logic, 55*, 207–238.

Quine, W. V. (1990). *Pursuit of truth*. Cambridge: Mass. Harvard University Press.

Quine, W. V. (1984). Review of Parsons. *Journal of Philosophy, 81*, 783–794.

Schatz, J. R. (2019). Axiom selection by maximization: V = Ultimate L vs. forcing axioms. Ph.D. thesis, UC Irvine.

Steel, J. R. (2000). Mathematics needs new axioms. *Bulletin of Symbolic Logic, 6*, 422–433.

Steel, J. R. (2014). Gödel's program. In *Interpreting Gödel*, J. Kennedy ed., Cambridge University Press, pp. 153–179.

Steel, J. R. (1982). A classification of jump operators. *Journal of Symbolic Logic, 47*, 347–358.

Steel, J. R. (2016). HOD in models of determinacy. In *Ordinal definability and recursion theory: the Cabal Seminar*, volume III, A.S Kechris et. al. editors, volume 43 of Lecture Notes in Logic, ASL, pp. 3–48.

Steel, J. R., A comparison process for mouse pairs. To appear in Lecture Notes in Logic (ASL, Cambridge University Press).

Steel, J. R., The comparison lemma. Annals of Pure and Applied Logic, to appear.

Usuba, T. (2017). The downward directed grounds hypothesis and very large cardinals. Journal of Mathematical Logic, vol. 17.

Usuba, T. (2019). Extendible cardinals and the mantle. *Archive for Mathematical Logic, 58*, 71–75.

Woodin, W. H. (2017). In search of ultimate-L. The 19th Midrasha Mathematicae lectures. The Bulletin of Symbolic Logic, vol. 23, pp. 1–109.

Woodin, W. H. (2001). The continuum hypo thesis II. *Notices of the AMS, 48*, 681–690.

Chapter 3
Reply to Steel

Penelope Maddy

Keywords Set theory · Maximize · Unify · Foundation · Generous arena · Shared standard · Meaning · Quine · Multiverse · Burgess

Professor Steel's essay is focused on an attempt by my colleague, Toby Meadows, and me (Maddy and Meadows, 2020) to reconstruct the multiverse project of Steel's 'Gödel's program' (Steel, 2014). Our goal in that piece was to spell out various turns in his line of thought more explicitly than in the original paper and to remove troublesome commitments that seemed to us unnecessary. Here Steel begins with his understanding of the philosophical background to all this, then gives voice to his reservations about our efforts before concluding with a welcome insider's assessment of the current state of play on $V = Ult{-}L$. Most of what I would have to say about Steel's multiverse project is already contained in my paper with Meadows, so I focus here on the philosophical background and Steel's own philosophical posture.

The story begins with Quine. As Steel notes, we've both been heavily influenced by Quine's writings, but in my case, this has taken the form of my gradual rejection of many central Quinean theses, beginning with his confirmational holism. Touching on the leading case study in my argument, Steel takes the early twentieth century experiments on Brownian motion to be just one more piece in the accumulating stock of evidence, albeit the one that finally tipped the scale into confirmation. Elsewhere, I've laid out my reasons for thinking—in company with most leading scientists at the time—that this evidence is of a qualitatively different sort,[1] so I won't repeat that

[1] See Maddy (1997), pp. 135–143, Maddy (2007), pp. 404–407.

P. Maddy (✉)
Department of Logic and Philosophy of Science, University of California, Irvine, CA, USA
e-mail: pjmaddy@uci.edu

S. Arbeiter and J. Kennedy (eds.), *The Philosophy of Penelope Maddy*,
Outstanding Contributions to Logic 31, https://doi.org/10.1007/978-3-031-58425-1_3

story here. What is worth noting is Steel's treatment of the indispensability argument that derives from Quinean holism.

Steel's characterization of the argument begins, 'mathematics is essential to the rest of science, so let's not dispense with it' (p. 15). I readily accept this and, in company with John Burgess, go further: the usefulness and fruitfulness of mathematics in natural science have earned it its place, even if it isn't strictly essential (in whatever sense). But I step off the path of both Burgess and Steel, and of course, of Quine, when they draw ontological conclusions. Steel boldly declares, 'as a set theorist who believes some quite strong set existence principles, I am a realist' (p. 22). This contrasts with my Arealism, which doesn't so much deny this ontology—as a nominalist would—as it denies that matters of truth and existence, pro or con, are what's at issue in the pursuit of a theory of sets.[2]

Given this metaphysical divergence, it's disorienting when Steel takes my attempted characterization of Maximize to be too focused on ontology, not enough on theories (pp. 20–21). This seems to me untrue: my formal characterization of what it is for one theory to 'maximize' over another is a purely syntactic matter of what the first theory can prove.[3] What apparently bothers Steel is that I allow myself, in another context, to speak the ordinary language of mathematics, to speak, for example, of an 'entity of classical mathematics' (p. 23). Thinking in terms of objects with properties standing in relation to one another is absolutely fundamental to human cognition, present already in pre-linguistic infants,[4] so it's natural for us to form our mathematical language and our mathematical theories in that vernacular, but the fact that we frame a discourse in this way doesn't force us to regard it as a truth-seeking enterprise or descriptive of some subject matter.

Steel is also bothered by what he sees as an anomaly in my treatment of the maxim Unify: though central to *Naturalism in Mathematics* (1997), it doesn't appear in 'Set-theoretic foundations' (Maddy 2017b) or 'What do we want a foundation to do?' (Maddy 2019a). The reason for this is simple, but perhaps worth pointing out. In the book, the foundational goal is called upon to justify methodological maxims for the proper pursuit of set theory: if you're out to provide a foundation, you should have a single theory (Unify) that's not restrictive (Maximize).[5] In the more recent papers, the question isn't—how should set theory be done if it is to serve as a foundation?—but—what foundational jobs does set theory, as currently conducted (as Unified and Maximized), manage to do? Among those are Generous Arena and Shared Standard, but these don't replace Unify, as Steel imagines (p. 11). The connection is more like this: if you want a Generous Arena, you should Maximize; if you want a Shared

[2] Steel characterizes Arealism as 'a form of instrumentalism' (p. 15), but the Arealist's methodological decisions aren't justified in terms of their consequences for some presumably 'concrete' basis. A closely allied position is labeled 'enhanced if-thenism' in Maddy (2022b). I resist the label 'fictionalism' (see Maddy (2011), pp. 98–99), but it's fairly apt.

[3] See Maddy (1997), pp. 220–224.

[4] See, e.g., Maddy (2007), §III.5, or Maddy (2018b).

[5] For the record, the methodological maxim that emerges from my discussion of maximization isn't to embrace maximizing theories (too broad) but to avoid restrictive theories. Schatz (2019) argues that $V = Ult{-}L$ is restrictive.

Standard, you should Unify. But since set-theoretic method isn't the topic of the more recent papers, Unify and Maximize don't come up.

Returning to Steel's Quineanism, there's the haunting question of unapplied mathematics. Steel's fellow indispensability theorist, John Burgess, is willing to loosen the connection between the role of mathematics in natural science and the confirmation of its claims: the efficacy of mathematics in application has earned it the right to its own autonomous modes of evidence.[6] Steel, in contrast, adheres to strict Quinean orthodoxy, or rather, he does Quine one better, embracing a strikingly elastic notion of holistic evidence: pure mathematics is empirically confirmed by its interconnections, at points, with applied mathematics; set theory is empirically confirmed by its interconnections, at points, with pure mathematics. The problem of unapplied mathematics disappears because there is no unapplied mathematics in the relevant sense.[7] Though I, too, regard it as methodologically advantageous to pursue unfettered pure mathematics and, for that reason, to Maximize in the practice of set theory (Steel would say 'maximize interpretive power'), it seems to me a stretch to count these methodological imperatives as conferring empirical confirmation. But again, these matters are covered elsewhere.[8]

Steel moves on from this philosophical backdrop to his discomforts with Meadows' and my reconstruction of his multiverse project—in particular, to our 'criticism' of his 'use of the notion of *meaning*' (p. 18). While we do find it difficult to explicate his notion of meaning, our guiding position, as I understand it (not to speak for Meadows), isn't that meaning is fundamentally problematic in some way[9]—it may be, but that's not our debate to engage—it's that meaning is a subtle and elusive notion, and if we don't need it, better to do without.[10] We hope to have shown that Steel's project doesn't need it, to have shown how to do without. In some passages, at least, Steel seems to think we haven't strayed overly far[11]:

> Meaning is at the center of it ... In this set theoretic context, our framework theory should be such that all others can be translated into it. Maddy's focus on methodology is not in conflict with this idea; after all, the logical positivists' slogan was that the meaning is the method of verification. If we understand 'method of verification' in a suitably broad sense, this seems ok to me. (p. 29)

---

[6] See 'Reply to Burgess'. My Arealist believes something analogous: if mathematics hadn't grown out of natural science as it did and if it didn't continue to intertwine with natural science as it does, its autonomous methods wouldn't be subject to rational analysis and evaluation.

[7] Thanks to Professor Steel for clarifying his position in our exchange over an earlier draft of this reply (cited with permission).

[8] See Maddy (2008) or Chap. 1 of Maddy (2011) in addition to references in footnote 1.

[9] As Steel's section title 'meaning skepticism' may suggest.

[10] Steel acknowledges that 'In the context of new axioms, it can be hard to distinguish sharpening the meaning of the language from discovering new truths' (p. 19). This may difficult be because there's no fact of the matter here, but our point is only that there is no need to take this on.

[11] See the discussion of *WRT* on p. 26 for another point of near agreement.

Leaving the positivists and verificationism aside, the sentiment here seems to be that thinking about what methods are appropriate and why isn't all that different from thinking about meaning (as Steel understands it).[12]

Let me close by noting Steel's response to a related paper by Meadows (Meadows, 2021). There Meadows demonstrates the equivalence (in a specific sense) of two theories, one an extension of Steel's *MV* in the multiverse language, the other the corresponding extension of *ZFC*. Given this equivalence, Meadows concludes that, for all practical purposes, we might as well continue to use the simpler and more familiar theory in the language of set theory, that the multiverse language and theory can be set aside without loss. Steel appears to agree, to a point: 'That is certainly true for elementary set theory' (p. 32). One senses the 'but', and sure enough: 'convenience isn't what motivates' the multiverse theory; it.

> is useful in isolating generically invariant set theory, and that helps understand what might justify [the corresponding extension of *ZFC*], and the extension $V = Ult{-}L \ldots$ that we shall describe in the next section. (p. 32)

This is what makes Steel's project so fascinating and potentially so important. For this, I can do no better than refer the reader to the closing sections of Steel's current paper.[13]

References

Maddy, P., & Meadows, T. (2020). A reconstruction of Steel's multiverse project. Bulletin of Symbolic Logic, 26, 118–169.

Maddy, P. (1997). *Naturalism in mathematics.* Oxford University Press.

Maddy, P. (2007). *Second Philosophy.* Oxford University Press.

Maddy, P. (2008). How applied mathematics became pure. *Review of Symbolic Logic 1*, 16–41.

Maddy, P. (2011). *Defending the axioms.* Oxford University Press.

Maddy, P. (2017b). Set-theoretic foundations. In A. Caicedo, J. Cummings, P. Koellner, & P. Larson, eds., *Foundations of mathematics: Essays in Honor of W. Hugh Woodin's 60th birthday* (pp. 289–322). American Mathematical Society.

Maddy, P. (2018b). Psychology and the a priori sciences. In S. Bangu (Ed.), *Naturalizing logico-mathematical knowledge: approaches from psychology and cognitive science* (pp. 15–29). Routledge.

[12] I skip over Steel's more localized criticism of the way Meadows and I understand L_{MV} and its relationship to L_ε (pp. 24–25). For my part, the puzzle posed in footnote 53 of Meadows and Maddy (2020) remains. Steel thinks there's an important subset of the sentences of L_ε that can be identified as the range of a certain function from L_{MV} to L_ε, but he also insists that L_{MV} is meaningless syntax until the function is defined (regarding it as a 'translation' assigns meaning to the sentences L_{MV}). It's hard to see how the unconstrained choice of a meaningless syntax and a mapping of that syntax to L_ε could be trusted to tell us anything about the sentences that happen to fall in its range. As Meadows and I reconstruct the argument, set theorists easily understand the talk of universes and sets in L_{MV}, there's good reason to regard the relevant mapping to L_ε as a translation, and a particular virtue of L_{MV} sentences is shared by the synonymous L_ε sentences in its range.

[13] Thanks to Professor Steel for helpful comments on a previous draft (see footnote 7).

Maddy, P. (2019a).What do we want a foundation to do? In S. Centrone, D. Kant, D. Sarikaya (Eds.), *Reflections on the foundations of mathematics* (pp. 293–311). Springer.

Maddy, P. (2022b). 'Enhanced if-thenism'. In *A plea for natural philosophy and other essays* (pp. 262–293). Oxford University Press.

Maddy, P. (2022h). *A plea for natural philosophy*. Oxford University Press.

Meadows, T. (2021). Two arguments against the generic multiverse. *Review of Symbolic Logic, 14*, 347–379.

Schatz, J. (2019). *Axiom selection by maximization: V=Ultimate L vs. forcing axioms*, Ph.D. dissertation, UC Irvine.

Steel, J. (2014). 'Gödel's program'. In J. Kennedy (Ed.), *Interpreting Gödel* (pp. 153–179). Cambridge University Press.

Chapter 4
Intrinsic and Extrinsic Evidence in Set Theory

Donald A. Martin

Abstract This paper discusses intrinsic and extrinsic justification and evidence in Maddy's work. I agree that extrinsic justification is at present more important than intrinsic justification, but I am more optimistic than she is about intrinsic evidence in set-theory. The main source of our disagreement is our view about the concept of set, which I explain in this paper.

Keywords Intrinsic evidence · Extrinsic evidence · Set theory · Iterative concept of set · Cantor · Global reflection principle · Categoricity · Zermelo

4.1 Introduction

The latter part of Chap. V of Maddy (2011) is devoted to a comparison of intrinsic and extrinsic justification in set theory and mathematics in general. At the end of the chapter, Maddy says that she has cast doubt on "the common opinion that intrinsic justifications are the grand aristocracy and extrinsic justifications are the poor cousins. The truth may well be reversed."

I agree with her about the importance of extrinsic justification and extrinsic evidence, and I think that they are at present more important than intrinsic justification and evidence in the search for new axioms for set theory. I think her examples of extrinsic justification are good examples. I disagree about what makes her examples good examples and about what some of her examples are examples of. Also I am more optimistic than she is about intrinsic evidence in set theory.

The main source of our disagreement is our views about the concept of set. She thinks that the concept's role is to provide "a workable heuristic picture of the sort of thing we're investigating mathematically." I think the concept is the proper subject of set theory.

D. A. Martin (✉)
University of California, Los Angeles, CA, USA
e-mail: dam@math.ucla.edu

S. Arbeiter and J. Kennedy (eds.), *The Philosophy of Penelope Maddy*,
Outstanding Contributions to Logic 31, https://doi.org/10.1007/978-3-031-58425-1_4

What is intrinsic justification? On page 124, Maddy says:

> When a principle is defended in terms customarily classified as intrinsic, various descriptors typically appear: the principle is intuitive, self-evident, obvious; it's part of the meaning of the word 'set'; it's implicit in the very concept of set; and so on.

Some of the examples in her discussion of intrinsic justification are about whether various sentences of the first-order language of set theory can be seen to be "implicit in the very concept of set"—or, as I would say, can be directly seen to be implied by the iterative concept of set. When I say that we have intrinsic justification for believing such a sentence, I will mean that we can show directly from the concept of set that the sentence would have to be true in any structure instantiating the concept. I say "would have to be true in any structure instantiating the concept" instead of "is true in every model of the concept" because: (a) I think that set theory is about the concept of a universe of sets and that it doesn't matter whether there are objects that instantiate this concept; (b) I assume that a structure fully instantiating the concept of set would be too large to be a set, so couldn't be a model. By intrinsic evidence for the truth of a sentence, I will mean evidence gotten directly from the concept of set.

4.2 The Iterative Concept of Set

The concepts I will discuss will be concepts for some first order language. For the iterative concept of set, the language is the language of set theory, the language whose only non-logical symbol is $\in$.

Maddy explains the iterative concept by saying:

> This well-known picture of the set-theoretic universe was introduced by Zermelo (1930) and subsequently entered into the fabric of set-theoretic pedagogy and practice; following Gödel's description, sets are understood as generated in a series of stages ('a set is something obtainable ... by iterated application of the operation "set of"'); at each stage we take every possible combination of the available elements[1] 'no matter whether we can define it in a finite number of words (so that random sets are not excluded)'; this process is indefinitely repeated into the transfinite (so, for example,'the totality of sets obtained by finite iteration is to be considered to be itself a set and the basis for further applications of the operation "set of"').[2]

Some find questionable the notion of "every possible combination" of the sets already generated. There is no way to avoid such notions. The concept of set is a fundamental concept, one that is not directly definable from more basic concepts. The same is true of the concept of natural numbers. Gödel's account is rather sketchy about the length of the iteration process. This isn't surprising, since that length is perhaps the most problematic aspect of the iterative concept of set. Maddy says that Gödel's

---

[1] I.e., every such combination of the sets already generated is generated as a set.

[2] Maddy (2011), page 124.

account of the iterative concept shouldn't be taken literally. She says, "Of course the language of 'generation' and 'iteration' is regarded as metaphorical, merely a colorful way of describing the structure of the set-theoretical universe, V." "For each stage we take" is literally metaphorical, but the language of generation and iteration occurs widely in mathematics, and it does not need to be understood metaphorically. (I do not mean to imply that Maddy thinks it needs to be so understood.) Explaining it in the case at hand involves explaining the notion of a well-ordering. One way to do this is by using the notion of "every possible combination."

Though there isn't a general problem with transfinite iteration, specifying a transfinite iteration involves specifying the length of the iteration. In the case at hand, this length is the length of the sequence of all ordinal numbers. For each ordinal α, the sets generated before the αth stage form a set called V_α. (Thus I am assuming that V_0 is the empty set.) In the quoted passage, the only thing said about the process's length is that there should be transfinite stages, though Gödel says more elsewhere about the ways that the process should not be too short. On pages 124–126 of Maddy (2011), Maddy discusses several possible non-shortness conditions.

The main attempt to specify the length of the iteration that generates sets comes from Cantor. Cantor probably didn't have an iterative concept of set, but he did have an iterative concept of ordinal number. In his 1883 Grundlagen[3] he presented an account of how the ordinal numbers are generated, one by one, in what he called an absolutely infinite process. In Grundlagen, ordinal numbers, which he called simply numbers (Zahlen) are the objects generated in this process. Being such objects is what they are. They are not defined as the order types of well-orderings, though for each well-ordering there is an ordinal number that is what Cantor called the Anzahl of the well-ordering. Cantor said that the ordinal numbers form an absolutely infinite sequence. He regarded absolute infinity as the true infinity, and he thought that ordinary infinity, transfiniteness, has much in common with finiteness. Each of his Grundlagen ordinal numbers has only finitely or transfinitely many predecessors. He made it clear that for him sets are precisely those multiplicities that are finite or transfinite, although his official definition of a set was "a multiplicity which can be thought of as one." Later in his career, he also called absolutely infinite multiplicities inconsistent multiplicities.

Cantor specified the process of generating the ordinal numbers by giving two principles of generation. The first principle asserts that there is no last stage. What the second principle asserts is that every unbounded transfinite sequence of ordinal numbers has a least upper bound (so there is no unbounded set-size sequence of stages). Cantor tacitly used this—essentially a Replacement principle—to get what he said was a theorem that for every ordinal γ there is a γth number class.[4]

---

[3] See Cantor (1883), or the translation, Cantor (2007).

[4] At the time of the Grundlagen, he seemed to think that a clearer specification was needed of when one of the two generating principles applies at given stage. He introduced a third principle of generation, a restrictive condition that gives an unambiguous necessary condition for the process to continue at a given stage. This condition was too restrictive, ruling out ordinal numbers that he did not want to rule out.

Today limitation of size is standardly treated as what determines sethood. Calling the totalities that are not sets the "absolutely infinite" ones is not standard, but there are examples of it. Later I will discuss one of them: the work of Welch and Horsten on Global Reflection Principles. Cantor did not make the notion clear, and he mixed it up with theology. In the Grundlagen, at least, he didn't even say whether all absolutely infinite multiplicities have the same size. Since "absolute" and "absolutely infinite" were synonyms for him, one would think that the answer is "yes." What he said was that the absolute is "in God," that it is "not determined," and that it "cannot be numbered." In his 1899 letters to Dedekind, he said that the multiplicities that are not sets count both as inconsistent multiplicities and as absolutely infinite multiplicities. In these letters, he asserts that two equivalent (same size) multiplicities are either both inconsistent or both sets. This statement presupposes that it makes sense to compare the sizes of inconsistent multiplicities, but Cantor doesn't say whether there can be inequivalent inconsistent multiplicities. Both the early and the late Cantor regarded the multiplicity of all ordinal numbers as absolutely infinite.

I will take the following to be part of the iterative concept of set: There are absolutely infinitely many ordinal numbers and no ordinal number has absolutely infinitely many predecessors. This seems the best attempt we have of asserting the non-smallness of the length of the iterative hierarchy. Note that the iterative concept, as I have described it, neither affirms nor denies that there are absolutely infinite multiplicities or whether such multiplicities would be actual—as opposed to potential—things. What it implies is that such objects would not be sets.

4.3 Zermelo's Concept of Set and His Categoricity Theorem

Maddy[5] says that the iterative concept of set was introduced in Zermelo (1930). It is fair to say this. Zermelo's concept agrees with the modern concept of set in that Zermelo's sets are generated by a transfinite process like the one Gödel describes. His theorems in Zermelo (1930) readily adapt to yield related theorems about the modern concept. Nevertheless, his concept differs from the modern iterative concept that I have been discussing in important ways. In particular, it is not categorical. His concept could be thought of as a result of trying to avoid problems about multiplicities that are too large to be sets (and perhaps going too far in trying to avoid those problems).

In Zermelo (1930) Zermelo described a concept by—in effect—axiomatizing it. I will present his axiomatization, which he calls ZF', by saying how it differs from first-order ZFC. Models of ZF' have at least one urelement, and how many urelements there are varies from model to model. There is no Axiom of Choice and there is no Axiom of Infinity. In ZF' the Axioms of Foundation, Separation, and Replacement are not stated in the first-order language of set theory. They are second-order, i.e., their conclusions are supposed to hold for all propositional functions (i.e., for all

[5] Maddy (2011).

definite properties and relations), not just those given by first-order formulas. The reason for the absence of Choice is that Zermelo, like Cantor, considered Choice to be a logical principle. The reason for the absence of Infinity is that Infinity's sole effect is to keep the hereditarily finite sets from counting as a model of the axioms, and Zermelo thought doing this is unnatural, though he did not regard the hereditarily finite sets as a model of "Cantorian set theory." I will ignore the absence of Choice and the existence of urelements: By "Zermelo's Axioms" I will mean ZF' plus Choice and without urelements.

One can doubt that the second-order axioms have a definite meaning. What one should not doubt is Zermelo's isomorphism theorem:

> Let $\mathfrak{M}$ and $\mathfrak{N}$ be models for the language of set theory that satisfy Zermelo's axioms. Either $\mathfrak{M}$ and $\mathfrak{N}$ are isomorphic or else one of them is isomorphic to some V_α of the other one.

If $\mathfrak{M}$ and $\mathfrak{N}$ are models of Zermelo's axioms, then the second order objects used in the proof that the two models are isomorphic are definable from those models.

Zermelo suggested a "meta-axiom" implying that the sequence of ordinal numbers is isomorphic to the sequence of *boundary numbers*, where a boundary number is the ordinal number of the well-ordered collection of objects that in some model of the axioms represent von Neumann ordinal numbers.[6]

Zermelo had the ingredients for putting together all the models of his axioms to get a structure that could be considered the universe of sets. This, however, is not what he did. He thought this would yield paradox. He did not think there is such a model, just as he didn't think there is a totality whose members are the ordinal numbers. Yet he seemed to think that it was legitimate to talk about all the ordinal numbers or about all the models of his axioms.

For him, the concept of set is a non-categorical concept whose instantiations are the models of his axioms. Each of these models instantiates what I am calling the iterative concept of set, except that the requirement about absolute infinity is replaced by the weaker second-order Axiom of Replacement. Zermelo thus avoided the problem of distinguishing sets from totalities that are too large to be sets. The latter do not exist. Zermelo's way of regarding the totality of all ordinals and the totality of all models of his axioms is reminiscent of Cantor's view in Grundlagen that such totalities are absolutely infinite and can't be "thought of as one." But Zermelo goes further, in a way that seems incoherent. He quantifies—or meta-quantifies—over the ordinal numbers and over the models of his axioms, and he talks of the sequence of all Cantor ordinals and the sequence of all models of his axioms, but he does not treat either as forming a totality, even an "inconsistent one." I understand why he won't consider a model of his axioms gotten by amalgamation, but I don't understand why he won't quantify over its elements in the same way that we quantify over sets in ZFC.

---

[6] Whereas what the sets are is relative to a model of the axioms, the ordinal numbers seem to be just "Cantor ordinals."

4.4 Intrinsic Evidence

The modern iterative concept of set is not the concept of a model of Zermelo's axioms. One way to describe what I am calling the iterative concept is to say that it is the concept of a structure that satisfies Zermelo's axioms plus the axiom saying that there are absolutely infinitely many ordinal numbers and no ordinal number has absolutely infinitely many predecessors. Let us call this extended set of axioms the Strong Zermelo Axioms.

Whatever problems there may be with the notion of absolute infinity, it is clear that no structure satisfying the Strong Zermelo Axioms can be the V_α of another such structure. Hence the Strong Zermelo Axioms are categorical. Zermelo's argument also shows that the concept of V_α is categorical (for any α clearly enough specifiable that it makes sense to talk about the concept of V_α).

As I indicated near the beginning of this paper, I think that the concept of set should be regarded as the main subject matter of set theory. What it should mean for a set-theoretic sentence to be true should be that it is—in Gödel's words—implied by the concept of set, where I mean by this that the sentence would have to be true in any structure instantiating the iterative concept (whether or not we know that it is so implied). It seems to me that this deals with the Benacerraf problem about mathematical knowledge better than Maddy's ways for solving it. To me, anyway, it is easier to understand. It also seems more objective. The question of whether a sentence of the language follows from the iterative concept of set or the concept of natural numbers may be difficult or impossible to answer, but it avoids the kind of subjectivity that could infect judgements of mathematical fruitfulness or depth. Finally, it seems more plausible to say that Maddy's proper set-theoretic method justifies introducing and studying a concept, than to say that it justifies adopting an axiom.

I'll now start to consider Maddy's view that "intrinsic justifications are secondary to the extrinsic."

At one point she imagines that some new axiom candidate for set theory is found, an axiom that conflicts with our current concept of set but is "mathematically fruitful" in the ways she has been considering.[7] She suggests that we would accept this axiom.[8] In spite of all the talk about paradoxes, the main line of the development of the concept from Cantor until now has been pretty free of this phenomenon. The concept changed by growing. Of course, set theorists have studied—and will study in the future—axioms that are incompatible with the standard concept. For example, Quine's New Foundations is still sometimes studied, but not with the idea of replacing the iterative concept by it.

Suppose Maddy's scenario occurs in the future. What part of the current concept will set theorists be willing to give up? I can't imagine an example that would affect

[7] See the last paragraph of page 135.

[8] I wonder what Maddy would say about the possibility of an analogous situation involving the concept of the natural numbers.

V_α for small α. The change would seemingly have to concern only the higher reaches of the iterative hierarchy.

A more plausible situation would be a set-theoretic analogue of the historical example that Maddy discusses: changing the axioms for groups by requiring the existence of inverses.[9] There are two possible versions of that example. The new concept could be for the same language as the old, or it could add a unary function symbol. Let's choose the latter version.

The new concept is more restrictive than the old one. If a structure instantiates the new concept, then its reduct instantiates the old concept, but there are structures that instantiate the old concept and are not the reducts of any structures that instantiate the new concept.[10]

Consider a set-theoretic analogue of the group-theoretic example in which the iterative concept of set plays the role of the old group concept. Assume then that the language for the new concept is an expansion of the language of set theory, and assume that the reduct of any structure that instantiates the new concept instantiates the iterative concept. It follows, by the categoricity of the iterative concept, that every structure that instantiates the iterative concept is the reduct of a structure that instantiates the new concept. Thus there isn't any full set-theoretic analogue of the group-theoretic example.

Before discussing Maddy's examples of proper set-theoretic method, I'll make a side remark. If one pays attention to actual current set-theoretic practice, one might conclude that trying to find new axioms and justify them is only a small part of that practice.

In Sect. 2.2, Maddy gives four examples of proper set-theoretic method. These examples are (1 and 2) Cantor's and Dedekind's introductions and uses of sets as single objects, (3) Zermelo's defense of his 1908 axioms for set theory, especially his defense of the inclusion of Choice among the axioms for set theory, and (4) the late 20th century arguments for $AD^{L(\mathbb{R})}$. I have no problem with Maddy's discussion of how these are examples of good mathematical practice. What I want to stress is the following. There is a difference between

(a) choosing what to study: introducing, sharpening, otherwise modifying, or replacing concepts, and
(b) studying the chosen concepts, in particular studying what sentences are implied by them.

[9] This can be thought of as adding a new concept and using what had been the name of an old concept to name the new one.

[10] Perhaps I should have said (to avoid ontological commitment), Our concept of set implies that if a structure instantiates the new concept ...".

(1) and (2) (Cantor's and Dedekind's introductions of sets) are stages in the creation of the iterative concept of set, and they should be regarded as examples of (a).

Zermelo's 1908 axioms for set theory constitute another stage in this process.[11] One might think of Zermelo as making proposals that certain statements should—and others should not—be part of the concept of set. One such proposal is that the concept should be purely about sets, not about sets and other objects. Not only did he avoid having official symbols for ordered pairs and functions, he didn't even introduce symbols for ordered pairs and functions by defining them to be certain sets. Most of his defense of the Axiom of Choice can be thought of as giving extrinsic evidence that Choice should be accepted as part of the concept of set. The only thing that doesn't fit well with this is his citing the self-evidence of Choice. Here he seems to be saying that we can see that Choice is already part of our concept of set. Maddy's first three examples of proper set-theoretic method are thus examples of (a). They—like the group-theoretic example—are examples of mathematical concepts' being improved or replaced in a way conforming to proper mathematical practice. Another example of (a) is Cantor's extending the concept of counting numbers by adding transfinite numbers. A good part of his justification of this is very much like what Maddy's justification of it would be.

Maddy's last example of proper set-theoretic method feels different from the others. By the time determinacy and $AD^{L(\mathbb{R})}$ became subjects of research, we already had the full iterative concept of set. To me it seems more natural to say that we have strong extrinsic evidence that $AD^{L(\mathbb{R})}$ is true of the iterative concept than it is to say that, because $AD^{L(\mathbb{R})}$ has properties such as leading to deep mathematics, we are justified in altering the concept of set by taking $AD^{L(\mathbb{R})}$ as an axiom. In any case, nothing would prevent us from studying the concept that would result if it were taken as an axiom.

To what degree can we use the categorical axiomatizations of the iterative concept of set and the categorical concepts of the levels V_α to discover which first-order sentences are implied by these concepts?

Gödel (1990a) says, "That new mathematical intuitions leading to a decision of such problems as Cantor's continuum hypothesis are perfectly possible was pointed out earlier." After discussing large cardinal axioms, he says on page 261, "there may exist, besides the usual axioms, the axioms of infinity, and the axiom mentioned in footnote 18, other (hitherto unknown) axioms of set theory which a more profound understanding of the concepts underlying logic and mathematics would enable us to recognize as implied by these concepts." Gödel is talking about possible intrinsic justification of axioms that imply or refute the Continuum Hypothesis. I take him to be expressing the hope that we might be able to discover the truth-value of the

[11] Most of Zermelo's axioms had already been used implicitly or explicitly by Cantor. Even in the early Grundlangen, he had used what amounted to special cases of Separation and Replacement. His 1899 letters to Dedekind had already asserted two of the Zermelo axioms: Union and Separation. In the same letters, he had formulated a principle equivalent to Replacement in the presence of Separation and Power Set. Zermelo didn't realize that Cantor had done this, partly because Cantor's formulations used proper classes, which Zermelo rejected.

Continuum Hypothesis directly from the concept of set (and other mathematical and logical concepts).

Why do we need extrinsic evidence in set theory? After all, every true sentence is implied by the concept of set. A similar statement holds for the concepts of the levels V_α. The answer, of course, is that there is no system of deduction for implication in these senses. The set of sentences implied by the concept of set is far from being recursively enumerable.

For the iterative concept of set and the concepts of the levels V_α, consider the following question. For what sentences σ of the language of set theory do we have at present intrinsic justification for believing that σ is implied by the concept?

Consider first the concept of the V_α's (or equivalent concepts). To avoid worries about whether it makes sense to talk about V_α, let us restrict ourselves to α less than, say, ω_1^{CK}. For the concept of the natural numbers (i.e., for the concept of V_ω) σ has to be something not far from being a theorem of first-order Peano Arithmetic.[12] For the concept of the natural numbers and the sets of natural numbers—i.e., for the concept of $V_{\omega+1}$—σ has to be something not far from being a theorem of the first-order theory called "Second-Order Arithmetic." And so it continues up the iterative hierarchy. For any ordinals α and $\beta > \alpha$ and any first-order sentence σ about V_α can be expressed as a first-order sentence σ' about V_β and the parameter α. In many cases, σ' is provable in the standard theory of V_β. When this happen, it provides intrinsic evidence that σ is implied by the concept of V_β, but it gives only extrinsic evidence that σ is implied by the concept of V_α.

Of course, it's possible that we might have "new mathematical intuitions" that would greatly increase the intrinsic evidence we have about these concepts. Until such a miracle occurs, we will have to rely on ZFC and extrinsic evidence to learn about the V_α's.

Now let's consider the full iterative concept of set. Here the absolute infinity requirement makes a significant difference. There is intrinsic justification for a good deal more than the sentences provable in ZFC. Also intrinsically justifiable are most known large cardinal axioms consistent with $V = L$. The evidence that stronger large cardinal axioms (e.g., measurable cardinals and Woodin cardinals) are implied by the iterative concept has been—at least until very recently—almost all extrinsic.

It is not impossible, though, that more can be intrinsically justified. Some evidence for this is an interesting idea of Philip Welch for getting intrinsic evidence for such strong large cardinal axioms. This idea is presented and studied in Welch and Horsten (2013) and Welch (2017).

The idea involves a change in the concept of set of the sort I discussed earlier. Actually the iterative concept of set is unchanged, but the new concept involves sets and other objects. These objects are treated syntactically like proper classes, but they are not supposed to be taken as mathematical objects. They are to be understood mereologically as actual objects that are Cantorian absolutely infinite parts of the

[12] Why not just the theorems of PA? One sort of example is arithmetic sentences expressing that PA is consistent, since any instantiation of the concept would have to be a standard model of PA.

universe V of sets. Their mereological nature is supposed to imply that they couldn't be elements of further set-like objects.

The new concept can be taken as a concept for the language of NBG set theory, and we may assume that it satisfies the axioms of NBG.

In any instantiation of the concept, let $\mathcal{C}$ be the collection of all classes.

The Global Reflection Principle (GRP) asserts that there is an ordinal κ and an elementary embedding

$$j : (V_\kappa, \in, V_{\kappa+1}) \to (V, \in, \mathcal{C})$$

with critical point κ. "Elementary" means that all quantifiers are over sets, but class parameters are allowed.

The authors of Welch and Horsten (2013) say, "The principle GRP says that the universe with its parts is, to a certain degree, indistinguishable from at least one of its initial segments V_κ and its parts. It says that the whole set-theoretic universe with all its proper classes is mirrored in a set-sized initial segment $(V_\kappa, \in, V_{\kappa+1})$, where the first-order quantifiers range over V_κ, and where the reflection of a proper class X is obtained by cutting it off at level V_κ." They say that this confers intrinsic plausibility on the principle. Reflection principles are often thought of as saying that the length of the iterative process is so great that there must be stages V_κ that closely resemble V, so their assertion is not unreasonable.

GRP implies, for example, the existence of a proper class of Woodin cardinals. Thus it does what earlier reflection principles have not done: implying the large cardinal axioms that imply $\mathrm{AD}^{L(\mathbb{R})}$ and its absoluteness under forcing. It is deducible from stronger large cardinal axioms.

I have mentioned GRP not to advocate its adoption but to suggest that intrinsic evidence is more promising than it may seem to be.

References

Benacerraf, P., & Putnam, H. (Eds.). (1964). *Philosophy of mathematics, selected readings*. Prentice Hall.

Browder, F. (Ed.). (1976). Mathematical developments arising from Hilbert problems. In *Proceedings of the Symposia in Pure Mathematics* (Vol. 28). American Mathematical Society.

Cantor, G. (1883). Grundlagen einer algemeinen Mannigfaltigkeitslehre: Ein mathematisch-philosophischer Versuch in der Lehre des Unendlichen. Teubner.

Cantor, G. (2007). Foundations of a general theory of manifolds: A mathematico-philosophical investigation into the theory of the infinite. In [Ewald, 2007] (pp. 881–920).

Ewald, W. (2007). *From Kant to Hilbert* (Vol. 2). Oxford University Press.

Feferman, S., Dawson, Jr., J. W., Goldfarb, W., Parsons, C., & Solovay, R. M. (Eds.) (1995). *Kurt Gödel: Collected works, volume 3: Unpublished essays and letters*. Oxford University Press.

Feferman, S., Dawson, Jr., J. W., Kleene, S. C., Moore, G. H., Solovay, R. M., & van Heijenoort, J. (Eds.). (1990). *Kurt Gödel: Collected works, volume 2: Publications 1938–1974*. Oxford University Press.

Gödel, K. (1990a). What is Cantor's continuum problem? In [Feferman et al., 1990] (pp. 254–270). Reprinted from [Benacerraf and Putnam, 1964], 258–273, which is a revised and expanded version of [Gödel, 1990b].
Gödel, K. (1990b). What is Cantor's continuum problem? In [Feferman et al., 1990] (pp. 176–187). Reprinted from American mathematical monthly (1947) (Vol. 54, pp. 515–525).
Maddy, P. (2011). *Defending the axioms*. Oxford University Press.
Welch, P. (2015). Global reflection principles. In *Logic, methodology and philosophy of science proceedings of the 15th international congress*.
Welch, P., & Horsten, L. (2013). Reflecting on absolute infinity. *Journal of Philosophy, 113*, 89–111.
Zermelo, E. (1930). Über Grenzzahlen und Mengenbereiche: Neue Untersuchungen über die Grundlagen der Mengenlehre. *Fundamenta Mathematicae, 16*, 29–47.

Chapter 5
Reply to Martin

Penelope Maddy

Keywords Set theory · Intrinsic/extrinsic · Mathematical depth · Iterative conception · Conceptualism · Absolute infinity

Professor Martin takes issue with my claim that extrinsic considerations are what serve to justify methodological decisions in set theory—the adoption of new concepts or new axioms—and that intrinsic considerations are of value only instrumentally, insofar as they guide or facilitate the search for extrinsic supports.[1] My thought is that extrinsic appeals to mathematical advantage embody a familiar type of means/ends reasoning: if these are your goals, and a proposed method facilitates those goals without interfering with others, then it's rational to adopt that method. But these extrinsic considerations, however well-suited to their goals, only amount to justification if the goals themselves are appropriate, that is, if those goals manage to track deep mathematics (or 'fruitful' or 'effective' or just 'good' mathematics).[2] Set theory, on this view, isn't the study of an abstract subject matter, but the construction of a theory intended to track good mathematics.[3]

When I speak of depth, I have in mind judgments like the one Martin mentions: the concept of group with inverses is preferable to—deeper than—one with cancellation laws alone, because it more effectively furthers the goals of abstract algebra, that is, it accommodates more structures from more diverse areas of mathematics and

[1] See Maddy (2011), V.4.

[2] Cf. Maddy (2011), p. 81. See also Ernst et al. (2015) for more on 'depth'.

[3] I speak here with the Arealist of Maddy (2011), IV. At one point, Martin wonders whether I would treat number theory in the same way as set theory (p. 51, footnote 10). The answer is no (see Maddy (2014b), (2018b)).

P. Maddy (✉)
Department of Logic and Philosophy of Science, University of California, Irvine, CA, USA
e-mail: pjmaddy@uci.edu

S. Arbeiter and J. Kennedy (eds.), *The Philosophy of Penelope Maddy*,
Outstanding Contributions to Logic 31, https://doi.org/10.1007/978-3-031-58425-1_5

facilitates a rich and powerful theory of features they all share simply because of their group structure.[4] In other words, the group concept is mining a strain of deep mathematics. Similarly, Cantor's theory of transfinite numbers is rich and productive, which provides extrinsic justification for adopting his axiom of infinity.

As for intrinsic considerations, it's obvious that the iterative conception has been an immensely powerful tool, an intuitive picture that makes the theory more accessible, guides its development and suggests new avenues, like large cardinal axioms or reflection principles. This is the kind of work that intrinsic considerations do so well—as a heuristic, an inspiration. But the ultimate test, the true measure of worth, for anything so generated, or for the concept itself, lies in its extrinsic success.

In opposition to this view, Martin continues a conceptualist line of thought he's been pursuing for some years now,[5] this time making a firm commitment: 'I think the concept [of set] is the proper subject of set theory' (p. 45).[6] According to Martin, a set-theoretic sentence is intrinsically justified if it can be shown

> directly from the concept of set that the sentence would have to be true in any structure instantiating the concept. (p. 46)

As Martin understands it, the relevant concept consists of ZFC^2 (second-order ZFC) plus

> the axiom saying that there are absolutely infinitely many ordinal numbers and no ordinal number has absolutely infinitely many predecessors. (p. 50)

The key notion of 'absolute infinity' here derives from Cantor. And finally, whether or not a sentence is implied (in Martin's sense) by the concept of set is objective (p. 50), which I take to mean that it's neither subjective—a matter of what I think—nor intersubjective—a matter of what we think.

Let me isolate three features of this account. The first has to do, not directly with the concept itself, but with its implications. When Martin requires that an implied sentence 'would have to be true in any structure instantiating the concept', he obviously doesn't mean structures in the sense of model theory—which consists of theorems straightforwardly provable from our official theory of sets. He's operating instead in an arena prior to our official theory and its theorems, where judgments are made about the propriety of that theory and its extensions. Presumably the formulation 'would be true' is intended to avoid the assumption that any such structures actually exist, but there are still modal facts in play—facts about what structures instantiating the concept *could* exist. In other writings, Martin appears to reject Gödel's abstract ontology of concepts,[7] but it's not clear that his modal commitments are any less substantive.

---

[4] Early on, the focus was on finite groups, so cancellation laws were enough to guarantee the existence of inverses. When attention turned to infinite groups, inverses had to be posited directly. See Maddy (2011), pp. 7, 79–80, 134–135.

[5] See, for example, Martin (2018).

[6] Unaccompanied page numbers refer to Martin, this volume.

[7] See, e.g., Martin (2018), pp. 1–2: 'Feferman takes mathematical concepts to be human creations. He calls them objective, but the objectivity seems ultimately only intersubjective. Gödel thinks that

The other two features concern the concept itself. The first is its unabashedly second-order nature: each stage of the hierarchy includes 'every possible combination' of what's gone before, in the sense that Zermelo has shown to be quasi-categorical (pp. 46, 49). Those uncertain, for example, that anything currently settles CH will wonder what underlies this full combinatorial concept.[8] The second striking feature of Martin's concept is its reliance on the notion of 'absolute infinity', where this addition is said to make Martin's concept, unlike Zermelo's, fully categorical (p. 49). Asked to explain what grounds this fact of the matter about precisely how far 'the absolute' extends, Martin would most likely continue to reject any abstract existents, but something or other in the same ambient metaphysical arena where his modal facts about structures and 'all possible subsets' hold true must be at work here as well. Those of us insensitive to metaphysical distinctions between ontology and modality might be forgiven for wondering how this improves on simple positing the universe, V, with its determinate width and height.

Thus the metaphysics. Martin's epistemology deploys a mix of intrinsic and extrinsic evidence. On the former, ZFC^2 is included in the concept, so its axioms trivially enjoy intrinsic support. Given that many extrinsic considerations have been offered in support of the various axioms of ZFC,[9] Martin's strong appeal to intrinsic evidence here might seem surprising, but he neatly sidesteps these concerns: any extrinsic considerations apparently evoked in favor of ZFC are actually just part of deciding which concept study, of forming the concept of set. So, for example, where I see the rich and productive theory of transfinite numbers as extrinsic support for Cantor's adoption of the axiom of infinity, Martin sees those same considerations as recommending Cantor's concept, with the axiom of infinity, as preferable to study (pp. 51–52). I have no objection to this formulation; I'm happy to describe the process of forming our theory of sets in response to considerations of mathematical advantage as one of molding our concept. The only difference between Martin and me on this point is that I think this same process of extrinsically-guided concept-formation is what properly informs all our methodological decisions, while for him, there are also times when our concept is fixed, when we progress instead by mining its implications.

How, then, is this mining done? This is where we might expect an explanation of how we come to know, by intrinsic means, which statements *would* be true in every

mathematical concepts are genuine objects, part of the basic furniture of the world. For him they are non-spatio-temporal entities. I would say that I stand in between the two, but in fact I don't have much to say about the ontology of mathematical concepts. I think that it is correct to call them objective and that it is more correct to say that they were discovered than that they were created by us.'.

[8] Martin himself retains some uncertainty about the determinateness of CH (p. 50, footnote 7), though this theme is less prominent here than in Martin (2018). All of us have to live with some trace of uncertainty about the consistency of ZFC, but Martin's worry here is that even if ZFC is consistent, an instantiation of ZFC^2 is might be impossible, that is, in his terms, ZFC^2 might be 'incoherent'. (Presumably there's an objective fact on this, one way or the other, in Martin's modal metaphysics.) Given that this concern plays no role in the rest of Martin's discussion, I set it aside in the text (but see footnote 11, below).

[9] See Maddy (1988), §I, or Maddy (1997), I.3.

structure in the ambient metaphysics that instantiates the concept—of how we gain epistemic access to those modal facts or to the details of 'all possible subsets'—but again Martin surprises. Though we get a familiar nod to the idea that his concept's inclusion of 'absolute infinity' yields intrinsic justification for small large cardinals, beyond that, it looks as if all that's intrinsically justified is what's provable from first-order formulations (p. 53).[10] Indeed, the fact that 'there is no system of deduction' for his second-order concept, explains why 'we need extrinsic evidence in set theory' (p. 52). Under the circumstances, it's easy to see why Martin would think extrinsic considerations 'are at present more important than intrinsic justification ... in the search for new axioms for set theory' (p. 45).

The remaining question, then, for Martin's epistemology is how extrinsic considerations manage to inform us about what structures there could be, what subsets are possible, in his robust metaphysics. How do extrinsic considerations serve, not just to guide concept-formation, but to reveal implications of the current concept that aren't present among the first-order consequences of its first-order formulation?[11] His example is $AD^{L(R)}$:

> To me it seems more natural to say that we have strong extrinsic evidence that $AD^{L(R)}$ is true on the [concept of set] than it is to say that, because $AD^{L(R)}$ has properties such as leading to deep mathematics, we are justified in altering the concept of set by taking $AD^{L(R)}$ as an axiom. (p. 52)

Beyond the bald claim of 'naturalness',[12] we're given no account of how extrinsic evidence manages to give us this kind of insight into the inner workings of the concept. Presumably such an account would have to explain how those extrinsic considerations track the modal facts in Martin's ambient metaphysics. This is where his epistemology finally gets difficult, but Martin is silent. Without more than he gives us, it's hard to see how his account 'deals with the Benacerraf problem about mathematical knowledge' (p. 50).

(To be fair, I should note that I've foreshortened this remark of Martin's; what he actually says is that his position deals with the Benacerraf problem 'better than Maddy's ways for solving it' (p. 50). Perhaps it's worth remarking, at least in passing, that I don't so much solve Benacerraf's problem as reject the presupposition that sets it in motion in the first place, namely what I call Robust Realism.[13] My objection is that the Robust Realist could be confronted with overwhelming mathematical

---

[10] Plus a bit more: see p. 53, footnote 14.

[11] Notice that if ZFC^2 is incoherent, presumably we would have to find this out by extrinsic evidence. It's not clear how this would go. If, for example, we had extrinsic evidence that P was implied by the concept of set and likewise extrinsic evidence that not-P was so implied, wouldn't these two batches of evidence undercut one another?

[12] See also Martin (2018), p. 27: 'the question of whether the [extrinsic] evidence for determinacy axioms and large cardinal axioms justifies belief that these axioms are implied by the concept of set is a difficult one. I would like to point out, though, that a lot of the evidence for large cardinals and determinacy ... does indeed feel like evidence for *truth* and not just for satisfying methodological desiderata [perhaps a nod to my position].'.

[13] See Maddy (2011), II.4. (I did attempt to solve it in Maddy (1990); Maddy (2011), p. ix explains why I soon rejected that approach.).

considerations in favor of axiom candidate and still refuse to adopt it until it had passed some additional test—like, for example, the requirement that it be implied by the concept of set. The fundamental point, it seems to me, is that the mathematical reasons should be enough on their own. This objection is prior to Benacerraf's worry that the additional test may be impossible to pass).

Surprisingly, Martin concludes his discussion of $AD^{L(R)}$ with the remark that

> In any case, nothing would prevent us from studying the concept that would result if [$AD^{L(R)}$] were taken as an axiom. (p. 9)

Here he reverts to the unproblematic use if extrinsic considerations to guide concept-formation, a use on which we agree. If Martin is open to this shared understanding of cases like this, the sharp contrast between us begins to blur.

My best guess is that the real disagreement hinges on objectivity; he writes that his account

> feels more objective. The question of whether a sentence of the language follows from the iterative concept of set ... may be difficult or impossible to answer, but it avoids the subjectivity that could infect judgements of mathematical fruitfulness or depth. (p. 7)

It's certainly true that our judgments about the depth of various bits of mathematics can be distorted by factors like fashion or self-interest or the preferences of funding agencies, but my claim in *Defending the Axioms* is that mathematical depth itself is nevertheless an objective matter[14]: some concepts, like group, simply do generate a richer array of interconnections; some assumptions, like the axiom of infinity, simply do generate richer theories. As with any objective matter, our judgments can, and sometimes do, go wrong, and a radical skeptic can frighten us with the thought that they might do so in every case. But there's this difference between objective facts about depth and objective facts about Martin's modal metaphysics: the extrinsic considerations that shape our mathematical concepts and guide our selection of mathematical assumptions are directly designed to uncover areas of apparent depth, which (short of radical skepticism) we have reason to hope are often tracking real depth, but there's no obvious connection between those extrinsic considerations and Martin's modal reality (the same lacuna as noted above in connection with Benacerraf's problem).

Still, there remains another type of objectivity that Martin's account promises to deliver and mine does not, for example, in the case of CH: one of CH and not-CH follows from his concept, from ZFC^2, though of course we don't currently know which. For me, there may or may not be a deep way of developing ZFC that would settle the question—this, like Martin's conceptual implication, is an objective fact that we don't now know—but if there is no such deep development, then there is, for me, no further fact of the matter about CH. Notice, though, that Martin's hope of finding out about CH hinges on whether there are extrinsic considerations that would reveal his underlying conceptual implication—the very same extrinsic considerations that would, for me, decide whether there is a way of settling CH. So Martin's 'such-and-such extrinsic considerations reveal that CH is true (false)' translates to my

[14] See Maddy (2011), III.4. Also, for full disclosure, see Ernst et al. (2015), pp. 245–248.

'such-and-such extrinsic considerations reveal that we should extend our theory in such a way that CH turns out to be true (false)'. What I've lost in this translation is the right to say that CH is now, currently, either true or false. What I've gained is freedom from any abstract metaphysics imposing extra-mathematical constraints on the pursuit of pure mathematics (and secondarily, from Benacerraf).

To return, at last, to Martin's central theme, it remains true that I take intrinsic considerations to be secondary, but on examination, they don't seem to play a leading role for Martin, either! At the end of his paper, he floats the possibility that Philip Welch's Global Reflection Principal (GRP) might provide intrinsic justification for stronger hypotheses like the existence of Woodin cardinals (pp. 53–54). Linnebo has questioned Welch's mereological treatment of the 'absolute infinite',[15] but beyond that, doubts have also been raised as to whether the GRP even qualifies as a reflection principle at all[16]—jeopardizing its claim to being implied by the concept of set (or the concept of set and class) and thus its ability to provide intrinsic evidence.

Ironically, my demotion of intrinsic considerations from justifications to heuristics apparently allows me to exploit them more fully.[17] From my point of view, niceties about mereology or the nature of reflection don't matter—all we want from intrinsic considerations is that they generate methods and concepts and axiom candidates that go on to gain extrinsic support. In the case of the GRP, then, implying the existence of Woodin cardinals isn't enough by itself – Woodin already formulated the notion, and it has plenty of extrinsic support—so the question is whether it gives us anything more, perhaps a more intuitive way of thinking about these cardinals that makes them easier to use, or perhaps more evidence for their consistency, as is often claimed for inner models of large cardinals.[18] If so, I'd be happy to draw out anything new the principle might facilitate, hoping for more mathematical fruit, without worrying about whether it satisfies the further constraint of being implied by the concept of set.

Most intriguing of all here, Martin leaves himself some wiggle room: if a fruitful principle is in danger of failing to be implied by the concept of set, he can always reclassify embracing it an act of concept-formation instead. But notice that this leads us back once again to concept-formation guided by extrinsic considerations, a common ground between Martin and me. At this point, the difference between his view and mine begins to look slight. To summarize, we agree that extrinsic considerations shaped the concept of set eventually codified in ZFC; we both naturally embrace anything provable from those axioms; we agree that small large cardinals are implicit in the process that formed ZFC; and we agree to rely on extrinsic considerations when extending ZFC. The key difference is that he distinguishes extensions of ZFC that reveal previously hidden features of the existing concept of set from those that serve to form a new concept—which looks to me like a distinction without a difference. So in the end, I'm left wondering, why cling to metaphysics?

---

[15] See his 'Maddy on classes', this volume.

[16] See, e.g., Koellner's talk to the 2013 VIG at UCLA.

[17] As in the discussion of Reinhardt's arguments in my 'Reply to Linnebo', this volume.

[18] This is an argument I wish I understood better.

References

Ernst, M., Heis, J., Maddy, P., McNulty, M. B., & Weatherall, J. O. (Eds.). (2015). 'Mathematical depth', special issue of *Philosophia Mathematica, 23*(2).

Linnebo, Ø. 'Maddy on classes', this volume.

Maddy, P. 'Reply to Linnebo', this volume.

Maddy, P. (1988). Believing the axioms. *Journal of Symbolic Logic, 53*, 481–511, 736–764.

Maddy, P. (1990). *Realism in mathematics*. Oxford University Press.

Maddy, P. (1997). *Naturalism in mathematics*. Oxford University Press.

Maddy, P. (2011). *Defending the axioms*. Oxford University Press.

Maddy, P. (2014b). A second philosophy of arithmetic. *Review of Symbolic Logic, 7*, 222–249.

Maddy, P. (2018b). Psychology and the a priori sciences. In S. Bangu, (Ed.), *Naturalizing logico-mathematical knowledge: Approaches from psychology and cognitive science* (pp. 15–29). Routledge.

Martin, D. A. (1998). Mathematical evidence. In H. G. Dales & G. Oliveri (Eds.), *Truth in mathematics* (pp. 215–231). Oxford University Press.

Martin, D. A. (2018). 'Completeness or incompleteness of basic mathematical concepts', 5/6/18 revision of his paper for Koellner's EFI project.

Martin, D. A. 'Intrinsic and extrinsic evidence in set theory', this volume.

Chapter 6
Maddy on Classes

Øystein Linnebo

Abstract Classes have been a central and recurring theme in Penelope Maddy's work. This article reviews and critiques her discussion of these matters. She offers a very perceptive analysis of the problem posed by classes and argues convincingly in favor of a "logical conception" of them. But her attempt to develop such a conception suffers from various shortcomings. I therefore outline some attempts to do better.

Keywords Set theory · Proper classes · Penelope Maddy · Henri Poincaré · Saul Kripke · Groundedness

6.1 Introduction

Are sets all the collections there are? Many mathematicians and philosophers think not. In addition to the sets, they think, there are classes—including *proper* classes, defined as classes that are "too large" to form sets. Although there is no set of all von Neumann ordinals or of all sets, for example, the mentioned collections exist as proper classes.

Classes have been a central and recurring theme in Penelope Maddy's work, starting in some of her earliest work,[1] resulting in two articles specifically on the topic (Maddy 1983, 2000), and discussed at length in a number of her other works.[2] Throughout her contributions, Maddy has been admirably clear about the mathematical and philosophical need for a conception of classes to supplement our far more familiar conception of sets. She has also provided an insightful analysis of the challenge that classes pose—including my favorite formulation of the concern: Why

[1] See Maddy (1981), which, as she explains on p. 511, fn. 17, derives from a chapter of her 1979 Princeton PhD dissertation.
[2] See, especially, Maddy (1988, pp. 752–54) and Maddy (1990, §3.3).

Ø. Linnebo (✉)
University of Oslo, Oslo, Norway
e-mail: oystein.linnebo@ifikk.uio.no

S. Arbeiter and J. Kennedy (eds.), *The Philosophy of Penelope Maddy*,
Outstanding Contributions to Logic 31, https://doi.org/10.1007/978-3-031-58425-1_6

aren't classes just "sets that we forgot to include" (Maddy, 1983, p. 122)? Lastly, she has made some interesting attempts to make sense of classes.

The aim of this article is to review and critique Maddy's discussion of these matters. I argue that Maddy sets up the problem in a very useful way and is right about our need for what she calls a "logical conception" of classes. However, as she herself admits, her attempt to develop such a conception suffers from various shortcomings. I will therefore end by sketching an alternative attempt to make sense of classes, which is still broadly in the spirit of her proposal.

6.2 The Need for Classes

Suppose we accept set theory, more or less as understood on the familiar iterative conception and as codified in the standard axiomatization ZFC—perhaps supplemented with certain large cardinal axioms, which assert the existence of particularly large sets. Why should we take on an *additional* commitment to classes, in particular to proper classes?

Some interesting and good reasons can be found throughout Maddy's work. To begin with, Maddy (1988, §VI) explains how classes are appealed to in ordinary set theoretic practice, especially in connection with reflection principles and other large cardinal axioms.[3] Reflection principles typically consider properties of the entire universe V of sets and states that these properties are found already in a proper initial segment V_α. While the first-order reflection principle is provable in ordinary ZFC, there are second-order versions that imply the existence of large cardinal numbers that cannot be established in ZFC alone.

Next, let me mention some other reasons people give for wanting a theory of classes.[4] The familiar set theoretic principles of Separation and Replacement are really axiom schemes, not single axioms. For each of these principles, there seems to be a single idea behind the infinitely many individual axioms that ordinary set theory uses to capture the principle. As Boolos (1984, p. 441) points out, we accept all these axioms *because* we accept the idea or principle behind them. It would be good if we could properly express this single idea, thus turning the axiom scheme into a single axiom.[5]

Next, consider the question of what set theory is about.[6] It is tempting to answer: it is about V, the cumulative hierarchy of sets. But according to set theory itself, there is no such set as V. If V exists at all, it will have to be a proper class of some sort. Some closely related question arise when we try to do the semantics of the language of set theory on its intended interpretation. What is the range of its quantifiers? What is the semantic value of the membership predicate '$\in$'? Whatever the answer to these

[3] See also Maddy (1983, p. 117).

[4] See Schindler (2019) for a recent and useful overview.

[5] See Parsons (1974) for some closely related ideas.

[6] See Parsons (1974) and Linnebo (2006).

questions is, it cannot be any particular sets. There might, however, be proper classes such as V and $E := \{\langle x, y\rangle : x \in y\}$.

None of the reasons just mentioned for wanting a theory of classes will surprise readers familiar with the philosophy of set theory. But Maddy gives a very distinctive further reason, which one rarely encounters elsewhere. This reason arises in connection with a discussion of an ancient philosophical question—famously discussed in Frege (1953)—namely: *What is a number?*[7] Maddy argues that "[n]umbers are properties of sets" (Maddy, 1981, p. 502); for example, the number 2 is a property had by all pair sets. This view has considerable plausibility. Sets are definite enough to admit an answer to the question about their number or cardinality, and this number is naturally understood as a property that the set shares with all equinumerous sets.

What are these number properties? In particular, are there distinct but coextensive properties of this sort? Consider the properties of being equinumerous with {Rome, Madrid} or with {Berlin, Paris}. In arithmetic, we do not distinguish between coextensive number properties such as the two just mentioned. On the contrary, we want to identify coextensive number properties. The resulting properties are therefore much like proper classes. Indeed, each extensional number property corresponds to the proper class of its instances and vice versa. In her earlier work, Maddy therefore took the further step of *identifying* numbers with proper classes; for example, 2 is identified with the class of all pair sets. She thus concluded that

> the problem of understanding the role of number properties within formal set theory is a special case of the problem of understanding the role of proper classes. (ibid, p. 508)

Maddy later expresses reservations about this identification.[8] But even then, she embarks on a conditional investigation of how best to make sense of numbers *if one did want them to be objects*. She argues that, on this assumption, her earlier view would still be appropriate.

6.3 Combinatorial Versus Logical Conceptions of Collection

If classes are indeed needed to supplement sets, what are these extra collections? In a famous passage, George Boolos protests:

> Wait a minute! I thought that set theory was supposed to be a theory about all, 'absolutely' all, the collections that there were and that 'set' was synonymous with 'collection'. (Boolos, 1998, p. 35)

His protest is understandable. Given the soaring ambitions of set theory, how can there be room for another kind of collection alongside sets? This is prima facie puzzling.

[7] Maddy (1981, 1990, §§3.2–3), but also Maddy (1983), e.g. p. 117.

[8] See fn. 50 on p. 100 of Maddy (1990) and surrounding discussing, where she appears to deny that properties, and numbers specifically, are objects at all.

Maddy's response is less impatient and more constructive. She articulates two useful and reasonable desiderata for a theory of classes (Maddy, 1983, p. 123):

1. classes should be real, well-defined entities;
2. classes should be significantly different from sets.

The first desideratum speaks for itself, at least for someone with realist inclinations. The second desideratum promises to address Boolos's concern. We can come to terms with the fact that sets are not "all the collections" provided that we are told how these additional collections differ from sets.

What, then, is the difference between sets and classes? Maddy distinguishes between two conceptions of collection, having to do with how a collection is "determined".[9] First, there is the conception of a set, understood as a collection *combinatorially determined* by specifying each and every element. This is a mathematical conception, which Maddy traces back to Cantor.[10] The conception was first clearly articulated by Paul Bernays, who wrote that in contemporary classical mathematics we

> abstract from the possibility of giving definitions of sets, sequences, and functions. These notions are used in a "quasi-combinatorial" sense, by which I mean: in the sense of an analogy of the infinite to the finite. [...] one views a set of integers as the result of infinitely many independent acts deciding for each number whether it should be included or excluded. (Bernays, 1935, pp. 259–60)

The combinatorial conception was further explicated in the iterative conception of sets, as developed by Zermelo (1930), Gödel (1947), Boolos (1971), and others. Second, there is the conception of a class understood as *an extension of a property or a predicate*. This is a logical conception of collection, which has its roots in Frege's notion of the extension of a concept. For example, there is a universal class and a class of pair sets, determined by the conditions '$x = x$' and 'x is a pair set', respectively.

The difference between the two conceptions of collection emerges particularly clearly when we consider tense and modality. According to Maddy,

> a class can change its membership, while a set cannot. The collection of things with a certain property can vary in membership from time to time, but as long as it is identified as the extension of the appropriate predicate, it remains the same class. A set, on the other hand, is completely identified by its membership (Maddy, 1990, p. 103)

The same goes for modality. While a set has its elements by necessity, it is possible for a class to have different members. The reason sets and classes differ in these ways is clear. Since a set is determined by specifying each and every element, there is no provision for a set to have different elements at other times or in other circumstances. By contrast, since a class is determined by its membership condition, its members vary in accordance with which objects satisfy this condition at different times and in different circumstances.

---

[9] See also Maddy (1990, §3.3 and pp. 121–24). As Maddy acknowledges, she is here inspired by Parsons (1974) (but see also Parsons (1977)) and Martin 1974. See also Lavine (1994, §V.1).

[10] Klev (2018, p. 244, fn. 4) traces the combinatorial conception back to Bolzano, Schröder, and Cantor and provides useful references.

Classes on the logical conception are therefore well suited to serve as numbers in a way that combinatorial sets are not. Reflecting on her discussion of numbers, Maddy writes:

> What we had in mind was a collection whose membership is allowed to change from time to time while it stays the same, that is, the collection which collects, at any given time, the three-element sets in existence at that time. Such a collection, even if it is small enough, is not a set. (ibid., p. 102)

An analogous point can be made about the collections we choose as the semantic values of our predicates (Linnebo, 2006). For example, the semantic value of '…is a pair set' should be allowed to have different members depending on which (possibly impure) pair sets exist relative to a time and possible world.

Both the combinatorial and the logical conception of collection are obviously threatened by paradoxes such as Russell's. According to Gödel, however, there is a stark difference between the two conceptions in this regard. He writes of the combinatorial conception that

> this concept of set […] according to which a set is something obtainable from the integers (or some other well-defined objects) by iterated application of the operation 'set of', not something obtained by dividing the totality of existing things into two categories, has never led to any antinomy whatsoever (Gödel, 1947, pp. 474–75)

That is, the combinatorial conception of set, as explicated by the famous iterative conception, has never given rise to any paradox. According to Gödel, the same cannot be said about the logical conception—either in its extensional form, as class theory, or in its intensional form, as property theory. Indeed, as Myhill (1984, p. 129) writes[11]:

> Gödel said to me more than once "There never were any set-theoretic paradoxes, but the *property-theoretic* paradoxes are still unresolved"

To make progress, let us take a closer look at the logical conception. Maddy makes some useful observations, noting that on this conception

> the entire universe is simply divided into two piles, depending on whether or not each thing satisfies the given predicate. This contrasts twice with the iterative picture: we are free to collect absolutely any things, regardless of whether they are all available at some stage, but we are not free to collect combinatorially, without recourse to a rule of any kind. The Cantorian, mathematical, collection is a set; the Fregean, logical, collection is a class. (Maddy, 1990, p. 103)

Maddy here identifies a fundamental dichotomy for our thinking about collections. On the one hand, there are logical collections, which can be "large"—in the precise sense of having members that are not all available at any single stage of the cumulative hierarchy—but which require a rule or explicit definition. On the other hand, there are combinatorial collections, each of which must be "small"—in the sense that all

---

[11] See also the views ascribed to Gödel in Wang (1996, §8.5). (Thanks to Juliette Kennedy and Mark van Atten for discussion).

of its members are available at a single stage—but where the collection can be purely combinatorial in character, without any rule or explicit definition.

The dichotomy can be made more robust. Suppose a division of "the entire universe" "into two piles" is permissible. Then there is a symmetry between the two sides of the division: each is just as legitimate as the other. This suggests that logical collections should be closed under complements: if C exists, then so does its complement $\{x : x \notin C\}$. We can now prove a simple proposition.

Proposition 6.1 *Assume there is a collection. Then Separation and Complementation are incompatible.*

Proof Suppose not. Since there is a collection, we apply Separation to obtain an empty collection and Complementation to the to obtain a universal collection U. Now apply Separation to U to get a Russell collection $\{x \in U : x \notin x\}$. As usual, this yields a contradiction. □

Thus, we are forced to choose between Separation and Complementation, as applied to any single conception of collection. As Maddy explains in the passage just quoted, Separation and the rejection of Complementation go naturally with the combinatorial conception. On the logical conception, on the other hand, Complementation is non-negotiable, thus putting great pressure on Separation. Perhaps, appearances notwithstanding, there are conditions that fail to effect a determinate division of the universe "into two piles". Some examples of how this might happen will be explored shortly.

Although the proposition represents a dichotomy for any single conception of collection, there is a cunning way to escape its jaws. Suppose we follow Maddy and accept both the combinatorial and the logical conception. Then we can retain both principles: Separation governing combinatorial sets, and Complementation governing logical classes.

With the distinction between the two conceptions under our belts, let us return to the question of what proper classes could be. The two main options, of course, are that they are large combinatorial collections of some sort or large logical collections. Maddy's early work is highly critical of the first option[12]:

> The problem is that when proper classes are combinatorially determined just as sets are, it becomes very difficult to say why this layer of proper classes a top V is not just another stage of sets we forgot to include. It looks like just another rank; saying it is not seems arbitrary. (Maddy, 1983, p. 122)

This criticism certainly sounds plausible.

In the intervening years, however, some new and attractive combinatorial options have become available. One option is to invoke plural logic, which allows us to refer to and quantify over many objects simultaneously.[13] In particular, '$x \prec yy$' expresses

[12] William Reinhardt expresses the same concern: "if you are going to add a layer on top [of the cumulative hierarchy of sets] it looks like you forgot to finish the hierarchy" (Reinhardt, 1974, p. 32).

[13] See Boolos (1984) for a seminal discussion of plural logic and Linnebo (2022) for an overview.

that the object x is one of the objects yy. Equipped with this logic, it becomes an option to hold that singular talk about proper classes would better be understood as plural talk about many sets.[14] Instead of V, for example, we can just refer plurally to all the sets. More generally, we get an appealing way to interpret and justify Morse-Kelley class theory. In terms of our dichotomy, however, plural logic deals with combinatorial collections: the predicate '$\prec$' allows us to talk about members, and the members of a plurality—that is, the individual objects among some given objects—need not be determined by any rule or explicit definition.

A related but different option is to invoke mereology, that is, the logic of part-whole relations. A proper class can then be understood as a mereological whole, and its members, as the atomic parts of this whole.[15] Again, we get an appealing way to understand Morse-Kelley class theory. And again, we are operating with combinatorial collections: each such whole has members, namely its atomic parts, and these members need not be given by any rule or explicit definition.

These later developments raise the question of whether Maddy stands by her dismissive assessment of the prospects for a combinatorial conception of proper classes. I don't know the answer. My own view is that Maddy's assessment, quoted above, was dead right and that later work on plurals and mereology shouldn't make us lose our nerve. Indeed, I argue in Linnebo (2010) that any plurality defines a set. If I am right, then to understand a proper class as a plurality will be little different from understanding it as a set that "we forgot to include". My argument turns on what is required to define a mathematical objects such as a set. Given some objects, I argue, we know precisely which set these objects would form, thus making this attempted definition permissible.

Analogous considerations apply to mereological sums of sets, provided that these too are understood combinatorially.

Maddy may not accept my argument concerning the relation between pluralities and sets, perhaps thinking it smacks too much of the "neo-Aristotelianism of ill-defined, potential entities" that she disparages elsewhere (1983, p. 123). I believe any such reaction—whether Maddy's or not—would be rash. There is nothing "ill-defined" about the contested sets. And the argument can be developed in a way that eliminates any "neo-Aristotelianism" in favor of less controversial considerations about what it takes for a mathematical definition to be permissible. On this issue, I can do no better than to quote Cantor's defense of a liberal view of what it takes for a mathematical definition to be permissible.

> Mathematics is in its development entirely free and only bound in the self-evident respect that its concepts must both be consistent with each other and also stand in exact relationships, ordered by definitions, to those concepts which have previously been introduced and are already at hand and established. (Cantor, 1883, p. 896)

Since proper classes cannot be understood as combinatorial collections, the alternative is to understand them as logical collections. This is Maddy's favored response

---

[14] This application of plural logic was a major motivation for Boolos; see e.g. Boolos (1984). For a nice development and defense of the view, see Uzquiano (2003).

[15] See Lewis (1991) and Welch and Horsten (2016).

and my own firm commitment. But the devil is in the details. Unlike ZFC and the iterative conception of sets, there is no established theory or worked-out conception of logical collection. We do, of course, find various suggestions, such as Cantor's intriguing remarks about "inconsistent multiplicities". But Maddy is surely right that Cantor leaves this notion philosophically mysterious and logically underdeveloped. Clearly, we need to do better. And when we set out to do so, we are well advised to heed Maddy's two desiderata for a theory of classes, namely that classes should be "real, well-defined entities" and "significantly different from sets".

6.4 Maddy's Development of the Logical Conception

Maddy's approach to logical classes is inspired by Kripke (1975)'s seminal approach to truth.[16] Membership facts are to be accounted for successively, at each step presupposing only facts that have already been accounted for. In this way, the entire construction will be grounded, much as facts about truth are grounded on Kripke's account.

Since we want an account that subsumes both classes and sets, we face the question of whether to operate with a single membership relation applicable to both types of collection, or with a distinct membership relation reserved for each type of collection. Maddy chooses the former, although nothing substantial hangs on this choice.

We start with the hierarchy V of sets. For reasons of simplicity, Maddy assumes that for each set s there is a constant $\bar{s}$. As she notes, this assumption can, if desired, be avoided in favor of a slightly more complicated account using variable assignments. There is also a constant $\bar{V}$ for the hierarchy of sets. Finally, she defines a collection of class terms of the form $\hat{x}\varphi$, for "the class of x such that φ". This is done by a simultaneous recursion on formulas and terms, the two crucial clauses being the following:

(i) if t and t' are terms, then $t = t'$ and $t \in t'$ are formulas;
(ii) if φ is a formula with x free, then $\hat{x}\varphi$ is a term.

Let T be the class of all these set and class terms, and let T^* be the subclass of closed terms (that is, terms with no free variables).[17]

Just as Kripke successively builds up an extension and anti-extension for the truth-predicate, Maddy's strategy is successively to build up an extension and anti-extension of the membership predicate. I will represent each of these extensions and anti-extensions as a class of ordered pairs of closed terms, that is, as a subclass of $T^* \times T^*$. Here my bookkeeping deviates slightly from Maddy's, although the

---

[16] See Maddy (1983, 2000), especially the latter, which improves on the former in various respects. In fact, as Maddy observes, "Kripke-like constructions for the case of classes preceded Kripke's own work on truth" (ibid, p. 300); see Feferman (1984), esp. Sect. 14, for discussion and useful references.

[17] Notice that we here use classes in our metatheory; see Maddy (2000, p. 315, n. 10). An alternative would be to let V be an inaccessible rank of the metatheory.

substance of the account is unchanged. I do this in order to highlight the parallelism with Kripke's construction.

Let an *extension/anti-extension pair* (or *EAP* for short) be a pair $E = \langle E^+, E^- \rangle$ of classes of the mentioned type such that $E^+ \cap E^- = \varnothing$. We can use strong Kleene logic to regard an EAP E as a three-valued model. For literals, truth and falsity in the model are defined as follows:

(1) $E \models t = t'$ iff $t = t'$ (i.e. t and t' are the same term)
(2) $E \Dashv t = t'$ iff $t \neq t'$
(3) $E \models t \in t'$ iff $\langle t, t' \rangle \in E^+$
(4) $E \Dashv t \in t'$ iff $\langle t, t' \rangle \in E^-$.

For complex formulas, the clauses are the ordinary strong Kleene ones, for example, $E \models \varphi \vee \psi$ iff $E \models \varphi$ or $E \models \psi$, and $E \Dashv \varphi \vee \psi$ iff both $E \Dashv \varphi$ and $E \Dashv \psi$.

As in the case of Kripke's account of truth, there is a natural notion of monotonicity. Define $E_1 \sqsubseteq E_2$ as $E_1^+ \subseteq E_2^+$ and $E_1^- \subseteq E_2^-$. It is useful to think of this as ordering EAPs by their information: the pair E_2 is at least as informative as the pair E_1. It is easy to prove that our notion of satisfaction is monotonic. That is:

Proposition 6.2 *Let E_1 and E_2 be EAPs such that $E_1 \sqsubseteq E_2$. Then any formula that is true (or false) in E_1 is also true (or false) in E_2.*

Also as in the case of Kripke on truth, we can define a jump operation J that takes us from one EAP E to another $J(E) = \langle J^+(E), J^-(E) \rangle$ that is at least as informative. Our task is thus to define J^+ and J^-. The crucial idea is that if $E \models \varphi(t)$, then we add the ordered pair $\langle t, \hat{x}\varphi \rangle$ to the extension $J^+(E)$, and if $E \Dashv \varphi(t)$, then this pair is added to the anti-extension $J^-(E)$. The idea is quite intuitive. $E \models \varphi(t)$ means that the information encoded in E suffices to establish, once and for all, that $\varphi(t)$. This makes it safe to add $\langle t, \hat{x}\varphi \rangle$ to the extension of the membership predicate. Returning to formal considerations, we now prove:

Proposition 6.3 *Let E_1 and E_2 be EAPs such that $E_1 \sqsubseteq E_2$. Then applying the jump operation J yields two EAPs such that $J(E_1) \sqsubseteq J(E_2)$.*

This proposition ensures that it is possible to iterate applications of the jump operation.

We are now ready to describe Maddy's construction. We start at stage 0 with an EAP that encodes just the information we have from set theory and our explanation of the constant '$\bar{V}$'. Let:

- E_0^+ be the class of all ordered pairs $\langle \bar{a}, \bar{b} \rangle$ such that $a \in b$, as well as all ordered pairs $\langle \bar{a}, \bar{V} \rangle$;
- E_0^- be the class of all ordered pairs which are either (i) of the form $\langle \bar{a}, \bar{b} \rangle$ where $a \notin b$, or (ii) of the form $\langle t_1, t_2 \rangle$ where t_1 is a class term and t_2 is either a set term or '$\bar{V}$'.

We now build up a monotonic sequence of EAPs by successively applying the jump operation, taking unions at limits:

- $E_{\alpha+1} = J(E_\alpha)$
- $E_\lambda = \langle \bigcup_{\gamma<\lambda} E^+_\gamma, \bigcup_{\gamma<\lambda} E^-_\gamma \rangle$.

As Maddy (2000, p. 308) shows, however, the process has no fixed point among the ordinals of the class V with which we begin.[18]

How does Maddy's account fare? She is able to prove various interesting facts, such as that the class of infinite classes is a member of itself (p. 304), and to develop parts of a theory of Frege numbers (pp. 311–12). There are some serious problems, however, to which Maddy devotes substantial attention. The problems stem from the fact that the account is very "gappy": there is *a lot* of indeterminacy. She is remarkable candid about this:

> Still, the context of three-valued logic—true, false, and neither—makes the system awkward, so I think the jury must still be considered out on the question of a workable theory of classes in general and of Frege numbers in particular. (Maddy, 1990, pp. 105–6)

The widespread indeterminacy becomes particularly problematic in the case of universal generalizations, which in strong Kleene semantics are rendered indeterminate by a single indeterminate instance. This means that it becomes extremely hard for a universal generalization to be determinately true.

This push towards indeterminacy is problematic for at least two reasons. First, in a Fregean theory of numbers, the notion of equinumerosity is obviously crucial. To say that two classes are equinumerous is to say that there exists a bijection between their members. Thus, to say that two classes are *not* equinumerous is to say that every (code of a) relation fails to be such a bijection. Since this is a universal generalization, it faces the threat of indeterminacy. And indeed, as Maddy observes on p. 312, her construction fails to render true even the seemingly trivial claim that $\{\varnothing\}$ is not equinumerous with $\{\varnothing, \{\varnothing\}\}$. In the context of a Fregean theory of numbers, it is hard to tolerate such rampant indeterminacy.

A second problem, as Maddy observes in Section II, is that the theory individuates classes in an exceedingly fine-grained way: when two formulas φ and ψ differ, then so do their classes $\hat{x}\varphi$ and $\hat{x}\psi$. The obvious strategy for fixing this problem is to define a coarser equivalence by letting $\hat{x}\varphi \simeq \hat{x}\psi$ iff $\forall x(\varphi \leftrightarrow \psi)$. But again we run up against the problem that universal generalizations are subject to rampant indeterminacy. Indeed, it is easy to see that $\hat{x}\varphi \simeq \hat{x}\psi$ requires the two classes to be total, in the sense that every membership claim concerning each class is either true or false. This means that partial classes can never be equivalent. For example, we don't even have obvious equivalences such as $\hat{x}(x \notin x) \simeq \hat{x}(x \notin x \wedge x \notin x)$.[19]

Clearly, if possible, it would be good to do better.

---

[18] The situation will be the reverse if our V is just a set of the metatheory (cf. footnote 17) and we look for fixed points among the ordinals of the metatheory. Then a fixed point will exist, for reasons analogous to those familiar from Kripke's account of truth.

[19] As Sam Roberts pointed out to me, the rampant indeterminacy raises a third concern as well. Are the various classes required by Maddy's Kripke-inspired account of classes available according to this theory itself? However, this concern can be assuaged by working (as suggested in footnote 17) in a purely set-theoretic metatheory, where the V of the account of classes is represented by a set of the metatheory.

6.5 Stability as a Way to Do Better

I will now briefly describe two attempted improvements that have a lot in common with Maddy's—but which ultimately fail to deliver what we want—before turning to a different, and more successful, approach.[20]

Instead of successively building up extensions and anti-extensions of the membership relation, as Maddy does, Horsten and Linnebo (2016) build up extensions and anti-extensions of an extensional notion of class identity—what we just sought, but failed, to define in Maddy's setting. Since their investigation takes place in a second-order setting, membership can be handled *à la* Frege by defining '$x \in y$' as '$\exists F(y = \hat{u}Fu \wedge Fx)$'. Here is the key idea behind the step to be iterated. Assume that the model associated with one stage of the construction yields that $\forall x(\varphi(x) \leftrightarrow \psi(x))$ is true (or false). Then at the next stage the ordered pair $\langle \hat{x}\varphi(x), \hat{x}\psi(x)\rangle$ is added to the extension (or anti-extension) of the identity predicate. Horsten and Linnebo show that this approach has valuable applications to the theory of neo-Fregean abstraction. For our present purposes, however, the approach is subject to a severe limitation: it only works when '$\in$' is banned from occurring in the formulas whose class abstracts we consider.[21]

Kriener (2014) investigates whether we can do better by combining ideas from Maddy and Horsten and Linnebo (2016), namely, by a construction that successively, but in parallel, builds up extensions and anti-extensions for *both* membership *and* identity.[22] Regrettably, the verdict is again largely negative. For example, even the formation of a singleton class of a well-defined object turns out to be problematic.[23]

The picture that emerges is thus one of a failed research program. At the outset, Maddy's idea of building on Kripke's highly successful account of truth seemed very promising. And as we have seen, there are natural ways to transpose the account to the case of classes. Unfortunately, the resulting theories are not very attractive; and the situation does not improve materially on any of the more natural ways to modify Maddy's account.

Let us therefore change tack. I will end by outlining a different approach that does better. The approach is inspired by Fine (2005) and Linnebo (2006), but has recently received a far more streamlined development and probing analysis by Roberts (2022).[24] Once again, the idea is (as in Maddy) successively to build up an interpretation of the class membership predicate. But there are some striking differ-

---

[20] Since this last approach is clearly superior to the preceding three, there is no need, for present purposes, to compare the relative merits of these three approaches.

[21] More precisely, the severe limitation is that the second-order logic is predicative. Since the Fregean definition of membership is impredicative, this in effect bans the membership predicate from occurring in comprehension axioms.

[22] Despite the earlier publication date, Kriener (2014) was developed in parallel with Horsten and Linnebo (2016).

[23] See Kriener (2014, p. 266) and Horsten and Linnebo (2016, p. 21).

[24] In joint work in progress, Roberts and I attempt to explicate the underlying philosophical conception.

ences. A comparatively unimportant one is that we distinguish being a member of a class from being an element of a set. We write 'η' for the former, and '$\in$' for the latter. More importantly, there is no partiality: we build up a sequence of ever larger interpretations E of 'η'.

The guiding idea for how to avoid paradox is that, when defining a larger membership relation, we may presuppose only the previously available membership relation. Suppose we start with a base model with no class-theoretic structure; for example, the entire cumulative hierarchy of sets. Let $\mathcal{L}_0$ be the language of pure set theory. Since classes are determined by their membership condition, any condition in $\mathcal{L}_0$ should thus determine a class whose members are precisely the objects satisfying that condition. Let E_0 be the membership relation for these classes, and let $\mathcal{L}_1$ be $\mathcal{L}_0$ together with a predicate, 'η_0', for this relation. An *iterative conception of logical classes* can now be formulated based on the idea of iterating the simple step that took us from $\mathcal{L}_0$ to $\mathcal{L}_1$. Given $\mathcal{L}_1$, we consider classes defined by conditions that presuppose only what we've got—namely, the expressive resources of $\mathcal{L}_1$. Let E_1 be the membership relation for these classes, and let $\mathcal{L}_2$ be $\mathcal{L}_1$ together with a predicate, 'η_1', for this extended relation. Ultimately, we can repeat this process indefinitely. Paradox is averted because classes are only ever defined by conditions that presuppose a restricted and previously available membership relation.

To spell this out properly, one obviously needs to give precise technical content to the talk about presupposing only expressive resources that we have got. Here we take inspiration from the discoverer of predicativity, Henri Poincaré. In one of his writings on the subject, he provides a characterization of predicativity that is more liberal than the usual one in terms of the very strict vicious circle principle, which prohibits a definition from quantifying over a totality to which the defined entity would belong. Poincaré writes:

> From this we draw a distinction between two types of classifications applicable to the elements of infinite collections: the *predicative* classifications, which cannot be disordered by the introduction of new elements; the *non-predicative* classifications in which the introduction of new elements necessitates constant modification. (Poincaré, 1912, p. 47)

To develop this, we need to say what it means to "introduce new elements" and what it means for a condition to be "disordered" by such an introduction. Suppose we are given a membership relation E. We propose to understand the introduction of new elements as the adoption of an extended membership relation E', which reveals more classes. For example, suppose we start with a membership relation E with nothing but the universal class in its range: E thinks there is a class with everything as a member, but that's it. We can introduce new elements by adding new classes to E. For example, perhaps we add a class whose sole member is the universal class; in other words, we add the singleton of the universal class to E. The resulting extended relation E' would represent two classes: the universal class and its singleton.

Our notion of a condition being "disordered" by such extensions is now straightforward. We say that a condition is disordered by extensions of a given membership relation when its application conditions change as a result of moving from that relation to at least one of its extensions. More precisely, consider any first-order language

$\mathcal{L}_0$, and let $\mathcal{L}$ be the language that adds 'η' to $\mathcal{L}_0$. Let $\mathcal{M}$ be any model of $\mathcal{L}_0$, and let E be any relation over $\mathcal{M}$. Taking E to be our given membership relation, we let $\mathcal{M}[E]$ be the corresponding model of $\mathcal{L}$ that interprets 'η' according to E. We say that another relation E' *extends* E, in symbols $E' \sqsupseteq E$, when E' doesn't disagree with E about the application conditions of any of the classes represented by E (by being in its range).[25] Let $[\![\phi]\!]_E$ be the extension of ϕ in $\mathcal{M}[E]$; formally, $[\![\phi]\!]_E = \{x : \mathcal{M}[E] \models \phi\}$. Finally, we say that a condition ϕ is *stable* with respect to E just in case[26]:

$$\forall E' \sqsupseteq E([\![\phi]\!]_{E'} = [\![\phi]\!]_E)$$

Suppose we allow every condition that is stable relative to a given membership relation E to define a property relative to an extended membership relation E'. As explained, the idea is that these conditions presuppose only class membership facts encoded in the given relation E and should thus be allowed to define further classes. Suppose we start with an empty membership relation $E_0 = \varnothing$. We now define a sequence $\{E_\alpha\}$ of larger and larger membership relations by letting $E_{\alpha+1} = (E_\alpha)'$ and taking unions at limits. Moreover, the assumption of stability enables us to use a single membership predicate 'η' rather than the clumsy sequence of such predicates employed in the informal sketch above.[27] All that remains is to capture the model-theoretic notion of stability in the object language. In joint work in progress, Roberts and I show how this can be done by describing the construction in terms of a nice stage theory of classes, developed in Roberts (2022).

How does the resulting account fare? Let us consider its performance on the various questions that came up in our assessment of Maddy's account. First, there is no indeterminacy to worry about. Second, we obtain the desired Frege numbers, defined as the equivalence classes of equinumerous sets.[28] More precisely, a Frege

---

[25] Formally, '$E' \sqsupseteq E$' is defined as:

$$E \subseteq E' \wedge \forall y \in \mathsf{rng}(E) \forall x(\langle x, y\rangle \in E' \iff \langle x, y\rangle \in E)$$

We may want to add a clause to ensure that an object that represents the empty class relative to E continues to represent this class relative to E'.

[26] Stability corresponds to the notion of "determination" found in Linnebo (2009, Sect. 7). The importance of Poincaré's notion of stability is emphasized by Crosilla (2016) (under the name of "invariance").

[27] To see this, suppose we have $a\ \eta\ b$ relative to some membership relation E_α. Then b represents the class defined by some condition ϕ that is stable relative to E_γ for some $\gamma < \alpha$. The stability of ϕ ensures that $a\ \eta\ b$ will hold not only relative to E_α but also relative to E_β for any $\beta \geq \gamma$. And this means, in turn, that there is no need to index 'η' in a way that ties it specifically to the relation E_α.

[28] Fine (2005, p. 563) makes the same observation. To what extent does the account satisfy the other theoretical needs associated with classes that we identified in Sect. 6.2? Roberts (2022) shows that the needs arising from set theoretic practice are largely satisfied, provided the iteration is carried sufficiently far. Needs arising from semantics are shown to be satisfied in Linnebo (2006). Finally, using our classes it is straightforward to turn the axiom schemes of ZFC into single axioms.

number is any non-empty class n such that $\forall y(y \ \eta \ n \leftrightarrow x \approx y)$, for some member $x \ \eta \ n$. Finally, the classes are entirely extensional.[29]

I therefore submit that the outlined account provides a superior way to satisfy the desiderata that Maddy so usefully articulates for a theory of classes—namely, once again, that classes be "real, well-defined entities" and "significantly different from sets". Indeed, our account can be seen as an explication of the logical conception of class, as analyzed by Maddy. It even shares with Maddy's account the idea that paradoxes are to be avoided by successively generating the membership relation, at each stage presupposing only what we have got. The key difference is that this common idea is now explicated *à la* Poincaré rather than *à la* Kripke, thus avoiding all the troublesome indeterminacies.[30]

References

Benacerraf, P., & Putnam, H. (Eds.). (1983). *Philosophy of mathematics: Selected readings* (2nd ed.). Cambridge University Press.

Bernays, P. (1935). On Platonism in mathematics. Reprinted in Benacerraf and Putnam (1983).

Boolos, G. (1971). The iterative conception of set. *Journal of Philosophy*, *68*, 215–32. Reprinted in Boolos (1998).

Boolos, G. (1984). To be is to be a value of a variable (or to be some values of some variables). *Journal of Philosophy, 81*(8), 430–449.

Boolos, G. (1998). *Logic, logic, and logic*. Harvard University Press.

Cantor, G. (1883). *Grundlagen einer allgemeinen Mannigfaltigkeitslehre*. B.G. Teubner, Leipzig. Repr. and trans. in Ewald (1996).

Crosilla, L. M. (2016). *Constructivity and predicativity: Philosophical foundations*. Ph.D. thesis, University of Leeds.

Ewald, W. (1996). *From Kant to Hilbert: A source book in the foundations of mathematics* (Vol. 2). Oxford University Press.

Feferman, S. (1984). Toward useful type-free theories. I. *Journal of Symbolic Logic, 49*(1), 75–111.

Fine, K. (2005). Class and membership. *Journal of Philosophy, 102*(11), 547–572.

Frege, G. (1953). *Foundations of arithmetic* (Trans. by J. L. Austin). Blackwell.

Gödel, K. (1947). What is Cantor's continuum hypothesis? In *Benacerraf and Putnam* (1983).

Horsten, L., & Linnebo, Ø. (2016). Term models for abstraction principles. *Journal of Philosophical Logic, 45*(1), 1–23.

Klev, A. (2018). A road map of Dedekind's theorem 66. *Hopos: The Journal of the International Society for the History of Philosophy of Science*, *8*(2), 241–277.

Kriener, J. (2014). The groundedness approach to class theory. *Inquiry, 57*(2), 244–273.

Kripke, S. A. (1975). Outline of a theory of truth. *Journal of Philosophy, 72*(19), 690–716.

Lavine, S. (1994). *Understanding the infinite*. Harvard University Press.

Lewis, D. (1991). *Parts of classes*. Blackwell.

Linnebo, Ø. (2006). Sets, properties, and unrestricted quantification. In A. Rayo & G. Uzquiano (Eds.), *Absolute generality* (pp. 149–178). Oxford University Press.

Linnebo, Ø. (2009). Bad company tamed. *Synthese, 170*(3), 371–391.

Linnebo, Ø. (2010). Pluralities and sets. *Journal of Philosophy, 107*(3), 144–164.

[29] Though we could also, if desired, individuate them more finely; see Roberts (2022).

[30] Thanks to Penelope Maddy, Sam Roberts, and the members of the Oslo logic seminar for comments and discussion.

Linnebo, Ø. (2022). Plural quantification. In E. N. Zalta (Ed.), *The Stanford encyclopedia of philosophy* (Spring 2022 ed.). Metaphysics Research Lab, Stanford University.
Maddy, P. (1981). Sets and numbers. *Noûs, 15*(4), 495–511.
Maddy, P. (1983). Proper classes. *Journal of Symbolic Logic, 48*(1), 113–139.
Maddy, P. (1988). Believing the axioms, i and ii. *Journal of Symbolic Logic, 53*, 481–511 and 736–764.
Maddy, P. (1990). *Realism in mathematics*. Clarendon.
Maddy, P. (2000). A theory of sets and classes. In G. Sher & R. Tieszen (Eds.), *Between logic and intuition* (pp. 299–316). Cambridge University Press.
Myhill, J. (1984). Paradoxes. *Synthese, 60*(1), 129–143.
Parsons, C. (1974). Sets and classes. *Noûs, 8*, 1–12. Reprinted in Parsons (1983).
Parsons, C. (1977). What is the iterative conception of set? In R. Butts& J. Hintikka (Eds.), *Logic, foundations of mathematics, and computability theory* (pp. 335–367). Reidel. Reprinted in Benacerraf and Putnam (1983) and Parsons (1983).
Parsons, C. (1983). *Mathematics in philosophy*. Cornell University Press.
Poincaré, H. (1912). The logic of infinity. In H. Poincaré (Ed.), *Mathematics and science: Last essays* (pp. 45–64). Dover Publications.
Reinhardt, W. (1974). Set existence principles of Schoenfield, Ackermann and Powell. *Fundamenta Mathematicae, 83*, 12–41.
Roberts, S. (2022). The iterative conception of properties. Typescript.
Schindler, T. (2019). Classes, why and how. *Philosophical Studies, 176*(2), 407–435.
Uzquiano, G. (2003). Plural quantification and classes. *Philosophia Mathematica, 11*(1), 67–81.
Wang, H. (1996). *A logical journey: From Gödel to philosophy*. Bradford.
Welch, P., & Horsten, L. (2016). Reflecting on absolute infinity. *Journal of Philosophy, 113*(2), 89–111.
Zermelo, E. (1930). Über Grenzzahlen und Mengenbereiche. *Fundamenta Mathematicae, 16*, 29–47. Translated in Ewald (1996).

Chapter 7
Reply to Linnebo

Penelope Maddy

Keywords Set theory · Proper classes · W. N. Reinhardt · Extendable cardinals

It's kind of Professor Linnebo to resurrect this old work of mine from the 80s, when my younger self was still a set-theoretic realist.[1] My hope at the time was to develop a theory that could validate Reinhardt's arguments for large cardinals,[2] but by 'Believing the axioms' (Maddy 1988), I'd come reluctantly to the conclusion that his purported proper classes have to be characterized combinatorially at one point in the argument and logically at another.[3] My distinctly logical proper classes obviously couldn't underwrite this line of thought, and it's hard to see how any responsible account could have it both ways. So I gave up on the project.

What puzzles me now, looking back, is what kind of support I imagined this argument would have given to these cardinals if it had worked. Presumably an intrinsic justification of some kind was intended,[4] perhaps by conceptual analysis of the logical

[1] Maddy (1983), (2000b) (The publication date for Maddy (2000b) is misleading; the work dates to the mid-80s.)

[2] Traditionally, cardinals consistent with $V = L$ have been termed 'small large cardinals' and the rest 'large large cardinals', but these days.

'large large' may be reserved for those without traditional inner models, leaving measurables and Woodins perhaps as 'medium-sized large cardinals'. In any case, the cardinals Reinhardt was out to defend would be large large on either convention.

[3] See Maddy (1988), pp. 752–754. Others have come to similar conclusions, e.g., Roberts (2016). Koellner (2009), pp. 217–218, highlights other difficulties.

[4] The mathematical intuition of Maddy (1990), roughly analogous to sense perception, is much too weak, so I couldn't have had in mind 'intrinsic' in the official sense of the set-theoretic realist.

P. Maddy (✉)
Department of Logic and Philosophy of Science, University of California, Irvine, CA, USA
e-mail: pjmaddy@uci.edu

S. Arbeiter and J. Kennedy (eds.), *The Philosophy of Penelope Maddy*,
Outstanding Contributions to Logic 31, https://doi.org/10.1007/978-3-031-58425-1_7

(as opposed to combinatorial) notion of collection.[5] Linnebo sees both my project and his in these terms:

> ... our account can be seen as an explication of the logical conception of class, as analyzed by Maddy. (p. 78)

An axiom implicit in the concepts at the foundation of set theory should have the highest possible form of justification.

Reasonable as this may sound, the years since have gradually led me to doubt it, until *Defending the Axioms*,[6] whose explicit moral is that the fundamental justificatory force lies with extrinsic considerations instead: mathematical benefits like welcome consequences, theoretical clarification and unification, new and fruitful conceptual and axiomatic resources. To put the point starkly, what should we do if our mathematical concept X rules out a strain of good mathematics? My thought is that we should amend our concept or change to a new one.[7] This is precisely what happened in the case of the Axiom of Choice: it conflicted with the logical notion of collection, so the discipline eventually moved to the new combinatorial notion. This isn't to say that intrinsic considerations are of no use, as this example shows: the intuitive picture of an iterative hierarchy has been immensely fruitful as a heuristic—organizing and guiding our thinking, suggesting new avenues, and so on. The claim is that these virtues are strictly instrumental: an intrinsic consideration has justificatory force only in so far as it enables extrinsic gains.

This perspective prompts a reevaluation of particular intrinsic 'justifications', since they aren't justificatory in their own right, but to be judged by their extrinsic fruits. Returning to Reinhardt's argument in this spirit, its internal deficiencies seem less important than the fact that it generated the notion of an extendable cardinal—the conspicuous role of extendables in current set-theoretic practice casts Reinhardt's thinking in a more positive light. It hasn't had the heuristic impact of the iterative picture, it hasn't continued to be productive, but it has earned at least some respect. Following this line of thought to its logical extreme, we might question the motivation for projects that offer new intrinsic considerations in support of small large cardinals, given that these follow from medium-size large cardinals that already have strong extrinsic support. Similarly, the challenge for multiverse theorists would be to detail the mathematical benefits of embracing its alternative intuitive picture a range of worlds in place of the familiar iterative hierarchy.[8]

Viewed from this perspective, what contribution might my foray into proper classes have made if it had worked? It wouldn't have justified the positing of extendable cardinals, because they would already have earned their keep extrinsically. What

[5] I have in mind here is the kind of naturalist-friendly conceptual analysis described in Maddy (2017), pp. 214–218. Successful examples include Dedekind on 'continuity', Einstein on 'simultaneity', and Turing on 'computable'.

[6] Maddy (2011), §V.5.

[7] Drawing a distinction between these two moves is notoriously difficult (and I think pointless).

[8] On another reading, the multiversist would be proposing a formal theory to replace ZFC, not an intuitive picture to replace the iterative hierarchy (see, e.g., see Steel (2014), also Maddy and Meadows (2020), Maddy 'On multiversism', this volume).

it might have done was to elaborate Reinhardt's effective heuristic, to produce a sharper intuitive picture to support his argumentation, but that by itself would have been wasted effort unless that added clarity had been enough to generate new mathematical benefits.[9] Of course, none of that happened, but my point is that this is how success would be judged.[10]

Viewed through this lens, Linnebo's analysis[11] of the logical notion of collection, like mine, is intended to add an intuitive picture of logical collections to the intuitive picture of combinatorial collections provided by the iterative hierarchy. Of course, it can't do the job I had in mind—sharpening up Reinhardt's argumentfor the same reasons mine couldn't, but it's clearly superior to mine in many ways, and depending on specifics, it can deliver either NBG or MK in place of ZFC. My guess is that for many set theorists, the intuitive picture behind MK simply places another rank on the iterative hierarchy and labels them as 'classes'. Surely everyone knows this is a cheat,[12] and perhaps Linnebo's more responsible heuristic picture would make the system easier to work with and thereby suggest new avenues. This is a promissory note at this point, but extrinsic payoffs like these would be the relevant test on the analysis.

But, of course, Linnebo has a much broader lens! Beyond this possible heuristic role in set theory, he envisions applications in other fields, particularly in formal semantics where it has already produced advances[13] and further work is very much in progress. The project would obviously be worth pursuing even without significant contributions to set theory proper.

References

Koellner, P. (2009). On reflection principles'. *Annals of Pure and Applied Logic, 157*, 206–219.

Linnebo, Ø. (2006). Sets, properties, and unrestricted quantification. In A. Rayo & G. Uzquiano (Eds.), *Absolute generality* (pp. 149–178). Oxford University Press.

Linnebo, Ø. (2024). Maddy on classes. This volume.

Maddy, P. (1983). Proper classes. *Journal of Symbolic Logic, 48*, 113–139.

Maddy, P. (1988). Believing the axioms. *Journal of Symbolic Logic, 53*, 481–511, 736–764.

---

[9] It sometimes happens that when an intuitive picture is made clearer, it becomes possible to elaborate it further, and this in turn enables new extrinsic progress.

[10] One of the defining characteristics of the examples in footnote 5 is that each started with an antecedently familiar notion that had served well enough up to that point, but was inadequate to a specific new undertaking. The goal of the analysis then was explicit—do the new job—which provided a specific criterion of success. This criterion dovetails nicely with the requirement here of new extrinsic gains, though the precise nature of those extrinsic gains isn't always specifiable in advance.

[11] More accurately, this should be termed Linnebo and Roberts's analysis, as Linnebo clearly indicates. See Roberts (2024).

[12] I agree with Linnebo that recent developments based on plural logic and mereology don't change the situation, that they just provide a new layer of combinatorial collections under a different guise.

[13] See, e.g., Linnebo (2006).

Maddy, P. (1990). *Realism in mathematics*. Oxford University Press.
Maddy, P. (2000b). A theory of sets and classes. In G. Sher, & R. Tieszen (Eds.), *Logic and intuition* (pp. 299–316). Cambridge University Press.
Maddy, P. (2011). *Defending the axioms*. Oxford University Press.
Maddy, P. (2017). Set-theoretic foundations. In A. Caicedo, J. Cummings, P. Koellner, & P. Larson (Eds.), *Foundations of mathematics: Essays in honor of W. Hugh Woodin's 60th birthday* (pp. 289–322). American Mathematical Society.
Maddy, P. (2024). On multiversivism. This volume.
Maddy, P, & Meadows, T. (2020). A reconstruction of Steel's multiverse project. *Bulletin of Symbolic Logic, 26*, 118–169.
Martin, D. A. 'Intrinsic and extrinsic evidence in set theory', this volume.
Roberts, S. (2016). *Potentialism and Reflection*. PhD thesis, Birkbeck, University of London.
Roberts, S. (2024). The iterative conception of properties, unpublished.
Steel, J. (2014). Gödel's program. In J. Kennedy (Ed.), *Interpreting Gödel* (pp. 153–179). Cambridge University Press.

Chapter 8
A General Framework for a Second Philosophy Analysis of Set-Theoretic Methodology

Carolin Antos and Deborah Kant

Abstract Penelope Maddy's Second Philosophy is one of the most well-known approaches in recent philosophy of mathematics. She applies her second-philosophical method to analyze mathematical methodology by reconstructing historical cases in a setting of means-ends relations. However, outside of Maddy's own work, this kind of methodological analysis has not yet been extensively used and analyzed. In the present work, we will make a first step in this direction. We develop a general framework that allows us to clarify the procedure and aims of the Second Philosopher's investigation into set-theoretic methodology; provides a platform to analyze the Second Philosopher's methods themselves; and can be applied to further questions in the philosophy of set theory.

Keywords Generous Arena · Consensus-based criteria · Content-based criteria · Subject-based criteria · Methodology

8.1 Introduction

In *Defending the Axioms*, Maddy (2011) applies her position of Second Philosophy to set theory. In *Second Philosophy*, Maddy (2007) elaborates this position by describing the work of a certain idealized enquirer, the Second Philosopher persona:

> [The Second Philosopher has] good reason to pursue mathematics herself, as part of her investigation of the world, but she also recognizes that it is developed using methods that appear quite different from the sort of observation, experimentation, and theory formation that guide the rest of her research. (Maddy, 2011, p. 39)

Maddy notes that there are two types of questions for the Second Philosopher, one about set-theoretic methodology and the other about the nature of set-theoretic

C. Antos (✉)
FB Philosophie, Universitätsstrasse 10, 78454 Konstanz, Germany
e-mail: carolin.antos-kuby@uni-konstanz.de

D. Kant
FB Mathematik, Universität Hamburg, Bundesstrasse 55, 20146 Hamburg, Germany

S. Arbeiter and J. Kennedy (eds.), *The Philosophy of Penelope Maddy*,
Outstanding Contributions to Logic 31, https://doi.org/10.1007/978-3-031-58425-1_8

activity itself. The first question already encompasses quite a wide array of topics: The Second Philosopher wants to learn about set-theoretic methodology from a broad point of view and therefore considers not only localized methods (like specific theorem proving techniques), but also more general approaches (like adopting axioms). The second question, about the nature of set-theoretic activity itself, is based upon the results acquired by answering the first question. Maddy here develops the positions of Thin Realism and Arealism, which, she argues, are interchangeable and equally correct.

Although the questions of the second type seem to be the philosophically "richer" option, in this article we will concentrate on the first group of questions. One reason for this focus is that little has been done on this beside Maddy's own work.[1] More to the point, we know of no works that use Maddy's account of Second Philosophy to further investigations into the philosophy of set theory.[2] We feel that this is due to the underdeveloped status of central notions and motivations of the Second Philosopher's approach to set theory and we would like to contribute clarification to increase the usability of Maddy's approach. As we will show in Sect. 8.5, there is potential for further work in this direction.

Another reason for focusing on the first group of questions is that they are a crucial prerequisite when considering subsequent discussions, like Maddy's philosophical positions developed in the second type of questions, as well as her work on the universe/multiverse debate (as e.g. in Maddy, 2017). Here a meta-analysis of the Second Philosopher's methods will be a useful tool to get a clear picture on how moves, such as excluding heuristic reasoning, influence the Second Philosopher's analysis, as well as on what the limitations of her methodology are and how it can be changed.

In this article, we propose a *general procedural framework* for the Second Philosopher's investigation into set-theoretic methodology, i.e. a framework for her own methodology. We do so by developing Maddy's exemplifying descriptions in Maddy (2011) and (Maddy, 1997) into a general framework. This exercise has three aims. We want to

(A) provide *clarification* on the use of central notions and aims in Maddy's work;
(B) facilitate a general *analysis* of the second-philosophical investigation into set-theoretic methodology; and
(C) increase the *applicability* of this investigation for further work in the philosophy of set theory.

We will show that providing such a framework is not at variance with the Second Philosopher's stance and we will give examples on how the framework can be used to further investigate open questions in the philosophy of set theory. Also note that we

---

[1] The articles that refer to Maddy (2011) are mostly concerned with questions of the second type or more general considerations into the naturalistic nature of Second Philosophy.

[2] One exception may be Ternullo (2019) on the evaluation on multiverse accounts according to Maddy's philosophical approach. However, he does not focus on framing the philosophical methodology itself, but rather on Maddy's arguments in the so-called universe/multiverse debate.

will work on two different levels of analysis: Aims (A) and (B) are concerned with studying the very methods of the Second Philosopher (with respects to her inquiry into mathematics), whereas goal (C) is concerned with the Second Philosophy's actual application to set theory.

We address aim (A) in Sect. 8.2 by clarifying how we can provide a general procedural framework for Second Philosophy while at the same time clarifying central relevant notions. In Sect. 8.3 we examine in detail Maddy's case study of Cantor's introduction of sets as given in Maddy (2011), contrast the latter with her case study of Zermelo's defense of the Axiom of Choice and, finally, propose a first schema for her procedure. Here we are interested in the specific details of a second-philosophical analysis. We then use this first rundown to formulate the general procedural framework in Sect. 8.4. We show that it is indeed a framework for Second Philosophy and that it can accommodate possible changes in the Second Philosopher's methodology itself. We also address the possibility of separating Second Philosophy from Maddy's description of it. We therefore provide a platform for a general analysis, thereby addressing aim (B). Section 8.5 shows how we can use the framework to further applications of a second-philosophical analysis of set-theoretic methodology and to outline how this can impact discussions in the philosophy of set theory. This will a (first) answer to how we can achieve aim (C).

8.2 Some Comments on Meta-Philosophy

In this section we address aim (A) from above, namely, the clarification of central notions and issues in Maddy's work on Second Philosophy. This includes the general question of whether we should give a general procedure at all, as well as clarifying Maddy's terminology and introducing the central method of means-ends relations.

8.2.1 Is A Procedural Framework Desirable?

Here we want to address the question if it is at all in the interest of Second Philosophy to provide a generalized procedure as propose in this article. (Maddy, 2007, pp. 1–2) herself points out that Second Philosophy not only "has no theory", but also states that its methods will remain "without any definitive way of characterizing exactly what that term entails". Instead, she goes on to describe through examples how an idealized inquirer, the Second Philosopher persona, is supposed to work. She then concludes:

> Though 'Second Philosophy' is never explicitly defined in all this, I hope that Parts I and II (of (Maddy, 2007, pp. I, II)) provide enough guidance for at least some sympathetic readers to get the hang of how to carry on. (Maddy, 2007, p. 4)

Maddy (2011, p. 39) makes this even more explicit: “Indeed any attempt at a once-and-for-all characterization of our inquirer’s methods would run counter to the ever-improving, open-ended nature of her project.” But, Maddy also states that the Second Philosopher “looks into the matter of how and why the methods she and others use in their inquiries work when they do and don’t work when they don’t” (2011, p. 38). The possibility of evaluating the Second Philosopher’s methods presupposes the possibility of making them more explicit.

One main aim in the paper is exactly this: making the Second Philosopher’s methodology, as currently stands, more explicit and make it amenable to a general analysis. We will show in Sect. 8.4.3 that our framework provides a more explicit presentation of the second-philosophical methodology while at the same time being flexible enough not to impede the ever-improving, open-ended nature of the Second Philosopher’s project. It might even be possible to use the framework to distinguish between Second Philosophy and Maddy’s interpretation of it. Maddy (2007, p. 3) herself hints at such a possibility when writing that “one might sign on as a Second Philosopher while thinking I’ve gone astray in my pursuit of the particulars”.[3]

8.2.2 Methodology and Practice

First, it is crucial to note that Maddy’s understanding of the notion ’set-theoretic methodology’ differs from other such understandings in the philosophical literature. Maddy includes the whole setup of set theory in it, going back as far as set theory’s fundamental assumptions. While many set theorists would not talk about ‘adopting axioms’ as a method they use to pursue set-theoretic research, this is one of the most important parts of set-theoretic methodology in Maddy’s view, because the adopted axioms delineate what is allowed in a proof and what is not. Typical questions about set-theoretic methodology are: “[W]hat are the proper grounds on which to introduce sets, to justify set-theoretic practices, to adopt set-theoretic axioms?” (Maddy, 2011, p. 41) The reason for this very broad interpretation of set-theoretic methodology also lies in the goal of Maddy’s investigation, namely, to answer philosophical questions of the second type, as pointed out in Sect. 8.1. In the following, we will use the term ‘set-theoretic methodology’ to include mathematical methods of set theory (like forcing, transfinite induction, etc.) as well as the more foundational ones included by Maddy.

Second, it is also helpful to note that Maddy uses the term “set-theoretic practice” in a different way than the term “mathematical practice” is used for example in the context of the philosophy of mathematical practice. For one, when discussing case studies, she mainly refers to published material, therefore concentrating on what

[3] See also (Rittberg, 2016, Chap. 6), where he discusses the possibility of “stripping away the idiosyncrasies of Maddy’s investigations” (2016, p. 138).

could be called "the front" of mathematical practice.[4] This raises the question of the Second Philosopher's adequate description of practice. For example, (Rittberg, 2016, p. 137) argues "that the Second Philosophical argument for UNIFY does not lead to a faithful description of contemporary set-theoretic practice".

The difference does not lie only in the source material used, but also in how this material is used for (second-)philosophical considerations. The most obvious example here is probably the use of heuristic reasoning. As we will see in Sect. 8.4.2, Maddy excludes heuristic reasoning from set-theoretic methodology—not in the sense that set theorists should not use heuristic reasoning, but on the ground that such reasoning should not inform our philosophical investigations into the nature of set-theoretic activity.[5] This assessment would not be shared by authors such as Francois and Bendegem (2010).[6]

However, what she does share with many of the proponents of the philosophy of mathematical practice is the reliance on historical material and analysis. She is, for example, seen as furthering the practical turn in the philosophy of mathematics: "an exemplary case of the practical turn is the philosophy of Penelope Maddy" (Ferreirós 2016, p. 23), which Ferreirós bases on the following principle: "Faithfulness to actual practice, both presently and historically, is a key requirement for this kind of work" (Ferreirós 2016, p. 24). This, of course, stands in stark contrast to the aforementioned critique by Rittberg (2016).

We think that one reason for this divergence of opinions lies in the ambiguity of the relation between set-theoretic practice and set-theoretic methodology. From the outset it could seem that practice is simply the more general term, with methodology being a part of the practice. Then we could simply describe set-theoretic methodology by collecting together all the instances of methods that come up in set-theoretic practice and perhaps augmenting this with an ordering according to importance, simplicity or whatever other feature we are interested in. However, this is not the route Maddy takes. She still takes practice as the basis for her investigation, but instead of just amassing together case studies of methods, she evaluates them. Only methods that pass certain criteria are allowed to be considered part of set-theoretic methodology and then, later on, be used for further philosophical consideration.

---

[4] In contrast, many studies from the philosophy of mathematical practice focus on the so-called "back" of the practice, as for example here in relation to explanation in mathematics: "The mathematical practice which we consider is drawn from the 'back' of mathematics. This contrasts with the 'front' mathematical practice" (Pease etal. 2019, p. 17). The distinction goes back to ?.

[5] For further discussion on the role of heuristic reasoning see Sect. 8.4.2.

[6] Cf. (Francois and Bendegem, 2010, p. 118):

> From the proof viewpoint, it seems obvious that different layers are required: the search for a proof and related concepts involves heuristics, proof search methods. As in many cases such techniques become interiorized […]. Thus, this proof practice "feeds" a particular metaphysical view that supports and stimulates the mathematician's search and that definitely should be taken into account.

8.2.3 Means-Ends Relations

According to Maddy, she succeeds in isolating the proper set-theoretic methods, by finding "an array of new methods for justifying claims, methods that appear to be both rational and autonomous" (2011, p. 54). We want to know *how* she manages to identify those methods. Here Maddy gives advice:

> In the cases we've surveyed, the community eventually reached a consensus that the controversial method was admissible because it led to certain varieties of mathematics, that is, because it was an effective means to particular desirable ends. Thus the positive counsel of history is to frame a defence or critique of a given method in two parts: first, identify a goal (or goals) of the relevant practice, and second, argue that the method in question either is or isn't an effective means toward that goal. (Maddy, 1997, p. 194)

Here Maddy observes how the community accepts a certain method (because it is an effective means towards a desirable end) and she proposes to take this justificatory structure as a guideline on how we should advance our own inquiries into set-theoretic methodology. This guideline tells us to frame the investigations into methodology in terms of a means-ends relation and test the means in the relation against its effectiveness towards the goal. The formulation of the means-ends relation and the evaluation of the effectiveness of the means takes place through historical analysis and consideration of contemporary practice; for example through a consensus of the community.

Regarding the role of consensus: In the above quote it is presented as the outcome of some concerted reasoning in the community. However, in our general framework it can play two further roles: First, a consensus in the community can suggest a method that should be incorporated into set-theoretic methodology (for example the adoption of a certain axiom, as Maddy discusses it in regard to the Axiom of Choice). Such a suggestion does not imply that a method can be included directly into the methodology; we still have to frame it as part of a means-ends relation that we will then evaluate by some criteria. This also means that it is unproblematic if the method is only suggested by part of the community. Second, consensus can be used as a criterion in the evaluation of the means-ends relation. In contrast to the previous case, it is indeed important to figure out if the consensus is only partial and what it is comprised of. Further, consensus cannot be the only criterion for evaluation, for that would make the investigations trivial (a method is suggested for investigation by consensus and then passes the threshold of acceptance because of consensus). Thus it is important to keep these two roles distinct. In the following, we will call the former the *proposing role* and the latter the *evaluative role* of consensus.

A note on terminology: As far as we can see, the means in the means-ends relation is always a method, whereas the end in the relation is always a goal. We pointed out at the beginning of Sect. 8.2.2 the broad notion of methods at play; the same will hold for goals, meaning they can "range from relatively local problem-solving, to providing foundations, to more open-ended pursuit of promising mathematical avenues" (Maddy, 2011, p. 52). Moreover, being a method or a goal is not mutually exclusive; a specific part of practice can be either in different mean-ends relations:

"we should expect that some goals will take the shape of means toward higher goals, and that goals at various levels may conflict, requiring a subtle assessment of weights and balances. But the simple counsel remains: identify the goals and evaluate the methods by their relations to those goals" (Maddy, 1997, p. 194). Whether we want to consider something as a goal or as a method is then determined by the context and the role something plays in the means-ends relation under consideration.

At the end of this process, a method is accepted as part of the methodology if we can show that it is an effective means towards a goal. Maddy also calls such methods "rational" or claims that it is rational to include them into the methodology. As we want to avoid the discussion about what role standards of rationality play in this context,[7] we will use the effectiveness criterion for a specific mean in a means-ends relation to pass the Second Philosopher's test.

8.3 Two Case Studies From Maddy

Our strategy in this section is the following: Building on Maddy's own account of the second-philosophical procedure, we analyze what she does in the case study on *Cantor's introduction of sets* Sect. 8.3.1; we then propose a detailed schema for her methodology in this case study, from which we extract three key features. Subsequently, we test the schema in a further case study, *Zermelo's defense of his axiomatization* Sect. 8.3.2.

8.3.1 Cantor's Introduction of Sets

The analysis (Maddy, 2011, p. II.2.(i)) gives of Cantor's introduction of sets is typical for a second-philosophical analysis. The guiding question of the present section is how Maddy interprets the historical and mathematical material in order to finally identify a specific means-ends relation and then can conclude that specific methods and goals found in set-theoretic practice are part of the justificatory structure of set theory.

In her reconstruction of the historical case study, Maddy identifies the method with the introduction of sets as new entities. She draws this from historical analysis, i.e. Ferreirós interpretation of Cantors work. Later on she implicitly uses the proposing role of consensus, for instance when she assumes that Cantor's introduction of sets is a *typical* example for set-theoretic methodology (Maddy, 2011, p. 52). She is justified in the assumption insofar as the point can be made that today there is a consensus on sets as accepted entities in the mathematical community.[8]

---

[7] For elaboration, see Maddy (1997, p. 197).

[8] Though there is an ongoing debate about the foundational role of set theory with respect to other candidates such as category theory and homotopy type theory, the use of sets is widespread over

In investigating this method, she describes how Cantor was led to the introduction of his definition of derived point set (Cantor, 1872) and explains why this can be considered to be the moment of the introduction of sets as new entities. In the setting of this case study, she identifies a specific research goal of Cantor, based on her observation that "Cantor was engaged in a straightforward project in analysis: generalizing a theorem on representing functions by trigonometric series" (Maddy, 2011, p. 41). This suggests that, in general, the research goal of the involved agent(s) can be interpreted as a goal of the relevant practice.

She establishes a means-ends relation in the following way: "a new entity—a set—has been introduced as an effective means toward an explicit and concrete mathematical goal: extending our understanding of trigonometric representations" (Maddy, 2011, p. 42). Thus, when establishing the means-ends relation, she already evaluates the method as effective, based on her prior analysis in which she points out that Cantor could prove an important theorem about trigonometric series by using point sets in this specific way. Finally, the method of introducing sets as new entities is included into set-theoretic methodology.

We can now give a schema for Maddy's procedure which is extracted from our analysis of the Cantor example and her description of the second-philosophical method; it is to be understood as a coherent interpretation of Maddy's methodology. In the following, M denotes the means and X denotes a goal:

We start with the observation that some M, often suggested by consensus, is an established part of mathematical methodology. Two questions present themselves: What is the justification for M? (This is the research question.) When was M introduced and by whom? (This is the methodological question.) As a next step, we have to choose the material from which to draw conclusions: we consider historical reconstructions about and source material of (one of) the event(s) surrounding the introduction of M. Then we reconstruct the event and try to identify a goal X. The central evaluation is carried out by assessing whether there is evidence that M is an effective means for X. If not, we can try to identify a different goal and repeat the assessment. If we succeed, we have found an argument in favour of M, if we fail we should not consider M to be a justified part of set-theoretic methodology.

There are three key components of Maddy's methodology that we extract from this schema: First, the focus on a particular philosophical method, the analysis via means-ends relations. Second, an evaluative component, which is used when assessing the means-ends relation at hand and when identifying the goals. And, third, the use of data both from contemporary practice as well as historical analysis to argue for or against the choice of methods and goals and to investigate the overall effectiveness of the means towards the end.

Let us now turn to one of the other case studies Maddy gives in Maddy (2011) and see if we can find these three key components at work.

---

various (if not all) mathematical research fields, and the argument here is not about the importance of set theory as a foundation, but on the mathematical notion of set and its introduction into mathematics.

8.3.2 *Zermelo's Defense of his Axiomatization*

Maddy (2011, pp. 45–47) offers a case study of Zermelo's defense of the adoption of the Axiom of Choice, building on the evidence found in Zermelo (1908a) and Zermelo (1908b). Here we will not reconstruct the entirety of Maddy's procedure, as we did in the Cantor case, but only discuss the main differences between the two case studies.

The Zermelo case study adds two aspects to the conclusion drawn from the Cantor case: it introduces a new method, namely the adoption of axioms and it discusses criteria for choosing goals.

As to the first aspect, it may not be clear from the outset how the adoption of an axiom is to be framed as part of a means-ends relation. For one, it is not clear why the adoption of axioms is a method and not a goal that one aims to justify. To clarify this, let us consider how axiom acceptance is used in Maddy's case study. In the Zermelo case, the most convincing arguments according to Maddy are the mathematical outcomes the adoption of an axiom brings about. These can be framed in terms of clear mathematical goals: we aim to prove a certain theorem, we stipulate that a certain mathematical feature should hold or not, etc. How can we reach these goals? By adopting a certain axiom that allows us to prove the theorem, by getting the mathematical feature to hold, etc. In this sense, the adoption of an axiom can be framed in a means-ends relation: adopting an axiom is the means that allows us to reach a certain goal. As we mentioned in Sect. 8.2.3, this might not be the only way to consider the adoption of an axiom, as means in one case can be ends in another. But, if we want to assess whether they should be included in set-theoretic methodology, we have to consider them as means, because the final evaluation of the means-ends relation judges the effectiveness of means.[9] Note that when presenting the Zermelo case study, Maddy does not outright frame it via means-ends relations. However, she does so in comparable examples like (Maddy, 2017, p. 303).

Secondly, Maddy discusses criteria for choosing goals by introducing the well-known distinction between intrinsic and extrinsic justification. This is based on the observation that Zermelo proposed both kinds of arguments in order to defend his Axiom of Choice. Maddy assesses the two types of justification differently. While extrinsic justifications are proper justificatory devices, intrinsic reasons are not and may be devoid of justificatory power.[10]

---

[9] Note that we do not claim that this framework of methods and goals follows the actual developments of the practice. The mathematical work (proving theorems, etc) might come first and only later an axiom is put forward that will entail desirable mathematics. Nonetheless, we would then say that the axiom is justified because it is a means towards this desirable mathematics and then we consider again the same means-ends relation.

[10] In Maddy (2011), this can also be detected in her comment to the effect that

> [Zermelo's] claim is that Choice is [intrinsically justified] ...; we might now express this by saying it is part of the informal 'concept of set'. But, as we've seen, Zermelo despairs of defining [the concept of set] with a precision adequate to the development of set theory. Instead he appeals to a second standard of evidence [i.e. extrinsic justification]. (Maddy, 2011, p. 46)

For the method of adopting the Axiom of Choice in the Zermelo case, Maddy mentions several goals, that this leaves us with two types of extrinsic goals: A mathematical one, as we need the Axiom of Choice in order to prove certain theorems that mathematical practice (not just set theory!) wants to hold. The second type of goal is foundational, i.e. to create more "productive science" and choose an axiomatization that is "sufficiently wide to retain all that is valuable in this theory" (Zermelo, cited in Maddy, 2011, pp. 46–47; cf. Maddy, 2011, p. 52). In Sect. 8.4.2 we will include criteria connected to extrinsic justification under the content-based criteria.

After studying Maddy's representation of the Second Philosopher's work in this section, we will proceed to formulate a general framework in the next section.

8.4 A General Framework for a Second Philosophy Analysis of Set-Theoretic Methodology

From the work surveyed in the last two sections, let us propose the following procedural framework for a second-philosophical analysis of set-theoretic methodology: The *core procedure* is given by the method of analysing set-theoretic methodology via historically informed and contemporary means-ends relations. This includes the formulation of the method qua means, the identification of the goal(s), as well as the final evaluation of the effectiveness of the method as means to realize the goal. This core part is supported by an *evaluative apparatus* that contains the criteria with which the Second Philosopher evaluates the choice of the methods, goals and case studies. The evaluation is carried out with the help of historical analysis and consideration of contemporary practice. The framework is procedural in the sense that usually the Second Philosopher will formulate a certain means-ends relation and then test the methods and goals according to the criteria of the evaluative apparatus. If they turn out to be unsatisfactory, she may re-formulate them, resulting in a different means-ends relation that again is evaluated—and so on. In the end, she either reach a satisfactory formulation of the methods and goals, in which case we can proceed to evaluate the effectiveness of the means; or we won't find such a relation. According to the outcome, she can add the method under investigation to set-theoretic methodology or not.

In the following, we will elaborate on the two parts of the framework. We are especially interested in providing a way to analyze the Second Philosopher's methodology itself, therefore addressing goal (B) from Sect. 8.1. We will show that one major feature of the procedural framework is to provide a background in which we can analyze the methodology of the Second Philosopher herself on different levels, while keeping track of how changes in different parts of the framework will alter and influence the overall methodology. This will also show that the framework allows for enough

---

Here we have an example on how goals are evaluated, meaning that even before the final evaluation of the means-ends relation takes place, goals have to satisfy certain criteria. In this instance, they have to be extrinsically, not intrinsically, defined goals.

flexibility to accommodate the "ever-improving, open-ended nature of [the Second Philosopher's] project" (Maddy, 2011, p. 38) and therefore address the worry voiced in Sect. 8.2.1.

8.4.1 The Core Procedure

The core procedure contains the methods used by the Second Philosopher in her investigations into set theory. Building on Maddy (2011), we identified the analysis using means-ends relations as the method used to investigate set-theoretic methodology. We discussed in Sect. 8.2.3 what the method and goals consist in. For the final evaluation, the method has to be effective to reach the goal in question. Providing that the means and ends satisfy the criteria of the evaluative apparatus, this can be judged through the analysis of the historical case study and of contemporary practice.

It is important to note that, by describing the core procedure in this way, we do not claim that an analysis via means-ends relation is the only method the Second Philosopher should use. Instead, as the Second Philosopher constantly reflects on her own methodology, she is free to add other methods to the core procedure; change details of how the means-ends analysis works; or even discard this method entirely. Of course, a change in the core procedure would require a thorough analysis. In Sect. 8.4.3 we will develop this further and give some hints as to why some changes concerning the method of employing the analysis of means-ends relations might be necessary to investigate *contemporary* set theory.

8.4.2 The Evaluative Apparatus

As we mentioned before, the Second Philosopher's ultimate goal of answering questions about the very nature of set-theoretic activity requires a careful choice of means-ends relations. While the choice of the the relevant setting by fixing a method and goal could be attained by simply picking a method and goal from practice (perhaps using the proposing role of consensus), here an evaluative component comes in. For example, the Second Philosopher wants to avoid including tangential methods or methods that lead practice astray in some sense.[11] This refers back to the evaluative role of consensus given in Sect. 8.2.3. Additionally, we have seen that Maddy favours extrinsic over intrinsic justification. This gives rise to criteria that say something about mathematical productiveness or relevance. But it also hints at the possibility of a negative criterion on what methods and goals should not be.

---

[11] This also allows the Second Philosopher to reject examples of individual practices or the practice of small sub-communities as template for general set-theoretic methodology.

Maddy provides one such negative criterion under the label of *heuristic aid* or *heuristic device*. Let us give two examples to illustrate what this criterion is meant to subsume:

Example 1

> Dedekind's paper on the natural numbers includes his belief that they are 'free creations of the human mind' (Dedekind (1888), 791). Given the wide range of views mathematicians tend to hold on these matters, it seems unlikely that [they] would all agree on any single conception of the nature of mathematical objects in general, or of sets in particular; the Second Philosopher concludes that such remarks should be treated as colorful asides or *heuristic aides*, but not as part of the evidential structure of the subject. What matters for her methodological purposes is that all concerned do feel the force of the kinds of considerations we've been focusing on here; these are the shared convictions that actually drive the practice. (Maddy, 2011, pp. 52–53, our emphasis).

In this example Maddy uses the evaluative role of consensus in the community to reject a particular group of beliefs as relevant for set-theoretic methodology. Not only does she claim that there was no consensus about such beliefs, but even more, that such a consensus would likely have never come about. She then uses consensus to identify such heuristic aides, aiming at excluding such beliefs pertaining to individuals or subgroups of practitioners from the justificatory structure of set theory. In the next example, we can see that consensus is not the only aspect of the "heuristic aid" argument.

Example 2

> [...] Linnebo and Pettigrew propose that the iterative conception justifies ZFC [...]. My own view is that the iterative conception is a brilliant *heuristic device*, but that the justification for the axioms it suggests [...] rests on their power to further various mathematical goals of set theory, including its foundational goals. (Maddy, 2017, p. 303, our emphasis)

Embedding this example in a means-ends relation, Maddy claims that choosing the goal of "compliance with the iterative conception" for the method "adopting ZFC as axioms" will not give rise to a usable means-ends-relation because the ends are not of the right type; instead they should be mathematical and foundational goals. Here, Maddy contrasts heuristics with subject-based considerations, i.e. mathematical or foundational considerations in mathematics.

These two examples show that there are different flavours of the "no heuristics" criterion and we argue that it is not *one* criterion after all. Instead, different aspects of it can be subsumed under different criteria, depending on which flavour we want to address. In Example 1, we claim that the "no heuristics" argument is actually part of a consensus-based criterion. As we discussed the different roles of consensus before, let us turn to Example 2. Here, we claim that the argument is part of what we call, for lack of a better term, a subject-based criterion. Let us give a broader background for this criterion:

Judging from Example 2, one could think that the difference between heuristic and non-heuristic arguments lies in philosophical vs mathematical considerations. Maddy (1997) describes the role of historical analysis as including negative counsel that "advises us that certain typically philosophical issues are ultimately irrelevant to the defence or criticism of mathematical methods" (1997, p. 194). At the same time, she concedes that "[d]isappointing as this may be, I think there is no principled distinction to draw between mathematics and philosophy [...]. At least, this isn't the course I take" (Maddy, 1997, p. 193). Instead there are instances of philosophical considerations that can and should be included into our analysis. One example is her argument that questions, such as whether CH has a definite truth value, are still legitimate even if they are independent from ZFC. They are legitimate because they can be framed and answered in mathematical terms (Maddy, 1997, pp. 194–195). Also, Maddy (2011) views foundational goals, like providing a unified basis for mathematics, as legitimate goals and does not characterize them as part of heuristics.[12] In the end, it seems to come down to a fine-grained analysis of the considerations in question, with the guideline that considerations are the more allowable the more they are mathematically informed, i.e. tied to mathematical content or mathematical ways in which they can be phrased instead of only concerning philosophical thinking. Coming back to Example 2, this raises the question why the iterative conception should be so decisively disregarded. There are ways in framing the iterative conception more mathematically, for example as done in Boolos (1971).[13] As we hinted at in Sect. 8.2.2, there are also arguments to be made that heuristics indeed is an important part of mathematical methodology and should inform further philosophical debate.[14] However, in the following we will include the "no heuristics" argument, as it plays an important part in Maddy's work and the goal of the framework is to structure this work.

Finally, let us list three criteria which we identified in Maddy's work and which are at play in the evaluative apparatus:

Consensus-based criteria[15]: "typical" (Maddy, 2011, p. 56), "natural" (Maddy, 2011, p. 85) shared beliefs that actually "drive the practice" (Maddy, 2011, p. 53).

Content-based criteria: part of the "evidential structure of the subject "(Maddy, 2011, p. 53), "productive" (Maddy, 2011, p. 85), "methodologically relevant" (Maddy, 1997, p. 193).

---

[12] For a comprehensive discussion of what foundational goals are, see Maddy (2017).

[13] Many thanks to Neil Barton for pointing this out to us.

[14] Notably in her very early work, Maddy (1988) does not use a "no-heuristics"-criterion. To the contrary, she relates heuristics to extrinsic reasoning when she emphasises that also ZFC axioms are supported by "'extrinsic' (pragmatic, heuristic) justifications, stated in terms of their consequences, or intertheoretic connections, or explanatory power, for example" (1988, p. 483).

[15] Note that this refers to consensus in its evaluative role, not its proposing role.

Subject-based criteria: no heuristics[16], no "philosophical" but mathematically informed content (in the way detailed above).

Three remarks are in order here: First, the dividing line between these criteria are not sharp. The idea behind the distinction between the first two criteria is that consensus-based criteria rely on judgments of the practitioners, whereas content-based criteria rely on the content being produced by the practitioners, like theorems, proof methods etc. Of course, it could be argued that 'relevance' is also a consensus-based criterion, however such ambiguities will not damage the general structure of the framework.

Second, it would be interesting to find and analyse a method or goal on which the criteria disagree. For example, what if a strong consensus emerges in the community about a goal or method that the Second Philosopher dismissed as heuristic device? Without being able to elaborate here, we think that the previously discussed case of the iterative conception could be construed in this way. This goes to show that we do not believe that the outcome of the procedural schema is guaranteed to be conclusive. More than that, we perceive the possibility of an inconclusive outcome as a desirable property of our framework as scientific practice is in flux and sometimes inconclusive itself. It should therefore be reflected by an analysis into its methodology.

Finally, we can see all of the criteria at work in Maddy's case study of Cantor's introduction of sets. As we pointed out in Sect. 8.3.1, the consensus-based criteria are met because sets as entities are regarded as a standard method in set theory. The content-based criteria are also met: the whole mathematical system of set theory is evidence for this. And, finally, the subject-based criteria are met, as evidenced by the following observation: Maddy uses terms like "exists" or even "ontology" not in "any philosophically loaded way: I just mean what the practice asserts to exist, leaving the semantic or metaphysical issues open" (Maddy, 2017, p. 296, footnote 13). So, when analyzing the method of introducing sets as new entities, we do not attach a metaphysical claim to this method, in the sense that sets are introduced as objects that "really exist", a belief which seems not to be a potential candidate for overall consensus. In this sense, no heuristics is involved in the method.

The procedural framework we have developed in the last two sections is modeled after Maddy's presentation of Second Philosophy. Our choice of criteria is guided by Maddy's analysis. However, we have also seen that some of the criteria can be called into question. In the following section, we will show that the procedural framework does allow for changes, for example in the choice of criteria. This opens up the possibility to be a Second Philosopher without having to agree with the entirety of Maddy's interpretation of it.

---

[16] Maddy (2017, p. 303)

8.4.3 Analysing the Second Philosopher's Methodology

In Sect. 8.2, we raised the question whether it is in the interest of the Second Philosopher to provide a generalized framework in light of the open-ended and ever-improving nature of her inquiry. In reply to this worry, we claimed that our procedural framework would be able to supply a concrete guideline on how second-philosophical inquiries can be analyzed, while at the same time being flexible enough not to impede the nature of the Second Philosopher's project. In this section we want to argue that we have succeeded on both accounts.

Changes in the framework can occur on different levels: For one, the content of the criteria can change. For example, foundational goals could change, as indeed happened with Maddy's **Final court of appeal**[17] that later seemed too restrictive and were revised into **Shared Standard** and **Generous Arena**.[18] Similarly, practice could produce a number of new results, showing a method to be productive that beforehand was not.

We could also add or remove criteria or make the proposed criteria more fine-grained. What our procedural framework does not allow for, however, is to get rid of the evaluative apparatus altogether (as it is an integral part of the framework). But this restriction does not present a problem for the Second Philosopher. She is not only interested in describing the practice, but in evaluating it in a certain way, separating relevant from irrelevant parts of the practice; this separation has to follow some criteria, therefore there has to be an evaluative apparatus.

The same holds in an even stronger sense for the core procedure, as it contains the methods used in the Second Philosopher's inquiry. Still, we have the possibility of introducing changes to the core procedure. Following Maddy's interpretation of the Second Philosopher's inquiry, it only contains the method of analysis via means-ends relations that is conducted through historical analysis and consideration of contemporary practice. The way in which we choose the means and ends is directed by the evaluative apparatus; any changes here would reflect back to changes in the evaluative apparatus, which we discussed above. This still leaves us with the possibility of adding new methods. Indeed, in the case of science, the Second Philosopher already has more than one method available to him (observation, hypothesis making and testing etc.), so it is reasonable to assume that she could come up with more than one way of analysing set-theoretic methodology. In the following we will provide an argument outline for why adding a new method (or at least adapting the method of means-ends relations) might be necessary.

We start with the following observation: There is one disadvantage to the dependence on historical analysis, namely that we cannot use it when considering very recent developments in set theory. Looking at the literature, historical analysis of set

---

[17] "Provide decisive answers to questions of ontology and proof: if you want to know whether or not a so-and-so exists, see whether one can be found in V; if you want to know whether or not such-and-such is provable, see whether it can be derived from the axioms of set theory" (Maddy, 2017, p. 296).

[18] For a more detailed discussion, see Maddy (2017).

theory becomes rapidly scarcer for the time frame after the 1950s and 1960s.[19] This is certainly not a result of historical oversight. Instead, if events are too recent, historical analyses are not forthcoming. It might simply not be methodologically possible at this point, as many historical sources are not yet available to the historian.[20] For the perspective of an overall history of mathematics, this might not be a problem; from the perspective of longue durée, fifty years may not even be a proper historical subject matter. But if we want to consider contemporary practice, fifty years is enough to drastically change a mathematical area both in its methods and its research goals (for example, consider the time between the 1850s and the 1900s in algebra).

As we have seen, the correct evaluation of the means-ends relation relies heavily on historical analysis. But if we cannot address (roughly) the last fifty years, we lose much insight into contemporary practice (especially in a discipline that is only little more than one hundred years young). Can this restriction be overcome in some way? Can recent set theory be made accessible to a second-philosophical analysis? We want to give on outline for an affirmative answer—at least up to a certain degree.

The main role of the historical analysis in the procedural schema is to create a test case for our intended means-ends relations. Here we can use a historical account of the practice to search for suitable methods and goals and test them against the criteria of the evaluative apparatus. We also use contemporary practice as a control case, for example to check whether the criteria still hold, whether the historical practice is still relevant, and so on. But if we don't have historical analyses available, we lose our primary source of test cases. To counteract this, we could try to use contemporary practice as the test case. This is difficult, however, as we will lack historical hindsight. For example, it will be more difficult to judge consensus-based criteria, like naturalness, for very new methods, as naturalness is (at least to a degree) connected to the community getting familiar with a new method. Content-based criteria are also difficult to assess, as a method can show much promise but *de facto* has not produced relevant results yet. And subject-based criteria are even more difficult to judge: When mathematicians are in the process of proving and formulating hypotheses, they heavily rely on heuristics and intuitions; some distance to the work is needed to see what was actually mathematically necessary and what was a crutch for thought that can be thrown away afterwards.

Nonetheless, we speculate that there is some work that can be done starting from the vantage point of analysis of contemporary practice, namely by considering future developments of the practice. The most immediate example in which considerations regarding the future come up are mathematical hypotheses. To be sure, they don't have the same status as scientific predictions, as the eventual assessment of their truth or falsity does not confirm or refute mathematics as a theory. But they can help provide a future frame of reference to test means-ends relations.

---

[19] For example Ferreirós (2007) ends with a short outlook on the time immediately after World War II; work in the 1960s is already scarcer, exceptions are scholarly work by Kanamori (2008); Moore (1987).

[20] For example access to Nachlässe, correspondences, etc.

Let us consider the following example: in Maddy's case study of determinacy, she quotes a speculation of Yiannis Moschovakis (1980) about the future developments of determinacy hypotheses:

> In his 1980 state-of-the-art compendium on the subject, Moschovakis observed that 'no one claims direct intuitions …either for or against determinacy hypotheses', that 'those who have come to favor these hypotheses as plausible, argue from their consequences' (Moschovakis (1980), 610). At that time, he concluded:
>
> > At the present state of knowledge only few set theorists accept [$AD^{L(\mathbf{R})}$] as highly plausible and no one is quite ready to believe it beyond a reasonable doubt; and it is certainly possible that someone will simply refute [it] in ZFC. On the other hand, it is also possible that the web of implications involving determinacy hypotheses and relating them to large cardinals will grow steadily until it presents such a natural and compelling picture that more will succumb. (Moschovakis (1980), 610–611)
>
> Here Moschovakis displays impressive foresight, as more have succumbed in recent decades, on the basis of new discoveries.

We can interpret this as presenting the following means-ends relation: The method is identified with the adoption of (some version of) determinacy as an axiom; the goals are *future* mathematical consequences of this axiom. We will be able to use future content-based criteria ("argue[ing] from the consequences") and future consensus-based criteria ("natural and compelling picture") to judge this means-ends relation.

Given this setup, can we use contemporary practice to build a test case for a means-ends relation and then check it on the basis of a future test case? Unfortunately, this is not so simple, as building future test cases will always commit us to certain *ceteris paribus* assumptions: We have to assume that in the meantime nothing changes the general picture, for example that no different axiom with similar productive power gets accepted. Still, we have hope that such general issues can be overcome in some way. This hope relies on the fact that it is already part of mathematical practice to create such future test cases. One example for this is the recent work by Bagaria et al. (2019), in which the authors formulate two possible ways set theory can develop depending on which horn of the HOD dichotomy will hold. Also, in general mathematics it is common practice to explore the consequences of some mathematical hypothesis when a proof is still lacking.

This is certainly only an outline of the idea of using future developments of mathematics in the second-philosophical analysis. In order to develop a reliable method, this outline has to be worked out in much more detail. We will have to leave this as future work to be done.

In this section we described how the formulation of a general framework can be used to analyze the work of the Second Philosopher. We have shown that the framework is flexible enough to not impede her work and indeed provides a platform on which we can clearly formulate possible limitations of her methodology. Finally, we have given an outline of how the limitation of the reliance of historical analysis could be overcome.

Concluding, let us note that we motivated the changes that can be implemented in the procedural framework with the Second Philosopher's motivation to have the possibility to change her methodology. This also provides us with the possibility to analyze a separation of Maddy's interpretation from Second Philosophy itself. We could, for example, not accept the criteria that are associated with the "no heuristics" argument and try to argue that we are still conducting a second-philosophical analysis. Of course, such an argument would have to involve a detailed analysis into the contents of the procedural framework, but the fact remains that the framework provides a platform for this kind of analysis to occur.

8.5 Applying the General Framework

In this last section we want to show how the procedural framework can be applied to new cases of set-theoretic methodology. We will give a short and concise example in which we can watch our procedural framework at work. We will then conclude by pointing out directions of further work that could be done by applying the procedural framework in Second Philosophy and discussing how this can be used to address questions in the philosophy of set theory. In this section, we will therefore address point (C) of our general goals in Sect. 8.1.

The constructible universe L is a fundamental object in contemporary set theory. Introduced by Gödel to show the consistency of the Axiom of Choice and the Continuum Hypothesis, and further investigated by Jensen in the 1970s, generalizations of it are studied in inner model theory. L is sometimes seen as a very natural model of set theory, due to its canonical structure. Based on its importance for set theorists, the Second Philosopher is interested in L and wants to understand its role in set-theoretic methodology.

The natural historical case study for such an investigation is Gödel's own work at the end of the 1930s, such as Gödel (1939) and Gödel (1940). Of course, Gödel's work has been studied extensively in the philosophical literature, however here we will mainly follow (Kanamori, 2007) as we want to focus on his mathematical work from a historical perspective.

Having established the setting for the historical case study, we want to figure out which method to choose. When looking at Gödel's works, we have an abundance of possibilities: the method could be foundational, like "adopting $V = L$ as a new axiom" with a foundational or mathematical goal to match; the method could be more local, like "using Gödel's Condensation Lemma" with the goal to prove GCH in L; it could be more broad like "minimality arguments" from the fact that L is the smallest inner model of ZF used in an abundance of mathematical theorems; and so on. Not all of these methods will lead to a satisfactory means-ends relation. As Maddy points out repeatedly, the method to adopt $V = L$ is be excluded from her analysis of set-theoretic methodology, as it contradicts foundational goals that are connected to maximality considerations (in the form of large cardinals) (Cf. Maddy, 1997, p. III.6). The historical case study could suggest a method like "work in (some variant of) NBG set theory", as Gödel indeed used NBG in Gödel (1940). However, when referring to contemporary practice, it turns out that this method should not be considered as it is not methodologically relevant or part of the evidential structure of the subject, therefore not fulfilling crucial content-based criteria.

Following Kanamori, we want to look at the method that is

> arguable the central feature of the construction of L: [...] *L* is a class definable in set theory via a transfinite recursion that could be based on the formalizability of def(x) [...]. Though understated in Gödel's writing, his great achievement here [...] is the submergence of metamathematical notions into mathematics. (Kanamori, 2007, p. 161)

Let us abbreviate this method as "the definability construction"; the goal suggested by our case study is a mathematical one, namely "proving the consistency of AC and GCH". First, we have to check if we pass the criteria of the evaluative apparatus.

The historical case study presents us with the curious fact that Gödel did use this method in his 1939, but in the longer exposition 1940 he uses the method of Gödel operations. Without going too much into the interesting relation between these two accounts,[21] we see with respect to contemporary practice that the definability construction remained methodologically relevant, as it gave rise to a great number of similar constructions (relative constructibility, ordinal definability and much more general studies of inner models). It has therefore become an integral part of the evidential structure of set theory. Additionally to these content-based criteria, the method and goal conforms to consensus-based criteria (building a model up in the way of the definability construction certainly counts as a natural construction). The subject-based criteria seem to be fulfilled as well, as both method and goal are mathematical.[22]

In comparison with the historical case study and contemporary practice, it turns out that our candidate passes the criteria of the evaluative apparatus. Further, we see that the definability construction is an effective means towards the end of proving the consistency of AC and GCH (as well as other mathematical goals) and therefore should be included in set-theoretic methodology.

Starting from this observation, the Second Philosopher could now turn towards the general question of the role of metamathematics in set-theoretic methodology. A first clue on this was given by Kanamori in the quote above. Baldwin (2018) gives an account of a shift in model theory that is connected to the introduction to metamathematical methods. To see if the same holds for set theory, we could therefore use our previous observation as first evidence for a comparable shift in contemporary set theory. Such an investigation would certainly be of importance to the Second Philosopher when turning to her questions about the nature of set-theoretic activity.

We see a further point where it could be interesting to follow the work on metamathematical methods: Maddy (2011) wants to explain how the Second Philosopher inquires into mathematics (cf. Maddy, 2011, p. 39). She focuses on set theory instead of general mathematics because she wants to investigate *pure* mathematics (Maddy, 2011, Section I) and assumes set theory to be a good foundation for it. We think that

[21] For more on this, see Kanamori (2007).

[22] Though it is interesting to note that this has not always have been perceived in this manner, see for instance Cohen's impressions recounted in (Cohen, 2002, p. 1087).

the viewpoint of metamathematical methodology could be an interesting starting point to revisit the connection between mathematical and set-theoretical methodology, as metamathematical methods seem much more prevalent in the latter than the former. This will also be important if we want to transfer insights from the Second Philosopher's inquiry into the nature of set-theoretic activity to the nature of mathematical activity in general.

Last, in the current debate in philosophy of set theory, there are several examples that can be studied with the above developed procedural schema—the most prominent of them being connected to the universe/multiverse debate. Maddy (2017) defends a universist stance by arguing with the methods of Second Philosophy. Ternullo (2019) argues to the contrary, mainly by addressing Maddy's foundational goals that she develops in (Maddy, 2017). As is shown in (Antos, 2022), the procedural framework can be used to find some middle ground in the debate. Here the claim that techniques like forcing have led to an introduction of models as new entities in set theory is investigated by applying Second Philosophy. The method of introducing models as new entities is regarded in a historical case study of the development of forcing by Paul Cohen, where the chosen goal is proving the independence of the Continuum Hypothesis. Antos (2022) concludes that this passes the criteria in the evaluative apparatus and that the method is effective towards this goal. It is inferred that the method of introducing models of set theory as new entities should be included into set-theoretic methodology. Also an outlook is given on how this might influence further discussions of Maddy's foundational goals. We think that such an investigation has the potential to build a bridge between the principles from which Maddy draws her form of universism and the focus on models of set theory from the multiverse side.

In this section we have shown how our general framework can be applied to investigate set-theoretic methodology in a second-philosophical manner. We have also given hints on how such an investigation can be used for the discussion of broader philosophical questions. Combining this with the possibility of changing aspects of the Second Philosopher's methodology itself, as developed in Sect. 8.4.3, we hope that this article contributes to further applications of Second Philosophy to the study of set theory.

References

Antos, C. (2022). Models as fundamental entities in set theory: A naturalistic and practice-based approach. *Erkenntnis, 89*(4), 1683–1710.

Bagaria, J., Koellner, P., & Hugh Woodin, W. (2019). Large cardinals beyond choice. *Bulletin of Symbolic Logic, 25*(3), 283–318.

Baldwin, J. (2018). *Model theory and the philosophy of mathematical practice: Formalization without foundationalism*. Cambridge University Press.

Boolos, G. (1971). The iterative conception of set. *Journal of Philosophy, 68*(8), 215–231. https://doi.org/10.2307/2025204

Cantor, G. (1872). Über die Ausdehnung eines Satzes aus der Theorie der trigonometrischen Reihen. *Mathematische Annalen, 5*, 123–132.

Cohen, P. J. (2002). The discovery of forcing. *Rocky Mountain Journal of Mathematics, 32*(4), 1071–1100.

Ferreirós, J. (2007). *Labyrinth of thought: A history of set theory and its role in modern mathematics*. Berlin: Springer.

Ferreirós, J. (2016). *Mathematical knowledge and the interplay of practices*. Princeton, USA: Princeton University Press.

Francois, K., & Van Bendegem, J. P. (2010). Revolutions in mathematics. more than thirty years after crowe?s “ten laws". a new interpretation. *Philosophia Mathematica, 18*(3), 107–120.

Gödel, K. (1940). *The consistency of the axiom of choice and of the generalized continuum-hypothesis with the axioms of set theory*. Princeton University Press.

Gödel, K. (1939). Consistency-proof for the generalized continuum-hypothesis. *Proceedings of the National Academy of Sciences, 25*(4), 220–224.

Hamkins, J. D. (2012). The set-theoretic multiverse. *The Review of Symbolic Logic, 5*, 416–449.

Kanamori, A. (2007). Gödel and set theory. *Bulletin Symbolic Logic, 13*(2), 153–188.

Kanamori, A. (2008). Cohen and set theory. *The Bulletin of Symbolic Logic, 14*(3), 351–378.

Maddy, P. (1988). Believing the axioms. i. *Journal of Symbolic Logic, 53*(2), 481–511.

Maddy, P. (2017). Set-theoretic foundations. In A. E. Caicedo, J. Cummings, P. Koellner, & P. B. Larson, (eds.), *Foundations of mathematics. Essays in honour of W. Hugh Woodin's 60th birtsday*, pages 289–322. American Mathematical Society.

Maddy, P. (1997). *Naturalism in mathematics*. Oxford: Oxford University Press.

Maddy, P. (2007). *Second philosophy: A naturalistic method*. Oxford: Oxford University Press.

Maddy, P. (2011). *Defending the axioms: On the philosophical foundations of set theory*. Oxford: Oxford University Press.

Moore, G. H. (1987). The origins of forcing. In F. R. Drake & J. K. Truss, (eds.), *Logic Colloquium '86 Proceedings of the Colloquium held in Hull*, volume 124 of *Studies in Logic and the Foundations of Mathematics*, pp. 143–173. Elsevier.

Moschovakis, Y. N. (1980). *Descriptive set theory*. North-Holland Pub. Co. ; Sole distributors for the U.S.A. and Canada, Elsevier-North Holland Amsterdam ; New York : New York.

Pease, A., Aberdein, A., & Martin, U. (2019). Explanation in mathematical conversations: An empirical investigation. *Philosophical Transactions of the Royal Society A*, 377.

Rittberg, C. J. (2016). *Methods, goals and metaphysics in contemporary set theory*. Dissertation, University of Hertfordshire.

Ternullo, C. (2019). Maddy On The Multiverse. Set Theory and General ThoughtsIn S. Centrone, D. Kant, & D. Sarikaya (Eds.), *Reflections on the foundations of mathematics: Univalent foundations* (pp. 43–78). Cham: Springer International Publishing.

Zermelo, E. (1908). Untersuchungen über die mengenlehre. i. *Mathematische Annalen, 65*(2), 261–281.

Zermelo, E. (1908). Neuer beweis für die möglichkeit einer wohlordnung. *Mathematische Annalen, 65*, 107–128.

Chapter 9
Reply to Antos and Kant

Penelope Maddy

Keywords Methodology · Set-theoretic method · Iterative conception · Heuristics

Professors Antos and Kant spotlight an aspect of work in the philosophy of set theory that's been relatively unattended, in the first instance, by me. To locate this aspect, first recall the distinction between questions about how set theory is properly carried out and questions about what sort of activity set theory is—Antos and Kant's concern is with the former. Zeroing in further, note that when set theorists allow impredicative definitions or avoid restrictive theories, these are set-theoretic methods. The study of these methods, then, is set-theoretic methodology, where I've tried to isolate set-theoretic methods and assess their rationality in light both local and global set-theoretic goals.[1] This kind of means-ends analysis, in turn, is a method of set-theoretic methodology. What Antos and Kant set out to do is describe and assess my use of this 'core procedure' (§4.1) of set-theoretic methodology, an inquiry that must count as the methodology of set-theoretic methodology! I think they've successfully established that it's difficult to piece together a coherent account of how my various appeals to means-ends analysis fit together, from Maddy (1997) to Maddy (2011) and beyond. What I'll try to do here, with the benefit of hindsight, is embed the three principles they isolate—Consensus-based Criteria, Content-based Criteria, Subject-based Criteria—into a more comprehensive, and hopefully more compelling, line of thought.

[1] Antos and Kant prefer to avoid talk of rationality (p. 9), but they allow themselves to speak of justification, which introduces a similar normative element.

P. Maddy (✉)
Department of Logic and Philosophy of Science, University of California, Irvine, CA, USA
e-mail: pjmaddy@uci.edu

S. Arbeiter and J. Kennedy (eds.), *The Philosophy of Penelope Maddy*,
Outstanding Contributions to Logic 31, https://doi.org/10.1007/978-3-031-58425-1_9

But first, one more general point deserves comment—the matter of Second Philosophy itself. Here Antos and Kant 'address the possibility of separating Second Philosophy from Maddy's description of it' (p. 3).[2] One conspicuous target is my claim that the Second Philosopher's methods can't be given a once-and-for-all characterization, that they evolve in open-ended and unpredictable ways (§2.1). In contrast, Antos and Kant set out to give 'a general framework for a Second Philosophy analysis of set-theoretic methodology' (§4). I think this disagreement is only apparent. When I say that the Second Philosopher's approach can't be conclusively characterized, I'm speaking of her fundamental approach, starting with ordinary observation, moving on through deliberate experimentation, theory formation and testing, and so on. What I resist is the attempt to characterize this in something of the form 'trust only science' on the grounds that the methods of 'science' evolve in those open-ended and unpredictable ways.[3] But this doesn't preclude the possibility that one or another of the various areas of human endeavor that the Second Philosopher investigates will turn out to have a conclusive characterization. So when Antos and Kant attempt to corral the methods of set-theoretic methodology, a much narrower target, I have no objection; for that matter, I might be said to have given something like a general characterization of set-theoretic method myself, something like 'employ efficient means for deep ends'. So I think there's no fundamental disagreement between us on the nature of Second Philosophy itself.

Turning to the central point at issue, the 'core procedure' of means-ends analysis is introduced in *Naturalism in Mathematics* (Maddy 1997). Consider again the example of impredicative definitions. Though the historical debate involved various constructivist and realist claims about the nature of mathematical ontology, we find no less a leading light than Gödel himself arguing that they should be accepted because they generate a classical theory of real numbers,[4] among other welcome consequences. Eventually, this debate over set-theoretic method was resolved in favor of impredicative definitions, while the debate over mathematical ontology continues to this day.[5] The key point is: when we set aside the attention-grabbing ontological debate, the underlying mathematical facts reveal a perfectly rational means-ends case for

---

[2] I admit to some passing amusement at this formulation, since the sole motive for coining the new term 'Second Philosophy' was to gain a free hand in stipulating what I intended it to mean (see Maddy, 2007, p. 1). But I think what Antos and Kant actually have in mind is not to revise the fundamentals of Second Philosophy, but to question my implementation of it in particular cases (p. 5)—a possibility I've embraced from the beginning (Maddy, 2007, p. 3).

[3] This is to take a stand on what philosophers of science call the 'demarcation problem'. If it were possible to corral these under a general rubric—methods of type X—I could have dispensed with the conceit of the Second Philosopher and stated my position straightforwardly as 'trust only methods of type X'.

[4] Maddy (1997), p. 174. This isn't to say that other conceptions of the points on a line aren't mathematically valuable, just that a classical theory provides an effective foundation for classical analysis, a worthy goal in itself.

[5] Maddy (1997), pp. 190–191. By the term 'ontological', I intend debates like realism vs. constructivism, objects versus structures, fictional versus literal. I should say 'ontological and semantic', to include debates like the one over truth conditions versus proof conditions in discussions of intuitionism, but I hope this can be taken as implicit and won't so complicate the text.

allowing impredicative definitions. This historical case and others like it suggest a technique for approaching contemporary cases: look past the ontological debates and examine the mathematical details for signs of an straightforward appeal to an effective means in pursuit of a mathematically identifiable end.[6] I go on to apply this technique in assessing the contemporary case against V = L, with at least partial success.[7]

So far, Maddy (1997). What's lacking here, conspicuously, is any assessment of the legitimacy of the ends toward which those means are aimed.[8] This is where Antos and Kant detect Consensus-based Criteria: since the debate was resolved in favor of impredicative definitions, since there was consensus, that's enough to establish that the goal is worthy. The same impression might be drawn from other parts of Maddy (1997), for example, from the discussion of Definabilism (§II.4.ii), where I take for granted the reader's implicit agreement with the consensus view that Reimann's goals of founding analysis or providing tools for analytic number theory are legitimate and that the many mathematical consequences of the Axiom of Choice that in fact tipped opinion its favor are indeed mathematically beneficial. For that matter, though the foundational goal central to the case against V = L is defended in an early passage,[9] even this defense might be taken to rely on general agreement that the various foundational jobs set theory does so well are themselves mathematically valuable. So Antos and Kant are hardly unreasonable when they see Consensus-based Criteria as a principle at work in my methodological study of set-theoretic method.

Still, despite appearances, this was never what I intended, but I wasn't able to articulate what I did intend until Maddy (2011). The opening chapter of that book describes the historical rise of pure mathematics and highlights the new responsibilities that come with independence from natural science: pure mathematical concepts, methods, and assumptions could no longer draw their legitimacy from their success in application. Set theory was soon identified as a backstop for the coherence of new mathematical definitions—a foundational job[10]—but in addition, beyond this baseline, it now fell to mathematicians themselves to identify which among the coherent possibilities are mathematically worthy. We could rely here on Consensus-based Criteria, but instead I've proposed that there is an objective measure of specifically mathematical goodness, of mathematical value beyond applications, that abstract algebra for example is objectively good mathematics (fruitful, important, etc.), that 'group' is an objectively good concept, that providing a foundation for classical

---

[6] Maddy (1997), p. 194.

[7] I'm not sure why Antos and Kant think the method of means-ends analysis can't be applied to contemporary debates (pp. 20–21). If the case of V = L isn't contemporary enough, Maddy (2017b, 'On multiversism', this volume) take up current discussions of multiverse thinking, and Schatz (2019) brings the discussion of restrictiveness in Maddy (1997) to bear on the debate between V = Ultimate-L and forcing axioms.

[8] Thus, what I soon came to think of as the 'nihilistic footnote' (Maddy 1997, p. 198, footnote 9, Maddy 2007, pp. 350–351).

[9] See Maddy (1997), §I.2. The foundational goal is treated in more detail in Maddy (2017b, 2019a).

[10] The beginning of what I call Risk Assessment in Maddy (2017b, 2019a).

mathematics leads to more good mathematics, that set theory is a good mathematical undertaking. This specifically mathematical goodness is what I've called 'depth', for want of a better term.[11]

Elsewhere in these replies, I've touched on the delicate role of depth in the metaphysics of Thin Realism and Arealism[12]—here we see how it anchors the methodological account as well.[13] With depth in the picture, Consensus-based Criteria is no longer a fundamental principle: assuming, as I do, that leading mathematicians generally have an eye for good mathematics, the fact that a strong majority them agree on something isn't itself determinative, but it is an indicator for depth. And, of course, even good indicators can go wrong (Maddy, 2011, pp. 81–82).

As for Content-based Criteria, I take this as a new label for exactly the sort of means-ends justifications—rooted in extrinsic considerations, ultimately in the search for depth—that I do obviously intend to foreground. This leaves Subject-based Criteria, which divide roughly in two: 'no philosophy' and 'no heuristics'. On the former, it should now be clear that 'no philosophy' isn't a fixed principle of methodological analysis, but a technique suggested by past performance, a strategy for uncovering any means-ends rationality that might be present in debate. This is how the example from Dedekind (pp. 15–16) should be understood: once the ontological remarks are set aside, the argument for sets as a means toward the end of characterizing the natural numbers and founding arithmetic is revealed.[14]

That leaves 'no heuristics', and most centrally, the role of the iterative conception. It's worth noting that the 'philosophy' in 'no philosophy' and the 'heuristics' in 'no heuristics' occasionally overlap—for example, Gödel's ontological realism played an important heuristic role in his metamathematical work (see Maddy, 1997, p. 192)—but the iterative conception is less an ontology than a highly articulated intuitive picture of the set-theoretic universe that organizes our thinking and guides it in

[11] See Maddy (2011), pp. 79–83. For discussion of the nihilistic footnote (footnote 8, above), see Maddy (2011), p. 82, footnote 42.

[12] See, e.g.,'Reply to Soysal', this volume.

[13] So what happens here if depth turns out not to be objective? (In the metaphysical sphere—part of addressing the second sort of question, 'what kind of activity is set theory?'—we've seen (in 'Reply to Soysal') that a failure of objectivity would constitute a win for Arealism over Thin Realism.) There would still be means-ends rationality—it would still be rational to choose a means to a preferred end—but the legitimacy of that preference would presumably be grounded intersubjectively rather than objectively, renewing the concern over nihilism in footnote 8 above. This would apply only to the pure developments of mathematics; providing tools for application in natural science would remain an objective end, providing a brake on a thoroughgoing nihilism of the aforementioned footnote in Maddy (1997). (I don't quite say this in Maddy (1997), pp. 350–351, but come close). Beyond that, we'd have to rely on the collective good judgment of the mathematical community.

[14] A means-ends analysis of Dedekind's introduction of sets to facilitate a rigorous characterization of the natural numbers and a foundation for arithmetic appears on p. 45 of Maddy (2011), so the subsequent dismissal of his ontological stance that follows on pp. 52–53 (which Antos and Kant quote) presupposes that the success of the strategy has already been established.

new directions (for example, large cardinals).[15] I certainly don't mean to downplay the highly beneficial role of the iterative conception in set-theoretic thinking; what concerns Antos and Kant is my denial that it plays a justificatory role.

To see how this last piece fits, it's important to recognize where it falls in the overall line of thought. The counsel of history has suggested the technique of looking past ontological discussions and searching the mathematical details for means-ends reasoning; this strategy has revealed rational structures in a number of methodological discussions and been extrapolated with some success to the contemporary case against $V = L$; the metaphysical account of Thin Realism and Arealism has been developed based on this methodological account, and mathematical depth identified as underlying both the methodology and the metaphysics. At that point, in the closing pages of Maddy (2011),[16] I return to the other side of the intrinsic/extrinsic distinction and ask what role is left for intrinsic considerations, and in particular, those based on the iterative conception of set. Does a plausible case that a claim is based on the concept constitute evidence for that claim?

Assuming what's gone before is on the right track, it seems to me that the answer is no, at least not directly. It certainly is mathematically advantageous to have a compelling intuitive picture like the iterative conception, but the researches and directions it suggests are justified, if they are, by their extrinsic benefits—means-ends rationality is still fundamental. So the fact that a claim appears to be part of or implicit in the concept of set gives us some reason to explore it, given the track record of the iterative conception, but it still awaits means-ends validation.[17] From this perspective, the role of intrinsic considerations is merely heuristic, not epistemic, but this isn't an evaluative principal present from the start; it's taken to emerge from the completed picture.

In sum, then, Antos and Kant's Content-based Criteria—that is, extrinsic justifications, means-ends rationality—are fundamental to my methodological analyses of deliberations over set-theoretic method. Consensus-based Criteria are relevant only as indicators of mathematical depth, which underpins both my methodology and its accompanying metaphysics. And finally, Subject-based Criteria are either historically motivated suggestions for a methodological technique that has panned out rather well ('no philosophy') or assessments of the remaining role for intrinsic considerations ('mere heuristics'). I'm grateful to Antos and Kant for opening this line of inquiry and hope this reply has cleared up at least some of its mysteries.

---

[15] Effective as the iterative conception is, it might one day be rational to replace it, for example, with some version of a multiverse conception—if the new intuitive picture were to generate more good mathematics.

[16] Antos and Kant actually cite a passage from Maddy (2017b), which obviously postdates the conclusion of Maddy (2011), so there, too, I'm presupposing the theoretical backdrop just described.

[17] Maddy (2011), pp. 136–137. Incidentally, I do something along the lines of what Antos and Kant suggest: 'it would be interesting to find and analyse a method or goal on which the criteria disagree' (p. 18). Suppose extrinsic considerations, that is means-ends justifications, suggest p, but p conflicts with the iterative conception. I suggest that if those extrinsic considerations were strong enough, if the mathematical benefits were overwhelming, we'd swap out the iterative picture for something new (Maddy, 2011, pp. 134–135) (Cf. footnote 15).

References

Maddy, P. (1997). *Naturalism in mathematics*. Oxford University Press.
Maddy, P. (2007). *Second philosophy*. Oxford University Press.
Maddy, P. (2011). *Defending the axioms*. Oxford University Press.
Maddy, P. (2017b). Set-theoretic foundations. In A. Caicedo, J. Cummings, P. Koellner, & P. Larson (Eds.), *Foundations of mathematics: Essays in honor of W. Hugh Woodin's 60th birthday* (pp. 289–322). American Mathematical Society.
Maddy, P. (2019a). What do we want a foundation to do? In S. Centrone, D. Kant, & D. Sarikaya (Eds.), *Reflections on the foundations of mathematics* (pp. 293–311). Springer.
Schatz, J. (2019). *Axiom Selection by Maximization: V=Ultimate L vs. Forcing Axioms*, PhD dissertation, UC Irvine.

Chapter 10
Maddy's Notion of 'Maximize' and Strong Theories of Sets

Jeffrey R. Schatz

Abstract This paper examines contemporary approaches to maximization in axiom selection debates in set theory, applying these methods to the current debate between $V = Ult(L)$ and forcing axioms. The paper first surveys the motivations for maximization as a tool in axiom selection debates, focusing in particular on the formal approach of Maddy. It then examines an important current debate in set-theoretic practice between strong axiom candidates capable of settling the value of the continuum: the inner model axiom $V = Ult(L)$ and strong forcing axioms. Finally, the paper will apply the formal notions of maximize to these axiom candidates, showing that–under highly plausible conjectures–forcing axioms strictly maximize over $V = Ult(L)$. The paper will then briefly conclude with an examination of the implications of this result for axiom selection as a whole.

Keywords Maximization · Axiom selection · Ultimate L · Forcing Axioms · Set Theoretic methodology · Means-ends analysis

10.1 Introductory Remarks

The following paper is a condensed and edited version of the main results from the author's dissertation, written under the supervision of Penelope Maddy and defended in 2019.[1] On this occasion of Pen being honored with a volume in Springer's Outstanding Contributions to Logic series, it might therefore be appropriate to begin with a few remarks about Pen's remarkable abilities as an educator and thesis advisor in the field of logic.

[1] Pen has humorously referred to this sort of paper as the "freeze-dried" version of a dissertation. Note that in this initial section the author will refer to Professor Maddy as 'Pen'; in the following sections, she will instead be referred to as 'Maddy', in keeping with academic custom.

J. R. Schatz (✉)
Yale University, New Haven, CT, USA
e-mail: schatzjr23@gmail.com

S. Arbeiter and J. Kennedy (eds.), *The Philosophy of Penelope Maddy*,
Outstanding Contributions to Logic 31, https://doi.org/10.1007/978-3-031-58425-1_10

As a teacher, Pen demonstrates an impressive ability to explain and illuminate difficult concepts and issues, cutting through the details in order to reveal the heart of the matter. Beyond her expertise in logic, Pen has shown the proficiency to teach on myriad subjects in epistemology, philosophy of science, and the history of philosophy, with many of these areas becoming future focuses of her research. Perhaps the most remarkable facet of her teaching, however, is her ability to use seminars as an opportunity to illustrate sound philosophical method, showing students how to approach philosophical questions. In this way, no matter the subject of the seminar, her students would be given a front-row seat to see how the Second Philosopher works, providing a illustrative example of how to apply the method to their own work. As a result, the author firmly believes that Pen's seminars have influenced a generation of young philosophers in how to naturalistically approach research in areas well beyond logic itself.

One of the most favored memories most students of Pen fondly remember comes from her fall quarter seminar, offered on late Wednesday afternoons. Since the seminar went so late, many of the graduate students taking her classes would be unable to return home for Thanskgiving dinner the following day. Instead, Pen would treat her students to dinner on Thursday night at Irvine's local Mom and Pop vegan Thai restaurant. This tradition–referred to in the LPS community as 'Pensgiving'– became one of the most eagerly awaited days of the year, as students and visiting scholars would come together and enjoy one another's company. Over dinner, Pen would lead spirited discussions about important philosophical issues, such as the necessary and sufficient conditions for a 'sandwich', the relative importance of gangster vs cowboy films for American cinema, and the sad state of Pen's preferred San Diego Padres. These dinners were a great chance to interact with Pen outside of the classroom, and led to many cherished memories for myself and many of my peers. To this day, for many LPSers and former LPSers (including the present author), it truly does not feel like American Thanksgiving without a fried banana with green tea flavored ice cream.

With regards to her work as a thesis advisor, it is clear that Pen goes well above-and-beyond the call of duty to help her students both write the best possible version of their thesis, as well as develop into strong philosophers and logicians in their own right. In weekly meetings with the author, Pen would discuss the shape and direction of the thesis, arguing many different points in order to help shape the work into a stronger piece. Beyond advising the dissertation, Pen takes a strong interest in helping her students develop more generally as philosophers and human beings, providing both academic guidance and genuine kindness during the sometimes tumultuous process of being a PhD candidate. Even after the thesis is concluded, Pen continues to guide and advise her former students both inside and outside academia. In this way, the effects of Pen's approach to philosophy and logic will hopefully continue to be felt in the profession for years to come.

With that said, we now turn to the present paper, which has benefited immensely from Pen's guidance and teaching. In this paper, we will examine Maddy's novel approach to the notion of 'maximize', both as a foundational goal and as a formal tool in axiom selection debates. We will argue that this is both a useful and well-justified basis for deciding between competing axiom candidates. As a result, we will then seek to apply this concept to the current chocies for extending $ZFC + LCs$: the inner model and the forcing axioms programs. We will then present a novel result, suggesting that maximize provides a prima facie reason to prefer one of these two choices. We then conclude by discussing the ramifications of this result for axiom selection going forward.

10.2 Axiom Selection and Maddy's Notion of 'Maximize'

One of the central elements of set-theoretic practice since the discovery of the independence of CH has been *axiom selection*: finding candidates for new axioms to extend ZFC and determining ways to justify these candidates, thereby deciding between them. Given that historic and contemporary set theorists have given serious effort to justifying preferred axiom candidates, any successful methodology of set theory would need to come to terms with and evaluate the efficacy of these efforts. This is an especially salient question for current philosophical attempts to study set theory given the presence of two well-developed but mutually exclusive programs for developing new axioms: the inner model program and the forcing axioms program. So the questions arises: how do set theorists attempt to adjudicate between axiom candidates, and how should they do so?

In her *Naturalism in Mathematics*, Maddy begins by considering one of the goals of set theory as a mathematical discipline, namely providing a foundation for classical mathematics. She notes that achieving this goal requires following two fundamental maxims: maximize and unify.[2] 'Maximize' ensures that the domain of mathematical objects is as generous possible: "if set theory is to play the hoped-for foundational role, then set theory should not impose any limitations of its own: the set theoretic arena in which mathematics is to be modeled should be as generous as possible; the set theoretic axioms from which mathematical theorems are to be proved should be as powerful and fruitful as possible. Thus, the goal of founding mathematics without encumbering it generates the methodological admonition to MAXIMIZE".[3] In addition, UNIFY ensures that there is a single comprehensive theory of sets, enabling all the objects and methods of classical mathematics to be studied together.[4] Combining these two maxims, we find that the aim is to find a theory which provides "a single arena where all the various structures in all the various branches [of math] can co-exist side-by-side, where their interrelations can be studied" and a standard for

[2] Maddy (1997, p. 208).

[3] Maddy (1997, pp. 210–211).

[4] While unify is an important methodological principle in its own right, this paper will focus primarily on maximize. For this reason, we will mostly set unify to the side, and instead focus on the effects of maximize on the current debate in axiom selection.

"what counts as proof".[5] In playing these roles, set theory aims to provide a domain of objects that includes everything that could fall under serious mathematical study, and avoids curtailing any area of study in pure mathematics. For example, Maddy argues that the axiom of a measurable cardinal implies the existence of $0^\sharp$, a well-motivated object of mathematical inquiry, while $V = L$ blocks the existence of this set, and so curtails interesting mathematical work.[6] For this reason, maximize counts against the acceptance of V = L.

It is worthwhile to be clear on the precise nature of the justification of 'maximize' from these foundational goals. The foundational goals are justified by Maddy on firmly extrinsic grounds, through their ability to enable fruitful mathematical research by providing a single mathematical domain where one can study and compare the multitude of mathematical objects under consideration by the varied disciplines of contemporary mathematics. Considering these goals, Maddy notes that a notion of 'maximize' provides the most effective means of achieving the foundational goal: but this is simply a case of means/ends reasoning as found throughout scientific inquiry. 'Maximize' is thereby able to attain the extrinsic justification of the foundational goals, representing the methodological means for attaining them in practice.

In *Naturalism in Mathematics*, Maddy puts this methodological principle to work in considering the case of V = L and the existence of measurable cardinals, aiming to assess the rationality of the arguments leading set theorists to reject V = L as an axiom candidate on the basis of the mathematical goals of set theory. As a rough (and very simplified) outline of this argument, Maddy notes that 'maximize' encourages that an axiom candidate leading to the existence of some interesting mathematical objects should be preferred over an alternative candidate, provided that the former doesn't require giving up interesting mathematical content of the latter. The existence of a measurable cardinal implies the existence of non-constructible sets, and in particular the large cardinal $0^\sharp$, which conflicts with the axiom V = L. In fact, Maddy shows that V = L cannot even contain any object with the same isomorphism type as $0^\sharp$. Since this would curtail the use of the large cardinal hierarchy as a measure of consistency strength, as well as blocking the fruitful study of the higher large cardinal axioms, Maddy concludes that maximize counts against the acceptance of V = L. Finally, she notes there the axiom of a measurable cardinal does not require forgoing any of the mathematical benefits of working in ZFC + V = L, as L itself provides an interpretation of the theory that even a V = L proponent would have to accept as fair. Thus, Maddy concludes that 'maximize'–properly understood–is able to justify the set theoretic community's rejection of V = L.

---

[5] Maddy (2016, p. 16, 15). As Maddy notes, it is not obvious that maximize and unify can both be satisfied: for example, it might turn out that "ZFC can be extended in a number of incompatible ways... and that no mathematically defensible considerations allow us to choose between them" (Maddy, 1997, pp. 211–212). In such a case, we might instead turn to a multiverse approach to set theory. For the purposes of this work, we will focus on possible ways we might reconcile maximize and unify, and therefore will not consider multiverse approaches; see Maddy and Meadows (Frth) for more on how multiverse approaches fit into this view of set theory.

[6] See Maddy (1997, pp. 212–15) for an abbreviated outline of this argument; see Maddy (1997, pp. 216–32) for the argument in full technical detail.

In this paper we will focus solely on the maxim of 'maximize' as a tool for settling disputes in axiom selection. This is not to ignore that there are potentially other maxims that might be put to work in settling the current dispute between the inner model and forcing axiom programs. Nonetheless, 'maximize' is both rationally supported by the aims of set theory and justifies previous historically successful instances of extending ZFC. Thus, it is the obvious first method to be used in future instances of incompatible but desirable axioms.

In looking to use a formal explication of 'maximize' in the current axiom selection debate,[7] we note that Maddy's notion provides a ready-to-use formal tool capable of doing real work in axiom selection debates. Maddy's original explication of 'maximize'–the first such account proposed[8]–claims that maximization entails two distinct components; for one theory to maximize over another, the former must both provide a "fair" interpretation of the latter, and must also offer some genuinely new mathematical content. In particular, Maddy's account requires that there is some object in the maximizing theory that cannot be properly represented in any of the more restrictive theories.

The central idea behind Maddy's approach is that many different, even incompatible theories are all of genuine mathematical interest, and can be fruitfully studied for various mathematical purposes. Perhaps the clearest example of this is with the tradeoff between the axiom of choice and determinacy axioms; since Zermelo, choice has been recognized as essential for a broad swathe of work in core mathematics, and so plainly necessary for any good foundational theory,[9] and yet there have been many mathematical benefits of studying the consequences of determinacy axioms in restricted inner models of V.[10] Given the utility of studying rejected theories, a maximizing theory should avoid curtailing any of this well-motivated math, and so must provide domains for the theories which are more restrictive but still worthy of sustained development. In more plain terms, this means that a maximizing theory should prove the existence of some interpretation of a more restrictive theory. In

---

[7] Some will find the task of finding a formal explication for such a central methodological notion to be a hopeless task, fearing that any such precise counterpart will be doomed to supporting obviously unacceptable theories (false positives) and will miss other cases where a theory does intuitively maximize over another (false negatives). Maddy directly considers such problematic "dud" theories which meet the formal criteria for maximizing over a desirable theory can be artificially generated for a wide family of desirable theories (Maddy, 1997, pp. 229–31). In order to deal with these issues, we will follow Maddy's lead and informally restrict the scope of the notions of 'maximize' under consideration to theories which are genuine contenders for a strong theory of sets, considering only "natural theories". For our purposes, we simply note that both ZFC+LCs+V=Ult(L) and ZFC+LCs+MM are natural theories. Lacking a more firm criteria for "naturalness", we will use the informal guideline proposed by Steel: a natural theory should be a theory that is seriously proposed and considered by practicing set-theorists for its mathematical merits. See Steel (2014 Sects. 2 and 3) for his treatment of "naturalness".

[8] The other prominent account is a consistency strenght approach, found most prominently in John Steel's work, especially Steel (2004) and Steel (2014). See Schatz (2019) for a treatment of Steel's version of maximization in relation to these issues.

[9] See Moore (1982).

[10] See Maddy (1988) Sect. 5.

such a case, we do not face a genuine trade-off between the two theories, as the maximizing theory permits the exploration of the other within this interpretation, and so nothing is lost in shifting to the maximizing theory.

Maddy's approach as articulated in *Naturalism in Mathematics* can be understood as rationally starting from this idea of "fair" interpretations, with the existence of a "fair" interpretation being a necessary but not sufficient condition for a theory to be maximal. For Maddy, an interpretation is "fair" just in case it is a definable inner model of the base theory;[11] we will use the term *M-fair* interpretations to describe this class. Beyond merely having such an interpretation, however, Maddy's version of maximize demands that there also is a genuine mathematical benefit to accepting the maximizing theory. That is, there must be something of mathematical interest that is supplied by the maximizing theory that cannot be found within the original theory. In particular, there should be a pair (X, R) such that no object in the interpretation of the restrictive theory can have any pair (Y, S) that is isomorphic to (X, R).[12] Given that not all isomorphism types are equally of mathematical interest, we also intend that this isomorphism type be mathematically relevant: this is to exclude such objects as odd Gödel codings that have no appeal to practicing mathematicians, and so their existence does not present a genuine advantage for a theory. Unfortunately, there is no clear way to formalize the notion of "genuine mathematical interest", and given that a suitable foundation should aim to provide all the necessary objects for the future development of mathematics, it is desirable to refrain from hewing too closely to current understandings of mathematics. As a result, we will not officially include the demand that the isormophism type (X, R) be particularly useful. Given the focus on the existence of useful isomorphism types in the maximizing theory, we will refer to Maddy's approach to 'maximize' as a *isomorphism type* approach. We will call this approach to maximization Maddy-Maximization (M-Max for short).

Maddy defines a theory T' weakly M-maximizing over a theory T as follows:

$$T \trianglelefteq_M T' \text{ iff there is some } \varphi(x) \text{ st i). for all } \sigma \in T \; T' \vdash \sigma^{\varphi},$$
$$\text{(ii). } T' \text{ proves that } \varphi \text{ is an M-fair interpretation,}$$
$$\text{(iii). } T' \vdash \exists x \exists R \subseteq x^2 \forall y \forall S \subseteq y^2 (\varphi(y) \wedge \varphi(S) \rightarrow (x, R) \not\cong (y, S)).$$

We can then define a theory T' strongly M-maximizing over a theory T just in case T' weakly M-maximizes over T, but T does not weakly M-maximize over T'.

Note that a theory T' M-Maximizing over a theory T consists of two distinct conditions: in particular, conditions (i) and (ii) above together require that T' proves the existence of an M-Fair interpretation of T, while condition iii) requires that T' prove that there is some particular isomorphism type that exists but cannot be

---

[11] See Maddy (1997, pp. 220–221). Note that Maddy understands "inner model" in a slightly nonstandard way. In particular, an inner model includes both class sized transitive models of ZFC and the various V_κ's for κ an inaccessible cardinal. Throughout our discussion of maximization, "inner model" will instead be understood in the more typical way, including only class sized transitive models. We will suppress the inclusion of set sized models as "inner models" since nothing hangs on this difference in the context of considering the case of the inner model program and forcing axioms.

[12] See Maddy (1997, pp. 221–222).

captured by the interpretation of T. It is natural to wonder whether the isomorphism type condition, which is usually the difficult condition to verify in practice, might be simplified to a more tractable form. In particular, we ask whether the existence of a non-trivial M-Fair interpretation[13] suffices for the satisfaction of the isomorphism type condition?

We should note that, in complete generality, the answer to this question is a clear negative. To see this, consider the theory $ZFC^- + AFA$, where ZFC^- is ZFC without the axiom of foundation, and AFA is Aczel's anti-foundation axiom.[14] Since ZFC^- suffices to prove that the class of well-founded sets WF is a model of the full theory ZF, we find that $ZFC^- + AFA$ proves that there is an M-Fair interpretation of ZFC. One might naturally expect that $ZFC^- + AFA$ would witness many new isomorphism types, as it permits a development of an entirely new domain of non-well-founded sets; nonetheless, Maddy notes that any isomorphism type in $ZFC^- + AFA$ will in fact already be witnessed within the class WF.[15] In fact, it is precisely examples like these that motivated the necessity of Maddy's formulation of the isomorphism type condition in the first place.

Note, however, that in this example one of the theories–$ZFC^- + AFA$–is fairly weak, in the sense of not including all of the standard axioms of ZFC. In fact, this relative weakness is what makes the full isomorphism type condition necessary: if both T' and T are extensions of the full theory ZFC, then the existence of a non-trivial M-Fair interpretation of T within T' suffices to establish that there is an isomorphism type that fails to be witnessed in the interpretation of T. That is, for theories extending ZFC, any set outside of a proper inner model interpretation provides a novel isomorphism type.[16] As a result of this fact, we find that we can state an equivalent formulation of the conditions of M-Max for strong theories of sets extending ZFC:[17]

Re-Characterization Theorem:

Let $T, T' \supseteq ZFC$. Then $T \trianglelefteq_M T'$ iff there is some $\varphi(x)$ st
i). for all $\sigma \in T$ $T' \vdash \sigma^\varphi$,

---

[13] Here we just mean an M-Fair interpretation $\varphi(x)$ where $T' \vdash \exists X \neg\varphi(X)$.

[14] An accessible pointgraph is a directed graph with a distinguished vertex such that for any vertex in the graph, there is a path from the distinguished element to it. We say that an accessible pointgraph A with vertexes V decorates a set X if there is some map $f : X \mapsto V$ st $\forall y, z \in X$, $y \in z$ iff there is some path between $f(y)$ and $f(z)$ in A. The axiom AFA says that every accessible pointgraph decorates some set. This immediately suffices to imply the failure of the axiom of Foundation: to see this, note that a graph with a single point and self-directed edge is an accessible pointgraph; thus, any model of $ZF^- + AFA$ must include some x st $x \in x$. See Aczel (1988) for a development of this theory.

[15] See Maddy (1997, pp. 216–217), especially footnote 5, for this argument.

[16] This is just a bit too loose of a way of stating the result; a bit more precisely, any set outside of the proper inner model interpretation is coded by a collection of sets of ordinals, and the isomorphism type of the union of the singleton of this set of ordinals with its transitive closure cannot exist in the interpretation. See the proof below.

[17] See Theorem A.1.2 in Appendix A of Schatz (2019) for the proof of this claim.

ii). T' proves that φ is an M-fair interpretation,
iii). $T' \vdash \exists X(\neg\varphi(X))$.

Thus, when considering theories extending $ZFC + LCs$ we only need to be concerned with showing the existence of non-trivial M-Fair interpretations: the existence of a particular isomorphism type escaping the interpretation automatically follows from the existence of *any* set outside of the interpretation. For our present purposes, this will make it somewhat easier to apply M-Max to our theories, as there has already been much work on studying the existence of various interpretations between the proposed extensions of $ZFC + LCs$.

Returning to the original motivations for using 'maximize' as a tool in axiom selection, we may question how well-motivated M-Max is by the foundational goals of set theory. After all, it is not enough that a formal explication of 'maximize' gets the "right" answer in past cases of axiom selection: it needs to be that the formalization decides these cases on the basis of the "right reasons". M-max suggests that a theory with a greater consistency strength should be accepted only if it can provide some value to offset this risk: namely, some new isomorphism type that could not be studied in the less-risky theory. Combined with the informal guideline to focus on the isomorphism types that are genuinely mathematically interesting, M-Max amounts to a methodological imposition to choose a riskier theory when and only when there is mathematical benefits to be gained by doing so. Given the centrality of cost-benefit analysis to mathematical practice, and to set theoretic activity and axiom selection in particular, this seems to be a sustainable approach to increasing the consistency strength of our best theories of sets:[18] only incur risk when there is a corresponding mathematical benefit. In this way, M-Max appears to provide a principled way of explaining the motivations behind past axiom selection debates in set-theoretic practice. In the following section, we will introduce the main such debate in current set-theory; we will return to M-Max in the final section in hopes of providing a basis to potentially decide this issue.

10.3 Contemporary Axiom Candidates: A Fork in the Road

With the notion of M-Max in place, we can now turn to the contemporary literature regarding extensions of $ZFC + LCs$. While one might naturally expect a wide bevy of proposed axiom candidates for this purpose, it is somewhat surprising to see that most of the axiom selection literature has focused on two particular options. Even more ironically, these axioms can be seen as directly flowing out of the two historically important proofs leading to the proof of the independence of CH: Gödel's proof that $L \models ZFC + CH$[19], and Cohen's use of forcing to prove the consistency

[18] See Maddy (1997) for more on the important role of cost/benefit analyses in the methodology of set theory.
[19] Gödel (1940).

of $\neg CH$.[20] The former leads to the inner model program, and eventually the axiom $V = Ult(L)$, which provides an axiom implying CH; the latter leads to forcing axioms, especially the maximally strong (traditional) forcing axiom MM, implying $\neg CH$. Before we apply M-Max to this debate, it will be helpful to briefly survey both of these possible choices.

Let us first consider the latter–the forcing axioms program–as it proves to be a much simpler matter to state the related axiom candidate. In general, forcing axioms assert that for well-behaved classes of forcing posets, a generic set already exists in V: as a result, any sentence that can be forced to be true through forcing with that class of posets is already true. Initially, forcing axioms were proposed not as genuine candidates for extending ZFC, but merely as formal tools which simplified certain sorts of proofs involving iterated forcing. In 1988, working towards this goal, Foreman, Magidor, and Shelah isolated the class of stationary-set preserving posets: roughly, a forcing poset is ssp if every large subset of $\aleph_1$ in the ground model must remain large in any forcing extension through the poset.[21] A forcing axiom can then be stated regarding this particular class. Since Foreman, Magidor, and Shelah proved that no further extension of the class of forcings under consideration could be consistent–in particular, even extending the class of ssp forcings by a single additional forcing poset would generate a contradiction–this axiom became known as Martin's Maximum (henceforth, MM). With the formulation of MM, a strict upper limit to the program of developing traditional forcing axioms had been found.[22] Most important for present purposes, MM directly implies $\mathfrak{c} = \aleph_2$. For this reason, $ZFC + LCs + MM$ provides a natural and clear theory implying the falsity of CH.

The other main contemporary candidate for extending $ZFC + LCs$ was discovered and first defended by adherents of the inner model program, in particular the Harvard school,[23] who have sought to find a theory capturing $V = Ult(L)$. Unlike the case with forcing axioms, it is much less clear what particular axom candidate is intended by the Harvard school. There is some ambiguity in how $Ult(L)$ is used in the literature: it is sometimes used more broadly to mean the core or canonical inner model of a supercompact cardinal, but is sometimes used more strictly to refer to the unique structure satisfying some axiom candidate, especially Woodin's current candidate for $V = Ult(L)$. Throughout this section, we will use $Ult(L)$ in the looser sense, referring just to a uniquely specifiable (in some particular sense) weak extender model for the supercompactness of an extendible cardinal. While this ambiguity might be somewhat unsettling, by the end of this section, with some

[20] Cohen (1963) and Cohen (1964).

[21] Foreman et al. (1988). See Jech (2003), Chap. 37 for a contemporary presentation of the main results from this work.

[22] There are other ways of strengthening MM, for example by adding additional conditions on the poset which the forcing axiom asserts exists. We will put these strengthenings to the side for this paper: while they are an interesting subject in their own right, all of the claims made regarding $ZFC + LCs + MM$ also apply to these stronger axioms.

[23] By the Harvard school we mean the group of mathematicians and philosophers working on the $Ult(L)$ program at Harvard from roughly 2010 through the present. In particular, this group includes Hugh Woodin, Peter Koellner, and Gabriel Goldberg.

charitable assumptions made on behalf of the advocates of $Ult(L)$, a more sturdy understanding of the meaning of $Ult(L)$ will be possible.

The minimal requirement for a candidate for $Ult(L)$ is that of a weak extender model for the supercompactness of an extendible cardinal δ, for some cardinal δ.[24] A weak extender model, within some larger V, not only agrees with V on the supercompactness of δ, but does so for precisely the same reason: that is, the proper class of measures witnessing the supercompactness of δ in the weak extender is just the restriction of the proper class of measures witnessing this property in V. In this sense, a weak extender model is an inner model that provides a fair interpretation for the existence of a supercompact cardinal within V. Note that a weak-extender model need not provide any of the insight into supercompact cardinals typically associated with core or canonical inner models: in this way, the notion of a weak extender model is, in itself, a very loose "from above" notion of an inner model.

While providing a good inner model of a supercompact cardinal is an interesting feature of weak extender models, well worth studying in its own right, this proved to be just the first of a large number of surprising properties. The initial key discovery regarding weak extender models for the supercompactness of an extendible cardinal is that such a model must also provide a similarly fair interpretation of all known large cardinal assumptions above that of a supercompact: that is, any embedding witnessing a known large-cardinal property in V will also witness that property in the weak extender (through its restriction to that model).[25] Thus, unlike previous inner

---

[24] The current standard source for much of the material from this section is Woodin (2017), especially Sects. 3 and 7. While this provides a comprehensive account of much of the theory of weak extender models and of the axiom candidate for $V = Ult(L)$, the former appears to be in the process of being superseded by an alternative approach: it seems that much of the proofs in Woodin (2017) Sect. 3 can instead be proven in a somewhat simpler way directly from the δ-covering, -approximation, and -genericity properties. This alternative approach is mentioned briefly in Woodin (2019). Since the main resulting theorems themselves are unchanged, and this alternative approach has not been fully worked out in print, for this section we will follow Woodin (2017).

[25] One might naturally wonder how to evaluate the limitation to "known large cardinal assumptions" in Woodin's common statement of this result: how seriously should we take the possibility of a future large cardinal assumption that escapes this result? Such a large cardinal would necessarily fail to have a presentation in the form of the existence of an elementary embedding from V into some inner model M. Given the assumption of AC, there is only a single serious candidate for such a large cardinal axiom, namely the so-called HOD analogues of choiceless large cardinals. If these choiceless large cardinals prove to be consistent with ZF, then, there will be large cardinals that are unable to be captured in a weak extender model, and so there will in fact be an anti-inner model theorem for weak extender models. Such a theorem would, however, be only the tip of the iceberg regarding problems for the $Ult(L)$ program: this would imply that there is no weak extender model contained within HOD ((Woodin, 2017), p. 24), and that all of the friendly assumptions made on the behalf of proponents of $Ult(L)$ are false.

Thus, it seems that either there are no consistent candidates for large cardinal assumptions that cannot be captured in a weak extender model, or that the $Ult(L)$ program is thoroughly shattered. Much work remains to be done examining the plausibility of the choiceless cardinals: in particular, their consistency implies the consistency of a proper class of ω-strongly measurable cardinals in HOD, while it is currently unknown if even small finite numbers of these cardinals can consistenly exist in HOD (See Woodin (2017), p. 25, remark 3.43, for the current state-of-the-art on ω-strongly measurable cardinals in HOD.) For our purposes, we will set aside the possibility of choiceless large

models of large cardinals, any inner model that closely agrees with V on the nature of a single supercompact cardinal will not face an anti-inner model theorem ruling out the existence of some even larger large cardinal notion. While a weak extender model for a supercompact cardinal need not be a core or canonical inner model, Woodin pointed out that any core or canonical inner model containing a supercompact would seemingly be required to be at least a weak extender model for a supercompact: in this way, the theory of weak extender models represents a minimal requirement for any future theory of a fine-structured inner model at the level of a supercompact. Given the surprising discovery, this implied that there would be no limits on the large cardinal strength of a fine-structured inner model of a supercompact, if such a model were discovered.[26]

In this way, a weak extender model for the supercompactness of an extendible cardinal δ must be close to V in the sense of agreeing with V on all known large cardinal notions. There are a wide variety of other ways in which such an inner model must be "close to V", however. First of all, the weak extender will agree with V on all singular cardinals, and the identity of their successors, above δ; as a result of this fact, the weak extender will agree with V on cardinal arithmetic involving singular cardinals above δ more broadly.[27] Secondly, a weak extender model will have strong covering properties in relation to V.[28] In fact, given the assumption of a proper class of supercompact cardinals in V, Woodin has reportedly shown that any inner model with such covering properties must be a weak-extender model for the supercompactness of some δ: that is, only weak extender models can be similarly close to V.[29] Finally, any elementary embedding from a weak extender model into itself is either trivial or pathological, in the sense of having a small critical point relative to δ.[30] In each of these ways, we find that a weak extender model–if such a model exists–is uniquely similar to V in a wide variety of mathematically useful ways.

---

cardinals in order to present the $Ult(L)$ program in its intended context; if these cardinal prove consistent in the end, then it is quite likely that this would prove the end of the $Ult(L)$ approach to the inner model program as we know it.

26 An important distinction must be made here: while such a hypothetical fine-structured approach to a supercompact would capture larger large cardinal assumptions, it would not necessarily provide a fine-structured, "from below" approach to these cardinals. In this way, the proposed $Ult(L)$ structure might provide only a coarse understanding of large cardinals above a supercompact, leaving the task of developing notions of strategic extender models capable of providing a fine-structured understanding of these very large cardinal notions. Whether this possibility holds or not cannot be known until the fine-structure of weak extender models is more fully developed; but it is worth noting that in this way a proof of the assumptions outlined in this section may not be a complete end to the inner model program.

27 See Woodin (2017), p. 10, Theorem 3.10.

28 In particular, the model will have γ-covering for all regular $\gamma > \delta$, where a model M has γ covering if for all $X \subseteq M$ where $|X| < \gamma$, there is some $Y \in M$ st $X \subseteq Y$ and $|Y| < \gamma$.

29 This theorem was presented in a series of seminar talks given by Peter Koellner at UCI in the Spring of 2019. A published version of this material is forthcoming.

30 See Woodin (2017), p. 14, Theorem 3.19.

Given the important and useful properties of weak extender models, then, a natural question arises of whether any such model must exist, given the assumption of sufficient large cardinal strength. At present, this question remains open. Nonetheless, work on the choiceless large cardinals has led to a somewhat plausible assumption (the "HOD hypothesis") capable of resolving this question: that there is a regular cardinal λ above an extendible cardinal which is not ω-strongly measurable in HOD. For present purposes, we need not be too concerned with the detailed machinery surrounding this assumption, but may instead focus on a particular implication. If the HOD hypothesis is true and an extendible cardinal κ exists, then there is a weak extender model for the supercompactness of κ: equivalently, HOD itself is such a model.[31] Importantly for our purposes, this means that the HOD hypothesis implies that there is a definable inner model that provably must be a weak extender model for the supercompactness of an extendible cardinal. Furthermore, if $V \neq HOD$, then there is a definable *proper* inner model that must be a weak extender model.

Note, however, two key problems preventing HOD under these assumptions from truly representing the hoped-for ultimate approach to L: first, there is no clear axiom or theory behind the informal conception of $Ult(L)$ as a definable weak extender model, and, secondly, this approach fails to properly characterize $Ult(L)$ as a canonical inner model. Thus, HOD alone does not suffice for providing a theory of $Ult(L)$ robust enough for the analysis of maximization notions in this paper–even assuming HOD is a weak extender model. Towards this end, we must look more directly to the notion of canonicity of an inner model itself. Unfortunately, it does not seem that much can likely be known about the properties of a canonical inner model of a supercompact cardinal until the precise nature of the extender-like models capable of generating a model of a supercompact cardinal is discovered.

Nonetheless, some progress had made been into these questions by reflecting on and abstracting properties from past canonical inner models. Central to past successes of the inner model program is the comparison lemma. The comparison lemma was inspired as an extension of Kunen's comparison lemma, which states that for any two iterable structures of a particular form, there eventually is a pair of their iterates with one an initial segment of the other.[32] Since it is then said that the structure with an iterate initial segment is no stronger than the other to which it is compared, this lemma shows that the strength of any two iterable structures of the proper form can be compared. The comparison lemma proper similarly states that any two mice[33] with sufficient iterability conditions permit a comparison through their eventual iterates,

[31] See Koellner (2017), p. 3222–3223 and the surrounding material for a list of several key equivalencies to the HOD hypothesis, and a discussion of their importance.

[32] More precisely: if $\mathcal{M}_0 = \langle L_\xi[U], \in, U\rangle$ and $\mathcal{N}_0 = \langle L_\eta[U'], \in, U'\rangle$ are iterable structures, then there is some α and some filter F st $\mathcal{M}_\alpha \equiv \langle L_{\xi'}[F], \in, F\rangle$ and $\mathcal{N}_\alpha \equiv \langle L_{\eta'}[F], \in, F\rangle$. See Steel (2010), p. 1611 or Jech (2003), pp. 348–352.

[33] A mouse is a particular type of well-behaved transitive, iterable structure which plays a key role in core model theory: in particular, the core model up to a measurable cardinal K is equivalent to L relatized to the collection of all mice. For present purposes, we simply note that mice represent the"from below" building blocks of the core model K. See Jech (2003), pp. 660–661.

though the necessary conditions are much more involved due to the heightened complexity of mice and iteration strategies compared to Kunen's iterated ultrapowers.[34]

While each version of a comparison lemma for a particular type of canonical inner model makes reference to the precise machinery of that iterable structure, Woodin noted that every known canonical inner model had a corresponding comparison lemma; furthermore, reflecting on the nature of these lemmas, Woodin was able to abstract the essential content of these lemmas as the weak comparison lemma, which allowed the notion of comparibility to be stated without involving particular fine-structural concepts.[35] Weak comparison follows from each form of a comparison lemma, and so Woodin notes that any acceptable fine-structure for a canonical inner model of a supercompact cardinal should also be expected to imply weak comparison. As a result, weak comparison can be used as a proxy for the future fine-structure that will result from further development of the $Ult(L)$ program.[36] Thus, weak comparison–together with a few necessary supporting assumptions needed to permit this principle to do much real mathematical work–presents an initial option for a theory of a canonical inner model of a supercompact.

Noting the difficulty of working with weak comparison and the supporting assumptions, however, in his dissertation Goldberg takes the process of abstracting from the particular details of the fine-structured versions of the comparison lemma one step further. Seeking a simpler principle more easily used in mathematical proofs, Golberg isolated the Ultrapower Axiom (henceforth, UA): the claim that every pair of ultrapower embeddings admits a comparison.[37] Given the supplemental assumptions used for working with weak comparison, Goldberg was able to show that weak comparison implies UA, and thereby the fact that UA holds in all previously discovered canonical inner models.[38] Thus, due to the centraility of the comparison lemmas

---

[34] The details of this lemma are quite involved and go beyond the purposes of this section. See Steel (2010) Sect. 3.2, especially Theorem 3.11, for a precise statement of this theorem.

[35] Woodin's weak comparison axiom states that for any two finitely generated transitive models of ZFC $\mathcal{M}_0$ and $\mathcal{M}_1$ that are Σ_2-embeddable in V with $\mathbb{R}^{\mathcal{M}_0} = \mathbb{R}^{\mathcal{M}_1}$, there must be a transitive N with embeddings $i_0 : \mathcal{M}_0 \mapsto N$, $i_1 : \mathcal{M}_1 \mapsto N$ with i_0 close to $\mathcal{M}_0$ and i_1 close to $\mathcal{M}_1$. See Woodin (2017) Sect. 6.5 or Goldberg (Frth) for more on the weak comparison principle.

[36] More precisely, the combination of ZFC with the existence of a supercompact, $V = HOD$, and the weak comparison principle can be studied as a proxy for a theory of a canonical model of a supercompact cardinal. It is an open question whether this theory is consistent, which is unlikely to be resolved until the theory of fine-structure for $Ult(L)$ is further developed: Woodin notes that this is a "natural test question for the existence of a generalization of L at the level of supercompact cardinals based on anything like the current methodology for the construction of such inner models" Woodin (2017), p. 88.

[37] For transitive models of ZFC P, Q and a cofinal elementary embedding $j : P \mapsto Q$ is an ultrapower embedding if there is some $a \in Q$ that generates Q from functions in P, meaning $Q = \{j(f)(a) | f \in P\}$. Note that any embedding generated from an ultafilter in the usual way can be shown to be an ultrapower embedding in this sense. For ultrapower embeddings $j_0 : V \mapsto M_0$, $j_1 : V \mapsto M_1$, a comparison of (j_0, j_1) is a model N and a pair of embeddings (i_0, i_1) st $i_0 : M_0 \mapsto N$ and $i_1 : M_1 \mapsto N$ are themselves ultrapower embeddings, and $i_0 \circ j_0 = i_1 \circ j_1$. See Goldberg (Frth) or Goldberg (2018) Sect. 2.3 for more details.

[38] A proof of this result is found in Goldberg (Frth); an early summary of the main results of the dissertation, including this proof, was shared with the author in the Fall of 2018.

("the central feature of modern inner model theory" (Goldberg, 2018, p. 3)), Goldberg notes that ZFC with the existence of a supercompact and UA serves as a falsifiable test case for the eventual fine structured model of a supercompact: "if one could rule out the Ultrapower Axiom from a supercompact cardinal, one would in fact rule out any sort of inner model theory from supercompact cardinals" Goldberg (2018), p. 3. Furthermore, in stark contrast to weak covering, UA has proven extremely fruitful to work with, enabling a large body of theorems to be developed from this assumption.[39] Of particular relevance for our present purposes, Goldberg was able to use UA to prove a number of the properties typically associated with coreness and canonicity, including that GCH must hold eventually.[40] Thus, $ZFC + \exists\kappa SC(\kappa) + UA$ provides our first serious candidate for a theory approximating the eventual theory of $Ult(L)$:[41] though $ZFC + \exists\kappa SC(\kappa) + UA$ will likely not be the eventual theory used in the $Ult(L)$ program, it provides a stable subtheory of the eventual resulting theory that can be used in analysis as this true theory continues to be developed.

We have therefore found our first candidate for a theory of $Ult(L)$ articulated with sufficient precision to be capable of being utilized in our analysis of formal maximization notions. The question naturally arises of whether interpretations of this theory are known or suspected to exist. Given Goldberg's above comments on the necessity of the consistency of UA and the existence of a supercompact cardinal for any hope of an extension of the inner model program to this level of the large cardinal hierarchy, it seems fair to conclude that any satisfactory canonical inner model of a supercompact cardinal must validate the UA axiom. Additionally, a canonical inner model must be uniquely identifiable–hence the reference to canonicity–usually by making reference to the fine-structural notions used to build the model from below. With these two facts in mind, there seems to be a clear conjecture underlying Golberg's work with UA: that there must eventually be some definable inner model of $ZFC + UA + \exists\kappa SC(\kappa)$. Let us refer to this implicit claim as the UA conjecture.

Note, though, that UA only aims to capture an abstract and somewhat removed picture of the consequences of canonicity, without dealing with the details of what an extender-based approach to an inner model of a supercompact might entail. Another approach to finding a theory capturing the $Ult(L)$ notion instead tries to engage with this problem more directly. Much of the difficulty of articulating a fine-structural approach for this level comes from the seeming inability of standard, (nonstrategic) extender models[42] of reaching the level of a strongly compact cardinal: Woodin has

[39] See Goldberg's dissertation Goldberg (Frth) for the vast majority of this literature.

[40] That is, for any λ st $\lambda > \kappa$ for some supercompact κ, $2^\lambda = \lambda^+$. See Goldberg (2018) Sect. 6.2 for more on this result.

[41] The assumption of a supercompact is necessary here: as UA is implied by previous versions of covering lemmas, its is true in each of the canonical inner models below a supercompact. Only with the additional assumption of a supercompact, ruling out these lower consistency strength models of UA, can Goldberg's axiom be seen as capturing the informal $Ult(L)$ idea.

[42] Here we mean any notion of extender models which uses a construction process defining the next model only from a single parameter (the previous model). This is in contrast to strategic extender models, which use a process defining the next strategic extender model from two parameters (the previous model and an iteration strategy which in effect codes the correct way to carry out this

shown that the least cardinal that is $\kappa^{+\omega}$ strongly-compact must fail to be $\kappa^{+\omega}$ supercompact in a nonstrategic extender model, while Goldberg showed that $UA + GCH$ implies just the opposite for the the least $\kappa^{+\omega}$ strongly-compact cardinal. Gven that $UA + GCH$ is seen as indicating the presence of fine structure, these results show that nonstrategic extenders are *not* the correct notion of fine structure at the level of a strongly compact cardinal.[43] Thus, there is a seeming incompatibility between fine-structure as underwritten by UA and fine-structure as generated from non-strategic extender models once as the level of a supercompact cardinal is approached. To explain this incompatibility, Woodin and Goldberg note the difficulties in proving the iterability[44] of non-strategic extender models past the finite levels of supercompactness, and suggest that this is due to a failure of iterability at this point.[45] Thus, a new notion of an extender-like model was needed which would be capable of preserving iterability to the level of a supercompact.

To this end, Woodin turned to the notion of strategic extender models, which supplement a series of extender-models with iteration strategies in effect coding the proper way to proceed in these constructions without violating iterability.[46] Crucially, the use of iteration strategies on particular infinitary games in these definitions allows Woodin to apply the fruits of the successful theory of AD^+ in $L(A, \mathbb{R})$ for amenable sets A under the assumption of a supercompact cardinal to the work of developing the theory of strategic extender models at this level.[47] While the state of this project is very much in flux as Woodin continues to work out the details of strategic extender models at higher infinitary levels, for our purposes we need only note that Woodin

process while preserving iterability). Note that partial extender models, first developed by Baldwin and Mitchell and used by Mitchell and Steel to develop inner model theory to the level of Woodin cardinals in the 1990's, are of the former sort. See Steel (2010) footnote 3, Mitchell and Steel (1994), and Woodin (2017) Sect. 5.1 for more on partial extender models; see below for more on strategic extenders.

[43] See Goldberg (2018) for a brief discussion of this result and the challenges it poses for nonstrategic extender models reaching very large cardinals. While Goldberg's result requires the additional assumption of GCH, there is some hope that the assumptions behind this and other related results will eventually be weakened to just UA.

[44] A model permitting iterated ultrapowers is said to be iterable if each of its iterates is itself well-founded. See Steel (2010) Sect. 3 for an approachable discussion of iterability, and see Schimmerling (2010) and Neeman (2010) for proofs of iterability for earlier notions of fine-structure.

[45] See Woodin (2017) Remark 7.1 and surrounding material for more on this failure of iterability.

[46] The bulk of Woodin (2017) is concerned with laying out the role and theory of strategic extender models: Sects. 4, 5, and 6 lay out the complications that prevent nonstrategic extender models from capturing a supercompact cardinal, and Sect. 7 develops the initial theory of strategic extenders. Note that until the construction process reaches the finite levels of a supercompact, beyond the level of a single Woodin cardinal, the addition of iteration strategies has no effect on the defined models, and so the nonstrategic and strategic hierarchies are identical; this is due to the unique branch theorem, which states that there is only one unique way of carrying this process out (See Woodin (2017) Thm 6.35). After the fist Woodin cardinal, these hierarchies diverge. It is currently conjectured that at some point iterability fails for the non-strategic hierarchy, while iterability continues for the strategic hierarchy up to and including the level of a full supercompact cardinal. See Woodin (2017) Sect. 7 for more on these conjectures.

[47] See Woodin (2017), pp. 90–92 for a brief outline of this work.

conjectures that applications of the theory of AD^+ in this way will permit a proof of the iterability of strategic extender models past the finite levels of a supercompact.[48] Let us refer to the conjecture that iterability will eventually be proved for the full strategic extender hierarchy as Woodin's Iteration Conjecture (henceforth, WIC).

Given the WIC, the eventual model constructed through the strategic extender hierarchy would provide a fully fine-structural, from below approach to a unique weak extender model of a supercompact cardinal, thereby fulfilling the promise of the informal $Ult(L)$ notion. But what would the theory of this canonical structure be? While one might naturally expect any answer to this question to depend on the precise details of the proof of iteration, and thereby be inaccessible for the time being, in another surprising development Woodin found that the connections between strategic extender models and the theory of AD^+ might already provide the answer: under the assumption of a proper class of Woodin cardinals, good approximations to the eventually generated structure exist and can be defined without any reference to fine-structural notions.[49] These determinacy based approximations to $Ult(L)$ form a hierarchy, and it has already been shown that the initial members of the hierarchy are also members of the strategic extender hierarchy. Furthermore, it is expected that there will eventually be a proof that all of the determinacy based approximations must be strategic extender models.[50] Given this expectation, the completion of each hierarchy will be extensionally equivalent, and so the structure ($Ult(L)$ itself) could be theorized through reference to determinacy notions instead of the more difficult fine-structural approach.[51]

It is precisely this possibility that motivated the current formulation of the axiom $V = Ult(L)$ by Woodin. This axiom states two distinct claims: first, that there is a proper class of Woodin cardinals, and, second, that for any true Σ_2 sentence φ there is

[48] See Woodin (2017), pp. 89–90 for a clear presentation of the recent state of the project of proving iterability. Note in particular the claim that "the only credible possibility that remains is that iterability is proved by induction and *not* on the basis of some general iteration hypothesis for V. Verifying that this is in fact what happens is the main task ahead." (Woodin, 2017, p. 90).

[49] See Woodin (2017), p. 90, especially "in the context of a proper class of Woodin cardinals, there are naturally defined approximations to Ultimate-L and the collection is rich enough to make a definition of the axiom, $V = Ultimate(L)$, possible without specifying the detailed level-by-level definition of Ultimate-L".

[50] Woodin (2017), p. 90: "The conjecture is of course that all the approximations are strategic-extender models and there is quite a bit of evidence for this conjecture... The key issue is whether the axiom $V = Ultimate(L)$ formulated in terms of these approximations must hold in *some* weak extender model for supercompactness assuming that there is an extendible cardinal. Presumably any proof of this must yield as a corollary that these approximations are all strategic-extender models".

[51] While it might seem initially strange that a canonical inner model could be axiomatized through a claim about close approximations to V, without reference to any fine-structural notions, it is worth noting that the same is true for earlier examples of canonical inner models. As a particularly salient example, Woodin notes that there is a very similar presentation of the axiom $V = L$ as the claim that any true Σ_2-sentence in V is true in an inner model of a particular form. Thus, it should not be too surprising that such an axiom could be found for $Ult(L)$. See Woodin (2017) Lemma 7.2.

a universally Baire set of reals A such that HOD in $L(A, \mathbb{R})$ models φ.[52] It is hoped that any model of this axiom would be so well-approximated by the determinacy based approximations to $Ult(L)$ that it would have to simply be the completion of the hierarchy of these approximations; furthermore, given the assumption that this determinacy hierarchy is a cofinal proper subset of the strategic extender hierarchy, any model of the axiom would also thereby have to be the completion of the strategic extender hierarchy. While there is a flurry of current activity by the Harvard school in developing the consequences of this axiom, some initial results have been quite promising. In particular, it has been shown that $V = Ult(L)$ implies CH[53], and there is much hope that other restricted versions of paridigmatically L-like properties will also be shown to follow from it.[54] Thus, with the candidate axiom for $V = Ult(L)$, we have a second possible theory aiming to capture the informal $Ult(L)$ idea: $ZFC + V = Ult(L)$.

We should note, however, that though there has been significant progress regarding Woodin's current preferred candidate axiom for $V = Ult(L)$ in recent years, and while proponents of the $Ult(L)$ program have great confidence that this will be the eventual axiom capturing $Ult(L)$, it may perhaps be a bit wise to refrain from a complete endorsement of $V = Ult(L)$ as the correct axiom. This is especially true given the extremely fast pace at which developments in the $Ult(L)$ program have happened. In particular, there has been a bit of shift regarding the axiom candidate for $V = Ult(L)$ in the past, which past proposed axioms having been found to be too strong to capture the informal $Ult(L)$ notion: for example, while the various axiom candidates attempt to specify a canonical example of a weak extender model of a supercompact, past proposed axioms have been shown to be incompatible with weak extender models.[55] At the time of writing it remains possible that either the

---

[52] The inner models $HOD^{L(A,\mathbb{R})}$ are the determinacy based approximations to $Ult(L)$ described in the above paragraph. So the candidate for $V = Ult(L)$ in effect states that the theory of AD^+ is sufficient to permit the development of these inner models, and that for any sufficiently simple sentence φ there is an approximation of this form that comes close to V, in the sense of agreeing regarding φ. While it has been shown for particular universally Baire sets of reals A that $HOD^{L(A,\mathbb{R})}$ is a strategic extender model, it is currently only conjectured that this is the case for any universally Baire A. A proof of this conjecture would likely be closely tied to a proof of the WIC. See Woodin (2017) Definition 7.14 for the official definition and surrounding discussion.

[53] See Woodin (2017) Theorem 7.26 part a). It is currently unknown whether $V = Ult(L)$ implies the full GCH, though Goldberg's proof of GCH above a strongly compact cardinal from UA is seen as providing some evidence that this implication does in fact hold.

[54] The existence of a supercompact cardinal poses a serious limitation for the extent to which the typical L-like properties can hold in $Ult(L)$. In particular, there is no hope for a "simple" well-ordering of the reals. For the other properties, however, these is some hope. Woodin has shown that $V = Ult(L)$ implies that $V = HOD$ and that V has no proper set-generic grounds (Woodin (2017) Theorem 7.26 parts b) and c)). Together with the HOD hypothesis, the former may lead to a (weak) covering theorem; the latter may underlie a notion of restricted absoluteness, together with further developments in set theoretic geology. As there is much that remains to be proven regarding these properties, however, we will leave the matter here, noting that it is an open question of how L-like the structure characterized by the $V = Ult(L)$ axiom is.

[55] For examples of this, see Axiom 1 and Axiom 2 in Woodin (2017), pp. 92–93. Note that–though they appear to have somewhat different forms–the current candidate for $V = Ult(L)$ is equivalent

current candidate axiom proves to be too strong–by proving incompatible with the notion of a weak extender model–or that it prove to be too weak–by failing to entail UA and thus falling short of a fine-structural theory of $Ult(L)$. For these reasons, we will be somewhat cautious with equating the informal $Ult(L)$ notion and models of this axiom. Nonetheless, given the strong confidence on the part of its proponents, as well as the non-trivial theory of its consequences that has already been developed, we will henceforth treat this axiom as the axiom of $Ult(L)$, and refer to it just as $V = Ult(L)$.

As with the theory $ZFC + \exists\kappa SC(\kappa) + UA$, the question similarly arises of whether a definable interpretation of the theory $ZFC + V = Ult(L)$ is known to exist. As with the earlier question, much remains to be seen through the development of the $Ult(L)$ program, but Woodin makes the expectation quite explicit through the $Ult(L)$ conjecture: $ZFC + LCs$ proves that for any extendible cardinal δ, there is an inner model N such that N is contained in HOD, N is definable from δ, and $N \models V = Ult(L)$.[56] The truth of the $Ult(L)$ conjecture would suffice to show the truth of the WIC, and thus that there is a fine-structural approach to a model of $V = Ult(L)$.[57] It seems fair to say that a proof of the $Ult(L)$ conjecture would constitute a completion and a validation of the entire $Ult(L)$ program up to this point.

In summary, in this section we have found three approaches to attempting to capture the informal notion of the Ultimate version of L: the notion of a weak extender model of a supercompact, the theory $ZFC + \exists\kappa SC(\kappa) + UA$, and the theory $ZFC + V = Ult(L)$. Additionally, each has a corresponding conjecture which implies the existence of an inner model representing that approach: the HOD hypothesis, the UA conjecture, and the $Ult(L)$ conjecture. Given these multiple options to use in our analysis of the maximality relations between the $Ult(L)$ program and forcing axioms, a choice must be made of which to pursue. Fortunately, it turns out that not much results from such a choice, due to the deep relationships between the

---

to a straightforward restriction of Axiom 2 to only Σ_2-sentences. In the end, Axioms 1 and 2 in fact insist on too much "closeness" between the determinacy inspired approximations and V. A similar incompatibility between the current candidate axiom and the notion of a weak extender model for a supercompact has not been found, but a proof of the consistency of $V = Ult(L)$ holding in a weak extender likely will only be possible after a further development of the associated fine-structural notions, if it is in fact true.

[56] There are many different formulations of the $Ult(L)$ conjecture. For standard treatments, see Woodin (2010b) (in its original form) or Woodin (2017). The most recent presentation in Woodin (2019) is based in the δ-approximation, -covering, and -genericity properties, but can be shown to be equivalent to earlier formulations. See also Koellner (2017), p. 3224 for a more explicitly arithmetical statement of the conjecture and ensuing discussion. Woodin sometimes distinguishes between weak, standard, and strong versions of this conjecture, regarding the large cardinal strength that must be assumed to prove the conjecture: the distinction between these versions of the conjectures need not make a difference for present purposes. If in fact only the weak $Ult(L)$ conjecture is true, the LCs assumption in the theories in Sects. 3.1 and 3.2 will need to be supplemented with the precise large cardinal assumptions required to prove the $Ult(L)$ conjecture.

[57] See Woodin (2017), p. 101: "Proving [the $Ult(L)$ conjecture] would show in a decisive fashion the transcendence of the strategic-extender hierarchy."

approaches. In particular, both the UA conjecture and the $Ult(L)$ conjecture imply the HOD hypothesis, and can only be satisfied in a weak extender model for a supercompact.[58] Additionally, given the WIC, $V = Ult(L)$ will entail that $Ult(L)$ is a fine-structural model of a supercompact in the traditional sense, and therefore that $UA + \exists\kappa SC(\kappa)$ holds.[59] In fact, the ability of $V = Ult(L)$ to imply $UA + \exists\kappa SC(\kappa)$ serves as a sort of verification that $V = Ult(L)$ is in fact the correct axiom candidate. With this in mind, not much hangs on whether $V = Ult(L)$ or $UA + \exists\kappa SC(\kappa)$ is used as our axiom of $Ult(L)$: given the focus of the Harvard school on $V = Ult(L)$, we will use this as our axiom in the remainder of the paper.

In conclusion, for our analysis of the maximality relations between the $Ult(L)$ notion and forcing axioms, we will use $V = Ult(L)$ as our axiom. Likewise, we will assume the $Ult(L)$ conjecture (and therefore also the HOD hypothesis and the UA conjecture) on the behalf of adherants of the $Ult(L)$ program, putting this program in its strongest possible position as a compelling and justified way of extending $ZFC + LCs$. We note, however, that not much would change if we were to instead assume only the UA conjecture and use $ZFC + UA + \exists\kappa SC(\kappa)$ as our theory of $Ult(L)$: in fact, all of the results in the following two sections similarly hold with $UA + \exists\kappa SC(\kappa)$ replacing every mention of $V = Ult(L)$, and the UA conjecture replacing every mention of the $Ult(L)$ conjecture. In fact, all that is required for this analysis is that there is some axiom capturing the $Ult(L)$ notion and that there is provably some interpretation of this axiom in $ZFC + LCs$ for some large cardinal strength assumption. But the existence of an axiom and theorem of this sort has been noted above as being a minimal prerequisite for a successful completion of the inner model program: if the assumptions of the $Ult(L)$ conjecture and the UA conjecture are eventually shown to be provably false,[60] this would constitute the end of the $Ult(L)$ program and of the inner model project as we know it. Thus, we note that the following results hold not just for the current axiom candidate $V = Ult(L)$, but for any reasonable alteration this axiom undergoes as the $Ult(L)$ program continues to evolve and develop.

[58] To see the implication, note that the HOD hypothesis is equivalent to the provable existence of any weak extender model for a supercompact contained within HOD (Woodin, 2017, Theorem 3.39).

[59] In fact, the connection between $V = Ult(L)$ and $UA + \exists\kappa SC(\kappa)$ might be even closer than this. Goldberg conjectures that $UA + GA + \exists\kappa Extendible(\kappa)$ in fact implies $V = Ult(L)$ (where GA is the claim that there are no proper set-generic grounds of V), thereby "recapturing" $V = Ult(L)$ from a collection of its most central consequences (See Goldberg, 2017). This is regarded as a highly speculative conjecture, with far more room for failure than the other conjectures and hypotheses articulated in this section. The truth of it, however, would represent the best possible situation for the $Ult(L)$ program.

[60] Woodin stresses that the $Ult(L)$ conjecture is in fact a simple arithmetical statement about the provability of a finite collection of sentences from another finite collection of sentences. As a result, it cannot be independent of ZFC, but instead must be either provably true or false. See Koellner (2017), pp. 3223–3224.

10.4 A New Maximality Result in Axiom Selection

We now return to the notion of Maddy-Maximization from Sect. 10.2. Recall that, by the Re-Characterization Theorem, for theories extending ZFC, T' M-Maxes over T if T proves that there is some non-trivial M-Fair interpretation of T, φ. As M-Fair interpretations only include definable inner models, forcing extension interpretations do not suffice for weak maximization. With these facts in mind, we now turn to the question of how to evaluate our two theories in terms of M-Max.

The first direction–regarding whether $ZFC + LCs + MM$ weakly M-Maxes over $ZFC + LCs + V = Ult(L)$–is extremely straight-forward. To show that $ZFC + LCs + MM$ weakly M-Maxes over $ZFC + LCs + V = Ult(L)$, it would suffice to prove in the former that a definable inner model interpretation of the latter. But the $Ult(L)$ conjecture states that $ZFC + LCs$ proves for any extendible cardinal δ, there is an inner model interpretation, definable from δ, of $ZFC + LCs + V = Ult(L)$. Letting δ be the smallest extendible cardinal, $ZFC + LCs$ therefore proves that there is an inner model interpretation of $ZFC + LCs + V = Ult(L)$. Furthermore, since $ZFC + LCs + V = Ult(L)$ proves $\neg MM$, there is at least one forcing poset in the inner model that does not have a generic in the inner model, and so the inner model interpretation must be proper. We thus find that $ZFC + LCs + MM$ proves that there is an M-Fair interpretation of $ZFC + LCs + V = Ult(L)$ under the assumption that the $Ult(L)$ conjecture is true: that is, it is immediate from our $Ult(L)$-friendly assumptions that $ZFC + LCs + MM$ weakly M-Maxes over $ZFC + LCs + V = Ult(L)$.

The question of the other direction–regarding whether $ZFC + LCs + V = Ult(L)$ also weakly M-Maxes over $ZFC + LCs + MM$–is much more intricately involved. The standard interpretation witnessing the consistency of $ZFC + LCs + MM$ within $ZFC + LCs + V = Ult(L)$ is a set-forcing interpretation, and therefore fails to provide an M-Fair interpretation. As a result, an additional method of defining interpretations of forcing axioms would be required for the two current theories to be M-Equivalent. So we turn to the question of whether there are any known inner model interpretations of forcing axioms, given sufficient large cardinal strength.

One natural possibility to consider when searching for such an inner model interpretation comes from the theory of the $\mathbb{P}_{max}$ forcing extensions. Originally developed by Woodin[61] in the early 1990's as a further development of the forcing axiom program, it was discovered that forcing with this poset on models of strong determinacy could provide natural interpretations of restricted versions of forcing axioms, assum-

[61] Prior to the discovery of the machinery of $Ult(L)$, Woodin's work focused prominently on the development of forcing axioms. Though he advocated for those axioms, and especially his axiom $(\star)$, his later proofs regarding weak extender models for supercompactness lead him to later reject this earlier line of argument. See Woodin (2010a) Definition 5.1 for the definition of $(\star)$ in terms of $\mathbb{P}_{max}$. Interestingly, these early arguments directly invoked maximality considerations, noting that $(\star)$ implied that any Π_2 over $H(\omega_2)$ which can be made true with set-forcing is already true in V. Though Woodin has rejected these arguments, other proponents of this line of reasoning persist: see Schindler (2017) for the current state-of-the art regarding these arguments.

ing the existence of a supercompact in V.[62] Woodin was able to show that under the assumption of a supercompact, the generic extension of $L(\mathbb{R})$ by this forcing poset exists as an inner model of V, and many of the standard consequences of forcing axioms at the level of $H(\omega_2)$ hold here.[63] Furthermore, for certain well-behaved pointclasses of reals Γ, the $\mathbb{P}_{max}$ extension of $L(\Gamma, \mathbb{R})$ can be seen to model restrictions of strong forcing axioms to small cardinals, including $MM(\mathfrak{c})$.[64] Additionally, these forcing extensions of inner models of determinacy are generically absolute, and therefore represent informally "canonical" inner models of restricted versions of forcing axioms.[65] Given this rich and structured theory regarding inner models of restricted versions of forcing axioms, one might naturally hope that a full definable inner model of $ZFC + LCs + MM$ could be found through further developments of this literature.

Alas, however, the prospects of a $\mathbb{P}_{max}$ extension providing an M-Fair interpretation of the full theory of $ZFC + LCs + MM$ seem highly implausible, at best. There are three distinct complications for this prospect. First, while the $\mathbb{P}_{max}$ extension of $L(\mathbb{R})$ itself provides a definable inner model, the more intricate $\mathbb{P}_{max}$ extensions of $L(\Gamma, \mathbb{R})$ for pointclasses Γ are only definable in terms of a parameter. The $\mathbb{P}_{max}$ extension of $L(\mathbb{R})$ only models a very limited selection of localized consequences of forcing axioms–in particular, the structure theory of $\mathbb{P}(\omega_1)$ under MM–and not the full axiom MM.[66] On the other hand, the $\mathbb{P}_{max}$ extensions of models of the form $L(\Gamma, \mathbb{R})$ are unable to provide M-Fair interpretations of forcing axioms. Thus, any

---

[62] The $\mathbb{P}_{max}$ extension of $L(\mathbb{R})$ was first introduced in Woodin (2010a), Chap. 4–6. A substantial theory of this forcing poset is developed and articulated throughout this textbook. In effect, forcing with $\mathbb{P}_{max}$ is designed to directly "lift" the structure theory of $\mathcal{P}(\omega)$ under $AD^{L(\mathbb{R})}$ directly to $H(\omega_2)$, preserving as much of these consequences as possible. See Koellner (2017), pp. 3212–3215 for an accessible presentation of this material.

[63] See Woodin (2010a) Theorem 4.54 for the existence of the generic extension, and that it models ZFC.

[64] See Woodin (2010a) Sect. 9.2. Note that the results are slightly stronger than stated above: in fact, $L(\Gamma, \mathbb{R})^{\mathbb{P}_{max}} \models MM^{++}(\mathfrak{c})$. Work on extending these results is ongoing: see Larson (2014) for a discussion of one such extension, to a restricted version of $MM(\mathfrak{c}^+)$.

[65] The term "canonicity" is frequently used in the literature regarding the $\mathbb{P}_{max}$ extensions. Note that this is a wholly separate usage of the term than is found in the literature regarding canonical inner models. To avoid confusion, we will not refer to the $\mathbb{P}_{max}$ extensions in this way throughout the remainder of the dissertation, instead directly stating the properties seen as providing a degree of "canonicity".

[66] It should be noted that these local consequences are particularly important to some advocates of forcing axioms. In particular, (Todorčević, 2012) presents these consequences for the structure theory of $\mathcal{P}(\omega_1)$ as the most important sources of justification for PFA and MM. In his envelope perspective argument, Koellner argues that since this structure theory concerns only the cumulative hierarchy up to H(ω_2), and so can be wholly captured in a proper inner model of V, these consequences can be fruitfully studied and developed regardless of whether forcing axioms hold globally in V or not. See Koellner (2017) Sect. 10.4 for a concise and focused treatment of this argument in print. A full-length treatment of the envelope perspective argument is also forthcoming. For present purpose, we note that regardless of whether the envelope perspective successfully nullifies Todorčević's particular argument for forcing axioms, it does not attempt to provide a response to the arguments for forcing axioms on the grounds of 'maximize' under consideration here.

possible interpretation arising from the theory of $\mathbb{P}_{max}$ extensions will seemingly either be too weak or fail to provide a genuinely M-Fair interpretation. Secondly, even if an interpretation only definable from a parameter were accepted, it currently seems unlikely that $ZFC + LCs + V = Ult(L)$ will prove that any $\mathbb{P}_{max}$ extension of an inner model of determinacy will model all of MM: the process of extending the strength of the forcing axioms true in these models to even $MM(\mathfrak{c}^+)$ has proved painstaking, requiring myriad assumptions and restrictions to be added to the axiom's formulation.[67] Given how artifical and strict the restrictions on MM must be to hold even for the smallest infinite cardinals in a $\mathbb{P}_{max}$ extension, it would be extremely surprising if such an extension were proved to model the full, unrestricted MM.[68] Lastly, even if a $\mathbb{P}_{max}$ extension of an inner model of determinacy modeling MM were to be found, this would still not suffice to provide an interpretation of $ZFC + LCs + MM$: as $\mathbb{P}_{max}$ is a homogenous forcing poset, the large cardinal strength of these extensions is limited to that of the original inner model $L(\Gamma, \mathbb{R})$, which fall well below that of a proper class of Woodins and an extendible. As a result, even if there were hiterto unprecedented developments in the theory of $\mathbb{P}_{max}$ extensions of inner models of determinacy, this would nonetheless fail to provide an interpretation of $ZFC + LCs + MM$. For these three reasons, the method of $\mathbb{P}_{max}$ extensions appears to be unable to provide an M-Fair interpretation of $ZFC + LCs + MM$ in $ZFC + LCs + V = Ult(L)$.

But what about other possible avenues for finding an inner model interpretation of $ZFC + LCs + MM$: might there be a method unrelated to the theory of $\mathbb{P}_{max}$ forcing extensions that could reveal the existence of such an inner model? This remains an open question: there are no known examples of inner model interpretations of MM, but there also is no general proof showing that such an interpretation is impossible. We note, however, that any such interpretation would have to be generated in a very novel fashion, as the usual methods of defining inner models cannot be used for interpreting $ZFC + LCs + MM$: neither inner models of determinacy[69], nor set-generic grounds[70], nor fine-structural inner models[71] can interpret $ZFC + LCs + MM$ within $ZFC + LCs + V = Ult(L)$. As a result, it is not even clear where one might start to look for such an interpretation of forcing axioms!

---

[67] The current strongest forcing axiom known to hold in a $\mathbb{P}_{max}$ extension is $MM(\mathfrak{c}^+)$ restricted to posets which a) do not reduce the cofinality of ω_3 to ω_1 and b) are stationary set preserving for posets which do not add ω_1 sequences of ordinals. See Larson (2014).

[68] Note that, unlike the case of MA, MM is not equivalent to any of its restrictions to particular infinite cardinals.

[69] As noted above, these models and their forcing extensions which provably are contained in V inherently lack the necessary large cardinal strength

[70] Woodin has shown that $V = Ult(L)$ implies that V is the minimum member of the generic-multiverse: see Woodin (2017) Theorem 7.26 3). For present purposes, we note that this implies that there are no proper set generic grounds of V.

[71] All known fine-structural inner models are models of CH, and so cannot satisfy strong forcing axioms: the same is true for all known coarse-structural inner models. It is believed that any future fine-structural inner model will also model CH, as well as a variety of combinatorial properties inconsistent with MM.

Given the inability of any of the established methods of defining inner model interpretations to properly interpret $ZFC + LCs + MM$ with an M-Fair interpretation, the prospects for the existence of such an interpretation might seem dubious. We further argue that this possibility is extremely unlikely. Towards this end, note that for all prior stages of the inner model program there has been a clear pattern: a canonical inner model will be found, representing the "smallest" possible class-sized inner model interpretation for a given large cardinal strength, and any other theories with that consistency strength can be interpreted in it through outer model interpretations (or patently unfair interpretations). If this pattern broke at the level of an inner model of a supercompact cardinal, this would represent yet another fashion in which the typical features of the inner model program are incompatible with the level of a supercompact cardinal. But, while earlier instances of failures of paradigmatic properties of the inner model program at the level of a supercompact–such as the lack of a projectively definable well-ordering of the reals–can be explained and understood as the direct result of central consequences of the existence of a supercompact, the existence of non-trivial class-sized inner model interpretations in $Ult(L)$ would seem to be entirely unrelated to the unique properties of supercompact cardinals. At the very least, this would represent one of the most surprising and seemingly inexplicable asymmetries between inner model theory below and above a supercompact cardinal.

Beyond merely being surprising, however, the existence of such an interpretation of forcing axioms within $Ult(L)$ would pose a serious challenge for the extent to which $Ult(L)$ could be seen as L-like in any meaningful way. If there were a definable inner model interpretation of $ZFC + LCs + MM$ then, given the $Ult(L)$ conjecture, the definition of $Ult(L)$ could be carried out in this proper inner model; as a result, there would be a definable proper inner model of the Ultimate version of L within the Ultimate version of L. Furthermore, this process could then be carried out in the interpretation of $V = Ult(L)$, leading to an infinite descending chain of Ultimate versions of L. This would be in stark contrast to the case of $V = L$, where there are no proper class-sized inner model interpretations whatsoever, and even to the case of canonical inner models of large cardinals below a supercompact, where any such proper class-sized inner model must lose large cardinal strength. Without knowing the exact nature of the definable inner model interpretations of $ZFC + LCs + MM$, it is unclear the extent to which these distinct inner models would necessarily resemble each other. Advocates of $V = Ult(L)$ explicitly hope that $ZFC + V = Ult(L)$ would suffice to settle all important undecidable questions modulo large cardinals[72], so any instances of disagreement over an independently motivated mathematical question would seriously diminish the prospects of $Ult(L)$ as a completion to the challenges of incompleteness. Without knowing more about the eventual details of these inner models, however, at this point we can only note that they may reveal a fundamental lack of even a minor degree of absoluteness.

[72] See Goldberg (Frth): "Woodin has proposed the axiom $V = UltimateL$... which, supplemented by large cardinal axioms, is expected to axiomatize the structure of the canonical inner model with a supercompact cardinal, *if it exists*. Therefore, supplemented with large cardinal axioms, the axiom $V = UltimateL$ should decide all questions of set theory."

Regardless of whether this possible descending infinite chain of Ultimate L's would prevent $Ult(L)$ from having the necessary degree of absoluteness expected of an L-like inner model, we argue that it reveals a more significant challenge for the justification of $V = Ult(L)$: a deep lack of canonicity for the $Ult(L)$ structure. Recall that an inner model is considered canonical if there are no arbitrary sets contained in it, and no arbitrary information introduced to it: Sargsyan eloquently explains the canonicity of L as meaning "*no random or artificial information is coded into the model.* Every set in L has a reason for being in it." (Sargsyan (2011), slide 9). In stark contrast to this, if there is a descending infinite chain of models of $V = Ult(L)$, then *any* model of $Ult(L)$ will contain arbitrary sets that fall out of the next proper inner model of $V = Ult(L)$. These sets will not be required by the fine-structural notions–the strategic extender models–used to construct a particular model of $V = Ult(L)$. In this way, $ZFC + LCs + V = Ult(L)$ would seem to clearly fail to identify a principled and robust collection of sets, instead varying significantly depending on the precise specifics of where it were constructed. In this way, the existence of a definable interpretation of $ZFC + LCs + MM$ within $ZFC + LCs + V = Ult(L)$ would appear to pose a serious challenge to treating the "Ultimate version of L" as a genuinely canonical inner model. In fact, given the myriad ways that such an interpretation would seem to prevent a significant degree of L-likeness for $Ult(L)$, its existence would raise many questions regarding the justification of $V = Ult(L)$ through the historical goals and methods of the inner model program.

Beyond its effects on the $Ult(L)$ program, the existence of an interpretation of $ZFC + LCs + V = Ult(L)$ within $ZFC + LCs + V = Ult(L)$ would also raise questions about the usefulness of M-Max as a methodological tool for evaluating extensions of $ZFC + LCs$. Recall that M-Max can be thought of as formally explicating a concept of having all of the mathematical content of a less-maximizing theory, and containing additional mathematical content that the other theory misses. But if there is an interpretation of $ZFC + LCs + MM$ in $ZFC + LCs + V = Ult(L)$, then we would find that some natural mathematical theories M-maximize over themselves: in terms of the informal gloss on M-Max, this would seemingly mean that some good mathematical theories "think" that they themselves miss out on genuine mathematical content! Such a state would be bizzare to say the least, and may well require a a rethinking of how to apply 'maximize' to theories extending $ZFC + LCs$. It is likely that further study of the specifics of these M-Fair interpretations would grant insight into how inner model interpretations broke with established patterns at the level of a supercompact cardinal, perhaps revealing how to better explicate 'maximize' for dealing with theories at such high levels of consistency strength.

But this all remains deeply speculative, given the complete lack of an obvious way to interpret $ZFC + LCs + MM$ with an inner model interpretation. Since such an interpretation would require novel and unprecedented methods to be developed and utilized, and since its existence would imply a wide bevy of surprising results regarding $Ult(L)$ which seem at odds with the significant theory as a fine-structured inner model so-far developed by Woodin, Koellner, and Goldberg, we here register strong skepticism towards the possibility that any such interpretation exists. For these reasons, we propose the following Conjecture:

Interpretation Non-Existence Conjecture ($INEC$): There is no M-Fair interpretation of $ZFC + LCs + MM$ within $ZFC + LCs + V = Ult(L)$.

With this conjecture in place, we are now ready to evaluate the maximization conditions between the theory of forcing axioms and of $Ult(L)$ in terms of M-Max. Given the $Ult(L)$ conjecture of the Harvard school, we find that $ZFC + LCs + MM$ weakly M-maximizes over $ZFC + LCs + V = Ult(L)$. Additionally, given our INEC, we find that $ZFC + LCs + V = Ult(L)$ cannot weakly M-maximize over $ZFC + LCs + MM$. We therefore find that the following implications hold:

If the $Ult(L)$ conjecture holds, then
$$ZFC + LCs + V = Ult(L) \trianglelefteq_M ZFC + LCs + MM$$
and
If the $INEC$ holds, then $ZFC + LCs + MM \ntrianglelefteq_M ZFC + LCs + V = Ult(L)$

Combining these, we find:

Maximality result: If the $Ult(L)$ conjecture and the $INEC$ are both true, then
$$ZFC + LCs + V = Ult(L) \triangleleft_M ZFC + LCs + MM$$

In conclusion, we note that the Harvard school strongly endorses the truth of the $Ult(L)$ conjecture, and the currently available evidence seems to suggest that the $INEC$ is true; and so, in the best possible justificatory situation for the $Ult(L)$ program, M-Max finds that forcing axioms are strictly more maximizing than $V = Ult(L)$.

10.5 Conclusion

In the previous section, we studied the maximization relationships between forcing axioms and the $Ult(L)$ program according to M-Max. In doing so, we showed that M-Max finds separation between the two theories, given our two conjectures. In particular, we found that $ZFC + LCs + MM$ strictly M-Maxes over $ZFC + LCs + V = Ult(L)$. In this final section, we will evaluate the consequences of this result for the justification of the $Ult(L)$ program.

In Sect. 10.2, it was noted that the 'maximize' maxim was closely related to the mathematical goals of set theory as a discipline, and so justified on the basis of means-ends analysis. Additionally, it was noted that 'maximize' was shown by Maddy to underlie the justification of $ZFC + LCs$ in the first place, and so seemed to be a natural source of justification for any acceptable extension of this theory. Finally, it was argued that M-Max is a well-justified formal explication of the informal 'maximize' notion, and therefore best positioned to be used in future justificatory endeavors. As a result, the likely strict maximization of another natural theory extending $ZFC + LCs$ over $ZFC + LCs + V = Ult(L)$ seems to pose a clear de facto challenge to the justification of the latter as the uniquely correct theory of the set-theoretic universe. But how should an advocate of $V = Ult(L)$ respond to this?

One response, given the dependence of our maximization result on two unproven conjectures, would be to reject one or both of these assumptions, thereby nullifying the result. Unfortunately, this does not seem to be a live option for an advocate of the $Ult(L)$ program. A rejection of the $Ult(L)$ conjecture would likely dash the hopes of the Harvard school to prove the WIC, leaving strategic extender models unable to provide a fine-structural presentation of a supercompact cardinal. Since Woodin notes that strategic extender models seem to be the only possible notion of fine-structure at this level,[73] this would leave the inner model program without any clear path forward. Additionally, we note that the analysis in Sect. 10.4 could instead be carried out with UA and the UA conjecture in place of $V = Ult(L)$ and the $Ult(L)$ conjecture: thus, this response would in fact require an even stronger rejection of the UA conjecture. But the lack of an inner model of UA with a supercompact cardinal has been described by the Harvard school as a potential falsification of the prospects of the inner model program at levels above a supercompact: as a result, its rejection would pose a much more signficant challenge for the justification of the $Ult(L)$ program than our maximization result! For these reasons, a rejection of the $Ult(L)$ conjecture (and also the UA conjecture) does not seem to be a live option for proponents of $V = Ult(L)$.

Another response would be to instead reject the $INEC$, arguing instead that there will eventually be a definable inner model interpretation of $ZFC + LCs + MM$ within $ZFC + LCs + V = Ult(L)$. By itself, a rejection of the $INEC$ would seem somewhat implausible: this type of response would naturally also require some sort of account of what novel methods could possibly generate this M-Fair interpretation. Additionally, as noted above, the existence of such an interpretation would raise serious questions regarding the extent to which $Ult(L)$ would represent a canonical inner model of a supercompact cardinal, thereby severing the supposedly close connection between $V = Ult(L)$ and previous axioms generated from the inner model program. Given the combination of the implausibility of the falsity of $INEC$ and the extent to which its falsity would raise new challenges for the justification of $V = Ult(L)$, rejecting the $INEC$ also does not seem to be a live option for supporters of the $Ult(L)$ program.

Thus, any satisfactory response should instead accept the maximality result, but instead question its significance for the current debate in axiom selection. In particular, a proponent of the $Ult(L)$ program could concede that other theories are better justified on the grounds of the 'maximize' maxim, but argue that $ZFC + LCs + V = Ult(L)$ is still the best choice for extending $ZFC + LCs$ on the basis of other methodological principles. Such a response may well be in keeping with the traditional arguments for $V = L$, which focused on the ease of working in

---

[73] See Woodin (2017), p. 89: "the hierarchy cannot consist of models constructed from *just* a sequence of partial extenders. The only alternative at present is an inner model theory based on *strategic* premice."

such a well-behaved and regulated model of set theory. An argument of this form for $V = Ult(L)$ would seemingly focus closely on the extent to which $Ult(L)$ can be shown to be L-like in some meaningful way(s), but without forcing an adherent to give up the use of large cardinals as a tool for measuring and comparing the consistency strength of theories. Without further development of the $Ult(L)$ program, and especially without a proof of the WIC and $Ult(L)$ conjecture yet in place, it is difficult to see how to evaluate the force that such an argument might have. Additionally, there has been relatively little work in the literature trying to identify what general methodological principles might underlie these lines of reasoning. As such, the burden at this time is on the $Ult(L)$ program to further articulate and defend such an argument.

Beyond arguments from L-likeness, there is one other argumentative route that can be taken for arguing that $V = Ult(L)$ is better justified than alternative theories in spite of the maximization result: that $Ult(L)$ is provably so "close" to V under any theory extending $ZFC + LCs$ to make the fruits of maximization negligible for set theoretic purposes. A particular instance of this "closeness" argument schema would then include a proof, for some particular explication of the notion of an inner model being close to V, that the interpretation of $V = Ult(L)$ must be close in that way, and a methodological argument that the particular sense of closeness allows all the mathematical fruits of the alternative theory to be achieved in the interpretation of $V = Ult(L)$. For some notions of closeness, a proof already exists: for example, the theory of weak extender models suffices to establish that the interpretation provided by the $Ult(L)$ conjecture–if it exists–will be close to any V in the sense of agreeing on large cardinal notions. Other proofs of closeness properties are promised by work on the HOD dichotomy and the HOD conjecture, which would show that the interpretation of $V = Ult(L)$ is close in the sense of being a class-generic ground of V under any theory extending $ZFC + LCs$. But, while there has been much work on proving that particular closeness properties hold between $Ult(L)$ and alternative models of $ZFC + LCs$, there has so-far been relatively little work on providing arguments for why such a notion of closeness might allow $Ult(L)$ to retain all of the important mathematical fruits of the alternative theory. While we note that such "closeness" arguments seem potentially quite promising as a source of justification for $Ult(L)$, it is difficult to evaluate them further without some particular instances being put forward and defended. Thus, as with the L-likeness arguments, the burden at this time is on the $Ult(L)$ program to explicitly provide such an argument.

In conclusion, in this chapter we used Maddy's formal tool of M-Max to evaluate the justification of the two foremost contemporary candidates for extending $ZFC + LCs$. In doing so, we provided an accessible presentation of the current $Ult(L)$ program, eventually isolating a particular formal theory as a precise formulation of the program. In addition, we highlighted a few crucial assumptions underlying the program that must be granted for argument's sake, especially the $Ult(L)$ conjecture. Then, in Sect. 10.4, we used M-max notion to show that, given the truth of plausible conjectures, the theory of forcing axioms strictly maximizes over that of $Ult(L)$. Finally, we noted some possible responses to this maximality result available to advocates of the $Ult(L)$ program.

In so doing, we hope to have accomplished a few distinct goals. First of all, we hope to have shown the usefulness and importance of M-Max as a formal tool for axiom selection debates. Though developed in the mid-90's, Maddy's concept continues to shed light on open questions in set theory almost 30 years later. Secondly, we hope to have provided an accessible explanation of the current state-of-the-art choices of extending $ZFC + LCs$, especially the extremely complex and occassionally opaque $V = Ult(L)$ axiom. Finally, we hope to have contributed directly to the current axiom selection debate with the Maximality result. In so doing, this paper aims to be a continuation of the project initially developed and defended in Maddy's groundbreaking *Naturalism in Mathematics*. This program continues to be one of the most compelling approaches in the philosophy of set theory, and the author is personally quite excited to see how Maddy's work continues to shape the debate around axiom selection for years to come.

References

Aczel, P. (1988). *Non-well-founded sets*, volume 14 of *CSLI Lecture Notes*. Stanford University, Center for the Study of Language and Information, Stanford, CA. ISBN: 0-937073-22-9. With a foreword by Jon Barwise [K. Jon Barwise].

Cohen, P. (1963). The independence of the continuum hypothesis. *Proceedings of the National Academy of Sciences of the USA, 50*, 1143–1148. ISSN 0027-8424. https://doi.org/10.1073/pnas.50.6.1143.

Cohen, P. J. (1964). The independence of the continuum hypothesis. II. *Proceedings of the National Academy of Sciences of the USA, 51*, 105–110. ISSN: 0027-8424. https://doi.org/10.1073/pnas.51.1.105.

Foreman, M., Magidor, M., & Shelah, S. (1988). Martin's maximum, saturated ideals, and nonregular ultrafilters. I. *Annals of Mathematics (2), 127*(1), 1–47. ISSN 0003-486X. https://doi.org/10.2307/1971415.

Gödel, K. (1940). *The Consistency of the Continuum Hypothesis*. Annals of Mathematics Studies, no. 3. Princeton University Press, Princeton, N. J.

Goldberg, G. (2017). Strong Compactness and the Ultrapower Axiom. arXiv e-prints, art. arXiv:1710.03586.

Goldberg, G. (2018). Compactness and comparison. arXiv e-prints, art. arXiv:1801.09153.

Goldberg, G. (2019). *The Ultrapower Axiom*. Ph.D. thesis, Harvard University, 2019.

Jech, T. (2003). *Set theory*. Berlin: Springer Monographs in Mathematics. Springer, 3-540-44085-2. The third millennium edition, revised and expanded.

Koellner, P. (2017). Themes from gödel: Some recent developments. *IFCoLog Journal of Logics and their Applications, 4*(10), 3171–3236.

Larson, P. (2014). Forcing axioms in $\mathbb{P}_{max}$ extensions. Colloquium talk given at the Harvard in the Spring of 2014, http://logic.harvard.edu/slides/Larson-2014-03-12.pdf.

Maddy, P. (1988). Believing the axioms. ii. *Journal of Symbolic Logic, 53*(3), 736–764. https://doi.org/10.2307/2274569.

Maddy, P. (1997). *Naturalism in mathematics*. Oxford University Press.

Maddy, P. (2016). Set-theoretic foundations. In A. E. Caicedo, J. Cummings, P. Koellner, & P. B. Larson, (eds.), *Foundations of Mathematics: Logic at Harvard Essays in Honor of W. Hugh Woodin's 60th Birthday*. American Mathematical Society.

Maddy, P., & Meadows, T., A philosophical reconstruction of steel's multiverse. Has since been published. *Bulletin of Symbolic Logic, 26*(2), 118–169.

Mitchell, W. J., & Steel, J. R. (1994). *Fine structure and iteration trees*. Lecture Notes in Logic (Vol. 3). Berlin: Springer, 3-540-57494-8. https://doi.org/10.1007/978-3-662-21903-4.

Moore, G. H. (1982). *Zermelo's axiom of choice*. Springer, New York. https://doi.org/10.1007/978-1-4613-9478-5.

Neeman, I. (2010). Determinacy in $L(\mathbb{R})$. In *Handbook of set theory. Vols. 1, 2, 3*, pp. 1877–1950. Springer, Dordrecht.

Sargsyan, G. (2011). An invitation to inner model theory. Talk given at the Fourth Young Researchers in Set Theory Workshop.

Schatz, J. R. (2019). *Axiom Selection by Maximization: V = Ultimate L vs Forcing Axioms*. Ph.D. thesis, University of California, at Irvine.

Schimmerling, E. (2010). A core model toolbox and guide. In *Handbook of set theory. Vols. 1, 2, 3*, pages 1685–1751. Springer, Dordrecht.

Schindler, R. (2017). Woodin's axiom $(*)$, or Martin's maximum, or both? In *Foundations of mathematics*, volume 690 of *Contemporary Mathematics*, pp. 177–204. American Mathematical Society, Providence, RI.

Steel, J. R. (2004). Generic absoluteness and the continuum problem. Talk given at the 2004 Laguna Workshop on Methodology of Pure and Applied Mathematics. https://www.lps.uci.edu/files/conferences/Laguna-Workshops/LagunaBeach2004/laguna1.pdf.

Steel, J. R. (2010). An outline of inner model theory. In *Handbook of set theory. Vols. 1, 2, 3*, pp. 1595–1684. Springer, Dordrecht.

Steel, J. R. (2014). Gödel's program. In *Interpreting Gödel*, pp. 153–179. Cambridge Univ. Press, Cambridge.

Todorčević, S. (2012). The powerset of ω_1 and the continuum problem. Talk given at the Harvard EFI Project in the Spring of 2012.

Woodin, W. H. (2010a). *The axiom of determinacy, forcing axioms, and the nonstationary ideal*, volume 1 of *De Gruyter Series in Logic and its Applications*. Walter de Gruyter GmbH & Co. KG, Berlin, revised edition, ISBN: 978-3-11-019702-0. https://doi.org/10.1515/9783110213171.

Woodin, W. H. (2010b). Multiverse views and the ω conjecture. Ziwet Lecture given at the University of Michigan. http://www.math.lsa.umich.edu/video/ziwet3.pdf.

Woodin, W. H. (2017). In search of Ultimate-L: the 19th Midrasha Mathematicae Lectures. *The Bulletin of Symbolic Logic, 23*(1), 1–109. ISSN: 1079-8986. https://doi.org/10.1017/bsl.2016.34.

Woodin, W. H. (2019). The continuum hypothesis. Colloquium Talk given at the University of Muenster, June 2019, https://www.uni-muenster.de/imperia/md/content/MathematicsMuenster/woodin-opening-colloquium.pdf.

Chapter 11
Exploring the Generous Arena

John T. Baldwin

Abstract We build on Maddy's, 'What is a foundation for?' to discuss the relationship of set theory and category theory. We propose to distinguish the goals of 'foundation' and 'organization', coining the term 'scaffold' for the latter. Then we argue that via various bi-interpretations set theory and category theory offer compatible foundations and scaffolds with advantages for each.

Keywords Generous Arena · Metamathematical corral · Category theory · Classification theory · Abstract elementary class

> So my suggestion is that we replace the claim that set theory is a (or 'the') foundation for mathematics with a handful of more precise observations: set theory provides *Risk Assessment* for mathematical theories, a *Generous Arena* where the branches of mathematics can be pursued in a unified setting with a *Shared Standard of Proof*, and a *Meta-mathematical Corral* so that formal techniques can be applied to all of mathematics at once.
>
> *What do we want a foundation to do?* Maddy (2019)

We view this analysis as a real step forward in foundational studies. Maddy considers several further criteria and concludes that set theory (specifically ZFC + large cardinals (LC)) serves as an ambient basis for mathematics that meets all the authentic criteria except *Proof Checking* and *Essential Guidance*. Mac Lane articulated a quite different view:

> It is an open scandal that the classical method of applying Zermelo-Fraenkel set theory as foundation for <u>all</u> practice of Mathematics is no longer adequate to the practice of Category Theory. The device of having both large and small categories in some Gödel-Bernays set theory was a convenient arrangement when it was first proposed by Eilenberg Maclane 23 years ago, but it no longer convenes[1] for functor categories (with large domain category)

[1] OED (1971) gives meaning II.9 of convene as 'to harmonize'.

Date: April 17, 2024. Research partially supported by Simons travel grant G3535. We acknowledge valuable conversations with Bousfield, Gorbow, Kennedy, Maddy, McLarty, Otto and others including a workshop at the University of Chicago organized by Thomas Pashby.

J. T. Baldwin (✉)
Department of Mathematics, Statistics and Computer Science, University of Illinois Chicago, Chicago, IL, USA
e-mail: jbaldwin@uic.edu

S. Arbeiter and J. Kennedy (eds.), *The Philosophy of Penelope Maddy*,
Outstanding Contributions to Logic 31, https://doi.org/10.1007/978-3-031-58425-1_11

> or the category of all categories as used in the theory of fibred categories or Benabou's profunctors. The alternative arrangement of categories in a Grothendieck Universe has been effective for getting on with the development of Mathematics, but it introduces assumptions as to inaccessible cardinals which palpably have nothing to do with the case, and it leaves unsettled (as yet) a variety of questions resulting from a shift of universe.
>
> What should we conclude? The happy security provided by one 'monolithic' foundation has been lost. First Principia Mathematica and then Zermelo-Fraenkel had this monolithic character, that all working Mathematics could be formulated within one system. This provided a convenient division of labor, between Mathematicians who just 'used' the system (usually in a naive form) and the Logician who investigated within the system various classical problems. The paradise is irretrievably lost; it is high time that open-minded young Mathematicians set to work to construct a new one—perhaps less monolithic. (MacLane, 1969, 130–131)

Maddy (2019, 22) thinks that whether set theory or category theory is 'more foundational' does not make for a productive debate. Rather she urges 'a concerted study of the methodological questions raised by category theory'. To begin such a study, we propose a distinction between Foundation and organization (Sect. 11.1). We consider category theory and model theory as 'scaffolds' and discuss their complementary virtues for specific mathematical areas (Sect. 11.2). We analyze in Sect. 11.3 comparable claims to be Foundations by material and structural set theories. Finally, in Sect. 11.4, we summarise the properties of the scaffolds and assess whether the Foundational 'paradise' is lost.

11.1 Foundations and Organization

Russell & Whitehead (1910) opine 'the chief reason in favour of any theory on the principles of mathematics must always be inductive, i.e., it must lie in the fact that the theory in question enables us to deduce ordinary mathematics.' Maddy agrees with that assessment and so we understand Russell and Maddy as taking Foundation with a capital F—a system for deducing ordinary mathematics.[2] And for Maddy, such a Foundation is justified not by an external consistency proof but on its success in grounding mathematics at large. As Maddy stresses in Maddy (2017, 2019), this grounding is not reflected in the daily practice of theorem proving. For her, it is a belief that mathematical research, vaguely thought of as carried out in naive set theory, can be reduced to a formal set theoretic foundation. She calls this belief *shared standard of proof*.

Secondly, Maddy requires a *Generous Arena* that encompasses all of mathematics. Maddy (2019, 13) asserts: 'the axioms of set theory imply the existence of (surrogates for) all the entities of classical mathematics—a simple affirmation of set theory's role as Generous Arena.' She stipulates that the surrogates are elements of the universe of sets, V.

---

[2] This f/F distinction arose on the Foundations of Mathematics listserve in the late 90's. Baldwin (2018) describes other variants (5, fn. 14) and studies small f foundations.

Although we focus on these two criteria, Maddy posits three others satisfied by ZFC + LC: risk assessment, meta-mathematical corral, and elucidation; two reasonable (essential guidance, proof checking) are not satisfied by ZFC, and rejects two (metaphysical insight, epistemic source). We generally agree with her judgements, but regard the two reasonable criteria as tasks for scaffolds, not Foundations.

In a section entitled 'Foundation or Organization' (Maclane, 1986, 406) is unsatisfied with ZFC (or well-pointed topos) as a **foundation**.[3] On the next page he describes both ZFC and category theory as **not wholly successful organizations** for mathematics. We adopt this distinction and use it to refine Maddy's criteria.

In contrast to a global Foundation, we describe a *scaffold* as an organization that includes both *local* foundations for various areas of mathematics and productive guidance in how to unify them. In a scaffold the unification does not take place by a common axiomatic basis but consists of a systematic ways of connecting results and proofs in various areas of mathematics. The scaffolds, model theory and category theory, provide local foundations including two flavors of set theory (Sect. 11.3.2).

Notation 11.1.0.1 The familiar *material* set theory takes the notions of element and set as fundamental; *structural* set theory takes function and set as fundamental (Shulman, 2019). In pure (without ur-elements) ZFC elements are sets. In contrast, the elements of a set X in structural set theory are not sets; they are functions from a terminal object 1 (For every object C, there is a unique morphism from C to 1.) to the object (set) X. Both material (vocabulary: ϵ, $=$) and structural (vocabulary has symbols for: objects, arrows, domain, codomain, equality, and composition) are first order theories.

Although there are many variants of either sort of set theory, all have an extensional notion of equality. The model theory scaffold gives local foundations for material set theory, ZFC. Various structural set theories extend ETCS (Elementary Theory of the Category of Sets Notation 11.3.2.1). We will examine whether these alternatives satisfy Maddy's criteria of *Generous Arena* and *Shared Standard of Proof*.

Maddy interprets Mac Lane's complaint (MacLane, 1986, 407) that there is too much 'sand[4]' in ZFC as a proposal that a Foundation should provide *Essential Guidance*, 'that would guide mathematicians toward the important structures and characterize them strictly in terms of their mathematically essential features' (Maddy, 2019, 19). She takes this as a plausible Foundational goal, though not a goal of ZFC. As she cogently argues, this goal conflicts with Generous Arena. So, with Mac Lane, we see guidance as organizational rather than Foundational. Indeed, in Mathias (1992, 119) Mac Lane asserts, 'But I see no need for a single foundation—on any one day it is a good assurance to know what the foundation of the day may be—with intuitionism, linear logic or whatever left for the morrow.'

[3] He argues on page 406 that the Gödel incompleteness theorems prevent any 'security blanket'. The longer quote better represents Mac Lane's long term skepticism of the relevance of ZFC to category theory.

[4] In other places (MacLane, 1988) objects that ZFC allows too many unproductive investigations such as the normal Moore Space conjecture.

11.2 Two Scaffolds

Bourbaki wrote,

> Today, we believe however that the internal evolution of mathematical science has, in spite of appearance, brought about a closer unity among its different parts, so as to create something like a central nucleus that is more coherent than it has ever been. The essential aspect of this evolution has been the systematic study of the relations existing between different mathematical theories, and which has led to what is generally known as the 'axiomatic method.' (Bourbaki, 1950, 223)

The Bourbaki treatise proceeds by the axiomatization of various fields (local foundations), establishes a common notation, and documents the links among them. This provides one component of a **scaffold** for mathematics. Rather than attempting a precise, full definition of scaffold, we discuss the goals of scaffolding with two examples: category theory and model theory. Along with *local foundations* a scaffold should promote *unity* across mathematics by providing a method for transporting concepts and results from one area to another. There is no requirement that the scaffold encompass all of mathematics but only that it makes connections among a number of topics. A successful scaffold contributes to generous arena and shared standard of proof. Proofs are built on prior assumptions. One does not prove a new result in mathematics directly from the material or structural set theoretic axioms; there must be a chain of arguments and constructions/definitions from the foundational axioms to the result at hand. A scaffold helps to organize the modules of proofs that involve various areas and provides *Productive Guidance*. I replace essential by productive as various scaffolds will give different advice depending on their framework. The success of the mathematics is the measure of the guidance.

Both scaffolds adopt local foundations by studying the properties of certain classes. The delineation of these areas differs among Bourbaki, model theory, and category theory. Bourbaki axiomatizes classes of structures in the informal way of Euclid or Hilbert (1971). Model theory posits fully formalized theories and Tarski semantics. Category theory gives mathematical definitions (commonly in naive set theory) of its basic classes and morphisms without semantics.

A *formalism-free* (Kennedy, 2013) approach to mathematics defines in *naive* set theory or natural language a class of objects. Often, the exact vocabulary of the class is unclear. In contrast the modern notion of an axiomatic system requires a distinction between a class of structures (defined set theoretically) and a formal language in which axioms are stated. Thus, Bourbaki and Category theory are formalism-free. While, model theory makes essential use of formalization.[5]

These are not the only scaffolds. E.g., Descriptive Set Theory scaffolds results in many subjects that are captured by Polish spaces, while the Langlands program expands on Weil's Rosetta stone to unify number theory with harmonic analysis (and points in between) (Frenkel, 2013).

---

[5] Remark 11.2.2.1, Baldwin (2018, Chap. 14), and Kennedy (2021) give (non-first order) exceptions to this claim. Section 11.3.2 shows that, for full strength, structural set theory must use formulas.

11.2.1 The Model Theoretic Approach

Benis-Sinaceur summarised the source of model theorists' ability to discover analogies across different fields of mathematics.

> Model theory deals specifically with *logical* analogies among mathematical procedures and theories. It proceeds by means of an analysis of the language of theories while exploring the reciprocal relations between this language and the mathematical models that satisfy it. (Benis-Sinaceur, 2000, 282)

In this section we (a) describe fundamental model theoretic ideas which clarify the role of ZFC as a generous arena and also as a metamathematical corral, (b) explain how the classification of theories serves as a unifying principle to treat different areas of mathematics, (c) provides a wild/tame distinction to give productive guidance and (d) (Sect. 11.3.1) demonstrates and uses properties of cardinality in exploring Cantor's paradise. See Baldwin (2018) for much more detail.

We concentrate on first order logic, as the so far most successful logic for organizing mathematics by providing and connecting formal local foundations. All the logics are based squarely in ZFC with occasional use of large cardinals and independence results. *Examining a particular mathematical topic*, the investigator selects certain concepts as fundamental. The vocabulary[6] is a set τ of relation symbols, function symbols, and constant symbols chosen to represent these basic concepts. A *τ-structure* with universe A assigns (e.g., to each n-ary relation symbol R an $R^A \subseteq A^n$).

Definition 11.2.1.1 A *full formalization* involves the following components.

(1) Vocabulary: specification of primitive notions.
(2) Logic:
    (a) Specify a class of well formed formulas.
    (b) Specify truth of a formula from this class in a structure.
    (c) Specify a formal deductive scheme for these sentences
(3) Axioms: specify the basic properties of the situation in question by sentences of the logic.

Remark 11.2.1.2 (*The value of formalization*) Formalization provides a number of clarifications and solutions for both imagined and real problems.

Category theorists (e.g. Leinster, 2014) often raise tendentious arguments that a mathematician might confusedly ask of two real numbers, 'Is $\pi \in 2$?' The notion of fixing a vocabulary describes exactly when this question makes sense—only if $\in$ is in the vocabulary. Mathematicians don't actually make this mistake; Shulman (2008) makes a more subtle distinction between two uses of $\in$.

---

[6] To avoid ambiguity, I have chosen the word 'vocabulary' rather than such rough synonyms as language, similarity type, signature or, even rougher, logic.

Shelah's 1974 proved the independence (from ZFC) of the famous topologist (J.H.C.) Whitehead's conjecture (motivated by complex analysis and algebraic topology): any Whitehead group[7] is free. The conjecture is stated strictly in the vocabulary $(+, 0)$ of Abelian groups. Shelah constructed a specific structure in the vocabulary $(\in, +, 0)$ that under $V = L$ is free as abelian group and under Martin's Axiom is not. Thus the *metamathematical corral* of independence results available in ZFC extends beyond technical problems about ZFC to problems arising in traditional mathematics. Shelah's construction is not directly available in category theory where objects don't have elements (Leinster, 2014, 403). As a further example, consistently, there are regular cardinals (e.g. $\kappa = \aleph_1$) and groups A ($|A| = \kappa$) such that every strictly smaller group is free but A is not free. Shelah's singular cardinal theorem show this fails for any singular cardinal κ and places the result in a general framework (Eklof & Mekler, 2002; Magidor & Shelah, 1994). Vasey (202x) transfers the result via internal size to accessible categories (Remark 11.2.2.1).

The mantra, 'model theory is the study of definable sets' depends on using ZFC to define structures. Fraïssé and Jónsson constructions further illustrate the significance of internal structure *in ZFC*. Although the properties of the construction are easily expressed in appropriate categories, translating the natural model theoretic proof into structural set theory would seem unnatural.[8] The existence of isomorphic structures that are *not* identical is essential. This framework underlies the notion of saturation (Sect. 11.3.1), the theory of random graphs (connection in Blass & Harary, 1979), even with edge probability $n^{-\alpha}$ (Baldwin & Shelahran, 1997), Abstract Elementary Classes, the study of exotic strongly minimal sets (Hrushovski, 1993), and Zilber's generalization to the complex exponential field.

An informal notion of *tame/wild mathematics* developed during the 20th century. Roughly, wild mathematics includes the 'wilderness' of point set topology; Pillay in Buss et al. (2001) includes any area exhibiting the Gödel phenomena, undecidability and coding of pairs (thus, no notion of dimension). Shelah's *Classification theory* divides complete[9] first order theories by syntactical conditions into a small number of classes. Theories in the same class share mathematically significant properties. Bourbaki (1950) posits three great mother-structures: groups, order, and topology and suggests vague notions of particularizing and combining them to study classical mathematics (e.g. topological groups). Shelah's classification refines the Bourbaki program by enabling (by meta-theory or more often by analogy) the transfer of results from one theory to another in the same class and provides guidance to distinguish the wild from the tame. The crudest distinction, between stable and unstable,

[7] Call A a Whitehead group if for any f, B such that $f: B \to A$ is a surjective (i.e. onto) group homomorphism whose kernel is isomorphic to the group of integers Z then B is isomorphic to the direct sum of Z and A. Any free Abelian group A is Whitehead (and conversely if A is countable).

[8] Either the notion of universally axiomatized class or the use of set theory to define 'class closed under substructure' is needed.

[9] Incomplete theories such as the theory of R-modules permit closure under natural operations such as product and homomorphism so interact with universal algebra. Concentrating on complete theories enables the classification.

echoes Bourbaki's mother-structure: order. Stable structures have no infinite subsets that are definably linearly ordered. Theories have increasing degrees of tameness as conditions such as superstability, ω-stability and strong minimality[10] hold. Among the unstable theories *o-minimality* yields a collection of dramatically tame theories. Instability divides into theories that satisfy the strict order property[11] (strop or sop) and those with the independence property (ip). Those with both (strop) and (ip) are thoroughly wild (e.g., Peano arithmetic and ZFC); despite the connotation, the meaning is: *very different methods will be needed to study these topics.*

The wildness of first order arithmetic did not prevent the great twentieth century advances in number theory. But these advances come not from working in the first order theory of arithmetic but embedding the natural numbers in tame structures, such as (stable expansions of) strongly minimal theories ACF_p. In Baldwin (2018), we describe the classification and in Chap. 6 stress two paths (stable/o-minimal) to pick out tame areas of mathematics. They share a common feature, missed in Bourbaki's list of mother-structures: combinatorial geometry (matroid theory) organizes dimensions of subsets of models of tame theories.

We look first at the stable side. The success of model theory in studying solutions of equations arises as equations define sets. And in theories with elimination of quantifiers (e.g. differentially or algebraically closed fields (*DCF, ACF*)) they control all definable sets. The use of first order axiomatizations, interpretations among theories, and stability theory for results in Diophantine geometry, number theory, and automorphic functions demonstrate the value of formalizing these theories (Bertrand & Pillay, 2010; Bouscaren, 1999; Freitag & Scanlon, 2018; Hrushovski et al., 2018). Note in particular, that while in algebra the domain of a structure is naively thought of as a set of points, the study of the (ω-stable) theory of differentially closed fields (*DCF*) focuses on models where the elements are functions. Model theory treats morphisms in two ways: (i) 'algebraic homomorphisms' between models (specializing to elementary embedding); (ii) definable maps between definable sets. Thus, in algebraic geometry *definable* rational functions correspond to the morphisms in the category: Zariski topologies on ACF with continuous maps (Poizat, 2001, 4.3). Recent work on DCF (Nagloo & Pillay, 2017) invokes stability theoretic tools to solve and refine one-hundred year old Painlevé problems on classification and transcendence of solutions of partial differential equations.

Turning to the unstable case, Grothendieck described his motivation (clarifying Mac Lane's reference to 'too much sand' (Sect. 11.2)) for a notion of 'tame topology'.

[10] We omit the easily available technical definitions of the stability hierarchy. A strongly minimal set is a definable set such that every definable subset is finite or co-finite. An algebraically closed field *ACF* is strongly minimal; so too is the set of solutions to the differential equation defining the Weierstrass j-function (Freitag & Scanlon, 2018). Some of the notions arising in abstract stability theory that are now applied across mathematics include: regular type, prime/minimal models, forking/rank, (non)-orthogonality, canonical base and (elimination) of imaginaries, and classification of combinatorial geometries (matroids).

[11] T is unstable and there is a formula $\phi(\mathbf{x}, \mathbf{y})$ that defines a chain on some (likely not definable) subset of n-tuples.

> I would now say, with hindsight, that 'general topology' was developed (during the thirties and forties) by analysts and in order to meet the needs of analysis, not for topology per se, i.e. the study of the topological properties of the various geometrical shapes. That the foundations of topology are inadequate is manifest from the very beginning, in the form of 'false problems' (at least from the point of view of the topological intuition of shapes) such as the 'invariance of domains', even if the solution to this problem by Brouwer led him to introduce new geometrical ideas. (Grothendieck, 1997, 258)

A theory of a linearly ordered structure is *o-minimal* if every definable subset (perhaps in a much richer language than order) is a finite union of intervals. Notably, Wilkie extended the ur-example, real closed fields, to the real exponential field. Wilkie (2007), Baldwin (2018, 160) argues that o-minimality is a direct response to Grothendieck's call because o-minimality:

(1) is flexible enough to carry out many geometrical and topological constructions on real functions and on subsets of real Euclidean spaces.
(2) builds in restrictions so that we are a priori guaranteed that pathological phenomena can never arise. In particular, there is a meaningful notion of dimension for all sets under consideration and any constructed by the means of 1)
(3) is able to prove finiteness theorems that are uniform over fibred collections.

Some of the successes of this program, beyond the basic (Dries, 1999), include a case of the Andre-Oort conjecture (Karp Prize 2013) (Pila, 2011) (relying on Peterzil & Starchenko (2010), Pila & Wilkie (2006), the developing subfields in o-minimal algebraic topology (Berarducci & Otero, 2002), and now o-minimal Hodge theory (Bakker et al., 2020). The 2018 Karp prize winning book, Aschenbrenner (2017), on asymptotic analysis brings to fruition notions of Hardy. The universal domain is the proper class of surreal numbers (Dries & Ehrlich, 2001; Ehrlich, 2012).

The developments outlined above are guided by the following strategies.

Remark 11.2.1.3 (*Model Theory Strategies*)

(1) Fix an appropriate vocabulary τ to study the subject.
(2) Give a (first-order) axiomatization T of the area involved.
(3) Study definable relations on the structure to obtain tameness.
(4) Modify your vocabulary to reduce quantifier complexity of formulas.
(5) Use syntactic conditions (stability hierarchy, *o*-minimality) and the dividing line strategy to guide your search for analogies.

Problem 1 *Give a philosophically meaningful definition of* tame. *What is sand? Does structural set theory, even as weak as ETCS, actually eliminate sand and if so how? What are examples?*

11.2.2 Category Theoretic Approach

Category theory describes classes of structures (which have no internal anatomy) by maps (morphisms) between them. It provides an axiomatic operational description of constructions which apply across mathematics and sufficient conditions for a category to be closed under them. We examine certain results and discuss why they are only awkwardly justified in ZFC + LC. Our analysis is limited to the 'first generation' ($\leq$1975) of general category theory. *Dependent type theory* and the HOTT program raise important issues (The Univalent Foundations Program (2015) and Voevodski (2014) lecture 3 slide 11) not dealt with here.

While model theory relies heavily on material set theory, the objects of a category have no elements. Thus a group is defined as a category[12] with one object and whose morphisms are all isomorphisms (i.e. invertible). Section 11.2.1 described the benefits of internal structure. The gain here is that the properties of constructions used in many areas are defined axiomatically without involving 'irrelevant' internal information.

We indicate some underlying causes of the wide influence of category theory in homological algebra, algebraic topology, and algebraic geometry.[13] Lou Kauffman explains,[14] 'We would never have regarded topological spaces as morphisms in a monoidal category, as is part and parcel of TQFT [topological quantum field theory] and Quantum Topology, without a category theoretic point of view.[15] *Categories make possible conceptual shifts that a strictly set theoretic point of view would not see.*'

Category theory began with the conceptual shift engendered by the realization that problems Eilenberg was studying on the continuous side (homology theories in topology) were reflected in problems Mac Lane was studying on the discrete side (group extensions). They described their goals:

> In a metamathematical sense our theory provides general concepts applicable to all branches of abstract mathematics, and so contributes to the current trend towards uniform treatment of different mathematical disciplines. In particular, it provides opportunities for the comparison of constructions and of the isomorphisms occurring in different branches of mathematics; in this way it may occasionally suggest new results by analogy (Eilenberg & MacLane, 1945, 236).

Category theory describes families of objects and morphisms between them. Morphisms implement the intuition of a function as a rule while hiding the actual rule but studying only the patterns of composition of morphisms like homotopy type. Thus f is a 1-1 function becomes: for every g, h: $fg = gh$ implies $g = h$ (monomorphism). De Toffoli (2017) argues that commutative diagrams and the technique of

---

[12] A usual group $(G, \times, 1)$ induces a category $\hat{G}$ and the elements of G acting as morphisms of $\hat{G}$ by left-multiplication. Conversely, take each morphism as an element and 1_G as the unit. Isomorphisms between material groups become functors.

[13] Other areas include K-theory, scheme theory, Langlands program, and model categories. Krömer (2007) and Marquis (2009) insightfully unite historical, philosophical and mathematical expositions of category theory.

[14] Foundations of Mathematics Listserve: January 23, 2020.

[15] Blass and Gurevich (2018) give a more detailed example. Lawvere and Schanuel (1997), a high school text written from a 'function-first' viewpoint, is built around age-appropriate 'real world' examples. The book is slow reading because 'almost familiar' ideas appear in a strange guise—conceptual shift.

'diagram chasing' are an effective hybrid of diagrammatic and linguistic elements that supports reasoning. The operational definition enables the discovery of common constructions (e.g. adjoint functor Sect. 11.3.1) for diverse subjects.

Problem 2 *Must formal reasoning be represented linearly? What is lost/gained in transforming diagramatic to linear reasoning?*

In contrast to model theory, equations are not studied as 'nice' definable sets. As, there are no definable sets. Rather, for example, the family of possible equations is considered as a vector space of linear transformations and the kernel (in the dual space) of such a transformation contains the solutions. The same tools are now available for both the 'syntax' and the semantics. Moreover, the same abstract formulation applies in many different areas.

The definitions of categories and functor exemplify *par excellence* 'formalism-free' mathematics (Sect. 11.2). Yet the *informal* Eilenberg-Steenrod axioms (ES-axioms) for homology required more complex formalizations than known in the 1940's. A *homology theory* on $\mathcal{A}$ (the category of pairs (X, A) of topological spaces with continuous maps as morphisms) assigns to each (X, A) a countable sequence of abelian groups, and group homomorphisms. There were at least 5 such theories by 1945 when (Eilenberg & Steenrod, 1952, Chap. I) announced the ES-axioms that *unify* those theories by provide a common basis, satisfied by each of the examples. That is, axiomatizing functors from $\mathcal{A}$ to (sequences from) **Abgrp** (abelian groups) and natural transformations among these functors. They can be formalized as a theorem schemata in ZFC (or a theory in Gödel-Bernays). A key theorem asserts that two homology theories fulfilling the axioms (and certain additional conditions) are naturally equivalent and yield isomorphic homology groups for a given space from an appropriate category of topological spaces. The ES axioms yield without calculation important results, such as Brouwer's fixed point theorem ('without any appeal to a concretely defined homology theory' (Eilenberg & Steenrod, 1952, 298)). They remark (page viii), 'Heretofore this [homology theory] has been an imprecise picture which the expert could use in his thinking but not in his exposition.' This exemplifies what Maddy later calls *Elucidation*.[16]

Maddy dubs as *surrogates*, the constructions described by Burgess:

> [this] *common, unified starting point will have to be such as to make provisions for all types of constructions by which new, auxiliary spaces, or number systems or whatever are manufactured out of old, traditional ones ....* There are a handful of basic types of constructions that keep being used over and over, in ever more elaborate combinations (Burgess, 2015, 62)

Why does Mac Lane doubt that ZFC is adequate to support categorical reasoning? A category is *concrete* if it is equipped with a forgetful functor[17] to the category of sets. Applying this functor allows one to think of the objects of the category as

[16] Elucidation replaces an *imprecise, pre-theoretic* notion which interferes with mathematical practice with a precise, 'set-theoretically defined one' (Maddy, 2017, 293).

[17] A forgetful functor is a map from a category $\boldsymbol{C}$ into sets that forgets the structure on the elements of $\boldsymbol{C}$ and maps the *morphisms to functions*. A functor is *faithful* if it is injective (which can defined in category theory terms) on morphisms.

sets with additional structure, and of its morphisms as structure-preserving *functions* (Adámek et al., 1990). The most naive interpretation of finding surrogates would consider only concrete categories. The fundamental classes studied in 20th century mathematics: groups, topological spaces, etc. are concrete categories that are usually defined in naive set theory. But there are categories which, *intrinsically*, are not concrete.

Let $\mathcal{H}ot_*$ be the category of pointed topologogical spaces (X, x). In the naive category, morphisms are homotopy classes of pointed maps. (The result hold for both the naive and the now standard notion.) Freyd wrote in 1970, '$\mathcal{H}ot_*$ is not concrete. There is no interpretation of the objects of $\mathcal{H}ot_*$ so that the maps may be interpreted as functions (in a functorial way, at least). $\mathcal{H}ot_*$ has always been the best example of an abstract category, historically and philosophically. Now we know that it was of necessity abstract, mathematically' (Freyd, 2004, 1).

Freyd defines an equivalence relation on the class of maps $f : A \to B$ for a fixed object B of $\mathcal{H}ot_*$ such that this equivalence relation has a *proper class* of equivalence classes. And he reaches a contradiction by showing no concrete category admits such an equivalence relation. See Mathew (2012).

Beyond products, unions, direct sums etc., Burgess points to quotients[18] as a tool for building the generous arena. Taking such quotients enables finding surrogates in Gödel-Bernays set theory. Kucera (1971), Adámek (1983, Chap. 6) proved that *every (possibly abstract) category can be represented as a quotient of a (possibly large) concrete category.* The equivalence classes may well be proper classes even though each object is a set. Suppose, as in MacLane (1971), we work in one 'universe', i.e. assume a single inaccessible[19] κ exists. Now one can modify the Kucera argument by considering categories to have objects in V_κ. Then, the surrogates for the morphisms in Kucera's argument become sets of cardinality κ. Thus, to guarantee surrogates are realized in a concrete category one uses ZFC + an inaccessible cardinal.

The issue is even more complicated. The forgetful functor $U : \mathbf{Hilb_r} \to \mathbf{Set}$ gives the category $\mathbf{Hilb_r}$ of Hilbert spaces with linear isometries the structure of a concrete category. However, the necessity of taking the completion means that the category is not closed under increasing unions. Moreover, Lieberman et al. (2023) prove there is no faithful functor from $\mathbf{Hilb_r}$ to $\mathbf{Set}$ which preserves directed co-limits (the appropriate abstraction of 'union'). The difficulty arises because, while one can choose representatives for the equivalences classes, one *cannot* select one representative of each class in such a way that the composition of two selected representatives is always a selected representative.

Remark 11.2.2.1 Accessible Categories/Abstract Elementary Classes (Adámek & Rosický, 1994; Baldwin, 2009) are treated by both scaffolds. AEC are Shelah's solution to the complications of syntax in infinitary logic. They provide a formalism-free axiomatic treatment of classes of structures. That is, there are (naive set-theoretic)

---

[18] Recall the quotient group of a group G by a normal subgroup H is constructed by putting a natural group structure on the collection of E-equivalence classes where aEb if $ab^{-1} \in H$.

[19] A cardinal κ is *inaccessible* if the V_κ is closed under power set and under unions indexed by member of V_κ (i.e. κ is regular).

axioms which describe the property of a pair $(\boldsymbol{K}, \leq)$, where $\boldsymbol{K}$ is a class of τ-structures for some vocabulary τ, $\leq$ is a notion of *strong substructures* where the properties of 'strong' is specified axiomatically but include being a subset. Although this definition depends heavily on the set theoretic notion of subset, the axioms specify conditions on closure under union of chains which are such that the class is closed under colimits and thus it becomes a special kind of *accessible* category[20] (Lieberman, 2013). Lieberman (2020) exploit the connection between (set theoretically defined) cardinality and *internal size*, defined in any accessible cardinal to propose an exciting new generalized 'eventual categoricity problem'. Lieberman et al. (2019) show how to generalize the fundamental study of abstract independence relations from model theory to category theory.

Riehl (2016, 11) emphasizes the unifying aspect of category theory, 'the action of packaging each variety of objects into a category shifts one's perspective from the particularities of each mathematical sub-discipline to potential commonalities between them.'

11.3 Foundations

In this section we explore how the category and model theoretic viewpoints differ regarding size/cardinality. Then we study two approaches to the formalization of set theory: material and structural set theory.

11.3.1 Size

Most mathematical theorems apply either to structures of bounded cardinality (e.g., classifying 3-manifolds) or to structures of arbitrary cardinality. Category theory aims at uniform results for objects (e.g. topological spaces) of all cardinalities. But the techniques require recourse to proper classes.[21] So, Eilenberg and MacLane (1945) introduced the concept of category theory with foundations in Gödel-Bernays set theory. 'Cardinality' does not appear in the index to Mac Lane's text, MacLane (1971). He considers three sizes: finite, small, and large. (Enayat et al., 2017, 3/4) distinguish relative and absolute solutions to the foundational issues arising from the small/large problem. An *absolute solution*, e.g. one inaccessible (MacLane, 1971) or Gödel-Bernays (Eilenberg & MacLane, 1945), fixes the distinction once and for all. Grothendieck adopted a *relative* view and built his general cohomology theory on

[20] For regular λ, a category is λ-accessible if it has λ-directed colimits, has only a set (up to isomorphism) of λ-presentable objects , and every object can be written as a λ-directed colimit of λ-presentable objects.

[21] Recall that in ZFC classes are definable pieces of the universe which are equinumerous with the universe. Gödel-Bernays set theory treats these as a different kind of object.

the basis of *universes*, essentially the existence of a (proper class) of inaccessible cardinals. Countably many universes suffices for most purposes. As argued in McLarty (2010), Grothendieck uses universes to provide a smooth general framework. Authors presume, but rarely detail, that this assumption is not needed in specific applications. In contrast, comparing cardinalities is a central feature of the model theoretic study of algebraic phenomena. In this section we study the necessity of the small/large distinction in category theory and the impact of cardinality on mathematics as mediated by model theory.

One of the main achievements of category theory is to prove theorems (not metatheorems), e.g., the *adjoint functor theorem*, that apply in many areas of mathematics. To see how proper classes arise in even formulating that theorem we need a few technical definitions but avoid others not essential here.

Definition 11.3.1.1 (Awodey, 2010, 24) A category is *small* if it has only a set of objects and there is only a set of morphisms between every pair of objects. Otherwise it is *large* and *locally small* if each 'hom-set' is a set. It is complete if it has all *small limits*. (*For the moment just think, 'all unions indexed by a set are included*[22]*'*.)

We do not define the notion[23]: 'the functor U has a left adjoint'. The three page list of examples in MacLane (1971, 85) includes: the left adjoint to the forgetful functor[24] from groups to sets is a free group; from fields to integral domains is the field of quotients, and from complete metric spaces to metric spaces is the completion. An *adjoint functor theorem* provides sufficient conditions on a category $\boldsymbol{C}$ and a functor $\boldsymbol{F}$ for the existence of a left adjoint to $\boldsymbol{F}$. A natural candidate for a general sufficient condition to obtain adjoints is a result of Freyd:

Theorem 11.3.1.2 *If* $\boldsymbol{C}$ *is a complete small category then every functor from* $\boldsymbol{C}$ *to another category* $\boldsymbol{D}$ *that preserves limits has a left adjunct.*

The *hypotheses* of Theorem 11.3.1.2 is too strong in two ways. First, as a short Cantor diagonal argument shows, any such category satisfies a quite restrictive condition.[25] (Otherwise a set of size $2^{|\boldsymbol{C}|}$ is imbedded in $\boldsymbol{C}$.) This well-known result is expounded in Enayat et al. (2017, Theorem 1) or in more detail (Awodey, 2010, Chap. 9). Moreover, the hypothesis doesn't apply to large categories such as Set or Grp. The standard solution is to weaken the hypothesis by considering *locally small* categories and replacing 'preserves limits' by a more technical *solution set condition* that allows multiple morphisms between a pair of objects. Further applicable sufficient conditions for left adjoints apply in many contexts. Yet, the small/large dichotomy is inevitable.

Baldwin (2018, Chap. 8) describes the intricate history of interactions among first order model theory, ZFC, cardinality, and cardinal arithmetic. One of the most

---

[22] See Sect. 11.2.2 and the text (Awodey, 2010) for details. Shulman (2008) and Enayat et al. (2017) give a fuller explanation of the category terminology and different resolutions of the issue.

[23] See Chaps. 4 and 5 of Marquis (2009) for the impact of Kan's work on adjoint functors.

[24] See footnote 17.

[25] It is a preorder; that is, it has at most one morphism between any two objects.

influential problems in modern model theory asked about the function, $I(T, \aleph_\alpha)$, the number of (non-isomorphic) models a countable theory T has with cardinality $\aleph_\alpha$. Morley conjectured that this function was non-decreasing for every T (except possibly from $\aleph_0$ to $\aleph_1$). Shelah (1990) solved the conjecture using his *dividing line strategy* (Baldwin, 2018, 13.3; Baldwin, 2021; Shelah, 2021). This strategy successively splits theories by syntactic properties until arriving at two classes: (i) *unclassifiable theories* have the maximal number of models in every uncountable cardinality and there is no uniform way to assign invariants determining isomorphism; (ii) *classifiable theories* have a local and sometimes global dimension theory; each model is determined by a tree of invariants with countable height; and $I(T, \aleph_\alpha)$ is bounded by a function of $|\alpha|$. Both the classification of theories and the crucial tools for the main gap proof permeate applications across mathematics.

The notion of a universal object for a diagram is fundamental in category theory (MacLane, 1971, page 2). MacLane (1971, 123) observes in a particular case that the size of the universal object is bounded in terms of the size of the diagram and the objects in it. But no actual cardinal calculations appear in the book. In contrast, Hausdorff (2005, H 1908) introduces a form of universality that is seminal for model theory: M is κ-universal for a class $\boldsymbol{K}$ if $|M| = \kappa$, if $N \in \boldsymbol{K}$ and $|N| \leq \kappa$ then there is an embedding of N into M. He proves the existence of $\aleph_{n+1}$-universal linear orders if $\aleph_{n+1} = 2^{\aleph_n}$.

Generalizing ideas of Hausdorff, Fraïssé, and Jónsson, Morley and Vaught (1962) proved the existence of saturated[26] models of any complete theory in all regular cardinals using the GCH. A minor variant (special) extends the result to singular cardinals. An important consequence of Shelah's classification theory is that GCH can be replaced by model theoretic hypotheses for 'tame' theories: A stable theory T has a saturated model in cardinality κ if and only if $\kappa^\omega = \kappa$ and a superstable one has saturated models in any cardinal $\geq 2^{\aleph_0}$. Intriguing problems in axiomatic set theory arise from determining which theories have universal models in which cardinals (Shelah, 2021; Baldwin, 2021).

The notion of a saturated (homogeneous-universal) model M plays a fundamental role in model theory for finding 'universal domains'. Model theory papers routinely begin, 'we work in a suitable *monster model*'. This means choose a domain (necessarily, for an arbitrary theory, whose cardinality is strongly inaccessible) large enough to encompass any construction from the objects actually under study (Baldwin, 2018, 5.2). As in the last paragraph, for 'sufficiently tame' theories, these inaccessible cardinals are not necessary.

Problem 3 *Further explore the distinctions between relative and absolute 'solutions of size'. What are the mathematical consequences of the choices; what are the philosophical justifications?*

[26] κ saturated means κ-universal and κ-homogeneneous: isomorphic strictly smaller submodels of M are automorphic in M. M is saturated means $|M|$-saturated.

11.3.2 Material and Structural Set Theories as Foundations

Notation 11.3.2.1 (*Weak set theories*) A well-known[27] family of bi-interpretable weak set theories (Mathias, 2001; Shulman, 2019) include these four.

(1) structural set theories:

  (a) a well-pointed topos with a natural numbers object and with the axiom of choice;
  (b) ETCS, Elementary Theory of the Category of Sets; Lawvere (1964).

(2) material set theories:

  (a) BZC, bounded Zermelo with choice[28];
  (b) Mac Lane set theory (finite order arithmetic (McLarty, 2020)).

Each of these weak theories omit the axiom of replacement. Thus they do not have the cardinals[29] $\aleph_\omega$ or $\beth_\omega$.

From a set theoretic standpoint ETCS is very weak. Mathias (2001) identifies $V_{\omega+\omega}$ as a natural model for Mac Lane set theory and gives the axioms in ϵ-style and a chart organizing various weak set theories. Shulman is a precise reference for the known ability to extend the axioms of ETCS by considering axioms which vastly strengthen ETCS: including full separation, collection, and replacement axioms. These extensions leave the formalism-free world and require explicit set theoretic syntax to formulate axiom schemes. He concludes that ETCS plus 'structural replacement', is equi-consistent and indeed mutually interpretable[30] with ZFC (Shulman, 2019, Cor. 8.53).

Remark 11.3.2.2 (*Large Cardinals*) As a first approximation, any cardinal whose existence implies the consistency of ZFC is termed 'large'. Thus, any inaccessible cardinal is large. From the set-theoretic standpoint merely inaccessible cardinals are quite small. Since large cardinal axioms (e.g. measurable cardinals) are properties of a particular cardinal κ, they can usually be easily rephrased as category theoretic properties of the cardinal κ. There is no real difficulty in *extending structural set theories by large cardinal axioms*.

Unlike category theory, set theory and model theory use the gradations of large cardinals. In particular, Maddy (2011, 47–51) expounds the equivalences between

---

[27] These equivalences date back to Mitchell (1972).

[28] BZC has comprehension only for Σ_0-formulas. We add C to any such abbreviation when choice is added. The Bourbaki foundation omitted the axioms of foundation and replacement.

[29] More algebraically, Mathias (2001, 9.32) notes that Mac Lane set theory *cannot prove* that for every n, one can iterate the process of taking the dual vector space n-times (starting with $\mathbf{R}[x]$). Mathias (1992) has an interesting dialog between Mathias and Mac Lane.

[30] MacLane and Moerdijk (1992, 343), citing (Mitchell, 1972) write that the basic interpretations are actually invertible and so a bi-interpretation of (a weak version) of replacement holds.

appropriate large cardinals, the projective hierarchy on first order definable subsets of $(\boldsymbol{R}, +, \times, 0, 1, \boldsymbol{N})$, and determinacy axioms.

In a little known example (Shelah, 1982; Baldwin, 2021, 2.3), Shelah invokes a smallish large cardinal (a beautiful cardinal[31]) to give an account of the main gap that, by counting non-mutually embeddible rather than non-isomorphic models, is more resistant to forcing. Several recent results on eventual categoricity in AEC, (e.g., Boney (2014), Shelah and Vasey (2018), Remark 11.2.2.1), require the existence of a proper class of strongly compact cardinals, while other AEC results require much smaller large cardinals. Thus, large cardinals are essential to placing all of model theory in the generous arena.

Remark 11.3.2.3 (*CCAF*) Lawvere (1966) introduces the 'category of categories' as a foundation for mathematics' as a first order theory with no ambient set theory. Enayat et al. (2017) lay out in detail two distinct goals for category theory, as Feferman had articulated in a series of works on Lawvere's proposal, labeled *R*: ('unrestricted existence', e.g. to have the category of all categories[32] (*CAT*) exist) and *S*: (the desire to contrast small and large sets: e.g. *Grp* versus *GRP*). The first suggests Russell's paradox (but to establish the paradox one needs information (included in *R*) about the ability to construct from given categories). The precise meaning of property *R* is central. In Ernst (2015)'s understanding, such a system must be inconsistent. Enayat et al. (2017) summarize and extend the different interpretations (Feferman, 2004 and Shulman, 2008) and show consistency of certain meanings of (R) with respect to various material set theories e.g. NFU (New Foundations with ur-elements), NFUA etc. Note that while the consistency of NF remains contentious, Jensen proved the extension NFU (by allowing ur-elements) is consistent relative to ZFC. NFU has a stratified notion of membership and admits a universal set. Thus Enayat et al. (2017) characterize various notions of *natural implementability* depending on the image of the usual category theoretic notions (Cat, Set) in the interpreting theory. Under their weakest notion of 'natural interpretability' they show NFUA interprets ZFC plus n Mahlo cardinals, for each standard n, on a part of its domain and a small/large distinction. Therefore, the category theory of MacLane (1971) can be implemented in NFUA on that subdomain. But the salience of the precise meaning of 'natural implementation' is emphasized as McLarty (1992) showed that under any 'reasonable definition of function', after such an interpretation the category *SET* of *all* sets is not closed under direct products and exponentiation. On the other hand, Forster (2007) points out that such issues have frequently appeared with NF and solved by finding a different definition of the troublesome concept (that is equivalent in ZFC).

[31] A beautiful cardinal is much larger than ω universes, but still smaller than a weakly compact.

[32] Expanding on Enayat et al. (2017), Gorbow regards the following definition of CAT as appropriate for (R): 'The category determined by (1) having the set of all categories as objects, (2) for any categories A, B, having the set of all functors from A to B as morphisms from A to B, (3) having identity functors as identity morphisms, and (4) having composition of morphisms defined by composition of functors.' Contra to Lawvere (1966), McLarty (1991), he does not require that CAT admit both finite products and exponentiation.

Gorbow (in correspondence) has emphasized the need for further investigation of the methodological gain for category theoretical consideration of CAT and SET.

Remark 11.3.2.4 (*Metamathematical Corral*) Most work in the corral is driven by the goal of finding implications among various principles and truly sufficient conditions which are not overly broad. Set theory itself is driven by internal problems such as the singular cardinals problem and problems arising in combinatorial set theory. Friedman's (1998) program produces concrete combinatorial properties of the natural numbers which require large cardinals. Examples arising from ordinary mathematical practice include the Whitehead problem and others (Sect. 11.2.1). Farah's ICM address (Farah, 2014) indicates the impact of model theory, descriptive set theory, and forcing constructions on the study of operator algebras. In these examples, material set theorists have the advantage; despite the interpretablity discussed above, ZFC provides a friendlier workplace.

11.4 Comparisons

We quickly assess how our two scaffolds meet three characteristics of a scaffold: (i) local foundations, (ii) unifying methods, (iii) productive guidance. Local foundations are found by specifying in first order logic either a complete theory or a category with appropriate morphisms. Unity[33] and Guidance are provided by the hierarchical classification[34] of first order theories or by Grothendieck style axiomatization of families of categories such as abelian category, additive category, topos.

Major differences are: (a) explicit syntax/semantics versus formalism-free and (b) use of cardinality versus large/small. Independence results in algebra and topology are naturally expressed through ZFC and model theoretic formalization (Remark 11.2.1.2). Reformulating these in category theory would not only require mimicking the forcing arguments in material set theory but formulating categories where the 'internal structure' of algebras was revealed.

Particular structural and material set theories are bi-interpretable at all levels from *BZ* to $ZFC + LC$ (Sect. 11.3.2). Thus, if we consider *Shared Standard of Proof* to be a question of which statements are theorems in the system,[35] and *Generous Arena* as giving surrogates for all mathematical entities there is nothing to choose between the approaches. The level within each approach makes a huge difference. Both model theory and set theory apply replacement, large cardinals, and forcing to obtain results in traditional mathematics (Sect. 11.2.1, Remark 11.3.2.4).

---

33 Unity arises in several ways: Bi-iterpretablity in model theory and natural equivalences in category theory connect different areas. Category theorem may view the same concept from different perspectives; Ashfaque (2020, 1.6) gives twelve categories for studying elliptic curves.

34 See the map of the universe at http://www.forkinganddividing.com.

35 This is a mathematical question as opposed to the 'sociological' (Maddy, 2017, 296) standard of *belief* in (§1) of proof from a common foundation (ZFC).

As noted in our introduction, Mac Lane proclaimed, 'inaccessible cardinals palpably have nothing to do with [category theory]'. We argued (Sect. 11.2.2) that Kucera's theorem exhibits the connection if the foundation is to be in set theory rather than class theory. Either material or structural set theories admit conservative extensions to class theory at each level. Recall that Mac Lane set theory sees no cardinal numbers $\geq \aleph_\omega$. As McLarty has observed, there is a proper class containing exactly the $\aleph_n$ in the natural class extension of Mac Lane set theory. So Mac Lane requires proper classes (Sects. 11.2.2 and 11.3.1) that are seen in ZFC as much smaller than inaccessibles. The McLarty (2020) proof that Fermat's conjecture is provable in Mac Lane set theory, using a suitable conservative class theory extension, demonstrates that only very weak class theory is needed for a classical problem. To obtain all of contemporary mathematics the *Generous Arena* requires a set theoretic bases of both replacement (Borel determinacy (Friedman, 1971)) and large cardinals.

Mathematicians of various stripes might object to the naturalness/accessibility of surrogates or proofs in either of the proposed foundations. Indeed, Tao suggests the Ax-Grothendieck theorem[36] supports the following:

> I have always been fond of the idea that model-theoretic connections between objects (e.g. relating two objects by comparing the sentences that they satisfy) are at least as important in mathematics as the more traditional category-theoretic connections (where morphisms are the fundamental connective tissue between objects) or topological connections (where the objects are gathered into some common topological space or metric space in order to compare them). Terry Tao[37]

A great advantage of *ZFC* is the simple set-up in terms of a single binary membership relation. In Lemma 11.2.1.2 we noted that fixing a vocabulary suitable to the topic studied resolves purported ambiguities about membership. This is a very soft and uniform form of typing. As Shulman (2013) points out, category theory provides a more rigid type system.

Alternatively, Scott (1974, 10) provocatively asserted, 'there is only one satisfactory way of avoiding the paradoxes, use some form of the *theory of types*. That was the basis of both Russell's and Zermelo's intuitions.' He goes on to outline a *cumulative* theory of types where all the sets up to a certain level form a partial universe which is a set. And the type of a set is the level at which it appears.

Problem 4 We have given descriptions of the mathematical and methodological differences between two scaffolds to engage the following problems. *Provide a more systematic philosophical analysis and justification of criteria for a successful scaffold. Find a notion of 'grounding' which accounts for the unease category theorists feel with ZFC (and model theorists with structural set theory) as the Foundation despite the equivalences between material and structural set theories.*

Celluci (2000) distinguishes between open and closed systems. He requires a closed system to have an immutable set of axioms. Maddy's notion of *metamathematical corral*, recognizes the essence of Gödel incompleteness; *there must be a*

---

[36] A 1-1 polynomial map between algebraic varieties is onto (e.g. Baldwin (2018, 102)).

[37] https://quomodocumque.wordpress.com/2012/09/03/mochizuki-on-abc/?#comment-10506.

continuing search for new axioms. All of mathematics may be an open system consisting of cooperating (and perhaps changing) closed systems.

We described two Foundational systems. They pursue different conceptions of 'set'; each allows revision of axioms. Why choose between them? Rather than a Foundationalist view, one may just consider several mutually reinforcing Foundations as an open system of Mathematics.

References

Adámek, J. (1983). *Theory of abstract structures* (Vol. 189). D. Reidel.
Adámek, J., Herrlich, H., & Strecher, G. (1990). *Abstract and concrete categories: The joy of cats*. Pure and Applied Mathematics. Wiley
Adámek, J., & Rosický, J. (1994). *Locally presentable and accessible categories, volume 189 of London mathematical society lecture notes*. Cambridge, New York.
Aschenbrenner, M., van den Dries, L., & van der Hoeven, J. (2017). *Asymptotic differential algebra and model theory of transseries*. Annals of math studies. Princeton University Press.
Ashfaque, J. (2020). An invitation to category theory. Lecture notes. https://www.academia.edu/37378620/An_Invitation_to_Category_Theory?email_work_card=interaction_paper
Awodey, S. (2010). *Category theory* (2nd ed.). Oxford logic guides. Oxford University Press.
Bakker, B., Klinger, B., & Tsimerman, J. (2020). Tame topology of arithmetic quotients and algebraicity of Hodge loci. *Journal of the American Math Society, 33*, 917–939.
Baldwin, J. T. (2009). *Categoricity*. Number 51 in University lecture notes. American Mathematical Society, Providence, USA.
Baldwin, J. T. (2018). *Model theory and the philosophy of mathematical practice: Formalization without foundationalism*. Cambridge University Press.
Baldwin, J. T. (2021). The dividing line methodology: Model theory motivating set theory. *Theoria, 87*, 361–393. Theoria: https://doi.org/10.1111/theo.12297. http://homepages.math.uic.edu/~jbaldwin/pub/atheoriabibjan619.pdf.
Baldwin, J. T., & Shelah, S. (1997). Randomness and semigenericity. *Transactions of the American Mathematical Society, 349*, 1359–1376.
Benis-Sinaceur, H. (2000). The nature of progress in mathematics. In E. Grosholz & H. Breger (Eds.), *The growth of mathematical knowledge* (pp. 281–293). Kluwer Academic Publishers.
Berarducci, A., & Otero, M. (2002). o-minimal fundamental group, homology and manifolds. *Journal of the London Mathematical Society, 2*(65), 257–270.
Bertrand, D., & Pillay, A. (2010). A Lindemann-Wheierstrass theorem for semiabelian varieties over function fields. *Journal of the American Mathematical Society, 23*, 419–453.
Blass, A., & Gurevich, Y. (2018). Who needs category theory? *The Bulletin of the European Association for Theoretical Computer Science, 124*. Math archive https://arxiv.org/abs/1807.11399.
Blass, A., & Harary, F. (1979). Properties of almost all graphs and complexes. *Journal of Graph Theory, 3*. Reprinted 2006. https://doi.org/10.1002/jgt.3190030305.
Boney, W. (2014). Tameness from large cardinal axioms. *Journal of Symbolic Logic, 79*, 1092–1119.
Bourbaki, N. (1950). The architecture of mathematics. *American Mathematical Monthly, 57*, 221–232.
Bouscaren, E. (Ed.). (1999). *Model theory and algebraic geometry: An introduction to E. Hrushovski's proof of the geometric Mordell-Lang conjecture*. Springer.
Burgess, J. P. (2015). *Rigor & structure*. Oxford University Press.
Buss, S., Kechris, A., Pillay, A., & Shore, R. (2001). The prospects for mathematical logic in the twenty-first century. *The Bulletin of Symbolic Logic, 7*, 169–196.

Celluci, C. (2000). The growth of mathematical knowledge: An open world view. In E. Grosholz & H. Breger (Eds.), *The growth of mathematical knowledge* (pp. 153–172). Kluwer Academic Publishers.

De Toffoli, S. (2017). 'Chasing' the diagram-the use of visualizations in algebraic reasoning. *Review of Symbolic Logic, 10*, 158–186.

Dries, L. V. D. (1999). *Tame topology and O-minimal structures*. London mathematical society lecture note series.

Dries, L. V. D., & Ehrlich, P. (2001). Fields of surreal numbers and exponentiation. *Fundamenta Mathematicae, 167*, 173–188. Errata: 2001, 169, 295–297.

Ehrlich, P. (2012). The absolute arithmetic continuum and the unification of all numbers great and small. *The Bulletin of Symbolic Logic, 18*, 1–45.

Eilenberg, S., & MacLane, S. (1945). General theory of natural equivalences. *Transactions of the American Mathematical Society, 58*, 231–294.

Eilenberg, S., & Steenrod, N. (1952). *Foundations of algebraic topology*. Princeton University Press.

Eklof, P., & Mekler, A. (2002). *Almost free modules: Set theoretic methods* (2nd ed.). North Holland.

Enayat, A., Gorbow, P., & McKenzie, Z. (2017). Feferman's forays into the foundations of category theory. In *Feferman on foundations* (pp. 315–346). Springer. Math https://arxiv.org/pdf/1603.03272.pdf.

Ernst, M. (2015). The prospects for unlimited category theory, doing what remains to be done. *Review of Symbolic Logic, 8*, 306–327.

Farah, I. (2014). Logic and operator algebras. In *Proceedings of the international congress of mathematicians—Seoul 2014* (Vol. II, pp. 15–39). Kyung Moon Sa, Seoul.

Feferman, S. (2004). Typical ambiguity: Trying to have your cake and eat it too. In *One hundred years of Russell's paradox mathematics, logic, philosophy* (pp. 135–152). de Gruyter.

Forster, T. (2007). Why the sets of NF do not form a cartesian-closed category. https://www.dpmms.cam.ac.uk/~tf/cartesian-closed.pdf

Freitag, J., & Scanlon, T. (2018). Strong minimality and the j-function. *Journal of the European Mathematical Society, 20*, 119–136.

Frenkel, E. (2013). *Love and math*. Basic Books, Perseus Group.

Freyd, P. (2004). Homotopy is not concrete. Reprints in theory and applications of categories, *6*, 1–10. First appeared: Streenrod's festschrift: the steenrod algebra and its applications. Lecture notes in mathematics (Vol. 168). Springer, Berlin 1970; http://www.tac.mta.ca/tac/reprints/articles/6/tr6.pdf

Friedman, H. (1971). Higher set theory and mathematical practice. *Annals of Mathematical Logic, 2*, 325–357.

Friedman, H. (1998). Finite functions and the necessary use of large cardinals. *Annals of Mathematics, 148*, 803–893.

Grothendieck, A. (1997). Equisse d'un programme. written 1984, English version. http://www.landsburg.com/grothendieck/EsquisseEng.pdf

Hausdorff, F. (2005). Hausdorff on ordered sets, volume 25 of History of Mathematics. American Mathematical Society, Providence, RI; London Mathematical Society, London. Translated from German, edited by J. M. Plotkin.

Hilbert, D. (1971). *Foundations of geometry*. Open Court Publishers. translation from 10th German edition by Harry Gosheen, edited by Bernays 1968.

Hrushovski, E. (1993). A new strongly minimal set. *Annals of Pure and Applied Logic, 62*, 147–166.

Hrushovski, E., Martin, B., Rideau, S., & Cluckers, R. (2018). Definable equivalence relations and zeta functions of groups. *Journal of the European Mathematical Society (JEMS), 20*, 2467–2537.

Kennedy, J. (2013). On formalism freeness: Implementing Gödel's 1946 Princeton bicentennial lecture. *Bulletin of Symbolic Logic, 19*, 351–393.

Kennedy, J. (2021). *Gödel, Tarski, and the lure of natural language*. Cambridge University Press.

Krömer, R. (2007). *Tool and object a history and philosophy of category theory*. Birkhauser.

Kucera, L. (1971). Every category is a factorization of a concrete one. *Journal of Pure and Applied Algebra, 1*, 373–376.
Lawvere, F. W. (1964). An elementary theory of the category of sets. *Proceedings of the National Academy of Science of the U.S.A, 52*, 1506–1511.
Lawvere, F. W. (1966). The category of categories as a foundation for mathematics. In S. E. A. Eilenberg (Ed.), *Proceedings of the conference on categorical algebra, La Jolla, 1965* (pp. 1–21). Springer.
Lawvere, F. W., & Schanuel, S. H. (1997). *Conceptual mathematics*. Cambridge University Press.
Leinster, T. (2014). Rethinking set theory. *The American Mathematical Monthly, 121*, 403–415.
Lieberman, M. (2013). Category-theoretic aspects of abstract elementary classes. *Annals of Pure and Applied Logic, 162*, 903–915.
Lieberman, M., Rosický, J., & Vasey, S. (2019). Forking independence from the categorical point of view. *Advances in Mathematics, 346*, 719–772.
Lieberman, M., Rosický, J., & Vasey, S. (2020). Sizes and filtrations in accessible categories. *Israel Journal of Mathematics, 238*, 243–278.
Lieberman, M., Rosický, J., & Vasey, S. (2023). Hilbert spaces and C*-algebras are not finitely concrete. *Journal of Pure and Applied Algebra, 227*. arXiv:arXiv:1908.10200
MacLane, S. (1969). Possible programs for categorists. Lecture Notes in MathematicsIn A. Dold & B. Eckman (Eds.), *Category theory, homology theory, and their applications* (Vol. 86, pp. 126–132). Springer.
MacLane, S. (1971). *Categories for the working mathematician*. Springer.
MacLane, S. (1986). *Mathematics: Form and function*. Springer.
MacLane, S. (1988). Towards the greater health of mathematics. *Mathematical Intelligencer, 10*, 17–20.
MacLane, S., & Moerdijk, I. (1992). *Sheaves in geometry and logic*. Springer.
Maddy, P. (2011). *Defending the axioms*. Oxford University Press.
Maddy, P. (2017). Set-theoretic foundations. *Contemporary Mathematics, 690*, 289–322. https://doi.org/10.1090/conm/690/13872
Maddy, P. (2019). What do we want a foundation to do? Comparing set-theoretic, category-theoretic, and univalent approaches. In S. Centrone, D. Kant, & D. Sarikaya (Eds.), *Reflections on foundations: Univalent foundations, set theory and general thoughts*. Springer.
Magidor, M., & Shelah, S. (1994). When does almost free imply free. *Journal of the American Mathematical Society, 7*, 769–830.
Marquis, J.-P. (2009). *From a geometrical point of view: A study of the history and philosophy of category theory*. Springer.
Mathew, A. (2012). Homotopy is not concrete. https://amathew.wordpress.com/2012/01/26/homotopy-is-not-concrete/
Mathias, A. (1992). What is Mac Lane missing? In H. Judah, W. Just, & H. Woodin (Eds.), *Set theory of the continuum* (pp. 113–117). Springer. https://www.dpmms.cam.ac.uk/~ardm/scdlistG.html contains Mac Lane's reply.
Mathias, A. (2001). The strength of MacLane set theory. *Annals of Pure and Applied Logic, 110*, 107–234.
McLarty, C. (1991). Axiomatizing a category of categories. *The Journal of Symbolic Logic, 56*, 1243–1260.
McLarty, C. (1992). Failure of cartesian closedness in NF. *Journal of Symbolic Logic, 57*, 555–556.
McLarty, C. (2010). What does it take to prove Fermat's last theorem? Grothendieck and the logic of number theory. *Bulletin of Symbolic Logic, 16*, 359–377.
McLarty, C. (2020). A finite order arithmetic foundation for cohomology. *Review of Symbolic Logic, 13*, 296–325. http://arxiv.org/pdf/1102.1773v3.pdf
Mitchell, W. (1972). Boolean topoi and the theory of sets. *Journal of Pure and Applied Algebra, 2*, 261–274.
Morley, M., & Vaught, R. (1962). Homogeneous universal models. *Mathematica Scandinavica, 11*, 37–57.

Nagloo, J., & Pillay, A. (2017). On algebraic relations between solutions of a generic Painlevè equation. *Journal für die reine und angewandete Mathematik (Crelle's journal)*, *726*, 1–27. https://doi.org/10.1515/crelle-2014-0082

OED. (1971). convene. In *Oxford English dictionary*. Oxford University Press.

Peterzil, Y., & Starchenko, S. (2010). Tame complex analysis and o-minimality. In *Proceedings of the international congress of mathematicians* (Vol. II, pp. 58–81) (New Delhi, 2010).

Pila, J. (2011). O-minimality and the Andre-Oort conjecture for C^n. *Annals of Mathematics*, *173*, 1779–1840.

Pila, J., & Wilkie, A. J. (2006). The rational points of a definable set. *Duke Mathematical Journal*, *133*, 591–616.

Poizat, B. (2001). *Stable groups*. American Mathematical Society, Providence. translated from Groupes Stables.

Riehl, E. (2016). Category Theory in Context. Dover, Minneola, New York. Kindle edition available. https://web.math.rochester.edu/people/faculty/doug/otherpapers/Riehl-CTC.pdf

Russell, B., & Whitehead, A. N. (1910). *Principia Mathematica*. Oxford University Press.

Scott, D. (1974). Axiomatizing set theory. In T. Jech (Ed.), *Axiomatic set theory, volume XIII.2 of Proceedings of the symposium in pure mathematics of the American mathematical society* (pp. 207–214). AMS.

Shelah, S. (1982). Better quasi-orders for uncountable cardinals. *Israel Journal of mathematics*, *42*, 177–226.

Shelah, S. (1990). *Classification theory and the number of nonisomorphic models* (2nd ed.). North-Holland.

Shelah, S. (2021). Divide and conquer: Dividing lines and universality. *Theoria*, *87*, 259–348.

Shelah, S., & Vasey, S. (2018). Categoricity and multidimensional diagrams. Submitted.

Shulman, M. (2008). Set theory for category theory. Math Arxiv, October 2008. https://arxiv.org/abs/0810.1279

Shulman, M. (2013). From set theory to type theory. *n*-category Cafe. https://golem.ph.utexas.edu/category/2013/01/from_set_theory_to_type_theory.html

Shulman, M. (2019). Comparing material and structural set theories. *Annals of Pure and Applied Logic*, *170*, 465–504. Math Arxiv, December 2018. https://arxiv.org/abs/1808.05204

The Univalent Foundations Program. (2015). *Homotopy type theory: Univalent foundations of mathematics*. Institute for Advanced Study.

Vasey, S. (202x). Accessible categories, set theory, and model theory: an invitation. https://arxiv.org/abs/1904.11307

Voevodski, V. (2014). Foundations of mathematics—their past, present and future. The 2014 Paul Bernays Lectures at ETH Zurich. September 9–10, 2014.; Website: https://www.math.ias.edu/vladimir/sites/math.ias.edu.vladimir/files/2014_09_Bernays_3%20presentation.pdf

Wilkie, A. (2007). O-minimal structures. Séminaire Bourbaki, 985. http://eprints.ma.man.ac.uk/1745/01/covered/MIMS_ep2012_3.pdf

Chapter 12
Reply to Baldwin

Penelope Maddy

Keywords Set theory · Category theory · Essential guidance · Generous arena · Methodology

In his essay, Professor Baldwin makes a valuable contribution to the methodology of mathematics. He frames this effort, in part, using the taxonomy of foundational jobs from Maddy (2017) and Maddy (2019), but he also introduces his own notion of a 'scaffold' and a distinction between 'foundations' and 'Foundations'. The latter appears to coincide more or less with what's sometimes called 'local' and 'global' foundations,[1] but the former strikes me as new. Often, the best way to understand new things is to put them into one's own words, so let me try to describe what I take him to have done.

Baldwin's title references Generous Arena—the goal of providing a single framework that encompasses all of mathematics, so that concepts and theorems in one branch can be freely exported to and imported from another[2]—but the goal he most thoroughly explores, amends, and extends is Essential Guidance. In Maddy (2017) and Maddy (2019), this notion is derived from a criticism that category theorists have lodged against set-theoretic foundations, namely, that that its surrogates for various items introduce irrelevant structure (what MacLane calls 'sand'), that this

[1] See, e.g., Kennedy (2021) and/or my review Maddy (2021).

[2] Cf. Steel (2014), p. 164: 'we want one framework theory, to be used by all, so that we can use each other's work. It is better for all our flowers to bloom in the same garden'.

P. Maddy (✉)
Department of Logic and Philosophy of Science, University of California, Irvine, CA, USA
e-mail: pjmaddy@uci.edu

S. Arbeiter and J. Kennedy (eds.), *The Philosophy of Penelope Maddy*,
Outstanding Contributions to Logic 31, https://doi.org/10.1007/978-3-031-58425-1_12

excess structure distracts from their mathematically significant features, and that its construction techniques don't distinguish between what's mathematically important and what's not. My contention is that the category theorists are right to point out that set-theoretic foundations don't provide Essential Guidance, but that this doesn't undercut their ability to do several other important jobs, including Generous Arena.[3]

In fact, as Baldwin himself remarks (p. 145), there's a tension between Essential Guidance and Generous Arena: it's highly unlikely that any single way of thinking, of formulating concepts, of spotting fruitful new avenues, could work for all areas of mathematics. Baldwin notes just these sorts of mismatches in his comparison of model theory and category theory in his §2, where he assesses both as potential scaffolds:

> along with *local foundations* a scaffold should promote *unity* across mathematics by providing a method for transporting concepts and results from one area to another. There is no requirement that a scaffold encompass all of mathematics but only that it makes connections among a number of topics. (pp. 144–145)

So, a scaffold delivers local foundations—capturing concepts and guiding research in its range of areas—but in the process, it also provides 'systematic ways of connecting results and proofs' across those areas (pp. 144–145).

In a second adjustment to Essential Guidance, Baldwin replaces 'essential' with 'productive':

> various scaffolds will give different advice depending on their framework. The success of the mathematics is the measure of the guidance. (p. 146)

My use of 'essential' was intended to capture the category theorist's goal of removing the 'sand' introduced by set theory and revealing the true nature of mathematical concepts and structures. My own view is that mathematical concepts should be crafted simply with an eye to generating good mathematics, so I take Baldwin's modification of 'essential' to 'productive' as a friendly amendment. The resulting foundational goal—find a fruitful scaffold—he calls, appropriately enough, Productive Guidance. An example would be the scaffold that descriptive set theory provides for theorizing about Polish spaces (p. 146).[4]

With this background in place, Baldwin sets out to assess the effectiveness of model theory[5] and category theory in providing Productive Guidance. Turning to

---

[3] Baldwin also notes Risk Assessment, Shared Standard, Meta-mathematical Corral, and Elucidation. For the record, I regard Essential Guidance and Proof Checking as laudable goals, just not what set-theoretic foundations are designed for, but Metaphysical Insight (the goal of determining what a mathematical item 'really is') and Epistemic Source (the goal of providing certainty) seem to me misguided.

[4] Polish spaces are found in several distinct areas of inquiry. Approaching each of these with the tools of Descriptive Set Theory captures its leading concepts and guides its research productively (local foundations)—and this allows methods and results to be transferred among the various areas (a scaffold).

[5] For the record, Baldwin sees his model theory as 'based squarely in ZFC with occasional use of large cardinals' (p. 147).

model theory first, he takes Shelah's Classification Theory as his leading example; in particular, he's out to

> explain how the [model-theoretic] classification of theories serves as a unifying principle to treat different areas of mathematics ... [and] provides a wild/tame distinction to give productive guidance. (p. 147)

Shelah's idea is to sort complete first-order theories by syntactic criteria in such a way that 'theories in the same class share mathematically significant properties', so results can be transferred 'from one theory to another in the same class' (p. 148)—in other words, each of Shelah's classifications forms a scaffold for the theories in that class. As a bonus, this classification 'provides guidance to distinguish the wild from the tame' among classes, which is crucial, since 'very different methods will be needed to study' the wild topics (ibid.).

This classification theory depends on internal structure, but Baldwin recognizes 'that the properties of constructions used in many areas are defined ... without involving "irrelevant" internal information' and acknowledges 'the wide influence of category theory in homological algebra, algebraic topology, and algebraic geometry' (p. 151). This time, category theory is the scaffold that provides a local foundation for each of these fields—captures its concepts, guides the work in productive ways—and also connects concepts and proofs from one to another. Moreover, just as a set-theoretic scaffold is less appropriate for the examples that don't depend on internal structure, a category-theoretic scaffold is less appropriate for those that do. By switching the focus from Foundations to foundations, Baldwin shows how different scaffolds, each designed for its own style of mathematics, can and do deliver Productive Guidance.[6]

Switching the focus back to Foundations, Baldwin compares the 'material' set theory of ZFC and its extensions with the 'structural' set theory of ETCS and its extensions.[7] ETCS escribes a natural category, but it's also quite weak, mutually interpretable with BZC; to reach the full strength of ZFC requires a Replacement schema, which in turn 'requires explicit set theoretic syntax to formulate' (p. 157). After that, though, 'there is no real difficulty in extending structural set theories by large cardinal axioms' (ibid.). Here we have two candidate Foundational system—ZFC + LCs and ETCS + R + LCs—based on two 'different conceptions of "set"' (p. 17).[8] How to choose?

Baldwin's bold suggestion is that we needn't:

---

[6] Baldwin generously notes that he's engaged here in just the sort of methodological project I recommend in Maddy (2019a), p. 304: about which 'ways of thinking' are effective for which areas of mathematics, about how they differ, about what makes them so effective where they are and ineffective where they aren't, and so on. See also Maddy (2017b), p. 307.

[7] Note that what's being compared here is two versions of set theory, not a theory of sets and a theory of categories. See footnote 10.

[8] Should we be concerned that the extensions of ETCS 'leave the formalism-free world [of category theory] and require explicit set theoretic syntax to formulate axiom schemes' (p. 157)? Is some element of material set theory present even in our structural set theory? Baldwin raises what may be a related question: 'does structural set theory ... actually eliminate sand and if so how? What are examples?' (p. 150).

> Why choose between them? Rather than a Foundationalist view, one may just consider several mutually reinforcing Foundations as an open system of Mathematics. (p. 161)

There seems to me much wisdom in this, though perhaps also a trace of exaggeration; what we have is not 'several reinforcing' candidates of starkly different origin, but two conceptions of set, material and structural. Baldwin's stance here echoes John Burgess in his *Rigor and Structure*:

> the most natural reaction to the 'sets vs. categories' debate seems to me to be 'Why not both?' In other areas of mathematics, the availability of different formulations of a problem, say geometrical and algebraic, is considered an advantage, since it means two different kinds of tools would be at hand in seeking a solution. The equivalence of [ETCS+Replacement] with ZFC ... suggests the possibility, in foundational work, of passing back and forth between different formulations, set-theoretic and category-theoretic. How could this not be an advantage? (Burgess (2015), p. 189)

Burgess goes on to suggest that the study of independent questions and the search for new axioms to settle them might stand to gain by drawing on the resources of both frameworks, a very enticing thought!

I think Baldwin and Burgess may well be right about this,[9] but I do have a reservation. It's one thing to propose category-theoretic set theory, ETCS + R + LCS, as an alternative to ZFC + LCs as Foundation, and quite another to propose a theory straightforwardly about categories, say CCAF (The Category of Categories as a Foundation), as an alternative to either one of the set theories in that role.[10] (The name CCAF suggests that the latter is the intent of at least some category theorists.) Recall Generous Arena. Baldwin and Burgess argue that ETCS + R + LCs can provide this as well as ZFC + LCs, presumably in very much the same way.[11] But ETCS + R + LCs is just one category in CCAF, or really, in CCAF + 'there is a category satisfying ETCS + R + LCs', because an axiom for each special category has to be added into CCAF by hand. Where, then, is the Generous Arena? Inside ETSC + R + LCS? What of all the other categories included in CCAF, each with its own axiom, and all the relations between them?[12]

But these are questions for another occasion. For now, let me simply thank Professor Baldwin for his ambitious and stimulating paper. I only hope that I may one day better understand some of the deep mathematics he calls to our attention along the way![13]

---

[9] I retain some unresolved qualms in the vicinity of the previous footnote.

[10] The second is the claim I discuss in Maddy (2017b) and Maddy (2019a).

[11] Hence Baldwin's question in footnote 8.

[12] In Maddy (2017b) and Maddy (2019a), I argue that CCAF is a problematic candidate for Generous Arena.

[13] Thanks to Professor Baldwin for his most helpful correspondence on an earlier draft.

References

Burgess, J. (2015). *Rigor and structure.* Oxford University Press.
Kennedy, J. (2021). *Gödel, tarski and the lure of natural language: Logical entanglement, formalism freeness.* Cambridge University Press.
Steel, J. (2014). Gödel's program. In J. Kennedy (Ed.), *Interpreting Gödel* (pp. 153–179). Cambridge University Press.

Part II
Realism, Anti-Realism and Thin Realism

Chapter 13
Second Philosophy and Realism

Patricia Blanchette

Abstract This essay discusses Penelope Maddy's "second-philosophical" understanding of the subject-matter of mathematics. It is suggested that the fundamental commitments of the second-philosophical stance might be better reflected by the adoption of a somewhat more-generous foundation, which will result in a somewhat more realist account of that subject-matter.

Keywords Mathematical Realism · Skepticism · Objectivity · Depth · Truth

13.1 Introduction

Penelope Maddy's Second Philosopher is an intellectual hero of mine. The Second Philosopher takes empirical science seriously, forming her beliefs about the physical world, in the usual sensible way, on the basis of careful reasoning about the results of experiment and observation. She is, as it's sometimes put, a "realist about" the usual things: tables and chairs, atoms and quarks—which is just to say that she, like most of the rest of us, believes that there are such things. The especially admirable part of her world-view has to do with the rationale she provides for this realism, and (especially) with the kinds of rationale that she eschews. We'll get to that shortly.

The Second Philosopher is, on the other hand, not (or not exactly) a realist about numbers, sets, and functions. Her views about mathematics and its (purported) entities come, interestingly, not from any skepticism about the general excellence of mathematics itself, but from a deep respect for the discipline. While SP's respect for the methods of empirical science leads her straightforwardly to existential beliefs, her respect for the methods of mathematics doesn't have quite this effect. The story here, especially the distinction between empirical science and mathematics as SP views them, is fascinating and illuminating. In what follows, I hope to sketch enough of this story to bring out some of its most tantalizing features. I will also venture to

P. Blanchette (✉)
Department of Philosophy, University of Notre Dame, 100 Malloy Hall, Notre Dame, IN 46556, USA
e-mail: Patricia.Blanchette.1@nd.edu

S. Arbeiter and J. Kennedy (eds.), *The Philosophy of Penelope Maddy*,
Outstanding Contributions to Logic 31, https://doi.org/10.1007/978-3-031-58425-1_13

suggest that the most admirable features of SP's position should lead her not quite in the direction that Maddy charts for her with respect to mathematics, but instead to a position on mathematical entities that more closely resembles her views about empirically-detectable things.

13.2 Second Philosophy and Skepticism

The Second Philosopher believes in atoms because of the strong experimental evidence for them, including for example Jean Perrin's observations of the Brownian motion of suspended particles. So she disagrees with those who hold that talk of atoms is merely a convenient fiction, and with those who hold that scientific theories are in principle merely good guides to prediction rather than accurate descriptions of the world around us. But SP's rationale for belief in atoms is also importantly different from the rationales that underly the most familiar responses to these fictionalist and instrumentalist positions. The familiar responses involve a realism that's grounded in a two-premise argument: (i) Our best (e.g. well-confirmed, strongly-tested) scientific theory says that there are atoms, and (ii) We have good reason (provided, for example, by its predictive accuracy) to believe that such a theory is true. A realist giving this response to the unbeliever is wading into waters that have nothing to do with the reasons SP herself has for believing in the entities in question. If we want to know whether there are atoms, says SP, we should pay close attention to the relevant experimental results; if the experiments are compelling, then that is the end of the story. We don't need experimental results *plus* further philosophical reasons—of the kind exemplified in (ii)—to believe that there are atoms.

The Second Philosopher is so-called because as compared with those who engage in the kinds of arguments just sketched, for or against the reliability of empirical methods, she inverts the order of the philosophical and empirical questions. Her methods for finding out about the world, including the human capacity to know things about one's surroundings, are empirical: the philosophical request for a justification of the claim that, say, we *really* know that there is a hand in front of our eyes is answered by noting the role of light, of the signaling from the retina via the optic nerve, etc. Her reliance on observation and experiment is bedrock, not in the sense that she takes these methods to provide absolute certainty, but in the sense that her reliance on them, carefully assessed and corrected where necessary, is not based on any philosophical claims that she regards as more fundamental.

SP's reaction to the external-world skeptic is instructive. Consider the skeptic who holds, perhaps for the kinds of reasons given in Descartes' First Meditation, that (a) our empirical methods are fallible in the sense that they may lead to false claims about the existence and the nature of external objects, and (b) that the fallibility of these methods undermines the rationality of our reliance on them. SP's response to (a) is to agree that of course any particular empirical method might in principle be faulty, and that therefore all such methods should be subjected to regular scrutiny, checking each method by comparing its results with the results of other empirical methods, investigating the theoretical underpinning of its predictions, and so on. But SP doesn't find in the sceptic's story any general reason to view the methods of observation and experiment as unreliable ways of finding out what the world around her is like. This is, again, not to say that SP thinks that empirical methods are guaranteed to produce absolute certainty; but since she makes no claim to absolute certainty—but only to ordinary knowledge about the external world—this is not a problem.

SP's response to the skeptic differs importantly, then, from two other responses. The first, noted earlier, is the response of the realist who provides the two-step argument above in support of our reliance on empirical methods. This realist stands, as it were, "outside" of empirical science, claiming to provide reason for faith in the entire enterprise. SP by contrast stands firmly within the realm of observation and experiment, which methods she uses to justify particular empirical claims. The second response to the skeptic with which we can helpfully compare SP's is the response of the phenomenalist, according to whom objects such as tables and chairs, and perhaps smaller things as well, are in some sense simply collections of sensations—so that the skeptic who admits sensations while denying external objects is simply incoherent. SP, by contrast, has the usual view that tables and chairs, and atoms, are radically different from sensations. She does not take the skeptic's position to be incoherent: it is in principle possible, as she sees it, that our beliefs about the external world are radically mistaken. She just doesn't view our inability to rule out the skeptical possibility as undermining the force of ordinary empirical evidence.

The spatial metaphor, according to which SP stands "within" empirical science, is a helpful way to characterize the central difference between her position and the assumptions shared by both the external-world skeptic and the typical anti-skeptic. While both of these last two "stand back" from empirical science and evaluate its methods (the first negatively and the second positively), SP does not. Her tools are the tools of observation and experiment, and there is, from this position, no "standing back" and evaluating her general reliance on those very tools. For the same reason, SP offers no evaluation of her own stance: she finds out about the world, again, via observation and experiment, and doesn't presume to offer a justification for doing so.

SP's silence with respect to the virtues of her own stance, however, doesn't stop us, her fans, from noting those virtues. The skeptic's demand for a wholesale justification of the methods of observation and experiment—a justification that would include e.g. a demonstration from first principles that our senses offer us information about the world—is arguably either incoherent or hopeless, for familiar reasons. Because justification itself requires appeal to principles and methods of some kind, the idea that we can without any presuppositions examine all such principles and methods in one overarching evaluative endeavor is self-contradictory. And the idea that we can provide an external justification by appeal to some minimal but rock-solid body of presuppositions has not panned out well. Getting on with our knowledge about the external world in SP's way, via careful and self-critical application of the best empirical methods we have, and without falling into the quicksand that comes when we try to pin the legitimacy of those knowledge-claims on an independent justification of empirical methods as a whole, is, from the external fan's point of view, the only sensible way to go.

13.3 Second Philosophy and Mathematics

13.3.1 Mathematics in General

SP's interest in mathematics is the scientist's interest. Because mathematics plays an essential role in empirical investigations of the world, her critical examination of her own tools must include a careful look at mathematical methods themselves. She is also interested in mathematics as a human activity, and seeks "an account of what pure mathematics is: what sort of activity is it? what is the nature of its subject matter? how and why does it intertwine so remarkably with her empirical investigations?" (Maddy 2011, p. 39).

Turning specifically to set theory, a careful look at its history provides, in part, the following rich picture of set-theoretic methodology:

> The Second Philosopher hoping to undertake an investigation of sets has access to a rich array of methods, both for introducing them in the first place and for determining their extent and their properties thereafter. We've seen that sets were originally posited in the service of explicit mathematical goals: a broader condition on uniqueness of trigonometric representations, representation-free definitions for abstract algebra, an account of continuity to found analysis and launch a general study of continuous structures, a foundation for arithmetic. In broad overview, these goals range from relatively local problem-solving, to providing foundations, to more open-ended pursuit of promising mathematical avenues. The arguments subsequently offered for and against set-theoretic hypotheses and axioms take many of the same forms: the Axiom of Choice solves an array of problems, helps found the theory of sets and promises more 'productive science' … ; determinacy hypotheses settle the questions once raised by the early analysts and produce a rich and deep extension of the classical theory; V=L presents an unproductive alternative theory of projective sets and conflicts with set theory's foundational role by restricting the universe of sets. And so on. Given what set theory is intended to do, relying on considerations of these sorts is a perfectly rational way to proceed: embrace effective means toward desired mathematical ends. (Maddy 2011, p. 52)

In short, SP notes that the development of set theory has been driven by clear mathematical goals, and that its methods have proven fruitful in the pursuit of those goals. These methods—e.g. the embrace of new entities in order to increase expressive power, the appeal to the intrinsic interest of their consequences when choosing amongst theories—are both straightforwardly rational, given the mathematical goals they serve, and also importantly different from the methods of observation and experiment that SP employs in the empirical arena. Faced with the question whether those empirical methods ought to be brought to bear in order to underwrite or correct the mathematical methods, SP's answer is negative: mathematical theories are generally not supported by empirical confirmation, and the imposition of non-mathematical restraints on mathematical goal-setting or methodology would be counter-productive.

If we want to judge the reasonableness of a mathematical method, holds SP, we should look to its effectiveness in furthering mathematical goals, rather than its conformity to external constraints. Mathematics is, as SP understands it, an *autonomous* science.

If mathematical methods are to be judged solely in terms of their conduciveness to mathematical goals, there is no role for stand-alone metaphysical or epistemological principles in such judgment. So an immediate implication of taking seriously the autonomy of mathematics is that one sees the powerlessness of a popular form of argument against mathematical realism. This is the form of argument according to which mathematical knowledge or reference would be impossible if realism were true, owing to our inability to interact causally with abstracta. That this line of argument presupposes a handful of relatively shaky principles about the causal origins of reference and knowledge is clear; respect for the autonomy of mathematics underscores their unsuitability as principles by means of which to leverage the rejection of the existential claims of that science. In short: the idea that we should reject or re-interpret the deliverances of well-established mathematical methods on the basis of epistemological or semantic dogma is one that SP does not find compelling.

Given this background, one might have expected SP to hold that the well-established claims of mathematics, especially its existential claims, should simply be straightforwardly accepted. Pursuing methods that SP regards as rational, the set-theorist claims that there are, for example, sets with uncountably many members; the number-theorist claims that there are infinitely many prime numbers, and so on; and one might have thought that SP's stance here would lead her to the conviction that the methods of mathematical research demonstrate the existence of mathematical objects in essentially the same way that the methods of empirical science demonstrate the existence of empirically-discoverable objects. But this is not the standpoint of Maddy's Second Philosopher. To lay out the Second Philosopher's attitude to mathematics and especially to its existence-claims, we'll follow Maddy's discussion of three different approaches to the subject-matter of mathematics, along with the Second Philosopher's reaction to them.

13.3.2 Robust Realism

The first of these views is what Maddy calls "Robust Realism," the view that number theory is about numbers, set theory is about sets, and so on: the objects in question are "objectively existing abstracta," (Maddy 2007, 363) and mathematics is the science of these objects in much the same way that physics is a science of a different collection of "objectively existing" things. The "objective" part is just a matter of the independence of the subject-matter from us: on the Robust Realist's view, numbers and sets don't owe their existence or their characteristics to our thoughts or proofs or activities of any kind; there would still be infinitely many numbers and sets even if nobody had ever thought about them. The reason to hold that mathematical objects are objective, from the Robust Realist's point of view, is that the methods of proof in mathematics—the means by which we demonstrate the existence and the characteristics of mathematical objects—make no reference to such human activities. And the reason to hold that the objects are "abstract" is that those methods make no reference to spatial, temporal, or causal features of the objects. Finally, the Robust Realist's idea that mathematics is essentially the attempt to truly describe such abstracta implies, as Maddy sees it, a commitment to the view that mathematical methods require for their justification a persuasive story about the connection between the abstracta and the methods. If mathematics is the attempt to say what's true about the denizens of an abstract realm, then a mathematician's methods are rational only if they are likely to lead to true claims about those abstracta. The rationality of mathematical methods turns, in short, on a "non-trivial epistemology" linking mathematical practice and mathematical objects.

But Robust Realism will not do for the Second Philosopher, as Maddy sees it. For this form of realism immediately conflicts with the autonomy of mathematics. Consider for example Cantor's rationale for introducing sets: that "they allow [him] to formulate and prove a stronger theorem on uniqueness" of trigonometric representations. The problem is that "if the Robust Realist is right, if the goal of set theory is to describe an independently-existing reality of some kind, then it appears that Cantor's evidence needs supplementation ...: we need an account of how the fact that sets serve this or that particular mathematical goal makes it more likely that they exist." (Maddy 2011, 58). Similarly, consider more recent arguments against the axiom-candidate $V = L$. The perfectly rational set-theoretic argument against $V = L$ is essentially that its adoption would unnecessarily circumscribe the universe of sets, in a way that conflicts with the goal of providing a setting for mathematics that is "as generous as possible." (Maddy 2007, 358) If the Robust Realist is right, however, pursuit of this straightforward mathematical goal is a reasonable methodological principle only if there is independent reason to think that it's a good guide to truth. In short, the Robust Realist cannot rest content with the straightforward, purely mathematical justification of the set-theorist's methods, and hence fails to respect the autonomy of mathematical methodology.

The problem SP sees for Robust Realism is reminiscent of the central difference between SP's stance and the standard response to external-world skepticism. That standard response, recall, is to provide a philosophical argument in defense of the reliability of empirical science. SP, by contrast, does not subject the methods of empirical science to critique, but simply uses them. In the case of mathematics, the Robust Realist offers a philosophical argument to the effect that our mathematical methods are reliable ways of coming to know about the existence and the nature of mathematical objects. SP's resistance to such an external defense of mathematics does not have the same genesis as her lack of interest in the external defense of empirical science: she is not an inhabitant of mathematical methodology in the sense that she is an inhabitant of empirical methodology. The question of the overall rationality of mathematical methodology is a coherent one for SP, in a way that the question of the overall rationality of empirical methodology is not. But an important thread that runs through both is the idea (adopted by her external admirers, rather than by SP herself) that empirical science is just fine as it is, and the idea (explicitly held by SP herself) that mathematics is just fine as it is.

Having rejected Robust Realism, Maddy's Second Philosopher is left to choose between two further views, Thin Realism and Arealism. And interestingly, despite their differences, Maddy's view is that SP can consistently hold either one of them.

13.3.3 Thin Realism

Thin Realism is the view that sets are just the kinds of things that can be known about by doing set theory. The realism is "thin" in the sense that the qualities of sets, on this account, are pegged to the methods of set theory: instead of aiming at an accurate characterization of objects whose qualities are sufficiently independent of our theorizing that our methods might in principle just be all wrong, this view takes sets just to be the things accurately characterized by those methods when carefully employed. If, for example, *maximization*—the provision of a set-theoretic arena for mathematics that's as rich as possible—is one of those methods, then ruling out $V = L$ because of its violation of maximization is automatically justified without appeal to any external standards of correctness. On the Thin Realist's account of matters, the realm of mathematical objects doesn't give rise to external standards of correctness by means of which to evaluate the methodology of mathematics, since that realm of objects is by its nature just the way those methods reveal them to be. There is, as Maddy puts it, no "gap" between the methods and the objects. (Maddy 2007, 370).

Some care is required here, however, about the lack of a 'gap.' Thin Realism is not a form of constructivism; it is not the view that mathematical objects are created by human mathematical activity. It shares with Robust Realism the view that set theory "aim[s] at describing the properties of an objectively-existing, non-spatiotemporal, acausal reality..." (Maddy 2011, 62). The essential difference between the Thin Realist and the Robust Realist is that, as noted above, the latter "requires a non-trivial account of the reliability of set-theoretic methods, an account that goes beyond what set theory tells us," while for the Thin Realist, "set theory itself gives the whole story; the reliability of its methods is a plain fact about what sets are." (Maddy 2011, 63) This difference comes out perhaps most clearly in the different responses provided by these realists to the radical skeptic who doubts that we are in a position to know that there are such objectively-existing objects. The Robust Realist proposes to respond to the skeptic by providing an account of the knowledge-conferring connection between sets and our cognitive processes, an account whose necessity the Second Philosopher views as incompatible with the autonomy of mathematics. The Thin Realist, on the other hand, views the mathematical skeptic's position as incoherent. Reflecting on the important mathematical virtues of the introduction of sets, the Thin Realist responds to the skeptic that

> to suggest that set theory could enjoy all these virtues and sets still not exist or be radically different than they seem is to misunderstand the nature of set theory and its subject matter. Recall the Thin Realist's credo: sets are the things set theory tells us about. ... [F]or the Thin Realist, sets simply are the sort of things we can find out about in these ways. From

> this point of view, there is no room for a radical epistemological gap between sets and set-theoretic methods; the skeptical challenge here, unlike the case of the external world, is simply ill-formed. (Maddy 2011, 75)

There is room for a worry here, though. The worry is that the response to the skeptic just outlined is incomplete, and perhaps incompletable from the Thin Realist's point of view. In order persuasively to say that the skeptic's challenge is "ill-formed," it will be necessary to say what, exactly, that skeptic has gotten wrong. And to say, as TR does, that sets "just are" the kinds of things that set theory tells us about is not enough, since the skeptic's question is whether there are, in fact, any things that set theory tells us about.

Recall the phenomenalist's response to the radical empirical skeptic. Here the incoherence of the skeptic's challenge, from the phenomenalist's point of view, is clear: the skeptic acknowledges the existence of sensations and collections thereof but questions the existence of tables and trees; and this is incoherent if, as the phenomenalist maintains, tables and trees are collections of sensations. The Thin Realist about sets doesn't have this easy account of the skeptic's incoherence, since TR rejects the constructivist account on which sets are reducible to something immediate—e.g. to mathematical ideas or proofs—the existence of which the skeptic grants. For TR insists, as one should, on the objectivity of mathematics. The close connection between the nature of sets and the methods of set-theoretic investigation is not due, on TR's view, to a conceptual or definitional connection between the two; it is instead the content of a "simple hypothesis" (Maddy 2011, 68) on TR's part. The difficulty here is that if the connection between objective sets and human methods can be the content of a non-trivial hypothesis, then the skeptic's denial of this connection is coherent after all. And this means that one cannot, in the way proposed by TR, simply brush aside the request for a justification of the claim that there are in fact objects accurately characterized by our set-theoretic claims.

The Thin Realist's resistance to a constructivist account of mathematical objects—and hence to one straightforward way of justifying the view that the radical skeptic's position is incoherent—rests on her conviction that mathematical research is getting at something objective. And here the Thin Realist's position is supported by the familiar phenomenon of the recalcitrance of the subject-matter:

> [J]ust as it's not up to us which bits of pure mathematics best serve the needs of natural science, just as it's not up to us that it would be counterproductive to insist that all 'groups' be commutative, it's also not up to us that appealing to sets and transfinite ordinals allows us to capture facts about the uniqueness of trigonometric representations, that the Axiom of Choice takes an amazing range of different forms and plays a fundamental role in many different areas, that large cardinals arrange themselves into a hierarchy that serves as an effective measure of consistency strength, that determinacy is the root regularity property for projective sets and interrelates with large cardinals, and so on. These are the facts that ... constrain our set-theoretic methods, and these facts ... are not traceable to ourselves as subjects. (Maddy 2011, 80-81)

Judgments about mathematical virtues of the kind just listed—about, as Maddy calls it, mathematical *depth*– are not subjective, but are instead objective features of the mathematical landscape that we aim to characterize, but can in principle get wrong.

> [T]he entire mathematical community could be blind to the virtues of a certain method or enamored of a merely fashionable pursuit without changing the underlying facts of which is and which isn't mathematically important. This is what anchors our various mathematical goals. ... [T]he overwhelming case for Dedekind's innovations depends in large part on the subsequent successes of the abstract algebra they helped produce. The key here is that mathematical fruitfulness isn't defined as 'that which allows us to meet our goals', irrespective of what these might be; rather, our mathematical goals are only proper insofar as satisfying them furthers our grasp of the underlying strains of mathematical fruitfulness. In other words, the goals are answerable to the facts of mathematical depth, not the other way 'round. (Maddy 2011, 81-2)

Our set-theoretic methods, in short, are objective because they "track ... these underlying contours of mathematical depth." (Maddy 2011, 82).

Maddy's beautiful account of the *contours of mathematical depth* in all their variety (of which I've represented only a small sample here) does, it seems to me, just what we want an account of mathematical objectivity to do. It meshes with the phenomena of mathematical recalcitrance, with the fact that we discover, rather than stipulate, which mathematical proposals will turn out to provide unifying, explanatory, fruitful avenues of investigation. And because our a priori methods of mathematical investigation are just the kind of tools one would expect to be well-suited for following these contours, it makes sense of the connection between our mathematical methods and the facts they uncover. Sets "mark off a mathematically rich vein within the indiscriminate network of logical possibilities," (Maddy 2011, 83) and the pursuit of the contours of this vein is the activity of set theory.

We can now see more clearly the structure of the Thin Realist's response to the radical skeptic, and her separation from the constructivist. On the Thin Realist's view, mathematical work aims at uncovering objective facts about mathematical depth. Although these are facts about which a mathematician can in principle be mistaken, they are the kinds of facts that mathematicians are trained to uncover, and for the discovery of which their tools are uniquely well-suited. The kind of skeptical claim that is, from the Thin Realist's point of view, incoherent is the claim that mathematical work might successfully uncover facts about mathematical depth while failing to uncover facts about numbers, functions, and sets. For as the Thin Realist conceives of it, mathematical objects are just those things that "track the contours" of mathematical depth.

> We can be uncertain whether or not a given set-theoretic posit will pay off, and therefore uncertain about whether or not it exists, but if it does pay off, there's no longer any room for doubt; we can be uncertain that we're getting at the deepest and most fruitful theory of sets, and therefore uncertain about whether or not our axiom candidate is true, but if we are succeeding, there's no further room to doubt that we're learning about sets. This is what defeats an Evil demon-style concern: the Demon might somehow induce in me all the experiences I'd have if there were an external world without there actually being such a world, but he can't present a set-theoretic posit that actually does a maximally-efficient job of tracking mathematical fruitfulness and yet doesn't exist – because the posit just is the sort of thing that does this sort of job. (Maddy 2011, 83)

Mathematics is *objective*, on the Thin Realist's account, because it's about the contours of mathematical depth, which contours hold independently of us. We can tell, perhaps just retrospectively, when we have done a good job of following those contours, because the mathematics that results in such cases is undeniably fruitful. So far, this sounds a lot like empirical science, in which we try to follow the contours of the physical world, and take ourselves to be doing this well when we get results that are explanatorily rich, empirically verifiable, and so on. But an important dissimilarity arises, from the Thin Realist's point of view, when we ask about the justification of the science's claim to accuracy. Faced with the skeptic's claim that our empirical methods *might* all be misleading, and the world radically different from our portrait of it, SP acknowledges that (remote) possibility. The skeptical possibility is coherent because the apparent explanatory richness and empirical verifiability of our empirical theories is compatible with their being entirely false. SP simply takes that possibility to be uninteresting, and sensibly rejects the idea that the justification of her empirical beliefs requires it to be ruled out. Faced with the parallel skeptical claim in the mathematical case, the Second Philosopher responds in an importantly different way, a way that distinguishes two versions of the skeptic's claim. If the skeptic claims that, for all we can tell, we are entirely mistaken about mathematical depth, then the Thin Realist agrees. The price of objectivity is, as ever, the possibility of error. It's just not a very interesting possibility. If on the other hand the skeptic claims it to be possible that we are *right* about mathematical depth but wrong about the existence of numbers, sets, and functions, then, as just outlined, the skeptic's position is incoherent.

Just as there is a difference between actually tracking the cause of a given sensation and merely thinking one is doing so, there is a difference between actually tracking the contours of mathematical depth and merely thinking one is doing so. The skeptic who insists on the importance of this gap can agree with the Thin Realist that the only thing needed to justify the positing of mathematical objects is a justified claim to have discovered rich veins of mathematical depth, while insisting, entirely coherently, that there is no guarantee that our mathematical methods track the contours of mathematical depth.

The purpose of raising the skeptic's worry is not to suggest that the Second Philosopher owes the skeptic a response. It's rather that if the worry noted above is coherent—if our mathematical methods as a whole can in principle be badly suited to tracking the contours of mathematical depth—then the Second Philosopher is not in a position to dismiss that worry as incoherent by characterizing mathematical objects as so "thin" that they're automatically tied to our mathematical methods. The fundamental concern is that there is a tension between the objectivity that the Thin Realist claims for mathematics and the essential connection she posits between the methods and the objects of mathematics. We'll return to this issue below. For now, let's turn to the second of the two positions Maddy takes to be open to the Second Philosopher, the position she calls "Arealism."

13.3.4 Arealism

Arealism is the view that there are no mathematical objects (no numbers, no sets), and that mathematics is not "in the business of discovering truths." (Maddy 2007, 377) With respect to the role of mathematics in empirical science, the Arealist's position is nicely illustrated by Maddy's example of the application of mathematical results to the prediction of a cannonball trajectory: the mathematics here plays its role entirely independently of whether one takes the structure described by the purely-mathematical part of our theorizing to exist, and hence independently of whether one takes the mathematics itself to be true (Maddy 2007, 379–80). More generally: from the Arealist's point of view, nothing about the role of mathematics in our scientific theorizing turns on the idea that mathematical statements are true descriptions of anything.

The primary problem with Robust Realism—its conflict with the autonomy of mathematics—is avoided straightforwardly by the Arealist, as Maddy sees it. For once we drop the idea that mathematical methods are intended to provide a route to the accurate description of a realm of independently-existing objects, there is no call for a defense of those methods via a connection between them and any such objects.

But the Arealist does not give up on the objectivity of mathematics. Indeed, she gives essentially the account sketched above on behalf of the Thin Realist: mathematics is responsive to the phenomenon of mathematical depth.

> [T]he Thin Realist holds that sets exist and set theory is a body of truths, and the Arealist denies both. But despite their disagreements over truth and existence, the Thin Realist and the Arealist are indistinguishable at the level of method. On grounds like those that motivated Cantor and Dedekind, both would elect to introduce sets into their pursuit of pure mathematics; both would regard Zermelo's defenses of his axioms as persuasive .... This methodological agreement reflects a deeper metaphysical bond: the objective facts that underlie these two positions are exactly the same, namely, the topography of mathematical depth ... For the Thin Realist, sets are the things that mark these contours; set-theoretic methods are designed to track them. For the Arealist, these same contours are what motivate and guide her elaboration of the theory of sets; she can go wrong as easily as the Thin Realist if she fails to detect the genuine mathematical virtues in play. For both positions, the development of set theory responds to an objective reality – and indeed to the very same objective reality." (Maddy 2011, 100)

Finally, in keeping with the Second Philosopher's recognition of the autonomy of mathematics, the Arealist intends her position to be entirely consistent with mathematics itself. This requires some explanation, since most mathematical theories contain a lot of existential theorems, and the Arealist's view would seem to contain the negation of those theorems. Maddy's view is that we can split the Arealist's relationship with the mathematician into two parts: the important agreement between the two of them when inside the realm of mathematics, and the relatively unimportant, or perhaps even nonexistent, disagreement between them when making statements about mathematics from the outside. Commenting on the claim that "the Arealist is disagreeing with what the mathematician says," Maddy remarks:

> In one sense, I've suggested that this isn't right: the Arealist is speaking as a Second Philosopher proposing an account of human mathematical activity; she isn't questioning the propriety of any aspect of the practice, including its assertions in the shape of existence claims. Of course, it's also clear that mathematicians themselves have philosophical views about the nature of mathematics, about what those assertions mean – recall ... Dedekind's remarks about 'free creations of the human mind' – and this part of what mathematicians believe *can* come into conflict with views like Arealism and Thin Realism. The Arealist doesn't disagree with what mathematicians say qua mathematicians, but when they branch out into questions of truth and existence external to mathematics proper – what is the nature of human mathematical activity? what is its subject matter and how do we come to know about it? and so on – then she reserves her right to differ. (Maddy 2011, 102-3)

The idea that we can split the mathematician's claims into the mathematical and the philosophical is straightforward, as is the idea that a recognition of the autonomy of mathematics rules out questioning the well-established members of the former camp, but provides no such insulation for members of the latter. But the important claim for the Arealist to make, as her position is outlined by Maddy, is that the mathematician's existence-claims fall into the latter camp, the camp over which mathematics itself has no authority. And this is not easy to understand. How can it be that, although it is a purely mathematical matter to provide, say, a proof of the existence of infinitely many prime numbers or of the existence of models of $V = L$, nevertheless the claim that there are such numbers or such models is not a *mathematical* claim?

The answer to this question, if I understand it correctly, is that the mathematical work itself delivers mathematical statements, e.g. statements of the form "there are infinitely many primes" or "there are models of $V = L$," and that it is a further philosophical question whether to take these statements to be true. The Second Philosopher, on top of her respect for the autonomy of mathematics, "may well wonder if its methods serve to establish truth and existence or actually do something else" (Maddy 2011, 105). An affirmation of the first disjunct results in Thin Realism, while the second leads to Arealism. And here, as Maddy sees it, there is no fact of the matter. Arealism and Thin Realism embody two different, but equally reasonable, decisions about how to use the words "true" and "exists."

The view that there can be such a thing as a "decision" in this case is a matter of applying to the words "true" and "exists" a view that is (or at least ought to be) uncontroversial as applied to many ordinary predicates. As the later Wittgenstein suggested, and as has been recently explored in detail by Mark Wilson, it is typical of many ordinary predicates that entrenched usage fixes some relatively straightforward criteria for inclusion in and for exclusion from the extension of the predicate, while leaving a good many cases indeterminate, subject to later stipulative decisions.[1] Competence with the use of the word "table" or "game" or "hard" need not involve appreciation of criteria that settle whether a newly-discovered thing falls into the term's extension. As one might put it, the well-entrenched usage of such a term in its home range need not decide the application of the term outside of that range. Maddy's view is that the ordinary use of the word "true" as applied in non-mathematical arenas, e.g. to statements about the physical world, does not determine its usage once we

---

[1] See Wilson (2006) *Wandering Significance*.

move to the more recherché realm of mathematics. Just as it is a matter of relatively-unimportant linguistic *fiat* that Pluto is (or is not) included in the extension of "planet," so too it is a matter of unimportant linguistic *fiat* whether one says that the sentences of our best mathematical theories are included in the extension of "true." For the Thin Realist, the answer is "yes:" those statements that are established by methods that meet the internal standards of the mathematical discipline in question are to be regarded as *true*, and the entities they posit are to be regarded as existing. For the Arealist, the answer is "no:" those statements that are established by methods that meet the internal standards of the mathematical discipline are to be regarded as just that: as meeting those rigorous standards, and therefore as doing all that we want a mathematical statement to do. The Arealist simply declines to use the word "true" to mark sentences meeting this standard.

But this, I think, is not a coherent position for the Second Philosopher to occupy. For SP is committed to the autonomy of mathematics; she does not take herself to be in any position to disagree with the well-established theorems, or the well-motivated axioms, of set theory and number theory. Consider one such theorem, e.g. the claim.

[U] Every pair of sets has a unique union.

Recognition of the autonomy of set theory requires not merely agreement with the metatheoretic statement that [U] meets the standards of the discipline (since this agreement is compatible with a complete rejection of that autonomy) but requires agreement with the set-theoretic statement that every pair of sets has a unique union. And this leaves the Second Philosopher in no position to reject the idea that [U] is true.

The general difficulty here is that the conditions for inclusion in the extension of "true" are not left open by the other semantic features of the language in the way that conditions for inclusion in the extensions of ordinary predicates can be.[2] Consider e.g. the word "alive," the well-entrenched use of which does not (I am told) determine whether viruses are in or out of its extension. We could in principle decide to extend the term in either way, consistent with past use. So the sentence [A] Virus V is alive could go either way: true or false. But once the decision is made with respect to the extension of "alive," the die is cast with respect to the inclusion of [A] in the extension of "true." If "alive" is firmed up by stipulating that satisfaction of condition C1, met by virus V, suffices for inclusion in its extension, then [A] is true; if the firming-up involves instead the stipulation that satisfaction of C2, not met by V, is necessary for inclusion in the extension of "alive," then [A] is false. In general: when a semantic stipulation fixes the truth-value of a previously indeterminate sentence, it does so by changing what the sentence says. Prior to such a stipulation, there is no sense to be made of the notion of "accepting" what the sentence says; in saying nothing true or false, it says nothing that can be accepted or rejected. And once it's been fixed what a sentence says, there is nothing left to stipulate with respect to truth and falsehood. If the Second Philosopher agrees with the things the mathematician says

[2] For the Frege aficionados, I take it that this is the central point behind Frege's claim that truth is not a property and "true" not a predicate. See e.g. "The Thought."

(*qua* mathematician), then she agrees that the sentences used to say those things are true.

Recall, however, that the Second Philosopher's focus is not just on truth, but—perhaps more importantly—on existence. And perhaps the best way to understand her position is to take the idea of an indeterminacy about the predicate "true" to be a red herring, and to focus on the more straightforward issue of an indeterminacy about the predicate "exists." The Second Philosopher holds that the paradigmatic use of the word "exists" in the empirical realm is insufficient to fix its extension in the mathematical realm. Consistently with that paradigm, one can take the so-called "existential" demonstrations of mathematics to demonstrate existence, or alternatively to demonstrate something only misleadingly so-called. One way to think of this is to take mathematical objects to occupy a gray area which might or might not, consistently with past use, be stipulated to belong or not to belong to the extension of the term "exists," in much the same way that minimally-hirsute people occupy a gray area which might or might not, consistently with past use, be stipulated to belong to the extension of "bald." The Thin Realist draws the line inclusively on this account, so that the shadowy area appears in the extension of "exists," while the Arealist draws it exclusively.

But here again, there are difficulties linked to the fact that the predicate in question, this time "exists," does not float free from the use of the other terms in the language, nor from the logic that the mathematician relies upon. The Second Philosopher is not a modern-day Meinong who thinks it makes sense to reserve the term "exists" for only some of the things she believes in. If she chooses Arealism, her view that there are no sets or numbers in the extension of "exists" is just the view that there are no sets or numbers. And this, I think, is not a coherent option for the Second Philosopher. When the number theorist shows via a quick calculation that $2^3 = 8$, the second philosopher who acknowledges the autonomy of mathematics (and so is in no position to reject the mathematical claim), and who eschews a re-interpretation of the mathematician's terminology (eschewing, in particular, a constructivist or fictionalist account of math), simply agrees that $2^3 = 8$. And when the number theorist remarks that this shows that 8 has a cube root, i.e. that there is a cube root of 8, there is no point in this brief line of reasoning at which the Second Philosopher can claim that the number theorist has stepped out of the role of mathematician. That there is a cube root of 8 is a mathematical claim. So the Second Philosopher, as far as I can tell, must agree not just that "there is a cube root of 8" is a statement that meets all of the internal standards of mathematics, but furthermore that there is a cube root of 8. Similarly of course for the claim that there are large cardinals. And if this is right, then the Second Philosopher can't be an Arealist.

13.4 Realism and Second Philosophy

The Thin Realist and the Arealist have the same goal: to recognize the autonomy and the objectivity of mathematics, and to resist the skeptic's claim that this recognition requires a metaphysics and an epistemology to support the view that the methods of mathematics deliver knowledge. Both the Thin Realist and the Arealist understand the objectivity of mathematics to rest on the ability of mathematical methods to uncover objective facts about mathematical depth. And both positions understand the relevant skeptic to be one who resists the inference from claims about mathematical depth to claims about the existence of mathematical objects. The Thin Realist's response to that skeptic is that the objects are nothing more than repositories of mathematical depth, which is to say (if I understand it correctly) that our talk of those objects is nothing more than talk of mathematical depth; the skeptic's position is therefore incoherent. The Arealist's response is to reject the target inference altogether: since the Arealist understands mathematical discourse as not really affirming the existence of mathematical objects, there is no point at which the skeptic (who agrees that we can track mathematical depth) can object to knowledge of mathematical claims so understood.

As we've seen in the discussion of Thin Realism, though, the gap between depth-detection and object-detection is not the only point at which skeptical worries might arise. The skeptic who asks for a guarantee that our mathematical methods in fact produce knowledge of facts about mathematical depth has not been answered, or effectively dismissed, by the replies given so far by the Thin Realist or the Arealist. Both of these positions take it for granted that our methods of proof, of calculation, of pursuit of purely mathematical goals in general, are ways to discover objective facts about mathematical depth. In this section, I would like to suggest that the Second Philosopher who adopts either Arealism or Thin Realism is not in a position to make this assumption. Or, to put the point differently: the Second Philosopher who—entirely correctly—*does* make this assumption is a slightly amended version of the Second Philosopher described by Maddy.

The facts about mathematical depth, to which the objectivity of mathematics is pinned, include for example the fact that there are "deep structural similarities between commutative and non-commutative groups," (Maddy 2011, 79) similarities that turn out to be important for our understanding of the interaction of groups with other mathematical structures; that the positing of transfinite ordinals opens up ways to successfully characterize facts about the uniqueness of trigonometric representations, and a wealth of facts about the fruitfulness of fundamental set-theoretic propositions like the Axioms of Choice and of Determinacy. These issues of *depth* go beyond merely logical connections between hypotheses, and involve a variety of different kinds of connection within and across mathematical disciplines. These are the facts, as Maddy sees it, that "play a role analogous to Kant's Euclidean space, the facts that constrain our set-theoretic methods, and these facts, unlike Kant's, are not traceable to ourselves as subjects." (Maddy 2011, 80–81) They are facts neither about logical implication nor about our own ways of constructing the subject-matter of mathematics; they are objective facts about that independent subject-matter.

Could we be systematically, drastically wrong about these kinds of facts? Could we make deep and systematic mistakes when reasoning about e.g. the uniqueness of trigonometric representations of functions, mistakes whose result is that our conception of the topography of mathematical depth is wildly wrong? Only the radical skeptic will answer in the affirmative; it would take a radical sort of confusion to be systematically mistaken here, just as it would take a radical sort of sensory-cognitive breakdown in order for us to be systematically, drastically wrong about ordinary empirical facts. Recall that SP's answer to that radical skeptic, in the case of empirical science, arises from an "internal" stance, one that takes it for granted that the way to find out about the external world is via experiment and observation. To the skeptic who suggests that the world might be radically different from the way we (seem to) perceive it to be, SP says: *I am not interested in the possibilities you lay out regarding radical error, and I don't take your bizarre scenarios to undermine my well-founded belief that there are atoms.* We who are watching SP's response can see that it is entirely rational: taking experiment and observation as bedrock, she can provide a clean justification of empirical knowledge, no matter what the skeptic says. But SP's stance with respect to mathematics is not internal: mathematical methods are not part of her bedrock toolkit for discovering objective facts. She examines mathematics from the outside, finds it rational, and finds that it should be evaluated with respect to its own goals, regarded as an autonomous science. Rather than taking mathematical methods as bedrock, SP thinks that she can *see,* by looking at the history of mathematics, that the mathematician's methods are successful in achieving mathematical goals. That is: she can tell that the mathematician's methods, when used critically, are good at uncovering objective facts about mathematical depth.

But something seems to have been slipped in quietly into this last step. If we characterize the goals of mathematics in subjective terms—e.g. as the goal of producing a line of reasoning that *seems* to show a fruitful connection between, say, transfinite ordinals and the unique representability of functions via trigonometric series, then we can certainly see, by looking at the history of mathematics, that mathematical methods are successful in achieving mathematical goals. But this gives us no assurance that those methods are successful in revealing objective facts. If on the other hand we characterize the goals of mathematics in objective terms—e.g. as the goal of actually providing (rather than seeming to provide) unique representations of functions, or as discovering (rather than seeming to discover) a broad array of results and connections provided by various set-theoretic hypotheses—then the history of mathematics cannot show us that the methods have helped to achieve the goals; it can merely show that they have seemed to do so. In either case, what's missing is a reply to the skeptic who wonders why we should take there to be any connection at all between the methods and the objective facts. SP cannot dismiss the skeptic's question in the way that she dismisses the external-world skeptic, and—if the above line of thought is on the right track—she cannot reply to that skeptic either. If that is right, SP is left without a clear reason to take mathematics to lead, after all, to objective knowledge.

It does not seem that we're wrong about the objective facts in question; there is no reason to suppose that careful mathematical work is deeply misleading. Our question wasn't *whether* we should trust the results of that careful mathematical work, but *on what basis*. The suggestion I've tried to press here is that SP, with the meager supply of methods that she takes for granted when investigating the world (all of which are suited to empirical investigation), does not have enough to go on. I will suggest, in what follows, that what's needed is a slightly-richer collection of bedrock methods, a collection that includes fundamental mathematical methodology as a means of discovering objective facts.

13.5 Conclusion

Robust realism, as Maddy has argued, is inconsistent with recognizing the autonomy of mathematics. Arealism has its own internal problems, or so I've argued. It also shares with Thin Realism the view that mathematical objectivity is a matter of the connection between mathematical methods and objective facts about mathematical depth: the methods track the contours of an independent reality. I have argued here that to agree with this last part of the shared view—which, in my view, one should do—is to occupy a position that's not available to the Second Philosopher. The difficulty is that the Second Philosopher is not in a position to say that the methods do in fact track objective facts. To put the point in terms of skepticism: the Second Philosopher is not in a position either to dismiss as incoherent, or to offer an argument in response to, the skeptic who doubts that seeming to track depth is an indication of tracking depth.

I would like to urge the Second Philosopher to jump more wholeheartedly into mathematics, to adopt the methods of mathematical investigation as part of her toolkit alongside the methods of observation and experiment. Just as it is *obvious* that the way to find out if there are atoms is to follow the results of careful experiment and observation, so too, with the amended toolkit, it is *obvious* that the way to find out about the contours of mathematical depth is to follow the results of careful proof, set-theoretic conjecture, and calculation. Such a Second Philosopher will say that we know about the deep structural similarities between commutative and non-commutative groups not because we have established these similarities via mathematical methods *and* we have reasons to take those methods to be scientifically respectable, but rather because we have established those similarities, via those methods, *simpliciter*. From this amended Second Philosophical standpoint, the response to the skeptic about mathematical knowledge is now exactly the response that the Second Philosopher gives to the empirical skeptic: We know these results because we've demonstrated them, and the possibility of extreme cognitive error, while coherent, provides no reason to doubt the results.

Such an amended Second Philosophical standpoint also gives us immediately a straightforward form of mathematical realism. From her expanded standpoint, SP will say for example that we know that $2^3 = 8$, and that there are infinitely many prime numbers, because we have gone through the easy calculations and proofs. Not (as the Robust Realist would have it) because we have calculated and proven *and* can give an independent justification of the veridicality of calculation and proof, the kind of justification that Descartes' demon would like us to provide. Rather: because calculation and proof, together with the rest of the rich array of methods employed when we pursue the goals of mathematics, are the ways we find out about numbers, sets, functions, and so on. The suggestion, then, is that instead of inhabiting empirical science while looking over the fence at mathematics as a science whose entire foundation can be examined from the outside, the Second Philosopher should take on the whole empirical-mathematical scientific enterprise as her home domain. She will then be a realist about numbers and sets in just the way that she is a realist about atoms, using the well-developed tools of the empirical-mathematical enterprise, critically and carefully, to learn about both.

Let me just say, in conclusion, that the worries I've tried to raise in this discussion are small. For the (or, at least, for this) fan of the Second Philosopher, the beautiful account of mathematical depth provided in *Defending the Axioms*, and the compelling route to respect for the autonomy of mathematics as described in both *Second Philosophy* and *Defending the Axioms,* are the cornerstones of an account of mathematics that manages to take the science seriously on its own terms, without falling into the twin errors on either side of the path: the error of succumbing to the metaphysician's or epistemologist's demand that we justify the methods of mathematics via stand-alone philosophical principles, or the error of responding to the difficulty of doing so by giving up on the objectivity of mathematics. The Second Philosopher shows us how to take mathematics to be fine just as it is. The central point of the above discussion is to argue that Maddy's persuasive account of the objectivity and the autonomy of mathematics cannot *quite* be adopted by the Second Philosopher, whose standpoint is insufficient to make sense of mathematical objectivity, but that a slightly amended Second Philosophy can straightforwardly fill the gap[3].

[3] Parts of this paper were read at the Paris Philosophy of Mathematics Workshop 2019. Many thanks to the audience members for helpful comments. Thanks also to Curtis Franks, Juliette Kennedy, and Mack Sullivan for comments on an earlier draft. Thanks especially to Pen Maddy for patient and helpful comments on an multiple earlier drafts.

Chapter 14
Reply to Blanchette

Penelope Maddy

Keywords Thin Realism/Arealism · Robust realism · Objectivity · Skepticism · Truth

Professor Blanchette's enthusiasm for Second Philosophy is heartening, and I'm grateful for it. She lodges serious objections to the way I think the second-philosophical approach plays out in the foundations of mathematics, but intramural disagreements of this sort between Second Philosophers are to be expected. What's less expected is her momentary tendency to go extramural—in the person of her Fan—with the hope of justifying the Second Philosopher's stance from the outside, against all comers.[1] However tempting it may be to confront the external world skeptic or van Fraassen's constructive empiricist head on—and I fully appreciate the temptation—these positions are epistemically airtight; for the Second Philosopher to engage with any one of them requires her to set aside all her hard-won methods for justifying her beliefs as price of entry, after which she inevitably finds herself at a loss.[2]

One response to the skeptic would be to declare her opponent's position incoherent[3]—to cut the problem off at the root—but in fact it makes perfect sense: she's simply being asked to produce a new, extraordinary sort of evidence, a demand

[1] This same impulse comes up toward the end of my exchange with Professor Paseau.

[2] For more how this plays out for the case of the external world skeptic, see Maddy (2017a), for the case of constructive empiricism, see Maddy (2022f).

[3] Blanchette appears sympathetic to this line: 'Because justification itself requires appeal to principles and methods of some kind, the idea that we can without any presuppositions examine all such principles and methods in one overarching evaluative endeavor is self-contradictory' (p. 175).

P. Maddy (✉)
Department of Logic and Philosophy of Science, University of California, Irvine, CA, USA
e-mail: pjmaddy@uci.edu

S. Arbeiter and J. Kennedy (eds.), *The Philosophy of Penelope Maddy*,
Outstanding Contributions to Logic 31, https://doi.org/10.1007/978-3-031-58425-1_14

that she understands, even respects, but doesn't know how to meet. The second-philosophical response, I argue, is to observe that her inability to produce extraordinary evidence leaves her ordinary evidence untouched. Perhaps the Fan imagines that the challenge, if allowed, would mean epistemic doom, when in fact, her ordinary evidence would still work just as well as it has all along.

But setting the Fan's ambitions aside, the centerpiece of Blanchette's paper is her discussion of the metaphysics and epistemology of mathematics. She raises objections to both Thin Realism and Arealism and offers a second-philosophical alternative to replace them. Let me take these up in turn.

14.1 On Thin Realism

Professor Blanchette well and rightly notes that Thin Realism faces two distinct skeptical challenges. First, the facts of mathematical depth are objective, so our beliefs about them are as vulnerable to radical skeptical challenge as our beliefs about the external world. Second, she notes:

> the kind of skeptical claim that is, from the Thin Realist's point of view, incoherent is the claim that mathematical work might successfully uncover facts about mathematical depth while failing to uncover facts about … sets. (p. 181)

Those clarifications in place, her objection to Thin Realism goes like this:

> if our mathematical methods as a whole can in principle be badly suited to tracking the contours of mathematical depth, then the Second Philosopher is not in a position to dismiss that worry as incoherent by characterizing mathematical objects as so 'thin' that they are automatically tied to our mathematical methods. (p. 182)

Phrased in terms of the two skepticisms, the claim here is: if the first form of skepticism is coherent, then it can't be turned away simply by noting that the second is incoherent. I certainly agree! So as far as I can tell, Blanchette's only objection to Thin Realism lies in the first skeptical challenge. Since depth is no less central to Arealism, this concern applies there as well, but Blanchette thinks Arealism also suffers from a special ailment of its own. Let's take that one first and return to the epistemology of depth in §3.

14.2 On Arealism

Blanchette's object to Arealism comes to this: the Arealist's claim that mathematics isn't in the business of establishing truth or existence stands in tension with ordinary set-theoretic assertions—Cantor's theorem is true, uncountable cardinals exist—and any effort to reform this aspect of the practice would violate her commitment to the autonomy of mathematical methods. How can the Arealist have it both ways on fundamental matters of truth and existence?

Now the Arealist certainly recognizes the historical, sociological, and psychological pressures that prompt us to extend the use of 'true' and 'exist' from ordinary applications in the physical world to new applications in the mathematical realm. She herself holds that elementary arithmetical claims like $2 + 2 = 4$ are straightforwardly true—rudimentary logical facts about the structure of our world—and that the standard model of arithmetic has deep psychological roots.[4] Descriptions of worldly situations using differential equations were once regarded as literal truths, only to give way when the particulate structure of matter was established in the early twentieth century.[5] When mathematical structures were shown not to correspond exactly to the world—the case of Euclidean geometry is a familiar example—mathematicians tended to opt for what Michael Resnik aptly terms 'Euclidean rescue'—for example, positing a realm of abstract spaces so that Euclidean geometry would remain true of something![6]

All this the Arealist regards as an instance of what Mark Wilson calls 'property-dragging'.[7] So, for example, the term 'weight' in its primary use here on Earth measures the effects of gravitational force, which leads us, quite reasonably, to conclude that a man who weighs 180 pounds on Earth would weigh only 33 pounds on the Moon. But when confronted with the case of an astronaut in Earth orbit, we 'drag' the property and our ordinary methods of detecting it—for example, how hard would it be to carry him?—to the new context, and we end up saying that he's weightless. Viewed from a broader perspective that includes both the Earth and the Moon, we immediately realize that he's really only weightless at a tipping point between the Earth's gravitational field and the Moon's, that in the orbiter, he isn't weightless at all, he actually weighs 176 pounds. Still, 'weightless' in the orbiter is established usage and does no harm as long as we don't confuse the orbiter's perspective with the broader perspective. In Wilson's terminology, 'weight' is operating on an uneven façade, and no harm is done as long as one doesn't freely import or export across separate patches of that façade.

The Arealist tells a similar story for the use of the term 'true' in mathematical contexts: those historical, sociological, and psychological forces have dragged it from the physical realm to the mathematical, but from the broader perspective of the physical and human sciences, she has her reasons for thinking that mathematics isn't the same sort of practice as the sciences, that the mathematical use of these terms is local to its context and shouldn't be exported to the broader perspective. So, just as the astronaut isn't both weightless and weighing 176 pounds, Cantor's theorem isn't both true and not a candidate for truth. (The term 'exist' comes along for the ride; to say a mathematical item exists is to say that the relevant existence claim is true.)

Of course, the analogy is imperfect, as all analogies are: the discourse around 'weight' in the orbiter is a relatively insignificant matter, while the discourse around 'true' in mathematics is mathematics itself, an immensely valuable discipline that the

[4] See Maddy (2007), part III, (2014c), (2018b).

[5] See Maddy (2008) or (2011), Chap. 1.

[6] See Resnik (1997), p. 130.

[7] See Maddy (2007), pp. 188–189, for a quick version, or Wilson (2006), 6.viii, for the whole story.

Arealist has good reason to practice. Though she ultimately regards it as a discipline that isn't in the business of establishing truth or existence—so Cantor's theorem isn't true (or false), uncountable cardinals don't exist (or fail to exist)—it remains entirely appropriate to say the usual things within the practice of set theory. This is where her respect for the autonomy of mathematics comes into play: she resists any effort to import inappropriate critiques from outside just as firmly as she resists efforts to export its truth and existence claims to the broader context. She respects and values mathematics for what it is; she takes it to have earned its right to its own methods.[8]

Thus the Arealist's defense; of course, the Thin Realist disagrees. She holds that mathematics has earned its right to 'true' and 'exist' even in the broader context. This is not simply to export such claims from within mathematics—as Blanchette seems inclined to do.[9] Rather, the Thin Realist makes appeal to the same considerations that the Arealist does when he concludes that mathematics has earned its rights, roughly, the way in which it's been intertwined with natural science over the centuries. The difference between them is that the Thin Realist takes mathematics to be enough like the natural sciences to count as one more development of her original inquiry, while the Arealist takes it to be different enough to qualify as something distinct. This is why the Arealist has to police import and export, and the Thin Realist does not.

Finally, the Second Philosopher's considered opinion is that both Thin Realism and Arealism are legitimate ways of describing the effort to uncover deep mathematics, that neither comes into conflict with any fact of the matter.[10] The only error is thinking that one or the other is uniquely correct, an error that both the Thin Realist and the Arealist tend to make: the Thin Realist insists that the Arealist is missing the fact that mathematics is a science and its claims are true; the Arealist sees mathematics as a different sort of undertaking and insists that its perfectly good uses of 'true' and 'exist' are not properly exported to the broader perspective. Eschewing both errors, the enlightened Second Philosopher's response to Blanchette's concerns would be: if you're more comfortable speaking the language of the Thin Realist, that's what you should do.

In the end, I suspect Blanchette's resistance to Arealism stems from a conviction that 'true' isn't just a word to be used and dragged at our convenience, but a term for a something more substantive, something with the nature of its own, something we theorize about philosophically.[11] Perhaps this is unsurprising in a distinguished interpreter of Frege! My own trajectory over the years has taken me in an Austinian direction that might be called 'deflationary' if that term didn't suggest 'deflationary

[8] Borrowing Professor Burgess's succinct formulation ('Reply to Burgess', p. 222).

[9] See p. 186.

[10] In Maddy (2011), V.1, I call the Second Philosopher's enlightened view 'Objectivism'. (In his (2024), Jeffrey Schatz observes that it's often clarifying to speak as an Arealist in discussing foundational contexts, and I think he's right about that.).

[11] Added later: Professor Blanchette agrees that this is the source of our disagreement about Arealism.

theory'.[12] But that's a topic for another venue—our topic here is the philosophy of mathematics.

14.3 On the Epistemology of Depth

Unlike the (purportedly) discredited Arealism, Thin Realism is consistent with both substantive and deflationary theories of truth,[13] so we return to Blanchette's fundamental epistemological concern: what reason does the Thin Realist have to think that her methods are reliably tracking what is and isn't mathematically fruitful or important (that is, mathematical depth)—or, as the Thin Realist herself might put it, what reason she have to think she's finding out the facts about sets?[14] Given that this radical skeptical challenge is coherent, is the Thin Realist really any better off than the Robust Realist? Doesn't she, too, need a non-trivial epistemology, the characteristic mark of Robustness?

In one sense, of course, the answer is no: she's entitled to a trivial epistemology in response to a skeptical challenge to the connection between her methods and sets; that is, assuming her methods are reliably tracking depth, there's no room to complain that they may not be delivering reliable information about sets. A Robust Realist using the same methods, methods reliably tracking depth, would still owe a substantive account of why depth is an indicator of truth, why depth-tracking methods also deliver reliable information about sets. But the Thin Realist still faces the other skeptical challenge, the question of that assumption: what a reason is there to think that her methods *are* reliably tracking depth?, what reason is there to think that our judgments of which mathematical avenues are most fruitful, productive, revealing, etc., are accurate?

In this sense, the Thin Realist and the Robust Realist *are* both subject to a radical skeptical challenge that runs parallel to concerns about our knowledge of the external world: in theory, the skeptic claims, an agent could have epistemically impeccable evidence, the best possible evidence, and still be entirely wrong. Blanchette sometimes suggests that this is the challenge she has in mind for the Robust Realist: 'the kind of justification that Descartes' demon would like us to provide' (p. 190). But the fact that a poor, put-upon agent can't provide such an extraordinary justification isn't the same as the traditional Benacerrafian worry about Robust Realism: Benacerraf isn't asking his Realist to provide evidence in the face of a mathematical Evil Demon; he's just asking for some sense of how the epistemic connection works, something or other that plays the role of our perceptual connection to the natural world. This is the relevant challenge for our purposes.

---

[12] See Maddy (2022g).

[13] See Maddy (2007), pp. 370–376.

[14] I phrase this as a challenge to Thin Realism because it's the remaining viable option for Blanchette, but of course Arealism is also susceptible.

To see this from another angle, recall what the Second Philosopher says in response to the radical external world skeptic: she falls back on her ordinary forms of evidence and her ordinary ways of assessing and defending their reliability—and holds that this ordinary evidence isn't undermined by her inability to provide extraordinary evidence. For the Robust Realist's purported knowledge of mathematics, the ordinary forms of evidence are proofs and other mathematical methods, but when it comes to assessing and defending their reliability as guides to the mathematical realm, his account breaks down. The Second Philosopher tells a story that begins with an examination of sense perception to evaluate its reliability, where it tends to err, and so on, and continues through scrutiny of her more advanced methods and their possible improvement. The problem for the Robust Realist is that he has no counterpart to this story, not even the beginnings of an account of how his methods match up with his metaphysics. The problem isn't that he can't answer the radical skeptic—the Second Philosopher can't, either—the problem is that he has no way to support the ordinary methods he hopes to fall back on.

The key point here in response to Blanchette is that the prospects for assessing and defending those ordinary mathematical methods are far better for the Thin Realist than for the Robust Realist. For her, the relevant ordinary methods are the kinds of considerations I've been isolating: Cantor's appeal to sets to enable his work on trigonometric representations; Dedekind's ideals as means toward a representation-free abstract algebra; the Cabal's support of determinacy hypotheses as a means toward a more complete theory of definable sets of reals; the defense of large cardinals as measures of consistency strength. In each case, assessing and defending reliability comes to arguing that the method proposed is actually effective toward its goal of tapping into depth: Cantor's and Dedekind's work was productive; determinacy has unlocked what certainly looks to be deep mathematics in a surprising range of ways; the ability to order mathematical hypotheses by their level of risk they pose is invaluable. Tracking depth in ways like these is difficult, obviously, because doing good mathematics is difficult—sometimes we go wrong, pursue what turns out to be a dead end, and have to circle back—but it's what mathematicians train to do and become adept at doing. The radical skeptic can still challenge them to set aside all these ordinary methods and provide some extraordinary new way of tracking depth, but their inability to meet this challenge doesn't undercut the soundness of their ordinary methods.

So the Thin Realist's response to Blanchette's concern is that the Thin Realist *can* assess and defend her ordinary methods. Meanwhile, the Robust Realist remains stymied by Benacerraf's challenge.

14.4 Blanchette's Alternative

This brings us, at last, to Blanchette's preferred alternative, her 'amended Second Philosophy':

> instead of inhabiting empirical science while looking over the fence at mathematics as a science whose entire foundation can be examined from the outside, the [amended] Second Philosopher should take the whole empirical-mathematical scientific enterprise as her home domain. She will then be a realist about numbers and sets in just the way she is a realist about atoms. (p. 190)

Here the Amended Second Philosopher begins her inquiry not just with perception and deliberate observation, but also with some mathematical tools: 'calculation and proof, together with the rest of the rich array of methods employed when we pursue the goals of mathematics' (p. 190).[15] Some of these (logic, elementary arithmetic) the ordinary Second Philosopher finds ready to hand as well, but peculiarly mathematical methods (beginning with the completed infinite of Peano Arithmetic, and extending into analysis and higher set theory) she will, at least initially, view as a new type of undertaking, 'over the fence', so to speak. (Any human practice can be viewed and investigated in this way, from an empirical perspective, including science itself—that's scientific methodology.) In contrast, the Amended Second Philosopher begins with some explicitly non-empirical methods that deliver evidence about mathematical abstracta.

With apologies, I feel I must pose an annoying type of question that's often been directed at me: should we also include from the start a kind of rational intuition that delivers information about concepts or propositions or whatever? What about a capacity for revelation of theological truths? A psychic power that detects auras?[16] If we allow one non-empirical method, how are we to disallow others? My own answer has always been straightforward: mathematics grew out of and is intertwined with natural science in a fundamental way that these other practices are not.[17] This isn't a Quinean indispensability argument—empirical methods aren't delivering ontological information about mathematics—but it is a reason for the Thin Realist to regard mathematics as a body of truths and to 'jump the fence', to participate directly herself, rather than merely investigating.[18]

This appeal to the role of mathematics in natural science isn't one Blanchette wants to make; she wants mathematics to be accepted from the get-go, not as a

---

[15] For simplicity, I'm assuming the Amended Second Philosopher is still to be understood as 'bottom-up' in the terminology of Maddy (2011), pp. 105–111, though it might be equally natural to see her as 'top-down'. I argue in those pages that this different is irrelevant to the issues that concern us here.

[16] Often, the example is astrology—'what about astrological naturalism?'—but the Second Philosopher has a straightforward answer in that case: the predictions of astrology aren't empirically confirmed (Maddy, 2007, pp. 108–109).

[17] See, e.g., Maddy (1997), pp. 203–205, Maddy (2007), pp. 345–347, Maddy (2011), p. 101.

[18] The Arealist also jumps the fence, but without taking 'true' with him. In Maddy (1997), p. 199, I call this 'going native'.

later consequence of empirical developments. Though she may have some other response to the problem of proliferation, for now, it seems to me that the Amended Second Philosopher stands at some disadvantage compared to my Thin Realist. But philosophical dialectics aside, would we regard mathematics with the same kind of respect if it weren't linked with empirical science in this way? What would such a 'mathematics' even look like?

On the other side of the ledger, the purported advantage of Amended Second Philosophy is epistemological. Blanchette frames this in terms of the radical skeptical challenge to our beliefs about depth, not the connection between methods and sets:

> the skeptic who insists on the importance of this gap [between actually tracking the contours of mathematical depth and merely thinking one is doing so] can agree with the Thin Realist that the only thing needed to justify the positing of mathematical objects is a justified claim to have discovered rich veins of mathematical depth, while insisting, entirely coherently, that there is no guarantee that our mathematical methods track the contours of mathematical depth. (p. 182)

But Blanchette can't really be challenging the Thin Realist to rebut this skeptical challenge because she grants that it runs parallel to radical skepticism about our knowledge of the external world and that no direct refutation is possible.

What is possible is the Second Philosopher's observation that her lack of extraordinary evidence doesn't undermine her ordinary evidence. For the Thin Realist, the ordinary evidence is the sort of thing appealed to above in the comparison with Robust Realism, so I'd disagree with Blanchette's claim that the Thin Realist 'take[s] it for granted that our methods of proof, of calculation, of pursuit of purely mathematical goals in general, are ways to discover objective facts about mathematical depth' (p. 187). For that matter, I'd disagree that the Second Philosopher 'takes it for granted that the way to find out about the external world is via experiment and observation' (ibid.). In both cases, we have ordinary tools to assess, and if necessary, to reevaluate and improve our methods. This, I take it, is what Blanchette is asking for: not a reply to an Evil-Demon-style challenge, but ordinary reasons to think we're successfully tracking depth. These I think we have.

Of course, Blanchette doesn't see it that way. As far as I can tell, what's troubling her is the fact that my Second Philosopher is looking at mathematics 'over the fence', that its methods 'are not part of her bedrock toolkit for discovering objective facts' (pp. 188, 190). On its face, the second observation doesn't strike me as an obstacle: many perfectly reliable methods aren't present from the start; it took patient empirical investigation to understand the evidentiary force of double-blind trials, hadron collider images, redshifts. On my telling, the Thin Realist, building on her expanding empirical investigation, gradually develops new methods of investigation that reveal the existence of mathematical abstracta, much as she develops sophisticated new methods in other areas of her inquiry.[19] That these new methods weren't present among her 'bedrock tools' doesn't make them any less reliable.

So I think Blanchette's real concern traces to that 'over the fence'—if the Second Philosopher is just examining mathematics from the outside, as she does with other

[19] See, e.g., Maddy (2011), §III.1.

human activities, from poetry to tiddlywinks, the methods she identifies never become her own, and for that reason, she isn't in a position to appeal to them herself. I agree that this would be a problem, but as should be clear by now, I understand the Thin Realist to have enthusiastically jumped that fence. She does mathematics herself and regards it as a pursuit of truths about sets, which are, given the nature of sets, truths about mathematical depth.

In the end, I think the Thin Realist has an answer to the epistemological challenge concerning mathematical depth—not an answer to the corresponding radical skepticism but a rich array of ordinary methods as she introduces, assesses, amends, and improves in appropriate ways. What does the Amended Second Philosopher say when asked to assess and defend her mathematical methods? Ironically, it seems she's the one who simply takes her methods for granted:

> The suggestion I've tried to press here is that the [Second Philosopher], with the meager supply of methods that she takes for granted when investigating the world ... does not have enough to go on. I will suggest, in what follows, that what's needed is a slightly-richer collection of bedrock methods, a collection that includes fundamental mathematical methodology as a means of discovering objective facts. (p. 189)

So it isn't clear to me that Blanchette's amendment has bought us any epistemic gain.

To sum up, it seems that Blanchette's understanding of truth rules out Arealism, but I don't see why she shouldn't opt for Thin Realism. For myself, I stick with the further thought that Thin Realism and Arealism are equally acceptable ways of describing the same phenomenon—and thank Professor Blanchette for her invigorating discussion![20]

References

Maddy, P. (1997). *Naturalism in mathematics*. Oxford University Press.
Maddy, P. (2007). *Second philosophy*. Oxford University Press.
Maddy, P. (2008). How applied mathematics became pure. *Review of Symbolic Logic, 1*, 16–41.
Maddy, P. (2011). *Defending the axioms*. Oxford University Press.
Maddy, P. (2014c). A second philosophy of logic. In P. Rush (Ed.), *The metaphysics of logic* (pp. 93–108). Cambridge University Press Reprinted in Maddy [2022].
Maddy, P. (2017a). *What do philosophers do?* Oxford University Press.
Maddy, P. (2018b). Psychology and the a priori sciences. In S. Bangu (Ed.), *Naturalizing logico-mathematical knowledge: Approaches from psychology and cognitive science* (pp. 15–29). Routledge (To be reprinted in [2024h]).
Maddy, P. (2022f). On the question of realism. In [2022h] (pp. 49–90).

[20] To repeat, I'm grateful to Professor Blanchette for a penetrating and productive exchange through more than one draft one each side. [Added later: From private correspondence (cited with permission), I can report that Blanchette and I now agree on where we disagree: on the coherence of the Arealist's view of truth (an issue set aside here) and on the epistemology of mathematical depth. She writes, 'while I think (of course!) that there's more to be said in defense of my preferred stance, I think that the issues now come out really nicely. (And I have to admit that I think the 'proliferation problem' is serious.)']

Maddy, P. (2022g). A note on truth and reference. In [2022h] (pp. 173–187).
Maddy, P. (2022h). *A plea for natural philosophy*. Oxford University Press
Resnick, M. (1997). *Mathematics as a science of patterns*. Oxford University Press.
Schatz, J. (2024). Arealism, thin realism, and the problem of extrinsic evidence In N. Barton, C. Antos, & G. Venturi (Eds.), *Palgrave companion to the philosophy of set theory*.

Chapter 15
Meditations on Second Philosophy: Anti-Nominalist Reflections on Maddy's Semi-Nominalism

John P. Burgess

Abstract Maddy's successive views on mathematical existence are reviewed, with emphasis on her final view, characterizable as semi-nominalist. This is critically examined and evaluated from a firmly anti-nominalist (though not in any serious sense "Platonist") standpoint. As might be expected from this characterization, the evaluation is mixed, with positive and with negative aspects, all disagreement (and there is enough of it) put forward against a background of massive agreement.

Keywords Second philosophy · Semi-nominalism

15.1 Introduction

Over the years, Penelope Maddy's philosophical writing has given as much attention to questions of mathematical ontology as to any other single subject. Constant throughout has been a concern with the bearing, or as the case may be, the lack of bearing, of this or that ontological view on the project of seeking new set-theoretic axioms to settle questions left open by the currently accepted ones. Constant also has been a commitment to naturalism, first in a broad, general sense, later in a narrower, more specific form as she developed what she calls *second philosophy*. Against this constant background there have been substantial changes, since she has always been ready to give up a thesis if, having given it a run for its money, she in the end finds it unsatisfactory.

All told she has given sympathetic attention to at least four ontological positions. If one compares the question of the existence of sets to that of the existence of a deity, the four doctrines might be roughly likened to *theism*, *deism*, *atheism*, and *agnosticism*. To the first three she gives names: *robust realism* and *thin realism* and *arealism*. The fourth position is that, though the first view is definitely wrong, there is nothing to choose between the second and the third, which have an equal bearing or

J. P. Burgess (✉)
Department of Philosophy, Princeton University, Princeton, NJ 08544-1006, USA
e-mail: jburgess@princeton.edu

S. Arbeiter and J. Kennedy (eds.), *The Philosophy of Penelope Maddy*,
Outstanding Contributions to Logic 31, https://doi.org/10.1007/978-3-031-58425-1_15

lack of bearing on the new axioms project, and an equal claim to acceptability from her kind of naturalist standpoint. The first and second views are plainly varieties of *anti-nominalism* and the third a variety of *nominalism*, and so the fourth view, which Maddy leaves nameless, I will call *semi-nominalism*.

The first, robust or thick realist view, the one the semi-nominalist takes to be definitely wrong, was defended by Maddy in her Princeton dissertation and her first book (1990). It was developed under the influence of the well-known passage where Gödel (1947) writes of our having something like a perception of the objects of set theory. The commitment to naturalism at this stage amounted to an attempt to develop this thought in a way fully in accord with cognitive science and without any trace of parapsychology. Gödel's discussion famously suggests how intrinsic or extrinsic grounds might be found for adopting new axioms, especially large cardinal axioms. Maddy undertook her own examination (1988) of many post-Gödelian heuristic arguments in support of such axioms, but the ultimate outcome was her abandonment of her original ontological position, partly on account of what she saw as a mismatch between the character of these arguments and the kind of realist metaphysics she had been endorsing.

Her change of view was also closely connected with the direction the development of her naturalism took through her second book (1997), her comparative study (2007a), and her third book (2007b), as she articulated second philosophy. All four ontological views are discussed in her well-known paper on mathematical existence (Maddy, 2005), and further elaborated in her fourth book (Maddy, 2011). Between the paper and the book there is a change of emphasis in the formulation of the fourth view. The claim that it is not merely the case that thin or anemic realism on one side and arealism on the other are equally defensible, but further that the difference between them is merely verbal, comes much more to the fore in the later treatment, which for present purposes I will take to represent Maddy's final position. One might expect a philosopher who has arrived at such a position to conclude that, though much has been learned through pursuing the ontological question, it is time to deposit that debate in the archives and turn to other matters; and turn to other matters is indeed what Maddy has largely done in her later books and any number of her later papers.

15.2 Routleyism

I myself have also come to believe time should be called on the debate over nominalism, though my reasons are rather different from any considerations put forward by Maddy. Nominalism seems to me to have degenerated sadly since the period when Charles Chihara, Hartry Field, Geoffrey Hellman, and others came forward with ambitious programs of reconstrual or reconstruction of mathematics or mathematically-formulated science, with technical aspects of interest to the logician. These are what drew me as into the on-going debates, to address Chihara and Field in (Burgess, 1983) and, with my colleague Gideon Rosen as coauthor, Hellman as well in (Burgess & Rosen, 1997). Nominalism's present-day advocates, in contrast to those just mentioned, seem mostly to embrace cheaper and easier, quicker and dirtier routes to the conclusion that numbers, functions, sets, and such do not exist. Nowadays one may encounter a lazy instrumentalism of a kind earlier nominalists scorned, one that speaks as if mathematics were true while doing science, speaks as if mathematics were false while doing philosophy, and makes no real effort to explain or excuse the flip-flop.

And one often encounters claims to discern a distinction in meaning between a pair of synonymous expressions, or what many of us would take to be such, followed by exploitation of the supposed distinction to justify a kind of doublespeak that makes ontologically liberal-sounding noises out of one side of the mouth, and ontologically conservative-sounding noises out of the other. The most egregious example of this latter tendency, the cheapest and easiest, the quickest and dirtiest of all, dates from the last century and Richard Routley's *noneism*, which Lewis (1990) remarked might with equal justice have been called *allism*. Speaking with a forked tongue, Routley says *yes, there are numbers and functions and sets*, or something of the sort, but also *no, there do not exist any numbers or functions or sets.*

It is a peculiarity of Routley's view that he makes no attempt to locate a sense of being or existence that would apply to the Cantor set but not the Russell set, or to the equilateral triangle but not the round square. This peculiarity, usually called *Meinongianism*, will not concern me here, but rather the general feature of seeing a distinction in meaning between "there exists" and some other expression that in ordinary usage is synonymous with it, or viewed by many of us as being so, which contrasting expression is claimed to mean something less, or weaker. If I call a view *Routleyist*, it is this last feature I will be attributing to it.

One form of Routleyism sees a distinction, not between the existence and the being of abstract mathematical objects, but between their actual and their possible existence. Now we *do* sometimes speak of possible existence in mathematics. We may say that it is possible that there is a largest twin prime and possible that there isn't. But the possibilities we speak of in such cases are merely *epistemic* possibilities, ways that *for all we know* things may actually be. The grounds for the assertion are that it has been proved neither that there is no largest twin prime, nor that there is one. The assertion is not that there are two alternate mathematical realities, in one of which the sequence of twin primes comes to an end, and in the other of which it doesn't. The kind of more-than-epistemic possibilities involved in an alternate realities claim, possibilities understood as (the way things are plus all the) ways things actually aren't but potentially could have been, is called *dynamic* by linguists and *metaphysical* by philosophers. It is what is involved in Routleyist modal nominalism.

The same kind of modality may be said to have been involved in the projects of Chihara and Hellman, but for them the distinction between actual and possible existence was applied only to concrete objects, not abstract ones. The project was to replace standard assumptions about the actual existence of mathematical objects with suitable assumptions about the possible existence of physical objects. Such a project required considerable effort, while the Routleyist modal nominalist project is effortless. One simply replaces standard assumptions about the actual existence of mathematical objects with the corresponding assumptions about the possible existence of those very same mathematical objects. I myself do not find in our ordinary usage any serious, sustained invocation of any non-epistemic kind of modal distinction in application to purely mathematical facts about purely mathematical objects; and finding the distinction without use, I find it without meaning.

Lewis's theme in his discussion of Routley is that productive debate over noneism or allism is virtually impossible, and I would say the same about later Routleyist views, and about instrumentalist flip-flopping, and about the quasi-Routleyist claim to detect a difference in meaning between *existential consequence* and W. V. Quine's pretentiously inflated synonymous coinage *ontological commitment*. For discussion of claims of this kind see Rosen (2006). In conversation Rosen has gone so far as to characterize them as "muddying waters previously clear." I now find the waters so muddy I do not want to step into them. But before pulling back from the water's edge, I want to put on record what are for me the most important points on which I continue to differ from Maddy. I hope I do not need to say that such differences occur against a background of massive agreement.

15.3 Naturalism and Arealism

Maddy contrasts her form of naturalism with others, and perhaps to begin with I should say something about the contrast between the flavor of naturalism involved in Maddy's second philosophy and that involved in my joint work with Rosen, from which she quotes (Maddy, 2011, p. 39n) the following snippet (Burgess & Rosen, 1997, p. 69):

> The naturalists' commitment is at most to the comparatively modest proposition that when science speaks with a firm and unified voice, the philosopher is either obliged to accept its conclusions or to offer what are recognizably scientific reasons for resisting them.

Maddy says, by contrast, of her second philosopher:

> My inquirer doesn't decide to place her faith in something called 'science'; she is simply one of those speaking with a firm voice, on the basis of the evidence.

There need be no mystery about what this "thing called 'science'" is: For present purposes we may take it to be that which is done university science departments, or rather, in university *natural* science departments. For we may set aside the so-called *social* sciences, which perhaps don't very often speak with a firm and unified voice, while we must set aside the so-called mathematical sciences, since counting them as among the sciences whose firm and unified voice the naturalist is not to gainsay would be considered question-begging by nominalists. The shift in Maddy's conception of naturalism from the mathematical naturalism of her second book to the second philosophy in her latest work, with the concurrent drift in the direction of nominalism, was in substantial part a matter of decreasing willingness to count mathematics as a member in good standing of the circle of sciences with the same rights as physics and the other members.

For me—I believe Rosen and I still after nearly two decades agree, but I won't presume to speak for him—a naturalist will, where there exists a stable scientific consensus, generally fall in with or acquiesce in it, unless in possession of some special information or expertise. In the discussion of nominalism, what needs emphasis is the passive or negative side of this acquiescence, non-dissent, the renunciation of any claim to have some higher philosophical standpoint from which to supply a critique of science. The active or positive side, assent, comes to fore in other contexts. For instance, in deciding whether to have their children vaccinated, naturalists of my kind would let themselves be guided by conventional science-based medicine rather than alternative who-knows-what-based medicine.

Bierce (1993, originally published 1911) defines "faith" as "belief without evidence in what is told by one who speaks without knowledge, of things without parallel." Is my naturalist's reliance and trust in conventional medicine "faith" in some such pejorative sense? Well, in renouncing the idea of an external philosophical critique of science one is also renouncing the idea of an external philosophical foundation for science. But this renunciation does not in itself amount to holding that there is *nothing* that can be said in favor of science and that we must simply make a leap of faith and accept it. There is a good deal that can be said in favor of science, but it is the sort of thing said by popularizers of science and debunkers of pseudoscience, not something distinctively philosophical.

By contrast, the passage quoted above is one of several that tend to make Maddy's second philosopher appear to be a kind of Wonder Woman, who never needs to rely on or trust in the expertise of others, and therefore never needs to decide which others to rely on and trust in, and so never needs to consider whether the possession of institutional scientific credentials should be taken into account in making such decisions about reliance and trust. Rather, the second philosopher has the energy and resources to investigate every issue for herself, and form her own opinion based on her own evaluation of the evidence, not anyone else's (except that at one point, faced with a question about chimpanzee nutrition, the second philosopher does seek outside advice; but why she on this occasion has faith in the opinion of a scientific botanist rather than a New Age herbalist we are not told). Maddy does, however, seem to expect the second philosopher's opinions will largely recapitulate those of actually existing natural science. The second philosopher presumably knows and feels qualified to judge the technical evidence, clinical and epidemiological, bearing on the safety and effectiveness of vaccines against childhood infectious diseases; whereas my kind of naturalist may well form an opinion based on pamphlets for the lay public from the usual professional medical sources, supplemented perhaps with journalistic accounts of recent controversies by debunkers of anti-vaxxer conspiracy theories.

Such differences in flavor of naturalism, however, are perhaps not the main source of differences on ontology. Maddy arrives at semi-nominalism by judging that, though the second philosopher is not required to reject mathematical existence claims, such rejection is fully compatible with her kind of naturalism. My kind of naturalism, with its acceptance of whatever stable consensus is to be found in physics, I believe to be, by contrast, incompatible with nominalism, owing to the heavy involvement of physics with mathematics. And the difference here between claims of incompatibility and claims of compatibility is less a matter of differences in flavors of naturalism, than of differences in conception of the role of mathematics in science, especially physics.

Indeed, I can largely confine my discussion of differences with Maddy to examination of one short passage where I find the crux of the issue between us to lie, a passage in which Maddy enunciates her arealist's understanding of the role of mathematics in empirical science. It consists of just two consecutive sentences from her discussion of arealism in the fourth chapter of her fourth book. Concerning what her second philosopher finds when she examines the practice of the application of mathematics in science, Maddy focuses on the role of mathematics as a source of models, and writes as follows:

(1) In all this, the scientist never asserts the existence of the abstract model; he merely holds that the world is like the model in some respects, not in others.
(2) For this, the model need only be well-described, just as one might illuminate a given social situation by comparing it to an imaginary … one, marking the similarities and dissimilarities.

(Maddy, 2011, p. 90, minor typos corrected)

These are the views I especially want to examine. But first a matter of terminology. Gauss remarks somewhere that a lot of confusion would have been avoided if instead of saying *positive, negative, imaginary* mathematicians had said *forward, backward, sideways*; but despite his remark, *real* versus *imaginary* numbers remains the standard terminology. To avoid a clash with this technical mathematical usage I will, therefore, when discussing such claims as (1) and (2), write of the *fictitious* where Maddy writes of the *imaginary*, and use *genuine* rather than *real* as the contrasting term.

15.4 Mathematical Existence Theorems in Physics

The epistemological question nominalists so often press is that of how anyone could arrive at a justified belief that there are abstract mathematical objects. If physicists believe in the existence of abstract mathematical models, then there is a very easy answer to this question for those who accept scientific standards of justification and do not suppose these can be trumped by some higher philosophical standards. The answer is this: Just look at how, historically, physicists arrived where they are today, and *that* is how one can justifiably arrive at belief that are such abstract objects as mathematical models. Hence the significance of the question whether physicists do in fact believe in the existence of mathematical models, or whether on the contrary, as (1) suggests, they don't. Is (1) factually correct or not?

In a perceptive early review of Field's project by a distinguished philosopher of physics, various mathematical results important in physics that Field's approach fails to cover are enumerated, and these notably include "propositions that establish the existence of models with special features" (Malament, 1982, p. 528, item B). We seem to have here an assertion from one who can be presumed to know his business that physicists do discuss, and sometimes assert, the existence of models. But the remark just quoted is not, of course, addressed directly to (1), which first appeared in print more than two decades later. So though taking a hint from the review cited, I decided that rather than rely on the authority of the reviewer in addressing the status of (1), I should proceed more directly, and see if I could find quotations from physicists asserting the existence of models.

Now as Maddy brings out in a concise survey of the history of relations between physics and mathematics in the first chapter of the book under discussion, the paradigmatic cases of mathematical modeling of physical situations involve the framing of differential equations. The equations are sometimes spoken of as if they themselves were the models, but properly speaking it is *solutions* to the equations that are the models, where a single system of ordinary or partial differential equations may have many solutions with different special features, all of physical interest. The solutions to differential equations are, of course, functions, real- or complex-valued, of one or of several variables, and as such paradigmatic cases of abstract mathematical objects, in modern set-theoretic mathematics identified with certain sets. So our question becomes: do physicists assert the existence of solutions to differential equations?

It was solutions to the Klein-Gordon equation $\Box\psi = 0$ that were at issue in the review cited earlier, but I expected it would be easier to find quotations from physicists dealing with the field equations of general relativity. I expected it would be easy to find physicists asserting that Gödel (1949) proved that there exist, or that there are, solutions with closed time-like curves, or at least that Gödel found or discovered such solutions, which implies there are such solutions, since one cannot find or discover what isn't there. But while I had no trouble accumulating examples of assertions of this kind, in every case I located the writer was one whom a captious nominalist might accuse of being a mere philosopher or mere logician or mere mathematician, and not a genuine physicist; or if a genuine physicist, not a genuine physicist engaged in genuine physics, but rather in some kind of popularizing, or worse, philosophizing.

Setting the specific Gödel example aside, however, I did quickly locate the sort of thing I was looking for. Princeton professor R. H. Dicke (1916–1997) was without cavil a physicist: an experimentalist and inventor of experimental apparatus as well as a theorist, and an astrophysicist and cosmologist as well as a gravitational and an atomic physicist. He was a physicist of high standing to boot, winner of the National Medal of Science among other awards. The *Astrophysical Journal* (ApJ) is unequivocally a technical scientific periodical publishing the results of technical scientific research, not the musings of professional scientists pursuing a philosophical hobby after work or after retirement. And I quickly found Dicke, in an ApJ paper on the cosmic background radiation, writing as follows: "From the above discussion it is evident that there are solutions to the field equations for which..." where then follows a technical enunciation of certain special features that need not detain us (Dicke, 1968, p. 12). Googling will turn up any number of further examples. And so, I submit, if Maddy's arealist hasn't found scientists asserting the existence of abstract mathematical models, she hasn't read widely enough.

It may be suggested at this point that though a physicist may *assert* that solution exist, he doesn't really *believe* it. After all, Maddy has remarked that "the attitudes of scientists toward their best theories are complex and nuanced: some posits are regarded as useful fictions, some aspects are suspected to be artifacts of the modeling, and explicit idealizations are freely employed" (Maddy, 2005, p. 363). And this is very true if it is taken to mean that when a mathematical model of a physical situation is proposed, the physicist seldom if ever supposes that every item in the mathematical model corresponds to some item in the physical situation, or that those that do correspond do so exactly. But it is one thing to say that nothing physical corresponds exactly to such-and-such items in the mathematical model, and quite another to say that the mathematical model itself does not exist, and that whatever mathematical existence theorems say it does are falsehoods, and that whatever mathematicians believe such theorems are deluded. Do physicists ever, in the course of doing physics, say anything like the latter? If so, let the arealist produce a quotation to counter the one I have produced from Dicke.

It may be suggested instead that though physicists *do* assert and believe that functions constituting solutions to certain differential equations exist, they may, not having taken to heart the lessons of Frege (especially 1904), suppose the functions to be mathematical formulas, and as such not set-theoretic but linguistic entities. This does not make very much difference to the issue of nominalism if linguistic entities are just another species of abstract object beyond numbers, functions, and sets. But it may be further suggested that physicists suppose linguistic expressions to be, not abstract types, but concrete tokens, physical items made of chalk or the like. Solutions to the field equations would be bits of calcium carbonate.

To be sure, physicists say many things that contradict such an understanding: for instance, that two researchers on opposite sides of the world have produced the *same* solution to certain equations, when they certainly have not produced the same chalk marks on the same blackboard. Do physicists simply hold contradictory beliefs here? One cannot refute the suggestion that they do by eliciting explicit statements from physicists through questioning, since whatever they say when questioned may be dismissed on the grounds that when addressing philosophers' questions they are not doing physics but philosophy. There may be no finally refuting such a suggestion at all, but I do not believe for a moment that this is what Maddy had in mind with the formulation (1), or that she would try to defend her formulation with such desperate suggestions. As for myself, I would modify or amplify my characterization of naturalism by making it clear that the naturalist is not limited to parroting things said by scientists, but may also draw out consequences that obviously follow if their statements are accepted, and presuppositions that must be accepted if the statements are to make coherent sense.

15.5 Fictions in Social Science

All this, however, may be considered but a preliminary skirmish. For even if scientists do as a matter of actual fact assert and not deny the existence of abstract mathematical models, contrary to (1), still if (2) is correct they *need* not do so, and that may well be enough for arealists' nominalistic purposes, at least as some of them view matters. Here a difference in flavor of naturalism may be important, since for my kind of naturalist what matters is how the scientific community actually does things, not how it potentially could have done them. The claim is that our way of doing things is all right, not that it is uniquely right, that doing things any other way would not have been all right.

Maddy and I both reject so-called indispensability arguments, but for different reasons. For Maddy it is, to put it very briefly, because indispensability arguments portray mathematics as more like or of a piece with physics than it is. For me, it is because indispensability arguments concede too much to nominalism. For me, the form of skepticism known as nominalism is to be repudiated if it turns out, as I have claimed in the preceding section it does, that the assumption of the existence of mathematical models is customary in our way of doing science; it need not be shown to be indispensably necessary.

Be all that as it may, (2) with its comparison between social and mathematical situations merits separate and close consideration. Literary writers present us with many examples not merely of fictitious social situations but even of whole fictitious societies, Swift's Lilliput and Brobdingnag being perhaps the best-known. Let me begin with the role of these fantasy lands in social thought before returning to mathematics and its role in physics. Three points stand out.

First, many issues about our society can indeed be illuminated by comparison with another society whether that other society is a genuine or a fictitious one. This may include virtually all theoretical or descriptive questions. For instance, our tendency to hold people responsible for things beyond their control in some cases but not in others, a tendency that is the topic of a voluminous philosophical literature on so-called *moral luck*, can be thrown into relief by comparison with Butler's *Erewhon* (2002, originally published 1872), an account of a fictitious society where people are blamed and even imprisoned for falling ill, with no pleas for leniency allowed on grounds of having been brought up in an impoverished and insalubrious environment, whereas those found committing what among us would be considered serious felonies, including the host of the narrator in the novel, who perpetrates a major embezzlement, are not on that account thought badly of, though they do have to undergo treatment at the hands of so-called *straighteners* that may be almost as painful as some of the therapies of Victorian medicine.

But second, many comparisons between our society and another are illuminating only if the other society is a genuine and not a fictitious one. This may include virtually all practical or prescriptive questions. It is one thing to argue that it would be feasible to improve some of our practices regarding some matter or other on the grounds that they order the matter better in France, and quite another to argue that it would be feasible to improve those practices on the grounds that they order the matter better in Utopia. France genuinely exists and it might be feasible to imitate here what is done there. The same cannot be said of Utopia. When discussing the prospects for prison reform in the United States, comparison with France or other developed countries, which have drastically lower rates of incarceration, is appropriate. Comparison with Erewhon, where one can be jailed only for disease, is not.

And third, the question whether a given description is of a fictitious or a genuine society has always a clear sense. Consider, for instance, Margaret Mead's famous account of adolescence in the South Seas (Mead, 1928), which became the subject of an academic controversy in the 1980s, a few years after Mead's death. Was the society depicted in her classic book that of the genuine Samoa of a hundred years or so ago, or was it, as Freeman (1983) alleged, that of a fictitious Samoa described to a gullible twenty-something anthropology doctoral candidate by mischievous teen informants who were hoaxing her? The answer to the question is relevant to whether Mead's Samoa shows how we might hope to make the teenage years in our own society less a time of storm and stress than it generally is. My impression is that her profession has generally sided with Mead on this contentious question, but we who are not up on the iterature may not be in a position to give a confident opinion. For all that, we understand perfectly well what is at issue.

15.6 Fictions in Mathematics

When we turn from social to mathematical models, we must note that literature provides far fewer examples of fictitious mathematical objects or structures than of fictitious human societies, though it does provide some. One may for instance cite, in contrast to the genuine *surreal* numbers of the genuine mathematician John Conway (see Knuth, 1974), the fictitious *supernatural* numbers of fictional character Noam Himmel (see Goldstein, 1983). Or if the interest is not in fictitious species of numbers but fictitious members of genuine species, one may contrast the least counterexample to Merten's conjecture, a conjecture genuinely proved to have counterexamples by Odlyzko and te Riele, with the least counterexample to Fermat's theorem, a theorem genuinely proved to have no counterexamples by Andrew Wiles. The Fermat example figures in a blog post of a well-known mathematical gadfly (Zeilberger, 2012) that apparently took in some readers who did not notice that the date of the post was April first.

If examples of fictional mathematics are few, straightforward examples of the application of modeling by fictitious mathematical objects or structures are fewer. The supernaturals will never be useful as models since all the novel where they appear tells us is that a character's reputation as a genius was established by his introducing them, not what they are like. The April fools' day blog post makes a serious point about a *social* practice, contemporary editorial policies at mathematical journals, by invocation of a fictitious *social* situation, the embarrassing computational discovery of an error in what had been accepted as a major and prize-worthy conceptual proof.

Other fictions about mathematics, from Swift's description of Laputa to the well-known play *Proof* (Auburn, 2001) are also illuminating about various social and psychological issues; but that is not the kind of modeling we are now looking for.

So let us keep looking. There are, as is only too well known, many cases where lapses in rigor by physicists working on a mathematical model have the result that the physicists end up describing something mathematical that not only does not but cannot exist. And I suppose any such case can be described as a case of the physicist using a fictitious model. The most famous example, always cited in this connection, is Dirac's use of a supposed function δ that takes the value 0 everywhere but at 0, yet has an integral of 1 over any interval containing 0. Mathematics recognizes no such "δ-function," the features ascribed to it being inconsistent with each other. But Dirac's rigor was fairly promptly patched up by Laurent Schwartz. If this example is accepted, then eighteenth century mathematical physics provides no end of further examples. One may well wonder, however, whether in these cases the models can be properly called "well described" until the rigor has been cleaned up.

Prospects may be better for the future application of another kind of fictitious number. In very weak formal systems of arithmetic, where the principle of mathematical induction is not unrestrictedly assumed, it may not be provable that 10^x exists for every x, even though the conditional is provable that tells us that 10^{y+1} exists whenever 10^y exists. By a hundred applications of the conditional one can prove the existence of 10^{100} or *googol*, but it would take googol applications to prove the existence of 10^{googol}, or *googolplex*, and a proof of this size would not be feasible to write down, googol probably exceeding the number of elementary particles in the visible universe. If therefore we add to our weak arithmetic the axiom that googolplex does *not* exist, we obtain a system that is inconsistent but for which it is not feasible to exhibit an explicit derivation of a contradiction: there can be no genuine species of number satisfying the theory, though it is not feasible to prove this fact directly. There have been hints in the writings of mathematicians and logicians and philosophers from Alexander Yessenin-Volpin on (see the last paragraphs of Nelson, 1986) that this sort of fictitious species of number may have applications.

The reader may at this point be impatiently muttering that the sort of cases I have been discussing are surely not what (2) is pointing towards, and this is quite correct: (2) seems to point to something cheaper and easier, quicker and dirtier, than exploitation of what are fictitious mathematical models with our present understanding of *fictitious*. What is being contemplated seems to be discarding such understanding as we currently have, an understanding on which the supernaturals are fictitious but the surreals are not, and declaring *all* numbers to be fictions, and all functions and sets along with them.

The distinction between the surreals and the supernaturals would presumably even then still have to be marked in some way, but it would have to be a different way from labeling one genuine and the other fictitious. As it happens there is precedent for distinguishing among items all deemed fictitious the genuine fictitious from the fictitious fictitious. What are in effect several cases of this kind are discussed in (Hayaki, 2005). One striking instance, brought to my attention by Saul Kripke, is that of the many nouns in Biblical literature that have been understood to be the names of fictitious gods, which become the names of demons in later tradition down to Milton. Here there is a distinction between genuine fictitious gods, such as Chemosh or Dagon, whom archeology reveals to have been once the object of cults now long defunct, and cases like those of Belial in the Hebrew Bible or Mammon in the New Testament, both high-ranking fallen angels in *Paradise Lost*, where the understanding of the noun involved as a proper name is generally regarded by scholars as a confusion. The kind of nominalism being contemplated would demote the supernaturals to the status of Belial, in order to have room to demote the surreals to the status of Chemosh, and more generally to declare all mathematical models used in physics, such as solutions to the field equations of general relativity, to be as fictitious as Dagon, if not quite as fictitious as Mammon.

But here we see a very sharp difference between the sociological and the mathematical cases contemplated in (2). If someone claims Mead's Samoa is fictitious, we know how its being so would contrast with its being genuine; and if someone claims Mead's Samoa is genuine, we know how its being so would contrast with its being fictitious. But is there an equally meaningful distinction of the kind the arealist professes to see, which is to say, a distinction between the assumption that a model genuinely exists, which is supposed to be superfluous, and the assumption that it is a *well-described* fiction, which is supposed to suffice for purposes of applications?

As the reader will doubtless have guessed, I find no more meaning in this supposed distinction than in that between Routley's supposedly ontologically loaded existential quantifier and his supposedly ontologically innocent particular quantifier, or between the actual and the metaphysically possible existence of pure numbers, functions, and sets, or between existential consequence and ontological commitment. The arealist, who seemed to me to be misstating the facts in (1), seems to me to be making a distinction without a difference in (2).

Such, then, are my differences with the arealist, whom I judge to be little if at all better than a Routleyist. Turning from the nominalist character to her semi-nominalist creator, my differences with Maddy herself might be thought to be exactly half as large. For Maddy may be described as in a sense doubting the meaningfulness of the arealist's and the thin realist's assertions and counter-assertions, and I may be described as in a sense sharing her view of the former, though not of the latter.

15.7 Convention

But is this not, perhaps, overstating our differences? Am I not, in the very act of accusing the arealist of Routleyism, agreeing with the spirit, even if not the letter, of Maddy's claim that there is no more than a verbal difference between the nominalist's and the anti-nominalist's ways of describing our mathematical and scientific theorizing? Let me try to formulate the maximum concession I find myself able to make, so as to bring out what residual disagreement there may be.

Let us start a long way back from the issue of ultimate concern, right back at the first reification of numbers in the tradition leading to contemporary cosmopolitan mathematics. This world-historic occurrence may plausibly be dated to the transition from the use of numerals as adjectives, as in *six days*, to their use as nouns, as in *six is a perfect number*, which appears to have occurred sometime during the period between Pythagoras and Plato. It was attended, we are told by historians of logic, with a good deal of grammatical confusion (see Kneale & Kneale, 1962, p. 392), and as we know it set off some metaphysical fireworks, speculations about the Monad and the Dyad and so forth. Strands of what we would accept as number theory and strands of what we would reject as numerology were almost inextricably intertwined. Centuries later there was still a lot of this sort of thing going on, as when Augustine ponders whether six is a perfect number because God created the world in six days, or whether God created the world in six days because six is a perfect number.

If things had never progressed beyond this stage, I think we can all agree there would have been no reason to consider mathematics a science. A naturalistic philosopher of any flavor would have no reason to make truth claims for the subject. Were we to try to produce an idealized representation of theorizing at this stage using first-order logical formalization, such science as there was would be codified in a theory with variables ranging only over concrete, physical objects. Such mathematics as there was would be codified in a theory with variables ranging only over abstract, mathematical objects. No scientific applications of mathematics would link the two, and a naturalist philosopher might regard the one as true, or probably largely approximately true, and the other as mythology.

If the claim that assumptions about the existence of solutions to differential equations are components of many areas of modern physical theory is accepted, as I have argued it should be, then science and mathematics cannot be separated in this way. At least one branch of mathematics, the theory of ordinary and partial differential equations, must be held to be true if a large part of physics is to be held to be true. But what about the branch of mathematics, number theory, we were just considering? As it happens, number theory, even after the mists of number mysticism dissipated, remained without many important scientific applications down through the period between the world wars. Though applications to cryptography lurked just over the horizon, the celebrated number theorist and pacifist G. H. Hardy was able to write at the end of interwar period (1940) that he took satisfaction in the fact that *his* kind of mathematics had no military applications. If number theory were still like that, would a naturalist philosopher today have any reason to regard it as a body of truths?

I think the answer may be affirmative, owing the applicability of existence theorems for ODEs and PDEs, together with the way in which modern mathematics is organized. Modern mathematics is, for reasons carefully discussed by Maddy in a late paper (2017), organized today in a framework of axiomatic set theory, where the axioms suffice both for the rigorous reconstruction of older useful theorems, and for the continuing deduction of new ones, as well as for the deduction of less applicable results, number-theoretic and other. If utility is taken to justify directly ascription of truth to theorems on the existence of solutions to differential equations, it must be regarded as justifying indirectly ascription of truth to the axioms from which spring an on-flowing stream of such theorems. And if truth is ascribed to the axioms, it must be ascribed to their logical consequences, to anything rigorously deducible from them, and hence to all theorems, even useless ones.

Be that as it may, a formal representation of our actually existent science would have to take account of the fact that it speaks both of physical and of mathematical objects. But it remains the case that the kind of questions it makes sense to ask about numbers and functions and so on are very different from those it makes sense to ask about physical objects. To reuse a favorite example of mine, no cosmologist would consider the suggestion that it is numbers rather than neutrinos that are the bearers of the notorious "missing mass" a plausible or even an intelligible hypothesis. (Numbers' lack of mass is indeed one of the features that make them count as abstract, and make nominalists suspicious of them.) For this reason, were we seeking to model our mathematico-scientific theorizing using first-order formalization, it would be best to use a formalization in a *two-sorted* first-order theory, with one style of variable, say x, for concrete, physical objects, and a different sort, say ξ, for abstract, mathematical objects, rather than with a single style ranging over both. (This is indeed what is done in my joint work with Rosen.) For each predicate it would be specified which sort of variable is to go in each of its places. For instance, the two-place predicate _ *is the mass in grams of* _ would take a ξ in the first blank and an x in the second blank, because it makes sense to say a number is the mass (in specified units) of a body, but not that anything is the mass of a number, or that a body is the mass of anything.

But what of the quantifiers that go with two sorts of variables? These are governed by the same logical introduction and elimination rules. Should they also be represented by the same symbol? Or should we use two different symbols, perhaps $\exists$ to go with x and € to go with ξ? Well, because our custom in unformalized theorizing is to use the same word "exists" in both cases, use of the same symbol $\exists$ in the formalization would make the formal modeling of informal argumentation one point more faithful. By contrast, if we systematically used "exist" in one case, and another word, say "extrudes", in the other, and spoke, for instance, not of existence theorems but of extrusion theorems in the theory of differential equations, then the other notational option, distinguishing an existential quantifier $\exists$ from an extrusive quantifier €, would be one point more faithful instead.

My maximum concession to the semi-Routleyist would be to grant that if we had it all to do over again, our purposes might have been equally well served by the opposite to our actual usage, by our contrasting "exist" and "extrude" rather than making do with the single word "exists". One may say, if one wishes, that there is no *intrinsic* reason to prefer one option to the other. It is in this "could equally well have done otherwise" sense, the sense developed in Lewis's dissertation and endorsed by his thesis director Quine (see Lewis, 1969 and the foreword by Quine thereto), that our actual usage may be called *conventional*. Our actually existing natural science contains many conventional features present in our theorizing as the result of something like historical accident. (And that is why the second philosopher, if she really undertook to re-examine personally every scientific question ab initio, could *not* be expected to recapitulate exactly the course actually existing natural science has taken.).

But having conceded this much, I would insist that as a matter of actual fact we do *not* have it all to do over again: we are deep into on-going projects. And the fact that the "exists" rather than the "extrudes" usage has become our actual established usage is a powerful *extrinsic* reason to uphold it. Possession is nine-tenths of the law. Why make a linguistically disruptive switch if none of our purposes would be better served by doing so? The long and the short of it is that to call something conventional is not to criticize it. It was as the result of a conventional decision or historical accident that drivers in most of the world keep to the right. But to say so is not to say that we should switch over and drive on the left instead, or refrain from driving at all. There are reasons why less driving would be a good thing, but the conventionality of the choice of which side of the road to drive on is not among them.

More importantly, suppose we did use one quantifier for the concrete and another for the abstract, as we may use *someone* for persons, but *somewhere* for places. Even then our usage would not be Routleyist, since it would be lacking the central Routleyist feature of claiming that the expression used for the concrete is in some sense *stronger* than the the one used for the abstract. It is not, according to Routleyism, just that chairs and tables *exist* while number and functions *are*, as we may say battlefields *exist* but wars *occur*; rather the Routleyist claim is that concrete items not only *are* but also *exist*, while abstract items only *are*; or on a variant version, seats not only *possibly exist* but *actually exist*, whereas sets only *possibly exist*. If it is claimed that some items both *are* and *exist*, while others only *are*, how would things have looked different if the latter also existed, as the former do? This is a question to which I, as an anti-Routleyist, find no intelligible answer in Routleyist writings.

15.8 Phonemes and Genotypes

But the main difference between Maddy and myself I now take to lie in contrary attitudes towards conventionalism and conventional features of our theories. I simply don't see the recognition of a feature as conventional as warranting skepticism or agnosticism, as was perhaps already clear from my statement earlier of my reasons for rejecting indispensability arguments. The issue of conventionality at work here is further illustrated and illuminated by some other instances of appeal to abstract objects in scientific theorizing that I would now like to take up briefly before closing.

I have been writing up to this point as if the questions of the ontological status of numbers, of functions, and of sets were all of a piece, and though I do not think it affects the cogency of my discussion above, I must now acknowledge that Maddy has come to make distinctions here. In some of her later work (see Maddy, 2014) she has turned from her special preoccupation with set theory to consider number theory as a distinct subject, and arrived at conclusions a point or two less skeptical. What she has not, to my knowledge, addressed at any length, is the question of the existence of *non*mathematical abstracta.

I am thinking especially of what I would call *equivalence types* of physical objects or events, though a variety of other terms have also been used in connection with them. Their equivalence type is what items that are equivalent in some respect thereby have in common, as figures that are geometrically similar have in common their *shape* and patches that are chromatically matching have in common their *color*. Talk of shapes and colors and the like of course long antedates the development of any sort of advanced science, and precedes the introduction of such phrases as *geometrically similar* or *chromatically matching*, for which the ordinary expressions are indeed simply *of the same shape* and *of the same color*. Such equivalence types will be with Maddy's second philosopher from the very beginning of her inquiries. But there have also been some significant cases of the historically late recognition of certain species of equivalence types in science.

For instance, the late nineteenth and early twentieth century saw the development in linguistics of the concept of *phoneme*, equivalence types of physical events of a certain kind, short segments of utterances, with respect to an equivalence relation whose behavioral characterization is somewhat differently explained by different authorities, and need not concern us here. The predicates it makes sense to apply to equivalence types correspond to those predicates of the items they are the types of that are invariant or preserved under substitution of equivalents for equivalents. In the case of phonemes, given two utterance segments exemplifying the same phoneme, the one will be sibilant if and only if the other is, and so a notion of sibilance can be applied to phonemes. By contrast, one of the two utterance events may take place in Helsinki and the other in Honolulu, and so the phoneme cannot be assigned a spatial location. Given a background of modern set theory, equivalence types can be identified with equivalence classes, or sets of equivalents, and so raise no additional ontological issues; but this identification is not open to nominalists.

There is no question but that talk of equivalence types could be systematically avoided, at the cost of introducing a certain prolixity into our discourse, in favor of talk of the things that are equivalence types of. Equivalence types *could* be dispensed with, but for me at least, that circumstance, or that plus the abstractness of types, is no reason why they *should* be dispensed with. And in actually existing science they are *not* dispensed with. For whatever may be the case with physicists and solutions to differential equations, linguists do unquestionably assert the existence of phonemes. They tell us, for instance, that among modern European languages Greek has few more than twenty, Lithuanian few less than sixty. For the nominalist, by contrast, both have zero, since for the nominalist there are no such things as phonemes. Nor are there genotypes or a whole wide range of other equivalence types recognized in science, including linguistic expression types as contrasted with tokens.

The case of equivalence types is one that especially invites consideration from the perspective of the particular variety of conventionalism associated with the name of Rudolf Carnap (see Carnap, 1950). The theory of phonemes involves the introduction of a new way of speaking, what Carnap would call a *linguistic framework*. It should be admitted at once that there may very well be good reasons for resisting the innovation. One may object that the new notion is obscure because the relevant equivalence relation has not been sufficiently clearly characterized; and there is in fact a degree of such unclarity in the case of phonemes, which is one reason I was only able to cite approximate numbers for different languages just now. Or one may object that talking of phonemes will encourage unjustified assumptions about the existence of some neural correlate underlying the behavioral facts. This is something that may not have happened in the case of phonemes, but something like it did happen with unfortunate results in the case of the reification of a mathematical factor, the *g* that arises in factor analysis of IQ test results, as a *general intelligence* supposed to be a physical characteristic of, or even a fluid in, the brain (see Gould, 1981).

According to the Carnapian, however, one reason for resistance to phonemes that is *not* legitimate would be doubt about whether there really are such things as phonemes. For on a Carnapian account it is the cluster of basic assumptions of the theory that constitute the linguistic conventions that guide the use and so constitute the meaning of the word *phoneme*. If one does not accept those assumptions the word remains meaningless and the question whether phonemes exist cannot even be enunciated; while if one does accept those assumptions, they trivially imply an immediate affirmative answer to the existence question. In this sense the theoretical question whether phonemes really exist is in principle meaningless: Asking in the material mode whether phonemes really exist may in practice serve as a way of putting what, expressed in the formal mode, would be the meaningful practical question whether phoneme theory should be accepted; but it is a misleading way.

Such is the Carnapian view, which I share. And what makes the status of phonemes something of which I think Maddy owes us an account is the fact that one constant through her various changes of ontological view has been a rejection or ignoring of Carnapianism. I am curious as to how, then, her second philosopher would regard phonemes and phones in phonology, genotypes and phenotypes in genetics, and so on, and I hope Maddy may sooner or later find time and occasion to address such questions. I anticipate that I will not agree with what she has to say, but I am confident it will be interesting.

References

Auburn, D. (2001). *Proof: A play*. Faber and Faber.

Bierce, A. (1993). *The Devil's dictionary*. Dover Thrift Editions.

Burgess, J. P. (1983). Why I am not a nominalist. *Notre Dame Journal of Formal Logic, 24*, 93–105.

Burgess, J. P., & Rosen, G. (1997). *A subject with no object: Strategies for nominalistic reconstrual of mathematics*. Oxford University Press.

Butler, S. (2002). *Erewhon*. Dover Thrift Editions.

Carnap, R. (1950). Empiricism, semantics, and ontology. *Revue Internationale De Philosophie, 4*, 20–40.

Dicke, R. H. (1968). Scalar-tensor gravitation and the cosmic fireball. *The Astrophysical Journal, 152*, 1–24.

Freeman, D. (1983). *Margaret Mead and Samoa: The making and unmaking of an anthropological myth*. Harvard University Press.

Frege, G. (1904). "Was ist eine Funktion?" In *Festschrift Ludwig Boltzmann gewidmet zum sechzigsten Geburstage* (pp. 656–666) Leipzig: Meyer.

Gödel, K. (1947). What is Cantor's continuum problem? *The American Mathematical Monthly, 54*, 515–525.
Gödel, K. (1949). An example of a new type of cosmological solutions to Einstein's field equations of gravitation. *Review of Modern Physics, 21*, 447–450.
Goldstein, R. (1983). *The mind-body problem*. Random House.
Gould, S. J. (1981). *The mismeasure of man*. W. W. Norton.
Hardy, G. H. (1940). *A mathematician's apology*. Cambridge University Press.
Hayaki, R. (2005). Fictions within fictions. *Philosophical Studies, 146*, 379–398.
Kneale, W., & Kneale, M. (1962). *The development of logic*. Clarendon Press.
Knuth, D. E. (1974). *Surreal numbers*. Addison-Wesley.
Lewis, D. K. (1969). *Convention*. Harvard University Press.
Lewis, D. K. (1990). Noneism or allism? *Mind, 99*, 23–31.
Maddy, P. (1988). Believing the axioms. *Journal of Symbolic Logic, 53*(481–511), 736–764.
Maddy, P. (1990). *Realism in mathematics*. Oxford University Press.
Maddy, P. (1997). *Naturalism in mathematics*. Oxford University Press.
Maddy, P. (2005). Mathematical existence. *Bulletin of Symbolic Logic, 11*, 352–376.
Maddy, P. (2007a). Three forms of naturalism. In S. Shapiro (Ed.), *Oxford handbook of mathematics and logic* (pp. 437–459). Oxford University Press.
Maddy, P. (2007b). *Second philosophy*. Oxford University Press.
Maddy, P. (2011). *Defending the axioms*. Oxford University Press.
Maddy, P. (2014). A second philosophy of arithmetic. *Review of Symbolic Logic, 7*, 1–28.
Maddy, P. (2017). Set-theoretic foundations. In A. E. Caicedo, et al. (Eds.), *Foundations of mathematics: essays in honor of W. Hugh Woodin's 60th birthday* (pp. 289–322). providence: American Mathematical Society.
Malament, D. (1982). Review of Field, Hartry, science without numbers. *Journal of Philosophy, 79*, 523–534.
Mead, M. (1928). *Coming of age in Samoa*. William Morrow.
Nelson, E. (1986). *Predicative arithmetic*. Princeton University Press.
Rosen, G. (2006). Review of Azzouni, Jody, deflating existential consequence: A case for nominalism. *Journal of Philosophy, 103*, 312–318.
Zeilberger, D. (2012). Doron Zeilberger's 123rd Opinion. http://sites.math.rutgers.edu/~zeilberg/

Chapter 16
Reply to Burgess

Penelope Maddy

Keywords Naturalism · Second philosophy · Thin Realism/Arealism · Indispensability arguments · Nominalism · Carnap · Einstein

Professor Burgess takes direct aim at fundamentals: the ontology of mathematics, its subject matter. Though the rhetorical contrasts between his writings and mine are often stark, his piece eventually becomes a search for 'what residual disagreement there may be' (p. 212). I confess that I, too, find our disagreement elusive, but let me do what I can to locate it.

Though my Second Philosopher isn't, as Burgess jokes, a Wonder Woman, she is a highly idealized inquirer[1]: she begins with simple observation and straightforward generalization, then theory formation and confirmation, working her way methodically toward a complete theory of the world and our place in it.[2] This device is designed to acknowledge and sidestep the hopeless task of characterizing 'science' or 'scientific method' once and for all. As empirical inquiry evolves, its reach and methods expand—what was once thought impossible becomes possible—and none of this is predictable in advance.

Burgess feels no such constraint; he's happy to characterize 'science' as what's done in 'university *natural* science departments', where practitioners 'speak with

[1] I'm sometimes asked whether I'm myself a Second Philosopher. I don't think I use any methods or pursue any inquiries that aren't open to her, but she's clearly ever so much smarter and knowledgeable than I am!

[2] In Maddy (2011), §IV.5, responding to Burgess and Rosen (1997), I consider a top-down (as opposed to the current bottom-up) way of describing the Second Philosopher and argue that it makes no difference.

P. Maddy (✉)
Department of Logic and Philosophy of Science, University of California, Irvine, CA, USA
e-mail: pjmaddy@uci.edu

S. Arbeiter and J. Kennedy (eds.), *The Philosophy of Penelope Maddy*,
Outstanding Contributions to Logic 31, https://doi.org/10.1007/978-3-031-58425-1_16

a firm and unified voice' (p. 204), and to characterize his naturalist as deferring to these experts. One difference this makes is that Burgess, confronted with something like astrology or Descartes's a priori Method of Doubt, simply dismisses them as 'unscientific', while the Second Philosopher hears them out before drawing her negative conclusions. It also means that the Second Philosopher won't preemptively restrict her attention to what's studied in university departments of natural science. Her ongoing inquiry into the world and our place in it will include human history, sociology, anthropology, and linguistics, and more—difficult as it may be to reach firm conclusions in these areas—and she may even pause to observe physical details that fall beneath the notice of the natural sciences. But I don't think these niceties account for our differences over mathematics.

On that score, the Second Philosopher's investigations soon lead her to develop mathematical methods, as a ready aid to her work. Will she see this undertaking as of-a-piece with the rest of her practices, as having uncovered the existence of a new and surprising type of thing, mathematical abstracta (Thin Realism), or will she see it as a different kind of practice, with its own goals and methods, eminently useful, but not in the business of establishing truth, evidence, or existence (Arealism)? My claim on her behalf isn't just that there can be no evidence to choose one view over the other or that both views are open to her, but that the two views are essentially the same, merely rhetorical variants.[3] Though Burgess is inclined to speak with the Thin Realist, as is his right, its relation to Arealism is where the disagreement between us becomes particularly elusive:

> Am I not ... agreeing with the spirit, even if not the letter, of Maddy's claim that there is no more than a verbal difference between the nominalist's and the anti-nominalist's ways of describing our mathematical and scientific theorizing? (p. 212)

After describing aspects of our current usage that tend toward Thin Realism, Burgess concludes:

> My maximum concession ... would be to grant that if we had it all to do over again, our purposes might have been equally well served by the opposite to our actual usage. (p. 214)

This would appear to concede that the Second Philosopher could speak either way without coming into conflict with any matter of fact, which sounds like pretty much all she wants to claim.

I think this appearance of convergence is deceptive, as does Burgess, but locating our disagreement is non-trivial. At first, it seems easy to find its source in his criticism of the Arealist—he points out, for example, that physicists tend to speak as Thin Realists (pp. 206–208)—but in the end, he appears to agree with me that there's no substance to the Arealist's quarrel with the Thin Realist (p. 212). It would seem to follow that there's no substance to the Thin Realist's quarrel with the Arealist, but this Burgess denies: 'I may be described as in a sense sharing her view of the

[3] As I suspect Burgess would agree, 'agnosticism' isn't a correct analog here: the Second Philosopher's claim isn't that she doesn't know which is correct, but that both are equally correct, that despite appearances there's no difference between them. Is there an analogous view on the existence of a deity? Unitarianism?

[Arealist], though not of the [Thin Realist]' (p. 212). This suggests one possible diagnosis of our disagreement, as I would put it: his anti-nominalist is not my Thin Realist; his anti-nominalist has a viable objection to the Arealist that's not available to the Thin Realist. If this is right, I take the source to be our contrasting takes on the indispensability arguments.

Burgess notes that we both reject these arguments, though for different reasons; for him, they 'concede too much to nominalism' (p. 209). His point is that it's enough for mathematical objects to appear customarily in our scientific theories; it shouldn't be required that they appear indispensably. Enough, that is, for ontological consequences to follow. My own reservations, on the Second Philosopher's behalf, run deeper; I don't think the appearance of mathematical objects in our scientific theories, customary or indispensable, has ontological implications.

To take an illustrative example, when the Law of Universal Gravitation[4]

$$F = G\frac{m_1 m_2}{r^2}$$

is applied to the earth and the sun, we don't end up with a statement involving both the concrete astronomical bodies and various abstract mathematical entities, as in the indispensability arguments and Burgess's argument from custom, seem to suggest. Rather, we describe a pair of perfect spheres with two numbers, m_1 and m_2, associated with their centers, r as the distance between those centers, plus a constant G and a vector F. To this purely abstract system we apply our mathematical tools. Drawing conclusions about the physical world from these results depends on the degree of resemblance between the mathematical model and the worldly situation it's intended to describe, in this case, on our degree of confidence that the various simplifications, idealizations, and outright assumptions involved are both benign and beneficial in the case at hand.[5] The crucial point for our purposes is that we needn't assume the existence of the abstract system for this practice to function effectively; we need only describe it with sufficient clarity to be able to do mathematics with it.

[4] The example is intended to recall the famous passage from Putnam's (1975), pp. 74–75: 'To sketch the [indispensability] argument in a nutshell: ... one wants to say that the Law of Universal Gravitation makes an objective statement about bodies—not just about sense data or meter readings. What is the statement? It is just that bodies behave in such a way that the quotient of two numbers *associated* with the bodies is equal to a third number *associated* with the bodies. But how can such a statement have any objective content at all if numbers and 'associations' (i.e., functions) are alike mere fictions? It is like trying to maintain that God does not exist and angels do not exist while maintaining at the very same time that it is an objective fact that God has put an angel in charge of each star and the angels in charge of each of a pair of binary stars were always created at the same time! If talk of numbers and 'associations' between masses, etc. and numbers is 'theology' (in the pejorative sense), then the Law of Universal Gravitation is likewise theology.'.

[5] This example is examined a bit more thoroughly in Maddy (2022b). A modern treatment of gravitational theory, Malament (2012), p. 119, begins by defining 'a *relativistic spacetime* to be a pair (M, g_{ab}), where M is a smooth, connected, four-dimensional manifold and g_{ab} is a smooth metric on M of Lorentz signature (1,3)'. This is obviously a purely mathematical object. The extent to which it can be thought of as resembling physical spacetime is a matter of some subtlety.

The upshot is that the empirical success of the theory of gravitation has no ontological implications for the abstract model involved, vindicating the Arealist.

Burgess's anti-nominalist appears to disagree, appears to draw ontological conclusions from the customary role of mathematics in science. But what about the Thin Realist? Does her confidence in the truth of mathematics and the existence of its entities rest on the particulars of its appearances in natural science? Perhaps surprisingly, the answer is no; she sees the application of mathematics just as the Arealist does. Both are inspired to engage in mathematics (as opposed to astrology, say) because of its benefits to science, but once engaged, they recognize that it has methods and goals of its own, that application of those methods and pursuit of those goals has generated the very subject they so value, and that it would be misguided to attempt to interfere with those practices or limit their reach. On this point, then, the Thin Realist and the Arealist stand together[6] in apparent opposition to Burgess's anti-nominalist, insofar as he lodges a direct objection to the Arealist based on the role of mathematics in science.

So far, this looks like a plausible account of the disagreement between Burgess and me. Pushing on, though, notice that this reliance on indispensability considerations, or on what's customary in science, would open Burgess's anti-nominalist to the question of how far his ontological generosity extends. In his paper, Burgess embraces 'the axioms [that] suffice both for the rigorous reconstruction of … useful theorems … as well as for the deduction of less applicable results' (p. 213). He doesn't say which axioms this covers, but ZFC would seem to be an outer limit. So what's the status of higher set theory and the mathematics that depends on it? The Thin Realist has no qualms—this work is generated by good mathematical methods—but it's hard to see how this anti-nominalist could avoid the prospect of pruning it away, as Wittgenstein once described the process.[7]

Faced with this prospect, Burgess demurs

> My inclination is to think that by producing so much applicable material mathematics earns a place in the circle of science and a right to conduct its affairs autonomously. (Personal communication, 2/24/20, quoted with permission)

Here Burgess's anti-nominalist says exactly what the Thin Realist would say[8]: the role of mathematics in science doesn't dictate its ontology; it earns math the right to settle its own ontology. So I take it that Burgess's apparent appeal to indispensability, or custom, is misleading, that his anti-nominalist and my Thin Realist in fact agree on the upshot of applied mathematics for mathematical ontology.

---

[6] As noted above, where the Thin Realist and the Arealist disagree is on the status of this valuable new practice: the Thin Realist concludes that it's so intertwined with her previous modes of inquiry and uses so many of the same techniques that it ought to be considered as more of the same; the Arealist finds the differences more salient and regards mathematics as an eminently worthy enterprise, but not one in the business of establishing truth and existence.

[7] Wittgenstein (1937), p. 432e: 'What will distinguish the mathematicians of the future from those of today will really be a greater sensitivity; and *that* will—as it were—prune mathematics; for then everyone will be more concerned with absolute clarity that with the invention of new games'.

[8] This shouldn't be a surprise, given that Burgess was one of the inspirations for the position in the first place (see Maddy (2011), §III).

Agreement is always welcome, of course, but with this one, my first attempt at locating our disagreement evaporates. Albeit by a different route, I'm led to Burgess's own conclusion that our fundamental dispute is over the status of Carnapian conventions (p. 215). I agree with the Carnapian[9] that some features of our overall theorizing differ in kind from the rest, that we apply different, more lenient standards to their introduction. Where I disagree is with the claim that these distinctive, shall we say conventional, parts of our theory enter by a special act of linguistic fiat that takes place prior to and separate from our ordinary empirical inquiry, and that once this act is executed and the new language adopted, the claims in those parts of our ordinary theory are trivially true (p. 217).

Consider, for example, Einstein's description of spacetime as a continuous manifold in general relativity.Einstein admits that

> adhering to the continuum originates with me not in a prejudice, but arises out of the fact that I have been unable to think up anything organic to take its place. How is one to conserve four-dimensionality in essence (or in near approximation) and [at the same time] surrender the continuum? (Einstein (1949), p. 686)

Here we have a conventional element included by custom (if not indispensably). Presumably the Carnapian would count this as a linguistic decision, without which the theory—Einstein's equation, for example—couldn't even be expressed, without which the truth or falsity of its claims would be meaningless, and after which the continuity of spacetime follows trivially.

The trouble with this picture is that it isolates evaluation and introduction of the conventional elements to a prior, pragmatic venue, when in fact this is a crucial undertaking within science itself, both drawing on and influencing ordinary scientific claims.[10] In some cases, there's legitimate internal debate over whether or not a given aspect is conventional (e.g., physical geometry). In others, an idealization or simplification is adopted despite being known to be false, a fact that's recognized and investigated; natural scientists and applied mathematicians are keen to know which worldly factors account for the theory's success, where it succeeds, and its failure, where it fails. And finally, in cases like relativity theory, the ultimate status of a currently conventional element remains an open question within science itself.

In a nutshell, then, why should Einstein, in electing to adopt his mathematical framework, thereby commit himself to the continuity of spacetime? Why should he, after making this choice, forget the weak grounds on which he's made it? My answer, obviously, is that he shouldn't (and didn't), that 'recognition of a feature as conventional' should warrant 'agnosticism' (p. 215). Maybe the fundamental disagreement between Burgess and me comes down to this.[11]

---

[9] By this term, I intend a face-value reading of Carnap (1950), what Burgess seems to have in mind. How this squares with Carnap's more sophisticated writings is the subject of considerable scholarly commentary.

[10] This point is developed at somewhat greater length in Maddy (2000a), pp. 109–114.

[11] Burgess asks my position on the status of phonemes. I don't know enough about linguistics to have an informed opinion, but here's where I'd start. If speech does come as a series of units

References

Benacerraf, P., & Putnam, H. (1983). *Philosophy of mathematics* (2nd ed.). Cambridge University Press.
Burgess, J., & Rosen, G. (1997). *A subject with no object*. Oxford University Press.
Carnap, R. (1950). *Empiricism, semantics and ontology*. Reprinted in Benacerraf and Putnam [1983] (pp. 241–257).
Einstein, A. (1949). Remarks concerning the essays brought together in this cooperative volume. In P. A. Schilpp (Ed.), *Albert Einstein: Philosopher-scientist* (Vol. 2, pp. 665–688). Open Court.
Gasparri, L. (2019). Phonetic segments in the organization of speech, *Philosophy of Science*86, 304–324.
Maddy, P. (2000a). Naturalism and the a prior. In P. Boghossian, & C. Peacocke (Eds.), *New essays on the a priori* (pp. 92–116). Oxford University Press.
Maddy, P. (2011). *Defending the axioms*. Oxford University Press.
Maddy, P. (2022b). Enhanced if-thenism. In [2022h] (pp. 262–293).
Maddy, P. (2022h). *A plea for natural philosophy and other essays*. Oxford University Press.
Malament, D. (2012). *Topics in the foundations of general relativity and Newtonian gravitation theory*. University of Chicago Press.
Putnam, H. (1975). *What is mathematical truth?* Reprinted in his [1979] (pp. 60–78).
Putnam, H. (1979). *Mathematics, matter and method, philosophical papers* (2nd ed., Vol. I). Cambridge University Press.
Wittgenstein, L. (1937). *The big typescript: TS 213* (Trans. and Eds. C. G. Luckhardt & M. A. E. Aue). Blackwell (2005).

or 'phones'—this is where the actual ontological debate occurs, as far as I can tell (see Gasparri (2019))—then those phones, like any other objects, enjoy properties; just as apples can be red or not, or the pips on the face of a die can be three or not, phones can be aspirated or not. Austin and Quine have alerted us to the dangers of supposing that those red apples imply the existence of Redness (see (2017), pp. 63–65), and it's only the demands of mathematical arithmetic that move us to positing 3 (see (2014b), (2018b), (2018c)). In the case of numbers, I've argued for the counterparts to Thin Realism and Arealism. I don't know whether the linguistic theory of phonemes as equivalence classes of phones arises in response to demands like those of mathematical arithmetic; if not, it may just be a fancy way of talking about the properties of phones.

Chapter 17
Enhanced 'If-Thenism', Fictionalism, and Realist Anti-Platonism

Mary Leng

Abstract 'If-thenism' sees mathematical practice as a matter of working out what follows from our mathematical hypotheses without regard to whether those hypotheses are true. *Enhanced* if-thenism, as Maddy envisages it, adds a further component to mathematical practice: the identification of mathematically valuable hypotheses to investigate (where axioms as hypotheses are "chosen with an eye to facilitating important mathematical goals", and only some amongst all the logically possible collections of axioms are supported as tracking "underlying contours of mathematical depth"). While accepting that Maddy is right that the identification of mathematically interesting axioms is a sufficiently important component of mathematical practice to warrant a departure from 'if-thenism' in its purest form (according to which any consistent axioms would be fair game for mathematical investigation), this chapter argues that empirical science can also provide support for mathematical axioms. Indeed, a plausible reading of Hilary Putnam's rather elusive indispensability argument is as an argument against pure 'if-thenism' and in favour of an enhanced if-thenism where axioms are chosen for scientific reasons.

Keywords If-thenism · Fictionalism · Realist anti-Platonism

Penelope Maddy (2023) offers an updated version of mathematical 'if-thenism', the thesis that "contemporary pure mathematics is simply a matter of determining what conclusions follow from which assumptions" (Maddy, 2023, p. 263). 'If-thenism' has a long history in the philosophy of mathematics, but so described it is a position more flirted with than committed to. Bertrand Russell's dalliance with 'if-thenism' led him to define mathematics gloriously as "the subject in which we never know what we are talking about, nor whether what we are saying is true" (Russell, 1901, p. 75), but his considered logicist position looked to provide content to the axioms of mathematics through reduction to logical truth. And Hilary Putnam indicates in the (1979) preface to his collection, *Mathematics, Matter, and Method,* that while the collection includes

M. Leng (✉)
Department of Philosophy, University of York, Heslington, UK
e-mail: mary.leng@york.ac.uk

S. Arbeiter and J. Kennedy (eds.), *The Philosophy of Penelope Maddy*,
Outstanding Contributions to Logic 31, https://doi.org/10.1007/978-3-031-58425-1_17

his (1967a) defence of 'if-thenism', 'The thesis that mathematics is logic', the central thesis of that paper is one with which he "now very largely disagree[s]" (Putnam, 1979, p. xiii). In contemporary philosophy of mathematics, however, the spirit of 'if-thenism' remains alive and well in the form of mathematical fictionalism (indeed arguably Hartry Field's *Science without Numbers* (1980) can be thought of the book length development of the abandoned proposal of Putnam (1967a)). So it is gratifying to see another giant of the philosophy of mathematics, Penelope Maddy, extolling the virtues of the position.

Maddy considers three reasons why 'if-thenism' has been rejected, considering only the third as "serious enough to demand some substantial rethinking of the view". I would like to suggest, through consideration of Putnam's reasons for rejecting the 'if-thenism' of (1967a), that the second reason Maddy considers—involving concerns around the indispensability of mathematics in empirical science—also requires some move from what Maddy calls 'simple' if-thenism to an alternative 'enhanced' version. Indeed, when Putnam presents his indispensability argument as an argument for the objectivity of mathematics and not for the existence of abstract mathematical objects, I argue that this is best understood as a plea for adopting an enhanced 'if-thenism' according to which the objectivity of our mathematical concepts is secured by their indispensability in empirical science. Thus, I will argue, the indispensability of mathematics in empirical science already supports a move to an 'enhanced' if-thenism, even before we consider internal mathematical reasons for enhancement.

The first of the reasons Maddy considers for rejecting 'if-thenism' concerns pre-axiomatic mathematics: how can mathematics be a matter of working out what follows from what when for much of the history of mathematics mathematical theories have developed outside of the context of axiomatics. Maddy gives this objection short shrift, noting that 'if-thenism' need not be a philosophy of mathematics suitable for the entire history of mathematics, and that if the aim is to describe the mathematics of the 20th and early 21st centuries, in that setting viewing mathematics as engaged in working out the consequences of axioms looks just fine. I suspect Maddy is a little too quick here, as even developments in contemporary mathematics involve important pre-axiomatic stages. I discuss this issue in Leng (2010a), where I argue that the spirit of the 'if-thenist' picture (which I discuss there as involving an 'algebraic' approach to mathematical theories, borrowing terminology from Hellman (2003)) can survive in the pre-axiomatic setting if we view this as enquiry into what would have to be true of any objects satisfying our pre-formal conception. This allows an important sense in which axioms themselves can be discoveries. One of the cases I consider in that paper is nineteenth century: Hamilton's famous discovery of the quaternions, the equations $i^2 = j^2 = k^2$ excitedly graffitied onto Dublin's Brougham Bridge. Far from putting these forward as an arbitrary choice of consistent assumptions, Hamilton reports that he "felt a *problem* to have been at that moment *solved*" (Tait, 1866, p. 57). While Hamilton's 1843 discovery of the quaternions is outside the time period that Maddy considers suitable for an 'if-thenist' understanding, a similar dynamic (new axioms being introduced as a solution to an existing problem) is found in a twentieth century case I discuss in my paper, the (1943) 'Gelfand-Naimark Theorem', which established the appropriate axioms for C^*-algebras. Indeed, the existence of this kind of work in contemporary mathematics: arguments that support a choice of axioms by showing that they characterise a concept that is already of independent interest, is something we will return to below.

The second challenge is from applied mathematics, and it is worth noting that it is this challenge that led Putnam to abandon his own 'if-thenism'. According to Putnam (1975, p. xiii),

> it was my realization of the literal falsity of one sentence in this paper ('applied mathematics consists of assertions to the effect that *if* anything is a model for certain system of axioms, *then* certain statements "if A then B" hold, where A and B are first order statements *not* about the hypothetical model but about real things') – a sentence which claims, in effect, that mathematics is only used to derive statements in nominalistic language from statements in nominalistic language – that led me to the view espoused in *Philosophy of Logic* and in the two papers mentioned.

The 'if-thenist' approach to applied mathematics that Putnam first outlined (in 1967a) and then rejected is to say that so long as a mathematical theory is consistent, we can use it in our scientific theories to uncover which nominalist conclusions follow from our nominalistic scientific premises, i.e. the picture of applications developed by Hartry Field in *Science without Numbers* (1980), which aims to secure the applicability of 'if-thenist' mathematics by showing first that the assumptions of our scientific theories can be nominalised, and arguing that augmenting our scientific theories by adding mathematical axioms will be safe even if these mathematical axioms are not themselves true (even if the 'if's of our mathematical 'if-thens' are false), given that mathematics is conservative over nominalistic theories.

Maddy again gives short shrift to this challenge to 'if-thenism', arguing that it rests on a mistaken picture of the role of mathematics in empirical science, and in particular a picture of science as involving the formulation of theories as bodies of truths which, if they involve indispensable reference to mathematical objects, must then commit us to the existence of such things. Maddy offers in place of this a picture of scientific theorising as the creation of layers of models, including simplifications and idealizations, whose purpose is to help us understand and predict phenomena, and whose success in this regard requires only that "the model resembles the physical situation well enough to be used for certain purposes" (Maddy, 2023, p. 271), not that the models are literally true.

On this point, Maddy remarks that it is perhaps surprising that in my *Mathematics and Reality* I make such heavy weather of this, given that it's so obvious that merely fictional models can be used to represent and find out about real situations. Maddy's diagnosis is that I too am

> thinking of applications as Putnam does, of the Law of Universal Gravitation, for example, as involving both real physical objects and fictional abstracta. To make sense of this, she appeals to Walton's theory of fictions with props, which allows her to import the actual earth and the sun as props into a fiction about real numbers and functions. That this instance of the law is a legitimate move in the fiction-with-props is an indirect way of saying something literally true about the physical situation. My suggestion is that the earth and the sun are actually replaced by abstracta in a purely fictional story, which is then compared with the purely physical situation. (Maddy, 2023, p. 271 n. 21)

I think in fact the distance between Maddy and I is not so far here—I too see much of the use of mathematics in science as arising out of comparison of fictional mathematical models to physical situations. The heavy weather I make of this in my book, making use of Kendall Walton's (1990) account of fiction as make-believe, is meant to explain how it is that these 'fictions' can be used for descriptive and explanatory purposes without triggering ontological commitment to fictional entities. Not all accounts of fiction are fictionalist, so saying that mathematics gets into our scientific theories by providing useful fictions needs supplementation by an anti-realist account of fictional representation (such as the Waltonian account I adopt in Leng (2010b)) in order to respond fully to the claim that the indispensable use of mathematics in science requires acceptance of the existence of abstract mathematical objects.

However, while I think Maddy is too quick in her discussion to presume that the use of mathematics to provide theoretical models does not require commitment to abstracta, the upshot of the account of Leng (2010b) is that such use is indeed as ontologically unproblematic as Maddy presumes. At this point, having set aside the application of mathematics as a challenge to if-thenism, Maddy moves on to what she takes to be the most serious challenge, arising from mathematical rather than scientific practice, which is to explain why, if mathematics is just the study of what follows from which axioms, so much of actual mathematical practice seems to involve investigation into which axioms are appropriate to adopt. To do justice to this important aspect of mathematical practice, Maddy thus proposes that we replace 'simple' if-thenism with an 'enhanced' version, according to which:

> mathematics is a matter of figuring out what follows from what, where the concepts and axioms in the 'if' part are chosen with an eye to facilitating important mathematical goals. (Maddy, 2023, p. 277)

That mathematical practice tells us that not all consistent axiom systems are equal, and some choices of axioms are the right choices for facilitating our mathematical goals, is something that my own aforementioned discussion of pre-axiomatic reasoning in mathematics supports. The development of formal theories of quaternions and of C^*-algebras were substantial achievements precisely because they were driven by internal mathematical goals to pin down concepts of that sort.

But is Maddy right that it is only mathematical practice that forces us away from 'simple' if-thenism, according to which all consistent theories are equally candidates for part of mathematics, to an 'enhanced' version where mathematics involves not just figuring out what follows from what, but also working out which amongst the possible consistent theories we could come up with are mathematically important? What I would like to do here is to return for a moment to the issue raised by applied mathematics, and in particular to a puzzle that arises when we consider Putnam's rejection of (simple) if-thenism on the basis of the indispensability argument. In Putnam's introduction to *Mathematics, Matter, and Method*, while he openly rejects the if-thenism of Putnam (1967a) on the basis of his difficulty dealing with applied mathematics, he does not similarly repudiate the modal structuralist position of his (1967b). But as Field (1980) is to Putnam (1967a), so Hellman (1989) is to Putnam (1967b): a book length elaboration of the position intriguingly sketched out in that paper. And what is interesting about Hellman's modal structuralism is that it too could be thought of as a development of the 'if-thenist' idea that mathematics is a matter of 'figuring out what follows from what' (indeed, in Leng (2017), responding to Stephen Yablo's (2017) defence of I suggest the spirit of 'if thenism' about mathematics is best developed along the lines of Hellman (1989)). So to the extent that Putnam thought by 1979 that the indispensability of mathematics in science defeated the 'if-thenism' of (1967a) but not the 'mathematics as modal logic' picture of (1967b) which could also be thought of as a form of mathematical 'if-thenism', this raises the question of what, precisely, Putnam thought the indispensability argument established.

The issue becomes even more puzzling when one looks at those two papers in light of Putnam's later clarification that his indispensability argument, unlike Quine's, was never meant to be an argument for mathematical platonism. Putnam writes,

> my "indispensability" argument was an argument for the objectivity of mathematics in a realist sense—that is, for the idea that mathematical truth must not be identified with provability. Quine's indispensability argument was an argument for "reluctant Platonism," which he himself characterized as accepting the existence of "intangible objects" (numbers and sets). The difference in our attitudes goes back at least to 1975. (Putnam, 2012, p. 183)

This distinction between the indispensability argument as an argument for platonism and as an argument for objectivity is also present elsewhere in Putnam's writings, e.g. here in (2004)'s *Ethics without Ontology:*

> Everything about the success of mathematics, and the deep dependence of much contemporary science, including physics, but not only physics, *on* mathematics, supports taking mathematical theorems as objective truths; but nothing supports taking mathematical theorems as descriptions of a special realm of "abstract entities," and nothing is gained, in philosophy of mathematics or elsewhere, by so doing. (Putnam, 2004, p. 67)

In these later writings (but also even in his (1975) presentation of the indispensability argument) Putnam emphasises his continued support of the 'mathematics as modal logic' picture as part of his explanation of why the indispensability of mathematics supports what he calls the 'objectivity' of mathematics without supporting the existence of mathematical objects.

What, then, of the relevance of the indispensability of mathematics to the denial of 'if-thenism'? Isn't modal structuralism after all, with its view of mathematics as engaged in uncovering what follows logically from consistent axioms, just another take on the 'if-thenist' view of mathematics? Why does the indispensability of mathematics challenge Field's Russellian 'if-thenist' picture of applications, but not the modal structural alternative?

A simplistic answer would focus on the fact that modal structural mathematics delivers the truth of mathematical claims (without also delivering an ontology), and note that scientific realism and the indispensability of mathematics in scientific theories requires us to account for mathematics as true (which is just what the modal structuralist interpretation does). This, however, is clearly unsatisfactory as it stands. The modal structuralist picture delivers us the truth of theorems of *pure* mathematical theories by recasting them as claims roughly of the form 'the axioms for this theory are consistent, and, if there were any objects satisfying these axioms, then this theorem would also hold of those objects'. But the mixed mathematical/empirical claims of empirical science cannot be so easily modalised. The claims of an empirical scientific theory are meant to be true of this world, not simply to tell us what would have to be true in any world satisfying the theory's laws, so what the modal structural account needs is a modalised account of the mixed mathematical/empirical claims of empirical science which preserves the requirement that their truth is indexted to how things are in the actual world.

Hellman (1989) takes up this challenge, of modalising the mixed mathematical/empirical claims of empirical science in such a way that they are still understood as claims about *our* world, by introducing what he calls a 'non-interference' proviso. Effectively, Hellman's modalisations of mixed mathematical/empirical claims say something like the following 'the mathematical theories used in empirical science are consistent, and *were* there any objects satisfying our mathematical theories that were such that they *did not interfere in any way with the objects there actually are*, then {mixed mathematical/empirical claim} *would* hold'. The apparent complexity of introducing into the world the possibility of 'non-interfering' objects satisfying our mathematical theories so as to say things about how objects in the actual world *would* be related to them *were* there any such things might hide the true simplicity in this account, as well as its relation to the 'fictional models' picture discussed above. For according to Hellman's picture, whether there really are any mathematical objects satisfying our mathematical axioms doesn't matter: we can get to the content of our scientific theories by imagining how things would be were there a realm of such objects (and were they to be causally and spatiotemporally isolated from any actual objects in the physical world, so as not to interfere in any way with their behaviour). But this is just the sort of picture of the role of mathematics in empirical science appealed to in the 'fictional models' picture that Maddy and I both endorse, whereby we avail ourselves of the descriptive and explanatory power of mathematical models by imagining that there are objects as described in our mathematical theories, and by comparing model systems to actual systems of physical objects.

So it looks like Hellman's modal structural picture of applied mathematics is friendly to the 'if-thenist' account of applications that Maddy herself prefers (involving imagined mathematical models which are used to help us reason about real physical things). But then what are we to make of Putnam's indispensability argument as an argument against 'if-thenism'? Given Putnam's continued adherence to the 'mathematics as modal logic' picture of (1967b), he clearly saw his indispensability argument as applying only to the Russellian if-thenism of (1967a) and not 'if-thenism' in its modal structuralist version, as developed by Hellman (1989). But what really is the difference between Hellman's modal structural 'if-thenist' account of empirical science (making use of the models as fictions picture to view the mixed claims of empirical science as claims about how things would be with physical objects were there mathematical objects as described by our axiomatic mathematical theories) and Field's Russellian 'if-thenism' (which dispenses with mathematics in formulating the claims of our empirical theories, but makes use of the 'if-thens' of our mathematical theories in order to justify claims about which nominalistic consequences follow from our nominalistic theoretical hypotheses)? Why was Putnam worried about the latter incarnation of 'if-thenism' but not the former?

A full answer to this question requires a much longer essay (I develop the idea at length in Leng ms). But in terms of Maddy's framework what I would like to suggest here is that Putnam's indispensability argument (with its intended target being the Russellian 'if-thenism' of (1967a) as developed by Field (1980), rather than the alternative 'if-thenism' of (1967b) as developed by Hellman (1978) is really an argument—from the indispensability of mathematical concepts—for an *enhanced* form of if-thenism. In particular, it is an argument for the objectivity of certain kinds of mathematical concepts via their indispensability to our best confirmed scientific theories. Amongst all the consistent assumptions that could take the role of 'ifs' in our mathematical 'if-thens', only some are vindicated as part of the objectively best understanding of the physical world, as only some turn out to be indispensable for formulating our scientific theories. Had Field's programme worked—had mathematics been dispensable in formulating those theories—then our mathematical 'if-thens' would just be useful, but inessential, tools for working out what follows from our empirical assumptions. But if mathematics is *required* in our best descriptions of the world, then the mathematical concepts that are essential for these descriptions receive confirmation as part of the reality described by empirical science, and this marks them out as different from any old consistent collections of axioms. Truth in mathematics is then not just provability from any old axioms, but a matter of what follows from the *right* (objectively vindicated) choices of axioms.

Support for this picture can be found in Putnam's own writings, where he is clear that, while contemporary 'easy road' versions of nominalism that do not aim to dispense with mathematics in empirical science do not suffice to answer his indispensability argument (when considered as an argument for the objectivity of mathematics and not for mathematical Platonism), Hartry Field's programme, if successful, would succeed as a response:

> Hartry Field understands very well what my arguments were, and he attempts to meet them in their own terms. (Putnam, 2012, p. 190)

On the other hand, Putnam is clear that the so-called 'easy road' nominalism I advocate in Leng (2010b), whose account of applications accepts the indispensability of mathematics in providing idealised models that can be used to describe the physical world, is not an adequate response to his indispensability argument.

To explain his concern about 'easy road' nominalism, Putnam considers the easy-roaders' 'trick' of accounting for the applications of mathematics by viewing our mathematically stated scientific theories as 'nominalistically adequate', i.e., as using some idealised mathematical descriptions to speak 'as if' there are mathematical objects, as a means to paint a picture that gets things right about the concrete world. The succcess of this 'trick', Putnam complains, would depend on our being able to describe the concrete world in purely non-mathematical terms:

> I suspect that the "trick" Rosen describes depends, in fact, on assuming that all possible things are good old fashioned "concrete objects," which we know how to describe in nominalistic

> language, or good old fashioned "abstract entities," which are not in spacetime, and which are causally inert. But the entities of modern physics are neither of the above. (Putnam, 2012, p. 197)

Take, for example, talk of 'electrons', as concrete objects. Putnam points out that our theoretical description of 'electrons' are really mathematical descriptions of quantized fields:

> This means that strictly speaking there aren't such "objects" as electrons, in the "ontology" of the theory; what there are are quantized fields, and talk of "electrons" is a useful way of talking of, for example, certain components of vectors in the Fock space that we use to describe those fields. (Putnam, 2012, p.196)

Putnam's worry about easy road nominalism might seem puzzling, given that easy road nominalists explicitly deny that we can always describe the concrete world in entirely non-mathematical terms. Why can't the nominalist agree that our only theoretical descriptions of electrons are fully mathematical, but still say that the work done by the mathematics is as a descriptive tool, not a literally true account of a realm of mathematical objects (which of course the anti-Platonist in Putnam also rejects)? I suspect the key to understanding why Putnam's apparent restatement of the 'easy road' position (i.e. his emphasis on the descriptive indispensability of mathematics) is considered by Putnam to be an *objection* to easy road nominalism is to be found in Putnam's meta-ethical writings.

Putnam's meta-ethical writings are peppered with discussions of realism and platonism in the philosophy of mathematics, and he clearly sees the issues of mathematical and moral realism as related. Putnam's argument for evaluative realism is also presented an argument for 'realism without Platonism', or 'objectivity without objects', and is also defended by a form of indispensability argument. Specifically, Putnam argues that the expressive indispensability of thick concepts—which combine 'factual' and 'evaluative' components, shows that we cannot meet our aims of describing the world of 'facts' without also invoking evaluative concepts. So to the extent that we wish to be realist about the realm of fact, we should be realist about the world of value too:

> so-called factual and so-called evaluative predicates are mutually "entangled" ... it is a fantasy to suppose that the predicates we use to give sensitive and relevant descriptions of human beings and human interactions can be disentangled into two components, a purely descriptive component and an evaluative component. (Putnam, 2011, p. 288)

Bearing in mind Putnam's worry about easy road nominalism, and the entangling of mathematical and empirical concepts in our descriptions of physical phenomena such as in quantum field theories, Putnam's (anti-Platonist) realism about mathematics could be expressed, *mutatis mutandis*, in just this way:

> so-called physical and so-called mathematical predicates are mutually "entangled" ... it is a fantasy to suppose that the predicates we use to give sensitive and relevant descriptions of the physical world and physical interactions can be disentangled into two components, a purely concrete 'physical' component and an abstract 'mathematical' component.

And given that the objective vindication of evaluative concepts (via their descriptive indispensability) is considered to be an argument for *realism* about the evaluative, it is not surprising in the light of this parallel that Putnam considers the analogous considerations concerning the descriptive indispensability of mathematics to be an argument for a form of mathematical realism (albeit not an argument for Platonism).

What I would like to suggest then is that, while Maddy is right that a 'simple' if-thenism according to which all consistent theories are metaphysically speaking on a par should give way to a more sophisticated 'enhanced' if-thenism according to which some, but not all, mathematical hypotheses are objectively vindicated, she is wrong in thinking that it is only mathematical practice that uncovers the objective contours of mathematical depth. Applied mathematics too vindicates some choices of mathematical concepts over others, an observation noted by Putnam and one that, I think, is the lasting contribution of his realist-but-not-Platonist indispensability argument.

References

De Caro, M., & Macarthur, D. (Eds.). (2012). *Philosophy in an age of science: Physics, mathematics and skepticism.* (Cambridge, MA: Harvard University Press).

Doran, R. S. (1994). *C*-Algebras: 1943–1993; A fifty year celebration (1993: San Antonio, Texas).* Vol. 167 of Contemporary Mathematics. (Providence, RI: American Mathematical Society).

Field, H. (1980). *Science without numbers* (Princeton, NJ: Princeton University Press).

Gelfand, I. M., & Naimark, M. A. (1943). 'On the imbedding of normed rings into the ring of operators in Hilbert space'. *Mat. Sbornik*, 12: 197–213. Reprinted in Doran, (1994), 3–20.

Hanna, G., Jahnke, H.-N., & Pulte, H. (Eds.). (2010). *Explanation and proof in mathematics: Philosophical and educational perspectives.* (New York, NY: Springer).

Hellman, G. (1989). *Mathematics without numbers: Towards a modal-structural interpretation.* (Oxford: Clarendon Press).

Hellman, G. (2003). Does category theory provide a framework for mathematical structuralism? *Philosophia Mathematica, 11*(2), 129–157.

Leng, M. (2010a). *Mathematics and reality.* (Oxford: OUP)

Leng, M. (2010b). Preaxiomatic mathematical reasoning: An algebraic approach. In Hanna et al. (2010), pp. 47–57.

Leng, M. (2017). Reasoning under a presupposition and the export problem: The case of applied mathematics. *Australasian Philosophical Review, 1*(2), 133–142.

Leng, M. (ms). The Prim and the Prime: Putnam, Thick Concepts, and the Collapse of the Mathematical/Empirical Dichotomy'

Maddy, P. (2023). Enhanced If-Thenism. In *A Plea for Natural Philosophy and Other Essays* (Oxford: OUP), pp. 262–293.

Putnam, H. (1967a). The thesis that mathematics is logic, reprinted in Putnam (1979), pp. 12–42.

Putnam, H. (1967b). Mathematics without foundations, reprinted in Putnam (1979), pp. 43–59.

Putnam, H. (1975). What is mathematical truth, reprinted in Putnam 1979, pp. 60–78.

Putnam, H. (1979). *Mathematics, matter and method: Philosophical papers, Vol 1* (2nd edition 1979) (Cambridge: CUP)

Putnam, H. (2004). *Ethics without ontology* (Cambridge, MA: Harvard University Press).

Putnam, H. (2011). The Fact/Value Dichotomy and its Critics, Chapter 18 in Putnam (2012)

Putnam, H. (2012). Indispensability arguments in the philosophy of mathematics. In De Caro and Macarthur (2012), pp. 181–201.

Russell, B. (1901). Recent work on the principles of mathematics. *The International Monthly* 4 (Jul 1901), 83–101 Repr. as 'Mathematics and the Metaphysicians', in *Mysticism and Logic and Other Essays*, London, Longmans, Green, and Co., pp. 74–96]
Tait, P. G. (1866). Sir William Rowan Hamilton. *North British Review, 14*, 35–54.
Walton, K. (1990). *Mimesis as Make-Believe* (Cambridge, MA: Harvard University Press).
Yablo, S. (2017). If-Thenism. *Australasian Philosophical Review, 1*(2), 115–132.

Chapter 18
Reply to Leng

Penelope Maddy

Keywords Enhanced if-thenism · Hilary Putnam · Geoffrey Hellman · Michael Resnik · Structuralism · Fictionalism · Modalism · Robust realism

Professor Leng engages with a forthcoming paper of mine, 'Enhanced if-thenism' (Maddy 2022b), that attempts to revive a version of if-thenism: what's revived is the simple idea that mathematics is a matter of determining what conclusions follow from which assumptions; the enhancement is an account of rational criteria for deciding what concepts and assumptions belong in the 'if' clause. The unstated goal of that paper was to approach something like the Arealism of *Defending the Axioms* (Maddy 2011) by a more familiar route, hoping, perhaps quixotically, to increase its appeal. I'm not sure this was accomplished, but the piece did at least draw the attention of Geoffrey Hellman, who published a stern rejection (Hellman, 2021), and now of Professor Leng. She proposes an amendment that strikes me as friendly on its face, perhaps not so much a change, but I think this apparent agreement masks fundamental disagreements—with Leng and her allies (Putnam and Hellman) on one side and myself on the other—over such fundamentals as the nature of applied mathematics and the underlying motivation for if-thenism. Let me try to explain.

My defense of if-thenism begins with three traditional objections; Leng begins her discussion with my response to the first of these. The objection is that there was mathematics before the advent of modern axiomatic methods, and my response is that I'm only attempting to characterize mathematics of the 20th and early twenty-first centuries, after the rise of pure mathematics produced the need for explicit existence

P. Maddy (✉)
Department of Logic and Philosophy of Science, University of California, Irvine, CA, USA
e-mail: pjmaddy@uci.edu

S. Arbeiter and J. Kennedy (eds.), *The Philosophy of Penelope Maddy*,
Outstanding Contributions to Logic 31, https://doi.org/10.1007/978-3-031-58425-1_18

assumptions to found the practice.[1] Leng's objection to this response is that the kind of concept-formation I have in mind also occurs outside of axiomatic mathematics, both in the pre-axiomatic past and today, and thus that the kind of enhancement I propose isn't as limited as I seem to think.

Leng is certainly right that important cases of concept-formation occurred in the past. One of my go-to examples is the gradual development of the concept 'group': it was first defined with cancellation laws, but these were later replaced by the existence of inverses when the mathematical advantages of including infinite groups became clear. Leng's examples are quaternions and C*-algebras, but the moral is the same: mathematical concepts are fine-tuned to do various mathematical jobs.[2] So far, we're in complete agreement, but where does if-thenism come into this story?

Leng's answer emerges from her understanding of what happens after these new concepts have been defined: 'an enquiry into what would have to be true of any objects satisfying our … conception' (p. 226). So, for an elementary example, we prove 'if x is (were) a group, then x has (would have) a unique identity element'. It seems churlish to note that this if–then isn't a claim at all, it's an open sentence; what gets proved is actually 'for all x, if x is a group, then x has a unique identity element'. Putnam notes this explicitly in his description of Russell's passing if-thenism:

> What he meant was not, of course, that all well[-]formed formulas in mathematics have a horseshoe as the main connective! but that mathematicians are in the business of showing that *if* there is any structure which satisfies such-and-such axioms (e.g. the axioms of group theory), *then* that structure satisfies such-and-such further statements (some theorems of group theory or other). (Putnam, 1967a, p. 20)

This characterization contrasts with a more straightforward alternative centered on actual conditionals—such-and-such axioms imply such-and-such theorems—as, for example, when we prove 'if ZFC + LCs,[3] then $V \neq L$'.

Focused as I have been on the search for and justification of the fundamental assumptions that found contemporary pure mathematics, I've taken if-thenism in the second sense: mathematics is a matter of which theorems follow from which axioms (or axiom candidates). For the case of groups, where Leng (and her allies) see a claim about structures satisfying the definition of 'group', I see a proof by the ordinary rule of universal-quantifier-introduction of a claim about all groups, but the actual argument is the same for us both: we start by assuming we have an arbitrary group G and prove that G has such-and-such properties. In 'Enhanced if-thenism', where my interest was in turning aside objections, I noted this difference but made nothing of it.[4] In fact, it was more important than I realized.

To see this, consider the second strand of Leng's critique, the one that emerges from the second traditional objection to if-thenism: the role of mathematics in science

---

[1] See Maddy (2008) or (2011), Chap. 1.

[2] Leng phrases this in terms of 'axiom-selection', but what's going on is the development of good definitions. I reserve the term 'axiom' for the fundamental foundational assumptions.

[3] That is, ZFC plus the existence of large cardinals, currently the leading foundational theory. The debate is over how it might be extended.

[4] See Maddy (2022b), §I, especially footnote 6.

requires that the consequents of the if-thens be unconditionally, not just conditionally, true. Here we're returned to Putnam's indispensability argument, his reason for giving up if-thenism. I've argued at some length elsewhere that indispensability considerations don't have the metaphysical implications their supporters imagine,[5] but in the paper Leng is considering, my interest is in diagnosing what leads philosophers to think that they do. This is where the fundamental disagreement between Leng and me begins to emerge.

Let me first note that Leng's ultimate conclusion from her brush with indispensability can be read in a way that I entirely condone. She writes, 'the objectivity of [some of][6] our mathematical concepts is secured by their indispensability in empirical science' (p. 226) and 'applied mathematics ... vindicates some choices of mathematical concepts over others' (p. 233). I can read this as saying that good concepts are those that facilitate deep mathematics and that attending to the needs of empirical science is a rich source of such concepts (as history amply demonstrates)[7]; the objectivity here is the objectivity of mathematical depth. But this isn't how Leng intends these remarks. For her and for Putnam, an indispensability argument yields a metaphysical conclusion of some kind and the objectivity of the relevant mathematical concepts follows from that; in Leng's words, this is the 'the lasting contribution of [Putnam's] realist-but-not-Platonist indispensability argument' (p. 233). This is where our contrasting visions of how mathematics works in application come into play, so let's look first at the two visions and then at the conclusions that Leng and Putnam draw from theirs.

Sticking with Putnam's well-worn example of the law of universal gravitation, consider the case of the sun and the earth. As I see it, we end up with a mathematical description involving not the physical sun and moon, but initially two perfect spheres and ultimately (in light of Newton's shell theorem) two points separated by a precise distance r in R^3.[8] Once the relevant features of this abstract mathematical model are derived, the empirical claim is that the physical system of sun and earth in three-dimensional space resembles the model in sufficient respects[9] and that a conclusion

---

[5] See Maddy (1997), §II.6.

[6] Though it may occasionally sound as if Leng is taking the strong indispensability theorist's line that the only mathematics that's justified is what's required by science, this clearly isn't what she means: e.g., 'the development of formal theories of quaternions and of C*-algebras were substantial achievements precisely because they were driven by internal mathematical goals to pin down concepts of that sort' [i.e., concepts 'facilitating our mathematical goals'] (p. 228).

[7] Cf. von Neumann (1947), p. 182: 'It is undeniable that some of the best inspirations in mathematics—in those parts of it which are as pure mathematics as one can imagine—have come from the natural sciences'. This was certainly true then, and it remains true today. It's even conceivable, in principle, that the development of science could suggest stronger existence axioms beyond ZFC + LCs, though I've argued that this is unlikely (Maddy, 2011, pp. 34–37).

[8] Oddly enough, Putnam sometimes seems to share this way of seeing applications. Leng remarks, 'Putnam points out that our theoretical descriptions of "electrons" are really mathematical descriptions of quantized fields' (p. 232); she then quotes Putnam (2012), p. 196: 'This means that strictly speaking there aren't such objects as "electrons", in the "ontology" of the theory"'.

[9] Spelling out what those respects are and why they are enough is the job of applied mathematicians. The story differs in each case, e.g., in fluid dynamics or in statistical mechanics, in general relativity

can be drawn about the actual gravitational force between the two actual bodies. On this reading, nothing follows about the existence of the abstracta employed in the description. Nor does it follow that the physical situation could be described in purely nominalistic terms; often enough, mathematization is the only means we have of characterizing the natural world.

But this isn't how Putnam and Leng see the situation. In Putnam's treatment, the sun and earth themselves somehow survive the introduction of perfect spheres and so on, and the theoretical description ends up involving both physical and mathematical items. Some take this as confirming the existence of the mathematical objects along with the physical ones, but Leng resists this conclusion with a kind of fictionalism (in her (2010)) and Putnam prefers his 'mathematics as modal logic' (in his (1967b)), subsequently developed by Hellman (in his (1989)). For present purposes, Leng takes the difference between herself and Putnam/Hellman to be insignificant: her fictionalism, too, is ultimately modal in character because the claims of mathematics, though literally false, still 'tell us about what *would* be the case *were* there any objects, predicates, and relations satisfying our mathematical axioms' (Leng, 2017, p. 134).[10] So the moral all three draw from indispensability considerations is not the existence of an abstract ontology, but a metaphysics of modal facts about what would be the case if so-and-so—and those ifs have to be second-order to fully specify the desired structures. So, for example, 'if there were a structure satisfying second-order ZFC + LCs, then such-and-such would be true therein'.[11]

What's happened here is that the kind of if-thenism Leng describes for concept-formation, the kind that seemed slightly off to me, has been extended to the foundational theory itself. From this perspective, ZFC + LCs isn't a set of existence claims that ultimately back up conceptual quasi-if-thens like 'any structure that satisfies the definition of group has a unique identity'; instead, ZFC + LCs is just another definition, on a par with that of 'group'.[12] From this perspective, the straightforward claim that there's a proof of Cantor's theorem or $V \neq L$ from ZFC + LCs is replaced with something like 'it's second-order logically possible that there's a structure satisfying second-order ZFC + LCs, and in any such structure, Cantor's theorem would be true and $V = L$ would be false' (paraphrasing from Leng (2017), p. 134). (Notice that, in

---

or in quantum mechanics (where the story of what worldly conditions make the theory work is almost entirely unknown).

[10] For Putnam and Hellman, this is what mathematical statements say, and they are literally true.

[11] Leng suggests that the distance between her fictionalism and my position 'is not so far here—I too see much of the use of mathematics in science as arising out of comparison of fictional mathematical models to physical situations' (p. 228) and that Hellman's modal-structuralism is 'just the sort of picture of the role of mathematics in empirical science appealed to in the "fictional models" picture that Maddy and I both endorse' (p. 230). Obviously, I think this is misleading. The picture of a theory that mixes the mathematical with the physical, the picture she shares with Putnam and Hellman, is what motivates her appeal to Walton's account of fictions to explain how 'mathematics gets into our scientific theories' (p. 228) without triggering ontological commitment to the abstracta involved, as well as Hellman's account of 'mixed-statements' with his 'non-interference' clause (Hellman, 1989, p. 99).

[12] Hellman is explicit about this: he regards his second-order ifs 'as defining structures of interest, not just as starting points of proofs' (Hellman, 2021, p. 253).

addition to the if-thens themselves, a second-order logical possibility assumption is required to remove the specter of vacuity.[13]) But now ZFC + LCs isn't playing its foundational role, as it did when backing up the claim about all groups; the possible structures satisfying second-order ZFC + LCs that back up the new quasi-if-thens can't just be matter of what follows from the foundational theory. This is the key conclusion: the existence of these possible structures for ZFC + LCs must be facts in some kind of pre-theoretic, ambient metaphysics.

Leng, Putnam, and Hellman have argued that this is not a metaphysics of objects—fine—but it must be a metaphysics of possible structures, or at least, a metaphysics of robust modal facts. To take just one dramatic example, either ZFC + LCs + CH is second-order logically possible or ZFC + LCs + not-CH is second-order logically possible, but not both. These thinkers have rejected Gödel's universe of sets in which CH is either true or false, but their metaphysics is really no poorer for that; you squeeze the balloon in one place and it pops out somewhere else. They still need a non-trivial epistemology to account for our access to these modal facts,[14] which makes this position a version of Robust Realism.[15]

To be clear, I'm not asking Leng for an epistemology that allows her to determine, right now, which of these two theories about CH is second-order logically possible. My point is that her position entails that there's a pre-theoretic fact of the matter about this in her ambient metaphysics. For that reason, she owes an explanation of how and why the kinds of methods we use in practice—proposing new axioms candidates and judging them by their extrinsic merits—are effective methods for finding out what that pre-existing fact actually is. This is my over-arching quarrel with any form of Robust Realism: even if it's been shown that an axiom candidate meets every test of mathematical advantage, this isn't enough; there remains the extra question of whether it tracks whatever metaphysics is in play. It seems to me that mathematical advantage should be enough.[16]

Thus, my take on the two visions of how mathematics works in application and on the metaphysical conclusions that Leng, Putnam, and Hellman draw from theirs. (To my mind, 'structuralism' is a better term for their view than 'if-thenism'.) Obviously, my if-thenism is quite different: the ifs are sentences, axioms, and the thens are theorems; the claim is just that the latter are provable from the former. Which brings us, at last, my final question: what's the point of if-thenism in the first place? What's its motivation?

---

[13] See, e.g., Hellman (1989), pp. 71–72, where he assumes the possibility of a structure satisfying of second-order ZFC (a V_{κ}, for some inaccessible κ).

[14] Cf. Hellman (1989), p. 72: 'How best to describe and assess our "evidence" for such a hypothesis [the possibility assumption of the previous footnote] remains one of the most difficult challenges confronting mathematical epistemology'.

[15] See Maddy (2022b), footnote 62, for some reaction to Hellman's Robust Realism and his (2021). How does my Thin Realist differ? She takes the axioms at face value—ZFC + LCs is a list of assertions, not a disguised definition—and posits that the sets they describe have an odd metaphysical feature: that facts about them can be tracked by following the lead of mathematical advantage. Thus her trivial epistemology—it falls out of her metaphysics.

[16] The previous footnote indicates how the Thin Realist achieves this result.

In his well-known discussion, Michael Resnik describes the attraction of if-thenism—he calls it 'deductivism'—in this way:

> Deductivism is a powerful and appealing philosophy of mathematics [that] avoid[s] commitments to abstract entities, and ... simplifies ... epistemology. (Resnik (1980), p. 136)

Though the if-thenism of Leng, Putnam, and Hellman does avoid abstract objects, it accomplishes this only by positing an equally robust metaphysics of second-order logical possibility that raises equally problematic epistemological questions. The if-thenism I've proposed seems closer to Resnik's vision: freed of robust metaphysics, epistemology reduced to that of logic.[17] In 'Enhanced if-thenism', I attempt to meet the series of objections that Resnik described as 'a difficult and pressing challenge for present and future deductivists' (Resnik (1980), p. 360). My hope is that this 'enhanced' version of if-thenism enjoys at least some faithful descendent of the allure that Resnik observed in the original![18]

References

Hellman, G. (1989). *Mathematics without numbers*. Oxford University Press.
Hellman, G. (2021). If "if-then" then what? In *Mathematics and its logics* (pp. 237–255). Cambridge University Press.
Maddy, P. (1997). *Naturalism in mathematics*. Oxford University Press.
Maddy, P. (2008). How applied mathematics became pure. *Review of Symbolic Logic, 1*, 16–41.
Maddy, P. (2011). *Defending the axioms*. Oxford University Press.
Maddy, P. (2022b). Enhanced if-thenism. In [2022h] (pp. 262–293).
Maddy, P. (2022h). *A plea for natural philosophy and other essays*. Oxford University Press.
Leng, M. (2010). *Mathematics and reality*. Oxford University Press.
Leng, M. (2017). Reasoning under a presupposition and the export problem: The case of applied mathematics. *Australasian Philosophical Review, 1*, 133–142.
Von Neumann, J. (1947). The mathematician. In R. B. Heywood (Ed.), *The works of the mind*, (pp. 180–196). University of Chicago Press.
Putnam, H. (1967a). *The thesis that mathematics is logic*. Reprinted in his [1979] (pp. 12–42).
Putnam, H. (1967b). *Mathematics without foundations*. Reprinted in his [1979] (pp. 43–59).
Putnam, H. (1979). *Mathematics, matter and method, philosophical papers* (2nd ed., Vol. I). Cambridge University Press.
Putnam, H. (2012). Indispensability arguments in the philosophy of mathematics. In M. De Caro, & D. Macarthur (Eds.), *Philosophy in an age of science* (pp. 181-201). Harvard University Press.
Resnik, M. D. (1980). *Frege and the Philosophy of Mathematics*. Cornell University Press

[17] The nature and epistemology of logic is obviously another important story. In Maddy (2022b), I gesture toward my other writings on this topic.

[18] Thanks to Professor Leng for her helpful responses during the composition of this reply.

Chapter 19
Logical Realism—A Tale of Two Theories*

Gila Sher

Abstract This paper compares two theories of the nature of logic: Maddy's, (2007, 2014a)—henceforth "Theory 1"—and Sher's (1991, 2016)—"Theory 2". The two theories share a significant element: they both diverge from the commonly held view that logic is grounded only in the mind (language, concepts, conventions, etc.). Instead, they argue that logic is crucially grounded in the world. But the two theories differ in significant ways as well. Most distinctly, one is an anti-holist, "austere naturalist" theory while the other is a non-naturalist "foundational holistic" theory. This methodological difference affects their questions, goals, orientations, the scope of their investigations, their logical realism (the way they ground logic in the world), their explanation of the modal force of logic, and their approach to the relation between logic and mathematics.

This paper compares two theories of the nature of logic: Maddy's (2007, 2014a)—henceforth "Theory 1"—and Sher's (1991, 2016)—"Theory 2".[1] The two theories share a significant element: they both diverge from the commonly held view that logic is grounded only in the mind (language, concepts, conventions, etc.). Instead, they argue that logic is crucially grounded in the world. But the two theories differ in significant ways as well. Most distinctly, one is an anti-holist, "austere naturalist" theory while the other is a non-naturalist "foundational- holistic" theory. This methodological difference affects their questions, goals, orientations, the scope of

---

[1] While Maddy's *The Logical Must* (2014a) and Sher's *The Bounds of Logic* (1991) are fully devoted to logic, Maddy's *Second Philosophy* (2007) and Sher's *Epistemic Friction* (2016) are more general philosophical works. Each, however, has a substantial part devoted to logic: "A Second Philosophy of Logic" (Part III of Maddy, 2007, pp. 197–302), and "An Outline of a Foundation for Logic" (Part IV of Sher, 2016, pp. 237–338). Both authors presented their views in a number of papers as well. I shall refer to these according to need. In addition to "theory 1" and "theory 2", I shall use "theorist 1" and "theorist 2".

---

* I would like to thank Pen Maddy for helpful comments.

G. Sher (✉)
University of California, San Diego, CA, USA
e-mail: gsher@ucsd.edu

S. Arbeiter and J. Kennedy (eds.), *The Philosophy of Penelope Maddy*,
Outstanding Contributions to Logic 31, https://doi.org/10.1007/978-3-031-58425-1_19

their investigations, their logical realism (the way they ground logic in the world), their explanation of the modal force of logic, and their approach to the relation between logic and mathematics.

The paper is not polemic. Its goal is *not* to compare the two theories with respect to their merits and deficiencies. One of its goal is a perspicuous description and analysis of the two theories, explaining their differences as well as commonalities. Another goal is showing that and how (i) a grounding of logic is possible, (ii) logical realism can be arrived at from different perspectives and using different methodologies, and (iii) grounding logic in the world is compatible with a central role for the human mind in logic. The discussion is divided into three parts: I. Basic Questions and Methodologies. II. Logic's Grounding in the World and Logical Necessity. III. Relation between Logic and Mathematics.

19.1 Basic Questions and Methodology

A. Similarities

(a) *Questions.* Both theorist 1 and theorist 2 aim at a foundation or a grounding of logic, where by this I mean an investigation of the principles underlying logic, the source of the truth and reliability of logic, the nature of logical knowledge, the conditions of acceptance, rejection, revision of logical theories, the modal force of logic, and so on. Among the questions they ask are: "What is logic?" (Sher, 1991, p. 8), "[W]hat is the subject matter of the science of logic?" (Maddy, 2012, p. 484), "Is logic grounded in the mind or in the world?" (Sher, 2016, p. 255),[2] "[W]hat is the ground of logical truth?" (Maddy, 2014a, p. 121), "What specific features of the world is logic grounded in?" (Sher, 2016, p. 271),[3] "[W]hat exactly are [the] logical facts about?" (Maddy, 2012, p. 484), What is the source of "the *veridicality* of logic"? (Sher, 2016, p. xiii), "[W]hat makes our logic ... reliable"? (Maddy, 2014a, p. 109), "[What] underlies (and explains) the generality and strong modal force of logic[?]" (Sher, 2016, p. 271),[4] What is "the so-called 'logical must'"? (Maddy, 2014a, p. 1), and so on. Both theorists focus on the contemporary heir(s) of Aristotelian syllogistic logic and Fregean quantificational logic, which I shall call here "mathematical logic".

(b) *The Philosopher's job.* An additional area of similarity concerns the more general aspects of the philosopher's job. Both theorists view themselves as conducting *theoretical inquiries.* More generally, as philosophers they view themselves as seeking the truth, or the facts, about whatever philosophical

---

[2] Capital letters replaced by low-case letters.

[3] See Footnote 2.

[4] Unitalicized.

subject-matter they study, just as scientists and mathematicians seek the truth, or the facts, about the subject-matters *they* study.

(c) *Methodology.* Methodologically, both theorists are non-traditionalist in a number of ways: they reject the foundationalist methodology, "first philosophy", pure apriorism, the view that philosophy is isolated from other branches of knowledge, the view that logic is analytic, and so on. Both are neither nominalists nor relativists, instrumentalists, or radical pragmatists. Both regard all knowledge, including logical and philosophical knowledge, as fallible, revisable, subject to critical examination, and capable of improvement. Both believe that humans can correctly describe the world. Both reject the view that there is only one way of correctly describing the world—only "one true theory" of the world. Both reject the Kantian duality of thing-in-itself and appearance. Both regard philosophy as studying certain aspects of the world. And both view themselves as investigating classical philosophical questions using new methods.

(d) *Holism.* Another methodological similarity has to do with their attitude toward certain kinds of holism. Both theorists reject the kinds of holism that are connected with radical pragmatism, indispensability arguments, the view that our system of knowledge is confirmed or refuted only as a whole ("one-unit" holism (Sher, 2016, p. 25)), the view that confirmation is of the same kind in all fields of knowledge and in all cases and that all items of knowledge equally partake in the confirmation of all items ("monolithic" and "homogeneous" holism (Maddy, 2007, p. 275, p. 315)), the view that thin evidence is sufficient for confirmation, and so on.

B. Differences

(a) *Questions and Orientation*

   (i) Theorist 1 seeks to understand logic primarily by understanding the most elementary part of logic, while theorist 2 does not distinguish between the elementary and advanced parts of logic.

   (ii) Theorist 1 says that she is not interested in questions concerning the distinctive features of logical truths/inferences (features that distinguish them from non-logical truths/inferences), questions concerning the scope of logic, the question of logicality as it applies to logical constants, and so on. Theorist 2 pursues all these questions.

   (iii) Theorist 1 seeks an understanding of logic by picking up an example that *most people would regard as logical* and talking about that. Theorist 2 seeks an understanding of logic by focusing primarily on how it *should be* in order to perform its designated role/function.

(b) *Methodology: naturalism vs. figuring out.* Theorist 1 is an "austere" naturalist (as described in the inner jacket of Maddy, 2014a). She identifies her naturalism as "methodological" (Maddy, 2014a, p. 2) and describes it as close to "scientism" (Maddy, 2007, p. 103). She follows Quine's lead in naturalizing all philosophical questions. For example, "[f]aced with the … question … how do we come to know about the world[, she] turns to contemporary cognitive science" (*ibid.*: 90), she "holds a roughly empirical account of logical truth" (Maddy, 2014a, p. ix), and so on. Given any question, philosophical or scientific, she follows the same procedure in investigating it: "beginning from her ordinary perceptual beliefs, [she] gradually develop[s] more sophisticated observational and experimental techniques and correctives, eventually ascending to theory formation and confirmation, all in the sorts of empirical ways usually labeled 'scientific'" (*Ibid.*: 2). "[H]er inquiry takes place within ordinary science" (*ibid.*: 119). She "is born native to the laboratory" (Maddy, 2007, p. 91). Science, for her, always comes first; philosophy—second. (She calls herself "Second Philosopher" and her philosophy "Second Philosophy" (*ibid.*: 19).) She is a bottom-top, rather than a top–bottom, theorist: "all standards, all truth must be laboriously built up from below" (*Ibid* .:174).

As a naturalist, theorist 1 pursues her understanding of logic by pursuing empirical investigations within, or pertaining to, physics and psychology (cognitive science, neuroscience, biology). These include examination of whether different branches of contemporary physics share the world-view of elementary logic, psychological investigations of children's inferential capacities, and so on.

Theorist 2 is not a naturalist. She is not an anti-naturalist either, but she does not conduct her philosophical investigations by conducting scientific investigations. She believes that there is room for questions of multiple kinds about the world, including philosophical questions, and there are multiple methods for investigating such questions, including philosophical methods.

Philosophical questions/methods are not superior to scientific questions/methods, but they are not inferior to them either. Philosophy is neither a "first" nor a "second" philosophy. It is simply philosophy. Theorist 2 appeals to scientific and/or mathematical developments when these contribute to her philosophical investigations, but she does not replace (what she takes to be) genuine philosophical questions by scientific questions or translate them to the latter. She does not limit herself to bottom-up investigations either. She is engaged in both bottom-up and top-down investigations.[5]

Theorist 2's main method of discovery/investigation is a general, widely used method she calls "figuring out" (Sher, 2016, p. x, p. 85). The expression "figuring out" means what it means in everyday discourse: "putting two and two together", "configuring", "finding out by looking and drawing connections", and so on. Figuring out requires active *epistemic freedom*.[6] Theorist 2 conceives of it as a broadly *intellectual* method of investigation, one that can take multiple forms, is open to input of various kinds (including empirical input), and is often incorporated in other methods. Although figuring out requires freedom, it is subject to exacting standards

---

[5] See, e.g., Sher (1991), Chap. 3 is a top-down account and Chap. 4 a bottom-up account of logicality.

[6] For epistemic freedom see Sher (2016, Sect. 1.3).

of truth and justification. The activity of figuring out takes place both in experimental and in abstract fields of knowledge. For example, an experimental scientist has to figure out which experiment (which experimental activity among all the experimental activities he can engage in) will test a given scientific hypothesis. A theoretical scientist has to figure out whether two theories (such as quantum mechanics and relativity theory) are compatible and if not, whether and how they can be revised so as to be compatible. A meta-logician figures out whether a given theory is complete and how to prove its completeness/incompleteness. A philosopher figures out how to justify or refute a given philosophical argument (e.g., Hume's skeptical argument), how to solve philosophical problems (e.g., the Liar-paradox problem), and so on.[7]

When it comes to logic, theorist 2 is engaged in figuring out how to ground logic. This she does in a way that connects with her functional approach to logic: figuring out, first, what role or function logic is designed to play in our search for knowledge; figuring out, next, what features logic has to have in order to perform this role; figuring out whether having these features requires that logic be grounded in the world; figuring out what facets of the world logic has to, or can, be grounded in; and so on.

(c) *Levels of Inquiry.* Another methodological difference concerns levels of investigation. Theorist 1 affirms only one level of investigation, while theorist 2 affirms multiple levels.

Theorist 1 contrasts a *one-level* approach to inquiry with a *two-level* approach. Her paradigm of a two-level approach is Kant's empirical and transcendental levels of inquiry. "Kant's system ... epitomizes ... a striking structural feature: it involves two distinct levels of inquiry" (Maddy, 2007, p. 47). The lower level is empirical and realistic, the higher level is apriori and idealistic. Theorist 1 "sees no point in the second [higher] level" (*Ibid.*: 81). "[O]ne [of her] recurrent theme[s] is the rejection of extra-scientific perspectives such as that of Kant's transcendental philosophy" (Maddy, 1999, p. 109). She is "deaf and blind to the lure of the Kantian transcendental project" (Maddy, 2007, p. 101). Her "investigations are pursued on one level, as part and parcel of the single mosaic of natural science" (*ibid.*: 47). She "makes no sense of transcendental analysis" (*Ibid.*: 225). "In place of Kant's two-level view, [she] seeks one unified scientific account" (*Ibid.*).

Theorist 2 affirms multiple levels of inquiry. Given humans' significant cognitive limitations, they have to make full use of their cognitive capacities. And two especially important capacities they do have—"immanent" and "transcendent" thinking (Sher, 2016, p. xi, 163, 166)—are associated with different levels of investigation. In *immanent* investigation we "direct... our mental gaze" (*ibid.*: 163) at the world in order to study it. This is what we do when we speak from *within* a theory. A theory studies some region of the

[7] For more on *figuring out* see Sher (2016, pp. 85–86, 192, 199, 243n, 254).

world, asks what properties objects in this region have (what relations they stand in), what laws govern these properties (relations), and so on. Scientific, mathematical, and philosophical theories are all immanent in this sense. Epistemology, for example, is immanent in directing its gaze at that region of reality in which human knowledge is located. In *transcendent* investigation we go outside a given immanent inquiry and investigate this inquiry itself. We may investigate its relation to its subject-matter, its relations to other inquiries, and so on. Since our inquiries are part of the world, transcendent inquiry is also immanent. Not every immanent inquiry is transcendent, but every transcendent inquiry is immanent. Accordingly, we have multiple levels of inquiry: immanence, immanent-transcendence, immanent-double-transcendence, and so on. Epistemology is, for the most part, immanent-transcendent.

The standpoint of transcendent inquiries, as conceived by theorist 2, is *factual* and *rational* rather than *ideal* or *magical*. It is also *human* rather than *Godly*. Theorist 2 calls it "*HH [human-human] transcendence*: [t]ranscending one *human* standpoint, X, to another, *human*, standpoint Y" (*ibid.*: 167). One paradigm of transcendence of this kind is transcendence to a Tarskian meta-theory. There is nothing Godly, or mysterious, or magical, or absolute, or ideal about a Tarskian meta-theory, but it has certain capacities that a Tarskian object-theory (immanence without transcendence) lacks.[8] We can transcend the same theory to multiple other theories. For example, we can transcend science both to the sociology of science and to the philosophy of science. Transcendence (of this kind) "reflects the dynamic nature of human ... knowledge; in particular, our ability to move up and down, back and forth, from one [level of inquiry] to another, in cognizing the world (including ourselves in it)" (*Ibid.*: 232–3).

(d) *Holism.* We have seen that certain types of holism are rejected both by theorist 1 and by theorist 2. But while theorist 1 regards these types as exhausting holism, and as a result views herself as an anti-holist, theorist 2 does not. Theorist 2 sees new possibilities for holism, and she develops a new type of holism, "foundational-holism" (*ibid.*: viii, 20). Foundational-holism is neither a one-unit holism nor a monolithic/homogeneous holism, neither a radically pragmatic holism nor an indispensabilist or a thin-evidence holism. Foundational-holism is motivated by the belief that the classical philosophical project of constructing a foundation/grounding for knowledge, truth, logic, and so on is worthwhile, but the traditional method for carrying it out, foundational*ism*, is self-defeating. In particular, foundationalism's requirement that the grounding of knowledge be a strict partial ordering with minimal elements (a tree or a pyramid), where each unit of knowledge is grounded in units lower than it in the foundationalist hierarchy, is self-defeating. The problem is that the entire foundation depends

[8] See Tarski (1933).

on the minimal elements, but the minimal elements themselves cannot be properly grounded.

Foundational-holism discards the strict ordering requirement. It says that "there is no inherent connection between grounding our system of knowledge in reality and doing so in a strictly ordered manner" (*ibid.*: 23). Instead, "[o]ur system of knowledge is connected to reality by a multi-faceted relation (or network of relations), strictly-ordered in some sections, not strictly-ordered in others. All branches of knowledge, qua branches of *knowledge,* must be grounded in reality, but their grounding need not follow a single, strict, and rigid pattern" (*Ibid.*). This, however, does not mean that foundational-holism's conception of the grounding of knowledge is thin or watered-down. On the contrary. It is designed for making substantial grounding possible, as we shall see in the case of logic below. We may say that "foundational holism puts the holistic method in [the] service of a robust, world-oriented, universal foundational project" (*ibid.*: ix).

One of the problems that undermines Quine's holism, according to theorist 1, is circularity: "the Quinean holist ... can[not] carry out his empirical testing of logical laws without presupposing at least some of those laws" (Maddy, 2007, p. 203). Foundational-holism is especially adept at handling circularity of this kind. Rejecting destructive as well as trivializing ("P; therefore P") circularities, it explains away circularities of the kind theorist 1 attributes to Quine as *partial* and *temporary*. They are partial since in revising logic we use other resources than just logic: we use mathematical resources, philosophical resources, scientific resources, and so on, so the overall revision is only partly circular. And this partial circularity is in principle temporary: every revision requires holding something fixed (assuming something), but we are not required to hold the same thing fixed, either in the course of a given revision or in future revisions. In critically revising logic we use some logical resources but we hold off others, we switch back and forth between assuming X and leaving X open to revision (while assuming Y), and so on. Furthermore, upon completing a given revision, we turn to examining its assumptions, using any resources that are available to us, including resources that are external to these assumptions and resources that we created based on earlier revisions of our theories. This may lead us to reexamine some of our past revisions, but this is as it should be. No illusory promise of a final, ultimate, infallible examination is given, but progress is in principle achieved.[9]

[9] For more on the foundational-holistic treatment of circularity, see (Sher, 2016, pp. 30–34, 323).

19.2 Grounding Logic in the World and the Standing of Logical Necessity

Both theory 1 and theory 2 ground mathematical logic in the world, and in a sense they both ground it in the *formal structure* of the world. But they differ in the scope of their grounding, their understanding of "formal structure", the way they ground logic in the world, and their treatment of the necessity of logic.

Theory 1. In grounding a basic discipline such as logic, we face a special challenge: finding an effective starting point. Although theorist 1's grounding of logic is supposed to be empirical, she starts with Kant. Why Kant?—Because

> [t]here's considerable appeal [both] to the suggestion that logic depends on very general structural features of the world, and to the quite different idea that logic is embodied in our most primitive forms of conceptualization. Kant's combination of transcendental idealism and empirical realism accomplishes the neat trick of giving us both these at once: transcendentally, logic is dictated by the forms of judgment and pure concepts of the discursive intellect; empirically, logic describes the underlying structure of the world. (Maddy, 2007, p. 225)

Now, theorist 1, as we have seen above, rejects large parts of the Kantian account, including Kant's entire transcendental approach. What she is after is a "second-philosophical [i.e., *naturalist*] version of the Kantian position that preserves its merits while bypassing the transcendental" (*ibid.*).

Theorist 1 divides logic into two parts: "*rudimentary logic*" (*ibid.*: 231, my italics) and what I shall call *full-scale logic*. Rudimentary logic is pre-logic, proto-logic, or the most elementary part of logic; full-scale logic is standard,[10] bivalent, 1st-order mathematical logic in its entirety. What is included and not included in rudimentary logic? There is a "considerable distance between rudimentary logic and modern, first-order predicate logic" (*ibid.*: 283). Rudimentary logic includes a fair number of the primitive rules applicable to disjunction, conjunction, and to some extent negation, identity, and the universal and existential quantifiers. But its conditional is the syllogistic rather than the truth-functional conditional. And it has no provisions for avoiding indeterminacy, non-denoting names, paradoxical predicates, and so on. As a result, rudimentary logic is a *trivalent* rather than a classical (bivalent) logic, and is to some degree fuzzy as well. Very few of the classical logical truths hold in rudimentary logic, and many classical logical inferences fail as well. For example, whereas Modus Ponens and Disjunctive Syllogism hold in rudimentary logic, Modus Tollens and Reductio ad Absurdum fail. Furthermore, rudimentary logic lacks the systematicity of full-scale logic.[11] Another distinctive feature of rudimentary logic

---

[10] "Standard" in the sense of being restricted to the standard logical constants (excluding other logical constants such as the quantifier "Most").

[11] See (*ibid.*: 229–231, 283).

is that it includes some basic arithmetic, such as very simple additions and multiplications.[12] Epistemically, rudimentary logic can be characterized as the kind of logic-mathematics that is available to infants and very young children.

In speaking about theory 1's grounding of rudimentary logic we may distinguish between its grounding of the *veridicality* of rudimentary logic and its grounding of its *necessity*. The *veridicality* of rudimentary logic is grounded, metaphysically, in the *physical world*, and epistemically in the *human mind*. What, in the *physical world*, is the veridicality of rudimentary logic grounded in? It is grounded in what theorist 1 calls, after Kant and Frege, the "KF- structure" (*ibid*.: 229) of the physical world, which consists of "a domain of objects that bear properties and stand in relations, perhaps some universal properties, plus compounds of these involving conjunctions, disjunctions and negations", where "some interconnections between these situations are robust ground-consequent dependencies". (*Ibid*.: 228–9) This structure behaves in a way that supports the logical inferences of rudimentary logic. For example, a physical object that has the property P or the property Q but does not have the property Q has the property P. The veridicality of Disjunctive Syllogism is grounded in this fact.

How does theorist 1 establish the claim that the world has this KF-structure? She establishes it in a process of the kind described in Part I, namely, one that begins with common sense and then turns more and more deeply into science. Here is how she establishes the reality of KF-objects:

> Common sense clearly endorses the idea that the world contains many medium-sized physical objects. Such things cohere, have boundaries, and move continuously as units; examples range from apples, chairs, and people, to boulders, books, and baseballs. When the Second Philosopher examines these beliefs more closely, she finds both confirmation and explication. Such objects are indeed distinct from their surroundings; they are composed of intricately arranged atoms dotted throughout largely empty space, and those atoms and arrangements differ starkly from the atoms and arrangements in the space nearby. Their cohesiveness comes from the bonds between their atoms; their solidity comes from the electromagnetic fields they generate; they move according to certain principles of motion, and so on. On further, more specialized investigation, she uncovers further such objects: planets, blood cells, spider mites. She concludes that ordinary physical objects are real structures in the world. (*ibid*.: 234)

But science also limits the reality of the KF-world. Theorist 1 learns from science that the KF-world is real on the *macro-level* yet not on the *micro-level*:

> Our examination of the extent to which the world's structuring into objects can be pushed into the microscopic, turns up a... breakdown... . ... [T]he micro-world is not structured into things of the familiar sort; though the world does contain numerous ordinary objects, it also contains phenomena that are not so structured. Despite our scientific predisposition to see the world in these terms, our pursuit of science itself has taught us that the world is not as we expect it to be, not in all its parts. This portion of our empirical hypothesis – that the world consists of coherent objects that move as units along continuous spatiotemporal paths – must be qualified. The world is structured into such objects at the macro-level, but at the micro-level, all current evidence suggests that it is not. ... Here not only the spatiotemporal features of objects are undermined, but the pure notion of an object as an individual thing. Thus

[12] See (*ibid*.: 318).

> it seems the micro-world cannot be said to display an abstract KF-structure of individual objects. (*Ibid.*: 236–7)

The veridicality of rudimentary logic is grounded only in the macro-world, hence is limited to it.

Turning to the grounding of the veridicality of rudimentary logic in the *human mind* (epistemic grounding), the underlying idea is that for us to have a logic that simulates the KF- world, our mind has to be appropriately structured. For this, theorist 1 appeals to contemporary psychology: "[t]here can be little doubt that ordinary adults see the world in terms of individual objects" (*ibid.*: 245). But "there is morethan this" (*ibid.*): contemporary psychology tells us that this way of seeing the world is not "acquired or learned"; it is "the product of [humans'] primitive conceptual mechanisms" (*ibid.*). Contemporary researchers of infant and child cognitive development, such as Elizabeth Spelke, have shown that infants cognize the world in terms of object unity:

> [I]nfants as young as 4 months are able to individuate and identify objects using spatiotemporal criteria: spatial contiguity, common fate, and continuous motion. No such object can be in two places at once; two such objects cannot occupy the same location; all such objects travel on continuous paths .... They are 'complete, connected, solid bodies that persist over occlusion and maintain their identity through time'. (*ibid.*: 253)
>
> [O]ther principles [include] continuity, solidity, and contact ...[;] cohesion is fundamental. (*ibid.*: 254)

Objects of this kind are often called in the literature "Spelke-objects". Theorist 1 says:

> [I]t seems humans are so configured, biologically, that they come to perceive a world of Spelke objects, without instruction, given ordinary maturation in a normal environment. Or, in the Second Philosopher's terms, the ability to perceive Spelke objects is part of a human being's most primitive cognitive equipment. (*ibid.*: 258)

What connects the epistemic grounding of rudimentary logic to its metaphysical grounding is: "humans are so configured *because* the macro-world is so structured" (*ibid.*: 264, my italics). In the case of objects:

> [T]he spatiotemporal information central to the notion of a Spelke object is generally more reliable than the information infants eschew. ... And the types of spatiotemporal information infants prefer – cohesion, solidity, continuity of motion – are more reliable than those they tend to ignore – similarity, good continuation, good form. (*ibid.*: 264–5)

Theorist 1 is aware of the fact that the new theory of child development may not be completely accurate. Nevertheless, she says, it is clear that

> human beings have primitive cognitive mechanisms capable of detecting and representing KF-structures because they live in a largely KF-world and interact almost exclusively with its KF-aspects. The evidence clearly suggests that we – like the chicks, pigeons, and monkeys – are responding directly to some of the world's most elementary features. (*Ibid.*: 270)

We may sum up theory 1's grounding of the veridicality of rudimentary logic in the world on the one hand and in the human mind on the other thus:

A. "[R]udimentary logic is true of the world insofar as it's a KF-world, and in many but not all respects it is".
B. "[H]uman beings believe the simple truths of rudimentary logic because their most primitive cognitive mechanisms allow them to detect and represent the KF-structure in the world".
C. "[T]he primitive cognitive mechanisms of human beings are this way because we live in a largely KF-world and interact almost exclusively with its KF-structures". (*Ibid.*: 271).

But while the KF-structure of the physical world, at least on the macro-level, grounds the veridicality of rudimentary logic, it does *not* ground its *necessity*, according to theorist 1. The reason is that it is a *contingent* physical fact that the macro-world is a KF-world.[13] Why do people regard the inferences of rudimentary logic as necessary? Theorist 1's answer to this question is similar to Hume's answer to the question why the laws of nature are usually regarded as necessary: our psychology. But while Hume's answer is given in terms of "habit" and "custom", theorist 1 appeals to the results obtained by contemporary developmental psychology. The logical inferences of rudimentary logic, contemporary psychology tell us, are built into the primitive mechanisms of human cognition. As a result, these inferences appear "obvious" to us, and this, in turn, leads us to view them as necessary.[14]

It should be noted that in a certain respect, theory 1's account of rudimentary logical inferences does not distinguish between logical and non-logical inferences. Depending on what contemporary physics and psychology say about the basic features of the world and the basic features of human cognition, it may very well be the case that there is no difference between the grounding of non-logical rudimentary inferences (possibly: "This is a dog; therefore it barks") and logical rudimentary inferences ("This dog barks; therefore, it barks or meows").[15] In this sense, it is not specifically a grounding of logic. This, however, would not be problematic for theorist 1 who, as we have noted above, has no interest in explaining the difference between logical and non-logical truths/inferences.

Turning to full-scale logic, i.e., standard classical (bivalent) 1st-order logic, the move to this logic, according to theory 1, is motivated by the theoretical limitations or shortcomings of rudimentary logic. "In light of these ... shortcomings, ... logicians have sought stronger, more manageable systems of logic. The considerable distance between rudimentary logic and modern, 1st-order predicate logic ('classical logic') [is] bridged by a number of restrictions and idealizations, all more or less traceable to Frege" (*ibid.*: 283). These include the devising of an idealized, "'... logically perfect language'" (*ibid.*) under which "the names nam[e], the predicates classify" (*ibid.*: 288), the conditional is truth-functional, there is no limit on the complexity of formulas, and so on. "Here, just as in the rest of science, we [are] sensitive to

[13] See (*ibid.*: 273).

[14] See (*ibid.*: 273–274).

[15] In this respect, theory 1's account of logic may be similar to Quine's account of analyticity in terms of *stimuli*—"stimulus analyticity"—which does not distinguish between "Dogs bark" and "Bachelors are unmarried".

the benefits and dangers of idealization, and satisfy ourselves that the idealization in question is appropriate to the case at hand" (*ibid.*: 287). The result is a "vastly more effective instrument" (*ibid.*: 288) than rudimentary logic. "The justification [of the idealizations underlying this instrument is] that they make it possible to achieve results that would otherwise be impossible or impractical, and that they do so without introducing any relevant distortions" (*ibid.*). Under these conditions "the familiar logic can be trusted" (*ibid.*). In this way, full-scale logic is "grounded in a rudimentary logic that's both true of the world and embedded in our most primitive modes of cognition and representation" (*ibid.*). In contexts where classical logic is inappropriate, we may resort to various "deviant logics" (*ibid.*).

Summing up we may say that theory 1 grounds the veridicality of logic both (metaphysically) in the world and (epistemically) in the mind, and its necessity it grounds only (epistemically) in the mind. The veridicality of logic is grounded metaphysically in the KF- structure of the physical world and epistemically in our primitive cognitive mechanisms. These groundings are direct in the case of rudimentary logic, indirect in the case of full-scale logic. But the veridicality of full-scale logic is also grounded in our theoretical abilities, including our ability to engage in appropriate idealization, systematization, and so on.

Since the KF-structure is limited to the macro-level of the physical world, the veridicality of logic is limited to that level.

The seeming necessity of logic, unlike its veridicality, is limited to rudimentary logic. Epistemically, it is grounded in the same primitive cognitive mechanisms as its veridicality. These mechanisms make rudimentary logic appear obvious to us, hence seemingly necessary. Metaphysically, however, the necessity of rudimentary logic is illusory since its veridicality is contingent. It is contingent on the basic structure of the physical world and on the primitive cognitive mechanisms of the human mind.

***Theory** 2*. While theorist 1 has one starting point—Kant—theorist 2 has two starting points. The first is the question with which she actually started her investigation of logic and which gave her a clue to the grounding of logic. The second is the first step in her subsequent rational presentation of the grounding itself. The first is the question of logical constants, which is the focus of Sher (1991). The second is the question of logic's task (function) in the human pursuit of knowledge, with which her functional outline of a foundation for logic in Sher (2016, Chap. 10) begins. Here I shall start with the latter.

A. *What is Logic's Task in the Pursuit of Knowledge?* To answer this question, theorist 2 begins with a few observations on the "basic human cognitive-epistemic situation" (Sher, 2017, p. 373). These include the observations that (i) for one reason or another, humans seek to know the world in its full complexity, and not just practically but also theoretically, (ii) their cognitive capacities are considerable in some respects, limited—and sometimes severely limited—in others, (iii) the latter makes the pursuit of knowledge highly problematic for them, while the former enables them to find solutions to, avoid, or compensate for, some of these problems. As a result of this situation, it is both worthwhile and feasible for humans to develop methods for expanding their knowledge, such as methods of inference that use knowledge they already have to obtain new knowledge (employing means within their reach).

What methods of inference would provide this benefit? Given humans' aspiration to know the world as it is, rather than as they imagine or wish it to be (or are told, or are manipulated to believe, it is), one of their most important standards of success is *truth.* They would greatly benefit from methods of inference that *transmit truth from premises to conclusions.* And a method of inference that is *highly general* and *modally-strong* (i.e., guarantees the transmission of truth in all fields of knowledge with a strong modal force) would be especially beneficial. This, theorist 2 suggests, is logic's task (or one of its main tasks):

> *Logic's task is to develop a method of inference which is both highly general and has an especially strong modal force. More specifically, its task is to develop a method for constructing inferences that transmit truth from sentences to sentences with an especially strong modal force and regardless of field of knowledge.* (Sher, 2016, p. 255)

B. *What constraints does this task set on an adequate logical system and its foundation/grounding?* Here, theorist 2 begins with the observation that the kind of truth that logical inference is required to transmit is the kind of truth that serves as a standard for successful knowledge. Since the aim is to know the world as it is, this truth is *truth-in-the- world*, i.e., truth as correspondence with the world (though not necessarily, and probably usually not, correspondence in the traditional, naive sense of copy, picture, mirror image, or even direct isomorphism).[16] But if logical inference is to transmit truth-in-the-world, it has to take the world (or something relevant in the world) into account. To see this, suppose someone comes to you and says: "I invented/constructed/put-together a beautiful logical system, L. You should adopt it as your logic." To adopt L, you have to ascertain that it adequately performs the task of logic, i.e., that its inferences transmit truth-in-the-world from premises to conclusion and that they do this with a strong modal force. Now, suppose L says that the sentences $S_1,\ldots,S_n$ logically imply the sentence S, but *the world is such* that the conjoined truth of $S_1,\ldots,S_n$ *rules out* the truth of S. In that case, you should not adopt the logic L. And even if the world is such that all of $S_1,\ldots,S_n$ and S are true or at least one of $S_1,\ldots,S_n$ is false, but this is merely an accidental matter (there is no strong modal connection between the circumstances in the world under which $S_1,\ldots,S_n$ are true and those under which S is true), you should not adopt L. Conclusion: for a logical system to be adequate, it cannot say that S follows logically from $S_1,\ldots,S_n$ *unless* there is an appropriate connection between the *worldly conditions* that would render all of $S_1,\ldots,S_n$ true and the *worldly conditions* that would render S true.

C. *Is there a logical system and a foundation/grounding for such a system that satisfy these constraints?* Theorist 2's answer to this question is positive: an appropriately built logical system satisfies these constraints. What is an appropriately built logical system? Assuming the common conception of logical inference as taking into account only the logical structure of the sentences involved, where

[16] For discussion of non-naive correspondence, see Sher (2016, Chap. 8).

this is a matter of the identity and distribution of logical constants, we may characterize an appropriately built logical system as one that satisfies the following conditions: (i) its logical constants denote *formal* properties (relations, functions), and (ii) its inferences are based on laws governing the formal properties denoted by its logical constants.[17]

To see why such a logical system is grounded in the world and how its special grounding ensures its strong modal force, we need to turn to theorist 2's systematic conception of *formality*.[18]

Formality as Strong Invariance. Theory 2 offers a systematic explanation of *formality* in terms of *invariance*, a highly fruitful concept that is widely used in science and mathematics, and has also been used in certain developments of meta-logic.[19] There are various types of invariance; theorist 2 employs a notion of invariance that may be called *property-invariance.*

The basic idea underlying property-invariance is that properties, in general, are selective in character, i.e., they "notice", "distinguish", or "are affected by" some differences between objects but not others. A property is *invariant* under all and only replacements of objects that it does not distinguish between. For example, the 1st-level property *is-a-human*[20] distinguishes between men and dogs but not between men and women or between dogs and horses. It is invariant under replacements of men by women (and *vice versa*) and of dogs by horses (and *vice versa*) but not under replacements of men by dogs (or *vice versa*). Precisifying, theory 2 says that *is-a-human* is invariant under *all 1–1 and onto replacements* of men by women[21] (and *vice versa*) and of dogs by cats (and *vice versa*) *in any equinumerous domains*, but not under any 1–1 and onto replacement of men by dogs (or *vice versa*) in any equinumerous domains. Other properties have different invariances.[22]

One significant result of the application of invariance to properties is that some properties have the distinctive feature of being *maximally-invariant* (Sher, 2021), i.e., invariant under *all* 1–1 and onto replacements of individuals—*actual and counterfactual*—in any (equinumerous) domains. For example, the 1st-level property (relation) *is-identical-to* is of this kind. Expanding this feature to 2nd-

[17] For more on an "appropriately built" logical system see Sher (1991, Chap. 3, 2016, Chap. 10). For the purpose of this paper we can think of standard 1st-order mathematical logic as a paradigmatic example of such a system.

[18] In a sense, theorist 1 also grounds logic—or at least its veridicality—in the formal structure of the world, since the KF-features she talks about are intuitively formal. And in one of her works (Maddy, 2014b) she talks about the "formal" structure of the world instead of its "KF" structure. But she does not offer a systematic characterization of formality and seems averse to such a characterization.

[19] For relevant examples, see Mostowski (1957), Lindström (1966), and Tarski (1966, Published 1986).

[20] I use italics for 1st-level properties.

[21] More precisely, replacements of individuals which are men by individuals which are women.

[22] For example, *is-a-woman*, unlike *is-a-human*, is not invariant under replacements of men by women (or vice versa).

(and higher-) level properties[23] and using current logical and mathematical terminology, it is easy to see that all the other properties denoted by the standard logical constants—IS-NOT-EMPTY, IS-UNIVERSAL, COMPLEMENT, INTERSECTION, UNION, INCLUSION (denoted by "∃", "∀", "~", "&", "∨", "⊃")[24]—are also maximally-invariant.

The notion of *maximal invariance* coincides with that of *invariance under all isomorphisms*.[25] Noting that the latter notion intuitively captures, or precisifies, the idea of objectual or worldly *formality*,[26] theorist 2 introduces the following definition or criterion of formality:

> *Criterion of Formality: A property is formal* iff *it is maximally-invariant* iff *it is invariant under all isomorphisms.*

Take IS-NOT-EMPTY (the denotation of "∃") as an example. Its argument-structures have the form < D, B > , where D is a domain of individuals and B is the extension of a 1-place 1st-level property P in D. IS-NOT-EMPTY is formal because it is invariant under any replacement of <D, B> by an isomorphic structure <D′, B′>: B is not empty (in D) iff B′ is not empty (in D′).

It follows from the above criterion that the properties denoted by the standard logical constants are *formal*. And it further follows that, given a well-built system of logical inference, its inferences are grounded in laws governing formal properties. As an example, take the logical inference

(1) $(\exists x)(Px \vee Qx), \sim(\exists x)Px \vDash (\exists x)Qx$

This inference is grounded in a law that connects the combination of (i) non-emptiness of a union of two properties and (ii) emptiness of one of these properties to (iii) non-emptiness of the other. I.e., this inference is grounded in a law that connects one combination of formal properties (of objects) to another. That is to say, it is grounded in a law governing formal properties, i.e., a *formal law* governing the world. More generally, logical inferences are grounded in *formal facets* of the world, or in the *formal structure* of the world.[27]

But this is not all. Not only is the veridicality (transmission of truth) of logical inferences, but also their necessity, is grounded in the formal structure of the world.

---

[23] A (1-place) 2nd-level property **P** is maximally-invariant iff for any 1–1 and onto replacement *r* of individuals from any domain, D_1, by individuals from any equinumerous domain, D_2, and any 1st-level property P on D_3, P has the property **P** iff its image under *r* has the property **P** (or, more precisely: the extension of P in D_1, B, has the property **P** iff its image under *r* has the property **P**).

[24] (i) I use small capital letters for 2nd-level properties. (ii) The listed denotations of the logical connectives are in typical predicative contexts (e.g., Φx & Ψx), but they are expandable to other contexts as well. (See Sher, 2016, Sect. 10.4).

[25] For "invariance under (all) isomorphisms" see Sher (1991, Chap. 3; 2016, Sect. 10.4). For its relation to "maximal invariance" see Sher (2021).

[26] See Sher (1991, Chap. 3; 2016, Sects. 10.4–10.5).

[27] For further details see (*ibid.*).

Formality implies Necessity. This is another result that theorist 2 arrives at based on the connection between formality and invariance. Formal properties are maximally-invariant, i.e., they do not distinguish between *any* individuals, *actual or counterfactual.* In other words, they behave in the same way regardless of whether the individuals they apply to are actual or counterfactual.[28] As a result, the laws governing/describing formal properties, i.e., *formal laws*, cannot distinguish between *any* individuals, *actual or counterfactual*, either. (If they did, they would not accurately describe the behavior of formal properties.) That is to say: formal laws hold in all domains, regardless of whether their individuals are *actual or counterfactual*, and as such *have a strong modal force* or are *necessary*. Indeed, they hold even of counterfactual individuals that are not physically possible, hence they have an *especially* strong modal force or are *highly* necessary.

Thus, the strong modal force of

(1) $(\exists x)(Px \vee Qx), \sim(\exists x)Px \models (\exists x)Qx$,

is grounded in a law that governs the formal properties of non-emptiness, union, and complementation. Since this law describes the behavior of these properties in any domain of any individuals, actual or counterfactual, it is highly necessary.

Theorist 2 sums up her grounding of logic as follows:

> *Logic (logical consequence) is grounded in formal laws governing reality – laws governing formal features of objects and properties – and it is the ... applicability and strong modal force of such laws that underlies (and explains) the [veridicality] and strong modal force of logic.* (Sher, 2016, p. 271)

Theorist 2, we see, regards the formal laws governing the world as necessary. How does she respond to theorist 1's claim that the KF-structure of the world, which is theorist 1's version of theorist 2's "formal structure of the world", is contingent? Theorist 2 does not say that the formal structure of the world, including its formal properties and formal laws, could not have been different than they are. (She leaves this an open question.) What she says is that regardless of what the formal laws are, they are laws that govern formal properties which, being formal, are maximally-invariant, and that, as a result, these laws have a larger actual-counterfactual scope than whatever laws govern the physical, biological, psychological properties of objects in the world, which are not maximally-invariant. It is in this sense that formal laws have a higher degree of necessity than physical, biological, psychological and other laws, and as such are highly necessary.[29]

Theory 2's grounding of logic is, thus, not dependent either on a particular logic or on particular formal properties and laws. Theory 2 says that whatever the formal

---

[28] Or, if they are higher-level properties, they behave in the same way with regard to any properties of any actual or counterfactual individuals, and so on.

[29] Figuratively, we can describe this in terms of formal "multiverses". Different formal multiverses have different formal laws, but in each formal multiverse, the formal laws have a stronger modal force (a larger actual-counterfactual scope) than any of its physical laws (or any of the physical laws within the physical multiverses that are included in this formal multiverse).

properties and their laws turn out to be, they will ground a logical system that satisfies the task of logic in a world governed by those laws.

One result of grounding logic in the formal structure of the world, where formality is defined in terms of maximal invariance, is a delineation of the *scope* and *distinguishing characteristics* of (mathematical) logic. Suppose the formal properties are (the 2nd-level correlates) of those definable by contemporary (1st-order) set-theory. Then the scope of (mathematical) logic extends to all logical systems with logical constants denoting such properties. This includes logical systems with such logical quantifiers as MOST, FINITELY-MANY, INDENUMERABLY-MANY, IS-WELL-ORDERED, etc., i.e., logics with so-called "generalized quantifiers" of the kind studied in Barwise and Feferman (1985). The distinguishing characteristic of all logics of this kind, including standard 1st-order logic, is their *formality*: their having modally-strong truth-transmitting inferences grounded in formal laws.

I have said that theory 2, like theory 1, grounds logic both in the world and in the human mind. What in the human mind is logic grounded in according to theory 2? It is grounded in (i) humans' interest in an especially strong method of inference centered on the transmission of truth from premises to conclusion, (ii) humans' cognitive abilities, in particular their ability to figure out what the formal laws are[30] and how to construct a logical system grounded in these laws,[31] and so on.

To sum up: Theory 2 grounds logic—full-scale mathematical logic—in the formal structure of the world, and this grounding encompasses both its veridicality and its modal force. Formality is explained in terms of maximal invariance, which is tantamount to invariance under isomorphisms. The formality of logical inferences distinguishes them from non-logical inferences and determines the scope (maximal extendability) of (mathematical) logic.[32]

19.3 Relation Between Logic and Mathematics

Theorist 1, as we have seen above, places the elementary parts of arithmetic under the same category as the elementary parts of logic:

> [W]e've seen ... that the world ... enjoy[s] a considerable amount of logical structuring, and we now note that this includes the facts underlying such elementary arithmetical claims as $2 + 2 = 4$. Likewise it appears that any KF-structure validates the worldly instances of simple multiplications, like $2 \times 3 = 6$.... Exponentiation is more complex, but still within the range of KF-structuring. Thus, insofar as it is KF-, the world reflects the structure of elementary arithmetical equalities and inequalities. (Maddy, 2007, pp.318–9)

But more advanced parts of arithmetic are beyond the scope of this connection:

> [E]lementary arithmetical claims like $2 + 2 = 4$ are answerable to the logical structure of the world – which means they have the same status as rudimentary logic – but ... the '...'

[30] See discussion of the relation between logic and mathematics in Part 19.3 below.

[31] See Sher (1991, 2016, Chap. 10).

[32] For theory 2's approach to deviant logics, see Sher (2016, Sect. 10.10).

> of mathematical number theory is another matter. ... [A]s soon as the '...' comes into play, we've entered the realm of higher mathematics. (*Ibid.*: 361–2)

Other parts of mathematics, such as higher set-theory (set-theory beyond number theory) are still farther away from logic. These parts of mathematics seem not to be grounded in the KF-world at all in the way rudimentary logic or rudimentary arithmetic are.

Leaving aside most of theorist 1's rich work on mathematics (see, e.g., Maddy, 1988, 1997), I focus here only on her view of the relation between logic and what she calls "deep mathematics", which includes higher set theory.

Theorist 1 emphasizes the "sharp contrast" between her "robust worldly" grounding of "logic" and her "objective but metaphysically neutral" grounding of "set theory" (Maddy, 2014c, pp. 222–3). In the case of logic her focus is on "the ground of logical *truth*", in the case of set theory on "the ground of set theoretic *practice*" (*ibid.*: 223, my emphasis). Whereas metaphysical questions come first (or at least early) in her understanding of logic, "the question of set theory's metaphysics comes last" (*ibid.*: 228). And while she is a *robust realist* with respect to rudimentary logic, she is a *thin realist* or even an *irrealist* with respect to set theory. But mostly she sees set theory as pursuing goals *internal* to mathematics, goals that are objective but have very little, if anything, to do with the world or with truth. Indeed, in her view, there is "no fact of the matter ... whether or not the notions of truth and existence... should be extended to set theory and the rest of [deep] mathematics" (*ibid.*: 230). What grounds deep mathematics is neither objects nor truth: "What we have here is a form of Objectivism without objects, and even without truth. ... [W]hat grounds mathematics isn't mathematical objects ... [or] even [mathematical] truth, it's mathematical depth." (*Ibid.*: 231).

So, while the main focus of theorist 1's investigation of logic is its connection to the basic structure of the world on the one hand and the human mind on the other, this is not the focus of her investigation of mathematics. And while she approaches logic from the outside so to speak, asking what physics and psychology—which are external to it—teach us about its grounding, she does not approach deep mathematics in this way. She does not draw a systematic connection between logic and mathematics.

Unlike theorist 1, theorist 2 does draw a systematic connection between logic and mathematics. She connects logic and mathematics, including set theory beyond number theory, *in principle*, based on their joint *formality*. She is a realist with respect to the formal structure of the world, which she views as central to both logic and mathematics. The formality of mathematics, like the formality of logic, is an *invariance result*: all higher-level mathematical properties, including all higher-level correlates of 1st-level mathematical properties, are maximally-invariant, or invariant under all isomorphisms, hence formal. This result systematically connects logic and mathematics. Logic and mathematics, however, are not identical. Neither is one reducible to the other. There is a division of labor between them. *Mathematics studies the laws governing formal properties while logic utilizes these laws to construct a powerful method of inference.* Theorist 2 calls this joint conception of logic and mathematics "semantic formalism" (Sher, 2016, p. 321) and "formal-structuralis[m]" (Sher, 2001, 2016, p. 321n). Logic and mathematics, on this view, develop in tandem: starting from a minimal logic-mathematics, we use its resources to develop a more advanced mathematics, which we use to develop a more advanced logic, which we use to develop a still more advanced mathematics, and so on.[33]

Theory 2's realist account of logic is thus connected to a realist account of mathematics. This realism, however, is not Platonic. It starts with regular objects and proceeds, based on straightforward invariance considerations, to their formal properties, the modally-strong laws governing these properties, and the modally-strong inferences based on these laws.

19.4 The World[34]

Let me conclude with a clarification of theorist 2's counterpart of theorist 1's KF-world, which I shall refer to simply as "the world". What the basic constituents of the world are is, for theorist 2, an open question, determined jointly by investigations of different disciplines. She does not distinguish between a "rudimentary" or "straightforward" structure of the world (which theorist 1 associates with KF-structure) and the rest of its structure. Whatever the structure of the world is, it contains both formal and non-formal elements, and in discussing logic theorist 2 is interested in its formal elements. Due to its formality, a correct description of the formal structure of the world is likely to be given by mathematics. One candidate for a mathematical theory of formal structure is ZFC. If ZFC, or another theory of a similar kind, is a correct theory of the formal structure, then the basic elements of the world (as far as its formal structure is concerned) are individuals and properties (where by properties she means also relations and functions). Given theorist 2's understanding of formality, the formal elements of the world are properties, not individuals. Their formality is explained in terms of invariance. Among its formal properties she does not distinguish between "simple" properties on the one hand, such as identity, finite cardinality, intersection, complementation, etc., and "higher" properties such as infinite cardinality, being an ordering relation, a well-ordering relation, well-founded, etc. All of these are on a par as far as the formal structure of the world and its ability to ground logical consequences are concerned.

But not only does theorist 2 include in her counterpart of theorist 1's KF-world "higher" formal properties, she does not rule out that this world is not straightforwardly connected to our language. For example, she does not rule out the possibility that 1st-order mathematical statements, that on the surface refer to mathematical

---

[33] For further discussion, including the way 1st-order mathematics is related to logic through 2nd-order mathematics, the apparent tension between the unbounded infinitude of mathematics and the (presumed) finitehood of the world, and the multiplicity of goals of mathematics of which tracking the formal structure of the world is only one, see Sher (2016, Sects. 8.4 and 10.8).

[34] I would like to thank Pen Maddy for a question that led me to add this section.

individuals, are true, yet the world does not contain mathematical individuals. For her, mathematical truth is grounded in the formal structure of the world, but since the formal structure of the world arises largely on the level of properties of properties, it does not require mathematical individuals. One possibility is that what we think of as mathematical individuals are posits representing higher-level formal properties. (For example, numerical individuals represent cardinality properties.) 1st-order arithmetic and set-theory are, thus, true or false of the world, but indirectly so. (The reason we introduce such posits may be that we, humans, figure out things better when we think about them in terms of individuals and their properties than in terms of properties and their properties.) This, from her perspective, does not conflict with robust mathematical realism: an indirect connection with the world is just as real and just as robust as a direct connection. By being a mathematical realist, though, theorist 2 does not say that the only goal of mathematics is to describe the formal structure of the world. Mathematics has a variety of goals, and describing the laws governing formal properties that hold of individuals and their properties in the world is just one of these, albeit a philosophically significant one.

Two additional notes: (a) Although theorist 2 talks in terms of individuals and properties, she does not rule out the possibility that the basic elements are of a different kind. In that case, the notions of formality and invariance will have to be adjusted. (b) For theorist 2 it is an open question whether the macroscopic and microscopic world have the same formal structure and if not, how much their formal structures differ, and whether their difference can be represented within a single theory of formal structure or not. She considers various possibilities here, including, but not limited to, ones that are similar to theorist 1's understanding of the situation. (c) Unlike theorist 1, theorist 2 is not committed to her counterpart of the KF-world being trivalent rather than bivalent. Whether the world is formally trivalent or bivalent is an open question, where the formal difference between the two is a matter of such things as whether the basic form of properties in a given domain of individuals is that of two complementary subsets of the domain, or three complementary subsets.

We have seen how two different approaches to logic, the one austerely naturalist, the other not, the one non-holist, the other foundational-holist, the one having a KF-conception of the world, the other another, lead to, and sanction, logical realism. Both ground the veridicality of logic in the formal structure of the world and one extends this grounding to its necessity as well. Both also emphasize the connection between the human mind and the world as central to this grounding. One focuses on evolutionary connections between the basic structure of the world and humans' cognitive mechanisms, the other on humans' interest in knowledge of the world and their ability to expand it by building a method of inference that tracks its formal structure. Logical realism is a minority view in contemporary philosophy. I hope that the tale of two theories that arrive at it from different directions will contribute to philosophers' understanding of the motivation, grounds, and richness of this view.

References

Barwise, J., & Feferman, S. (Eds.). (1985). *Model-Theoretic logics*. Springer.

Lindström, P. (1966). First order predicate logic with generalized quantifiers. *Theoria, 32*, 186–195.

Maddy, P. (1988). Believing the axioms. *Journal of Symbolic Logic, 53*, 481–511, 736–764.

Maddy, P. (1997). *Naturalism in Mathematics*. Oxford.

Maddy, P. (1999). Logic and the discursive intellect. *Notre Dame Journal of Formal Logic, 40*, 94–115.

Maddy, P. (2007). *Second Philosophy: A Naturalistic Method.* Oxford.

Maddy, P. (2012). The philosophy of logic. *Bulletin of Symbolic Logic, 18*, 481–504.

Maddy, P. (2014a). *The logical must: Wittgenstein on logic*. Oxford.

Maddy, P. (2014b). A second philosophy of logic. In P. Rush (ed.): *The Metaphysics of Logic*, pp. 93–108.

Maddy, P. (2014c). A second philosophy of arithmetic. *Review of Symbolic Logic*, 7, 222–249.

McGee, V. (1996). Logical operations. *Journal of Philosophical Logic, 25*, 567–580.

Mostowski, A. (1957). On a generalization of quantifiers. *Fundamenta Mathematicae, 44*, 12–36.

Sher, G. (1991). *The bounds of logic*. Cambridge: MIT.

Sher, G. (2001). The formal-structural view of logical consequence. *Philosophical Review, 110*, 241–261.

Sher, G. (2016). *Epistemic friction: An essay on knowledge, truth, and logic*. Oxford.

Sher, G. (2017). Truth and scientific change. *Journal of General Philosophy of Science, 48*, 371–394.

Sher, G. (2021). Invariance as a basis for necessity and laws. *Philosophical Studies, 178*, 3945–3974.

Tarski, A. (1933). The concept of truth in formalized languages. In: *Logic, Semantics, Metamathematics* (pp. 152–278). Indianapolis: Hackett.

Tarski, A. (1966). Published 1986. "What Are Logical Notions?" *History and Philosophy of Logic*, *7*, 143–154.

Chapter 20
Reply to Sher

Penelope Maddy

Keywords Logical realism · Naturalism · Transcendent/immanent · KF-structure/ possible world · Necessary/contingent

Professor Sher's piece provides a welcome opportunity to compare our two forms of logical realism, especially welcome, as she notes, given the rarity of positions in this general category. She begins by declaring philosophical questions and methods to be different in kind from scientific ones and embracing a 'transcendent' inquiry distinct from 'immanent' inquiry. This emphatically 'non-naturalist' approach she quite reasonably takes to distinguish her position from mine. The trouble here, at least for me, is that I can't find anything in her purportedly transcendent inquiry that doesn't fall well within the purview of the Second Philosopher. As Sher describes it, 'we go outside a given immanent inquiry and investigate this inquiry itself' (p. 246),[1] but surely the ordinary scientist does this sort of thing all the time when she steps back to critique her methods. So, for example, an individual scientist evaluates her experimental design, or early modern natural philosophers examine the assumption of corpuscular mechanism, or contemporary cosmologists ask what counts as cogent evidence for the existence of dark matter. Sher remarks that 'we can transcend science both to the sociology of science and to the philosophy of science' (p. 246), but sociology is itself an empirical science, and the Second Philosopher does, I would say, do philosophy of science. (Indeed, stepping back from the practice of mathematics to ask after its methods, its metaphysics, and its epistemology is exactly the topics of

[1] For the record, no God's-eye-view is intended here: we're only to transcend 'one *human* standpoint … to another, *human*, standpoint' (p. 246).

P. Maddy (✉)
Department of Logic and Philosophy of Science, University of California, Irvine, CA, USA
e-mail: pjmaddy@uci.edu

S. Arbeiter and J. Kennedy (eds.), *The Philosophy of Penelope Maddy*,
Outstanding Contributions to Logic 31, https://doi.org/10.1007/978-3-031-58425-1_20

Naturalism in Mathematics and *Defending the Axioms*!) So I have yet to see where this purported methodological divide comes into play.[2]

A second difference noted by Sher strikes me as both easier to grasp and more significant, namely, the goal of our respective logical projects. I'm out to consider extremely simple inferences—inferences we might be said to share with non-human animals—and to ask (1) what makes them reliable? and (2) how do we come to know this?—so, for example, if it's either red or green, and it's not red, why must it be, and how do we know it to be, green? In contrast, Sher begins from a design perspective, pointing out that it would be extremely handy for inquirers like us—and here she means sophisticated, exclusively human inquirers—to have

> a method of inference that is *highly general* and *modally-strong* ... [that is, that] guarantees the transmission of truth in all fields of knowledge with a strong modal force. (p. 253)

Her goal, as a logician, is to find such a method. These are obviously two quite different projects. To the Second Philosopher's eye, it isn't obvious that Sher's goal is attainable—still, she can appreciate that this would be a good thing to have, just as she appreciates that it would be a good thing to have Descartes's a priori method for finding out about the world,[3] so she's happy to sign onto the endeavor to see where it might lead.

Let me pause at this point to address what will seem an unrelated question, namely, why I often present my position by talking first about Kant.[4] There are a number of reasons for this, including a nod to my own intellectual path, but the crucial rhetorical motivation is to maneuver the reader into a perspective from which the division of the macro-world into individuated objects with properties isn't taken for granted.[5] Though I obviously don't mean to second Kant's claim that this division is the product of our transcendental psychology, I do mean to suggest that what I call KF-structure and its attendant logical laws are contingent, not necessary, and that our tendency to think of all 'possible worlds' as akin to the models of model theory is myopic, though understandable given our basic cognitive mechanisms.

Returning to Sher, we're seeking a highly general mode of inference with strong modal force. This, she argues, demands an 'appropriately built' system, that is, one whose 'logical constants denote *formal* properties (relations, functions)', and whose inferences track the laws governing such properties (p. 254). The pivotal notion of 'formality', in turn, is parsed in terms of 'maximal property-invariance': a property is *maximally property-invariant* if and only if for any domain, one-to-one replacement

---

[2] In his (2015), van Fraassen also claims that the Second Philosopher doesn't engage in an important 'transcendental' activity, in particular, what he calls 'interpretation'. In this case, however, the transcendental inquiry is explicitly outside all of science, outside of all the Second Philosopher's empirical inquiry; it takes place at the remove, as he puts it, of the philosophy seminar room. (See Maddy (2022f) for more.) I doubt that Sher has anything like this in mind.

[3] For the record, Descartes isn't a First Philosopher—he's a Second Philosopher whose method didn't work (see Maddy (2022c), footnote 56).

[4] This isn't essential (see Maddy (2014c)).

[5] For those comfortable with the expression, the division of the macro-world into individual objects with properties is being 'problematized'.

of all individuals[6] with those of an equinumerous domain doesn't change where the property holds or doesn't. So 'is human' is not maximally property-invariant because, when a human is replaced by a dog, it goes from holding to not holding; in contrast, 'is identical to' is maximally property-invariant.[7] The familiar logical constants—and, or, exists, and so on—are then second-order formal properties, for example, to say 'something is either red or green' is to say that the union of the property 'is red' with the property 'is green' is non-empty.[8]

What's happened here should be obvious: Sher is presupposing the division of the macro-world into objects with properties and its elaboration into KF-structures. With this backdrop in place, it's not surprising that she finds 'strong modal force' where I do not: if we assume that all 'possible worlds' have KF-structure, then the inferences validated by that structure will be necessary.[9] Surprisingly, though, late in her paper, Sher appears to call this assumption into question:

> Although [I talk] in terms of individuals and properties, [I do] not rule out the possibility that the basic elements are of a different kind. (p. 260)

Much as I failed to find KF-structure in the micro-world, Sher considers the structure of the 'microscopic world' to be 'an open question' (ibid.). Despite her earlier insistence on 'strong modal force', she seems satisfied to observe that

> formal laws have a higher degree of necessity than physical, biological, psychological and other laws, and as such are highly necessary. (p. 256, see also footnote 29)

But I, too, take the laws of logic to be contingent on fewer features of the world than the laws of other sciences—in Kantian terms, they depend only on the structure of any discursive intellect, not on the details of our particular forms of intuition—so here with metaphysics, as before with methodology, I find our differences less stark than might be imagined.

There are, however, two places where our differences are quite real. The first is the epistemology of rudimentary logic: how do we humans come to recognize the cogency of inferences like ours about red and green things? Though Sher's and my accounts of the metaphysical structures that underlie such inferences appear similar, our answers to the epistemic question are not. I point to the basic cognitive

---

[6] This includes both actual and possible individuals.

[7] Here it matters that the replacement is one-to-one.

[8] See pp. 254–255. Sher phrases this in such terms as the 'union of two properties', as in the text above, but perhaps this is shorthand for the 'union of the extensions of two properties' or something along these lines.

[9] A second modal element of her view is the inclusion of possible objects, in the original domain and among the relevant replacements (as noted in footnote 6). If all actual dogs bark, then for me, 'This is a dog' would imply 'This barks' (every situation in which this is a dog is one in which this barks), while for Sher, assuming non-barking dogs are possible, it would not. The example comes from p. 251 where she remarks that, for me, 'there is no difference between the grounding of non-logical rudimentary inferences' like this one and properly logical inferences. Not that I'm keen to distinguish 'logical' from 'non-logical' (as Sher also notes in this same passage), but still, I think this particular claim isn't quite right: I can distinguish between inferences that depend only on KF-structure and those that don't.

machinery—shared, as noted earlier, with apes, dogs, birds, and fish—that enables us to detect KF-structure where it's present.[10] For Sher, in contrast, our logical knowledge arises from our ability 'to figure out what the formal laws are and ... to construct a logical system grounded in these laws' (p. 257). However this is spelled out, it's a higher level, distinctly human activity.

This real difference may be the first step toward the second, our disagreement over the status of arithmetic and higher mathematics. So far, I've been glossing over the difference between saying the red/green inference is cogent (1) because any situation in which it's either red or green and simultaneously not red is also a situation in which it's green, and saying it's cogent because (2) the intersection of the union of the properties 'red' and 'green' with the complement of the property 'red' is contained in the property 'green'. Parsing this elementary bit of human knowledge as including notions of collection, intersection, complementation, and unionization—second-order properties, on Sher's reading—opens the door to higher-order properties, even to the theoretical posits of transfinite set theory. In contrast, (1) invites no such escalation.[11]

Ironically, my youthful claim (in *Realism in Mathematics* (1990)) that we have perceptual access to (some) sets of medium-size physical objects allowed a similar move to posits of higher-order, which I then enthusiastically embraced. That story rested on philosophical assumptions that no longer strike me as viable (indispensability arguments) and developmental psychology that's since been superseded (Piaget). Removing the former and updating the latter is what led me to the position discussed here on rudimentary logic—and also to new accounts of the nature of arithmetic and of higher set theory.[12] I now think it's a fundamental mistake to assume that the metaphysics and epistemologies of these very different undertakings—elementary logic and simple arithmetic identities, full Peano Arithmetic, ZFC and its extensions—should be uniform.

But that, as they say, is a whole nother story—our topic here is the philosophy of logic. On this front, given the rarity these days of straightforwardly realistic accounts, the disagreements noted in Sher's piece and this reply have to be counted as intramural skirmishes. I hope to have furthered the comparative exploration she was so good as to initiate.[13]

---

[10] Sher often describes this with locutions like 'the veridicality of logic is grounded ... epistemically in our primitive cognitive mechanisms' (p. 15) and concludes that she and I both ground 'logic both in the world and in the human mind' (p. 257). I'd prefer to say that rudimentary logic is grounded solely in the KF-structure of the world—this is what makes the inferences valid—and that, by the way, our minds are equipped (probably by evolution) to detect that structure. (For comparison, I wouldn't want to say that veridicality of our ordinary discourse about medium-sized objects is grounded in our minds, though we're also mentally equipped to detect them.) Including human psychology in the 'grounding' of logic opens me to a charge of psychologism (which Sher comes close to lodging in her (2016), pp. 240, footnote 3, and 256).

[11] See, e.g., Maddy (2018b), footnote 27.

[12] See Maddy (1997), (2011), (2014b), (2018b), (2018c).

[13] [Added later: Professor Sher kindly reviewed this reply and confirmed that 'it's right about our similarities and differences' (personal correspondence, quoted with permission)].

Chapter 21
Trumping Naturalism Revisited

Alexander Paseau

Abstract Whenever science returns a confident and univocal answer to a question, Trumping Naturalism enjoins us to accept it. The majority of contemporary philosophers are sympathetic to this sort of position. Indeed, several have endorsed it or something very close to it. Arguments for Trumping Naturalism, however, are scant, the principal one being the Track Record Argument. The argument is based on the premise that in cases of conflict, science has a better track record than non-scientific forms of inquiry such as philosophy. My chapter investigates this argument. I enumerate and assess several objections to the argument, from footling objections to stronger ones. My conclusion is that the Track Record Argument is not as strong as imagined.

Prologue

It is as much a pleasure as an honour to contribute to a Festschrift for Penelope Maddy. Her work's influence on my thinking in the philosophy of mathematics, as well as my conception of how it is best done, is immeasurable. Upon encountering *Naturalism in Mathematics* as a tyro graduate student, I felt compelled to work on 'Maddian themes' in general and on naturalism in particular. The present chapter is in this vein, even if it shies away from forging explicit connections with Maddy's more recent 'second philosophy', mainly for reasons of space. It is, however, un-Maddian in its willingness to try to pin down and assess the precise content of naturalism, and in its negative conclusion. But it is inspired by Maddy's work, informed by it and more generally indebted to it at every turn.

A. Paseau (✉)
University of Oxford, Oxford, UK
e-mail: alexander.paseau@philosophy.ox.ac.uk

S. Arbeiter and J. Kennedy (eds.), *The Philosophy of Penelope Maddy*,
Outstanding Contributions to Logic 31, https://doi.org/10.1007/978-3-031-58425-1_21

21.1 Beyond the Platitude

Science is, generally speaking, more credible than other forms of inquiry. Though the expression is a little crude, the sentiment is widely shared. Champions of a naturalist trend in philosophy go further, by advocating[1]

Trumping Naturalism (TN)

If science endorses *p* then one should believe *p*

We illustrate TN with an example. Take the statement that life on Earth arose about 3.7 billion years ago, which is supported by well-confirmed scientific techniques (carbon dating, analysis of the fossil record, etc.) and theories (geology, evolutionary theory, etc.). Despite its scientific backing, biblical literalists might disbelieve it. They might shrug off the scientific evidence and argue that we have no more reason to believe that life on Earth emerged billions of years ago than that it emerged a few millennia ago with the deceptive appearance of pre-existence. As the naturalist Philip Gosse put it, in an attempt to reconcile 19th-century advances in geology with the Bible, we might read the evidential record in a way that 'disturbs the conclusions of geologists respecting the antiquity of the earth'.[2] According to TN, however, the biblical account cannot cast doubt on the existing scientific consensus; it must yield to it.

In the past thirty-odd years, philosophers have endorsed TN or something like it in droves.[3] Yet arguments for TN are thin on the ground, notwithstanding the generally acknowledged need for naturalism in its various guises to justify itself.[4] The principal argument for TN is based on disciplinary track records. In a nutshell, the argument is that recent science has been very successful whereas, time and again, philosophy and other non-scientific disciplines have failed. And in cases of conflict, science has

[1] The word 'trumping' expresses the idea very clearly and accords with the terminology in Paseau (2008), which predates Donald Trump's political rise. I am loath to change it for these reasons, and because hopefully this essay will be read after Trump's political demise.

[2] Gosse (1857, p. iii). See also the more recent 'scientific creationist' movement typified by Henry Madison Morris.

[3] John Burgess and Gideon Rosen write: 'The naturalists' commitment is…to the comparatively modest proposition that when science speaks with a firm and unified voice, the philosopher is either obliged to accept its conclusions or to offer what are recognizably scientific reasons for resisting them' (1997, p. 65). Many other philosophers have backed TN, and several more have endorsed versions of it for conflicts between philosophy and disciplines other than science. In particular, Colyvan (2001, p. 13), Maddy (1997, p. 161), Burgess (2004), Hofweber (2009, p. 263), Schaffer (2009, p. 537 n.8) and Kit Fine (2009, p. 158) all endorse or are very sympathetic to the trumping claim that philosophy cannot overturn mathematics. Blome-Tillmann (2009, p. 307) believes that philosophy cannot overrule linguistics, as do others (see the references in Daly and Liggins (2011, p. 333)). Many others draw similar conclusions, e.g. Ladyman and Ross (2007, p. 16). The trend of the past few decades has deeper roots, some of the more proximate ones being Quine's naturalism (1981, p. 21), or the view that the philosophy of science should be descriptive rather than prescriptive (e.g. Holton, 1988). Prior to his later denunciation of naturalism (1983), Putnam endorsed TN (1971, p. 356). Critical discussions of Trumping Naturalism and related versions which inform this chapter include my (Paseau, 2005) and (Paseau, 2008) as well as Daly and Liggins (2011).

[4] For which, see for example my (2008) or Rosenberg (2012, pp. 248–9).

a better track record than other forms of inquiry. This argument can be regimented into premise-conclusion form, as follows: Science has an excellent (recent) track record; non-science has a much inferior track record; so: TN.

There is a much-quoted passage in which David Lewis sets outs his credo and articulates the Track Record Argument. I shall quote it once more, in full, the better to engage with Lewis's individual examples later.

> Mathematics is an established, going concern. Philosophy is as shaky as can be. To reject mathematics for philosophical reasons would be absurd. If we philosophers are sorely puzzled by the classes that constitute mathematical reality, that's our problem. We shouldn't expect mathematics to go away to make our life easier. Even if we reject mathematics gently—explaining how it can be a most useful fiction, 'good without being true'—we still reject it, and that's still absurd. Even if we hold onto some mutilated fragments of mathematics that can be reconstructed without classes, if we reject the bulk of mathematics, that's still absurd.
>
> That's not an argument, I know. Rather, I'm moved to laughter at the thought of how *presumptuous* it would be to reject mathematics for philosophical reasons. How would *you* like the job of telling the mathematicians that they must change their ways, and abjure countless errors, now that *philosophy* has discovered that there are no classes? Can you tell them, with a straight face, to follow philosophical argument wherever it may lead? If they challenge your credentials, will you boast of philosophy's other great discoveries: that motion is impossible, that a Being than which no greater can be conceived cannot be conceived not to exist, that it is unthinkable that anything exists outside the mind, that time is unreal, that no theory has ever been made at all probable by evidence (but on the other hand that an empirically ideal theory cannot possibly be false), that it is a wide-open scientific question whether anyone has ever believed anything, and so on, and on, *ad nauseam*? Not me! (Lewis, 1991, pp. 58–9).

Evidently, Lewis is aiming for comic effect here, and achieves it. He denied his examples the status of an argument, but that was a rhetorical ploy, since in the book the passage is from (*Parts of Classes*), he put them to use to rule out a structuralist interpretation of set theory which, as he saw it, clashes with mathematics. In our terminology, the examples are used to justify TN.[5]

The Track Record Argument seems to offer a direct and convincing route to TN. In this chapter, a sequel to Paseau (2005), I shall argue that it is not quite as powerful as it might initially seem. The conclusion I reach is very similar to my earlier one, but my reasons this time are slightly different.

21.2 Preliminaries

The term of art 'science endorses *p*' stands in need of clarification. One could read it as the claim that *p* follows from some core tenets of a particular science via a scientifically acceptable process of inference (which could include deduction, induction

---

[5] A similar argument is offered in Burgess (2004, p. 13). Along the same lines, Vargas (2009, p. 145) writes: 'armchair philosophical speculation about the construction of the brain has an uninspiring track record'. See also Ladyman and Ross (2007, p. 16).

and abduction).[6] This is a helpful first-pass elucidation, though it needs to be further sharpened. Since conceptual analysis is not our game here, we can rest content with an understanding of 'science endorses *p*' at a fairly intuitive level along these lines. It is important to appreciate, however, that (a) science could endorse *p* even if scientists do not personally believe *p*, e.g. if *p* is a consequence of an established scientific theory that scientists happen not to have considered; (b) science might endorse *p* even if scientists wrongly disbelieve *p* on scientific grounds, e.g. they have misjudged the balance of scientific evidence; (c) science might not endorse *p* even if all scientists believe *p*, e.g. all scientists might happen to believe that chocolate ice cream is tastier than vanilla or that God exists, but believe neither *qua* scientists; and (d) scientific confirmation comes in degrees, and endorsement implies a relatively high degree of confirmation. As an illustration of the last point, TN enjoins us to accept the core tenets of evolutionary theory, which are fairly settled, but has no implications for unsettled questions such as those fought over by phyletic gradualists and punctuated equilibrium theorists.

We may construe 'science' as including the traditional natural sciences: physics, chemistry, biology. A broader definition that includes, say, the established behavioural and social sciences, is possible, but would not significantly alter what follows. Since the argument is from apparent epistemic authority, we should also build mathematics into the definition of science. Mathematics is akin to the natural sciences in its manifest epistemic authority, even if its apparent ontology is different from that of the natural sciences. Indeed, mathematics plausibly has the most successful track record of all forms of inquiry. For this reason, most Trumping Naturalists (see fn. 3) count mathematics as a trumping discipline.

For the purposes of assessing the argument, we must assume that a boundary exists between science and non-science along the lines just given, even if it's fuzzy.[7] Our aim is to offer an 'internal' critique of the argument, which takes this distinction as given. This is not to close off the possibility of a critique that challenges the distinction's very existence. At any rate, the conventional line is that a clear distinction between science and non-science exists and that it emerged during the Scientific Revolution. This period stretches roughly from the mid-16th century to the late 17th century. Some historians of science play down the significance of the Scientific Revolution—and write books beginning 'There was no such thing as the Scientific Revolution, and this is a book about it'[8]—but it is the historian's stock in trade to emphasise continuities. Rhetoric aside, it is usually thought that a significant, albeit gradual, change in the way natural science was practised took place during this period, which

---

[6] As suggested to me by Tushar Menon.

[7] Burgess and Rosen (1997, p. 206) have claimed that there is a simple operational test for whether something is a scientific argument or a scientifically acceptable theory. They urge that a 'true test' would be to submit the theory to a scientific journal and see whether it is accepted and well-received by the scientific community. Elsewhere, I have criticised this proposal for being simplistic (2007, pp. 138–40). More promising would be a mildly idealised version of the test that prescinds from editorial limitations, editors' and readers' preconceptions about the kinds of discussions appropriate to journals, and so on.

[8] Shapin (1996, p. 1).

set it off more sharply from other areas of inquiry.[9] This is not to deny the monumental achievements of earlier times.[10] But even on such views, experimentation in its most mature form—'constructing artificial physical situations for the purpose of making specific observations and measurements' (Rovelli, 2011, p. 35)—had to wait until much later. In any case, the later we date the birth of science in the strict sense, the stronger the Track Record Argument becomes. Going along with the conventional 17th-century demarcation gives the argument its most charitable reading.

To be a Trumping Naturalist, must you explicitly express your allegiance to TN? No. If in practice you always side with the scientific consensus against an appeal to non-science when the two clash, you are *ipso facto* a Trumping Naturalist—irrespective of whether you embrace or disavow the label, accept or reject the science/non-science distinction, think that scientific standards are set in stone or instead evolve over time, are explicitly guided by TN or not, and so on. A Trumping Naturalist, in my book, is anyone who always judges in conformity with TN, whether they realise it or not, whether they avow it or not, whether it corresponds to their self-image or not, etc.

TN is plausibly an acceptable injunction only for someone with an antecedent desire to acquire many true beliefs even at the expense of acquiring a good number of false beliefs. To give the Track Record Argument a charitable hearing, we have to assume something like this background attitude. The reason is that current science surely furnishes us with a sizeable number of false claims, at least some of which are not even approximately true. As a result, TN would be unacceptable to anyone with a Cartesian terror of false beliefs.

I use the term of art 'norm' to mean a principle of scientific method, that is, an epistemic norm guiding or accurately describing belief-formation. Scientific norms are the norms currently underlying theory choice in the natural sciences.[11] For the

[9] Not coincidentally, it is to the thinkers who inaugurated this period—Francis Bacon, William Gilbert, Galileo and others—that we owe the scientific method's first proper articulation, and its definitive divorce from teleology and natural theology. Of course, many branches of science date the beginning of their 'modern era' to later than the 17th century. 'Modern' chemistry is marked by the passage from phlogiston theories to Lavoisier's theories of reduction and oxidation. The publication of Darwin's *On the Origin of Species* in 1859 irreversibly detached biology from philosophy and theology. The beginning of truly 'modern' biology is often dated, with remarkable precision, to the 25th of April 1953, the day on which the famous trio of publications on the structure of DNA (by Crick and Watson, Wilkins, and Franklin) appeared in *Nature*.

[10] For example, that of Alexandrian science and its practitioners such as Ptolemy (2nd century AD) and Hipparchus (2nd century BC). There is an argument to be had about whether some of these ancient Greek philosophers and mathematicians were scientists sensu stricto. Perhaps Anaximander, who flourished in the first half of the 6th century BC, was the first scientist (see Rovelli 2011 for an assessment); or perhaps science was born during the Hellenistic period, in about 300 BC (Russo, 2004).

[11] Feyerabend (1975) is one of the few to deny the stable existence of such norms. Even while registering important caveats, Kuhn does not dispute that 'accuracy, consistency, scope, simplicity, and fruitfulness…are all standard criteria for evaluating the adequacy of a theory' (1977, p. 322). How much leeway the scientific method allows in interpreting or weighting such norms remains a vexed question.

sake of simplicity, we may assume that some norms operate throughout science.[12] The paradigm of a scientific norm is empirical adequacy, i.e. agreement with empirical data.[13] Some form of the principle of simplicity, which in its most general version states that the simpler of two theories enjoys a theoretical advantage over the less simple one (in this respect), is also a scientific norm.[14] But not all epistemic norms are scientific; those stemming from religion, for example, tend to be non-scientific. Thus, compatibility with the sayings of some sacred person or text is not a scientific norm, nor is consistency with non-reproducible religious experience, and nor is consistency with various creeds and confessions of faith. Some philosophers also espouse norms that are neither scientific nor religious. In their 1947 manifesto, for instance, Nelson Goodman and W.V.O. Quine famously declared that nominalism is founded on a 'philosophical intuition' irreducible to scientific grounds.[15] The human sciences, it has been argued, also rely on different epistemic norms from the natural sciences: to make sense of other people, one must rely on *Verstehen*, a capacity to understand others from within. If that is right, these sciences are normatively different from the science of leptons or leptides or lepidoptera.

Scientific norms are a subset of all epistemic norms. So if they are a proper subset of all such norms, there is potential for divergence: scientific norms may endorse p yet other norms might endorse $\neg p$. In the hands of philosophers, the Track Record Argument typically pits scientific considerations against philosophical ones. The argument is more general, though, since it exalts science above any other forms of inquiry. It has implications not just for the contest between science and philosophy, but also for the contest between science and theology, common sense, and so-called pseudo-sciences such as astrology and parapsychology. That said, since this chapter is responding to a literature that concentrates on the clash between science and philosophy, a good deal of the critical discussion will focus on TN's anti-philosophical implications.

The tactic employed by the Track Record Argument may be illustrated by comparing two methods for predicting tomorrow's temperature.[16] The *roulette*

[12] If not, the Track Record Argument fragments into, say, the Track Record Argument from Physics, the Track Record Argument from Biology, etc.

[13] In practice, of course, most actual theories fail to mesh with all the empirical data: they contain 'anomalies', even from their inception. 'All theories, in this sense, are born refuted and die refuted', wrote Lakatos, because they contain 'unsolved problems' and 'undigested anomalies' (1978, p. 5). The better confirmed the theory, the fewer these anomalies and the easier they are to explain away.

[14] For Newton's attachment to the principle of simplicity, see the first of the *Regulae Philosophandi* in the third edition of his *Principia*. Paul Dirac later put it even more strongly: 'A theory that has some mathematical beauty is more likely to be correct than an ugly one that gives a detailed fit to some experiments' (cited in Holton, 1988, p. 15).

[15] 'We do not believe in abstract entities...We renounce them altogether...Fundamentally this refusal is based on a philosophical intuition that cannot be justified by appeal to anything more ultimate' (Goodman & Quine, 1947, p. 105). Goodman and Quine added that this fundamental rejection is fortified by certain a posteriori considerations. In his later and better-known incarnation, Quine repudiated this 'intuition' altogether.

[16] Say that the prediction applies to some specified location on Earth at noon local time, and that it is accurate if it is to within two degrees Celsius of the actual temperature at that time and place.

method involves spinning a normal (not meteorologically sensitive) roulette wheel and predicting that tomorrow's temperature is the number the ball lands on. The *same-as-today method* predicts that tomorrow's temperature will be the same as today's. The roulette method has a poor track record. The same-as-today method in contrast has a good track record, so good that even sophisticated meteorological techniques do not significantly improve upon it. The idea behind the Track Record Argument is that, in cases of conflict, scientific norms' track record is more akin to the same-as-today method's track record whereas unscientific norms' track record is more akin to that of the roulette method. The Track Record Argument is thus a meta-empirical argument for TN.

The locution 'getting things mostly right' is admittedly vague. However, further precision is not really necessary for our purposes, save to add that getting things mostly right is better understood as returning approximately true rather than unqualifiedly true answers to past questions. Judgements of past success rates are of course relative to some background beliefs; one cannot judge a method's success rate from a theoretically neutral perspective. In the temperature case, for example, one needs to take a view about past temperatures to judge that the same-as-today method has proved historically superior to the roulette method. This point will be very relevant later.

With these preliminaries behind us, we consider a series of objections to the Track Record Argument, broadly moving from less to more powerful. In roughly increasing order of strength:

The No Clash Objection (§3);

The Definitional Failure Objection (§4);

The Failure of Science Objection (§5);

The Good Philosophy is Scientific Objection (§6);

The Idle Wheel Objection (§7);

The No Dogmatism Objection (§8);

The Schools Objection (§9).

The last section—about the Schools Objection—in particular will show that the Track Record Argument depends on other anti-philosophical arguments for its plausibility. Our overall conclusion will be that the Track Record Argument is weaker than imagined.

21.3 First Objection: No Clash

This first objection to the Track Record Argument is that science and non-science cannot clash because they address different questions. The objection targets the Track Record Argument's premise by affirming that there are no clashes between science and non-science properly understood. From which it follows that there are no relevant cases over which to establish any sort of track record, superior or otherwise.

The idea that science is not in competition with other forms of inquiry has been around for a long time, as testified by Robert Hooke's draft preamble to the Royal Society statutes in the 17th century: 'The business of the Royal Society is: To improve the knowledge of natural things…(not meddling with Divinity, Metaphysics, Morals, Politics, Grammar, Rhetorick, or Logicks)' (cited in Holton, 1988, p. 7); note the parenthetical remark. More recently, the objection typically takes the form of denying that science and a particular non-science can clash. Some philosophers, for example, are fond of saying that science is in the business of making empirical claims about the world, whereas philosophy is not. Thus Wittgenstein:

> I am not interested in constructing a building, so much as in having a perspicuous view of the foundations of possible buildings. So I am not aiming at the same target as the scientists and my way of thinking is different from theirs. (1980, p. 7)

Wittgensteinian 'quietists' think of philosophy as not in competition with science, because philosophy dissolves rather than resolves problems.[17]

Perhaps a more common move is to contrast religion and science. Stephen Jay Gould was a great champion of what he called 'the principle of Non-Overlapping Magisteria':

> I…do not understand why the two enterprises should experience any conflict. Science tries to document the factual character of the natural world, and to develop theories that coordinate and explain these facts. Religion, on the other hand, operates in the equally important, but utterly different, realm of human purposes, meanings and values—subjects that the factual domain of science might illuminate, but can never resolve…I propose that we encapsulate this central principle of respectful non-interference…by enunciating the principle of NOMA, or Non-Overlapping Magisteria. (2001, pp. 4–5)

The 'magisterium of science' is the empirical realm, whereas the 'magisterium of religion' deals with ultimate questions. Similar claims may be found in the writings of Karl Barth and some other theologians.

Despite these impressive testimonials, the No Clash Objection is mistaken. Take common sense, religious claims, and philosophy as examples of non-science. Common sense is affected by science and by improvements in general scientific literacy. Nonetheless, common sense is distinct from science, and as history reveals, the two can clash. For instance, the special theory of relativity teaches that whether event *A* occurs before, simultaneously with, or after event *B* is not an absolute fact but depends on one's frame of reference. This conclusion is stubbornly at odds with common sense. Likewise for dozens of other instances that defied common sense at the time they were proposed or currently do so (or both). Examples include: wave-particle duality, the Copernican theory, Darwin's hypothesis that mankind is descended from 'lower' creatures, Newton's hypothesis of action at a distance—or indeed his first law of motion, which runs counter to common-sense impetus theory.[18]

---

[17] Compare Leiter (2004, p. 2).

[18] According to which a constant force must be applied to an object if is to continue moving at the same velocity.

Religion can also clash with science. Anyone who believes in the Bible's literal truth cannot reconcile their religious epistemic norms with current scientific ones. The example of the origins of life and the Earth's formation is a case in point. Traditional Christians, who base their belief system on some weighted combination of scripture, tradition, reason and religious experience,[19] similarly find their views in tension with science, be it regarding the doctrine of creation, original sin, Christ's nature, divine intervention in the form of miracles or otherwise, eschatology, and so on. Admittedly, Christian theology is more metaphysically loaded than the theology of other religions; but all that is required to see off the No Clash Objection is a single example.

As for philosophy, suppose for the sake of argument that a scientifically convincing quantum–mechanical argument can be developed to the effect that the external world is mind-dependent; see Part III of Wigner (1970) for one such attempt. Even so, philosophers might maintain that the external world is mind-independent, perhaps on the basis of a relatively a priori argument concerning the nature of language or mental content. If such a conflict arose, Trumping Naturalists would recommend that we side with science and accept that the world is mind-dependent. In such an event, we would have a clear-cut clash between philosophy and science—and a Trumping-Naturalist recommendation to follow the latter.

In general, the 'separate domains' conception of philosophy, according to which philosophy and science cannot clash because they are concerned with different domains, is belied by much philosophy. Relatively few philosophers maintain that philosophy is in the business of unempirical conceptual analysis, exploration, clarification of language, or intellectual therapy. On some conceptions of philosophy, science and philosophy may not clash; but on most recognisable ones, they can and sometimes do clash.

21.4 Second Objection: Definitional Failure

This next objection aims to deflate the Track Record Argument's significance by claiming that non-science must by definition fail. Most of the central theoretical disciplines have historically been intermingled with philosophy. As soon as they became more successful, they broke off from philosophy and emerged as their own subjects. A relatively recent example is the separation of psychology from philosophy at the end of the 19th century. If past non-scientific norms are simply those that turned out to be unsuccessful by our lights, then the Track Record Argument's premises and conclusion are true—but merely by definition. Sir John Harrington observed that

> Treason doth never prosper, what's the reason?
>
> For if it prosper, none dare call it treason.

[19] The so-called Wesleyan Quadrilateral—but here without any Methodist overtones.

The idea behind the present objection is that Harrington's ditty applies just as much to treason as to non-science. Neither ever prospers, *ex vi terminorum*—by dint of meaning.

This objection should be given short shrift. We do not hesitate to label norms as non-scientific even if we lack evidence for their untrustworthiness. In particular, if the objection were sound, no one engaged in the search for truth would call herself a philosopher as opposed to a scientist. Besides, the objection does not achieve what it thinks it does. The point of the pessimistic induction's conclusion is *not* that those of today's norms that will later be deemed unscientific cannot overrule those of today's norms that will later be deemed scientific. Otherwise, its cash value would be unclear: how can we tell which of today's norms will be considered scientific, and which not? It is of little interest to be told that non-science is untrustworthy if non-science is defined as whatever will come to be seen as untrustworthy. The thrust of TN is rather that the subset of norms we currently think of as non-scientific cannot overrule norms we currently think of as scientific. The Definitional Failure Objection therefore itself fails.

21.5 Third Objection: The Failure of Science

Another predictable objection to the Track Record Argument is that science itself has a poor track record. The items on Larry Laudan's (1981) familiar laundry list—the humoral theory of medicine, the hypothesis of the electromagnetic ether, the caloric theory of heat, and others—were all well-regarded in their time. Yet each of these theories turned out to be wrong. In a nutshell, Laudan's point was that observational success historically goes hand in hand with theoretical failure, that is, falsehood of the underlying theoretical assumptions. This supposedly defuses the realist argument that takes observational success as a reliable indicator of theoretical truth.

A familiar rejoinder to Laudan's argument, in a nutshell once more, is that its conclusion is blocked if we restrict attention to mature scientific theories that have accrued extensive experimental verification or to parts of theories that are truly pulling their weight in observational predictions. In fact, given the demographics of our planet, the great majority of scientists who have ever existed were born after 1900, and most of the science of the past century or so is remarkably stable. To take just one example, the physics of all but the very large and the very small and the way they fit together is generally held to be so secure that any conceivable final theory is bound to vindicate at least its approximate truth. This sort of response is compatible with a wide range of philosophical views about science, from various forms of structural realism to more traditional forms of realism.[20] In any case, Laudan's pessimistic meta-induction is really an argument that successful reference of theoretical terms is not necessary for successful science, however one construes 'success'. But since

[20] For a review of some of these positions, see. e.g. Chap. 2 of Ladyman and Ross (2007).

nothing about the Track Record Argument requires successful reference of theoretical terms, the two arguments are orthogonal to one another.

For our purposes, an even more direct reply to the pessimistic meta-induction is also available. The clashes between science and philosophy from which the Track Record Argument draws inspiration involve claims not questioned by the pessimistic meta-induction. In order to get off the ground, Laudan-inspired anti-realism must insist that science has been highly successful in its more observational claims. Yet some of the alleged cases of conflict—e.g. Lewis's example that motion is impossible—are straightforwardly observational. That science has a good track record at this relatively observational level is pretty much beyond dispute. (It is common ground, for example, in the debate between constructive empiricists such as van Fraassen and scientific realists.)[21] Another category consists of tenets central to all science, e.g. the claim that theories are made probable by evidence. Questioning such tenets is not necessarily beyond the scientific pale. But the tenets are evidently different from the theoretical hypotheses beloved by Laudan and other proponents of the pessimistic induction. It seems, then, that the alleged examples of observationally successful scientific theories eventually deemed to be false and the kinds of examples invoked by Trumping Naturalists do not overlap significantly, if at all.

Finally, even if we believe that parts of our current best scientific theories are unlikely to turn out to be approximately true, this thought is tempered by the knowledge that other parts are likely to be approximately preserved. Moreover, by assumption it is impossible to currently identify which is which. The overall credence reduction in currently successful science enjoined by a balanced assessment of Laudan's argument is therefore fairly low.

In sum, the Scientific Failure Objection does provide an opening against the Track Record Argument. But its effect is largely mitigated by the considerations mentioned. Of course, the more radical a view one takes of Laudan's and similar examples, the more askance one will look upon the Track Record Argument's premise. But it is hard to take so radical a view as to disarm the argument entirely.

21.6 Fourth Objection: Good Philosophy is Scientific

The No Clash Objection (§3) conceived of science and philosophy as distinct in kind, and thence incapable of clashing. A diametrically opposed objection takes science and philosophy to be so similar that they *never* clash. Some philosophers take philosophy to be entirely scientific, in the sense that exactly the same standards (or canons or norms) of reasoning apply in philosophy as in science. Since for them philosophical standards cannot diverge from scientific ones, philosophy and science

[21] Witness the experimental success of quantum electrodynamics. In Richard Feynman's words: 'To give you a feeling for the accuracy of these numbers, it comes out something like this: If you were to measure the distance from Los Angeles to New York to this accuracy, it would be exact to the thickness of a human hair. That's how delicately quantum electrodynamics has, in the past fifty years, been checked—both theoretically and experimentally' (1985, p. 7).

can never disagree. From this perspective, TN is true but has no anti-philosophical bite: if science endorses *p* then so does philosophy, so the two cannot clash.[22]

One may of course question the conception of philosophy this objection presupposes. Does philosophy really have no proprietary norms over and above the scientific? The most naturalist-friendly response, however, is to reformulate the argument in terms that are closer to the original intention behind it. We might let *p* be the sort of claim that contemporary natural science takes a direct and proprietorial interest in, or presupposes, rather than the sort of proposition contemporary philosophers argue over and publish papers respectively attacking or defending. For example, recalling Lewis's examples, science assumes that motion is possible, that things exist outside the mind, that time is real, that theory is made probable by evidence, and that people have beliefs. The revised version of TN is simply that if science endorses *p*, restricted to this sort of *p*, we should believe *p*. The anti-philosophical effect is exactly the same as the original argument, whilst acknowledging (for the sake of argument) that philosophy applies the same norms as science, only in the pursuit of answers to somewhat different questions—at least typically.

As we have seen, none of the first four objections to the Track Record Argument is particularly successful. Let's now turn to more threatening ones.

21.7 Fifth Objection: The Idle Wheel

There are propositions implied by current scientific theory that do not make a significant contribution to the theory in question. These 'idle wheels', though they seem to be part of the scientific theory in question, are not in fact spinning. As an illustration, take the debate between realists and idealists (Lewis's third example). Realists maintain that the universe and its constituents are mind-independent[23]; idealists maintain that they are mind-dependent. The Trumping Naturalist might simply read off realism from science, claiming that current science endorses realism. The idealist might respond that although science is in fact couched in realist terms, it need not be. Conceding that most current science adopts a realist language and conception, she—the idealist—denies that it endorses realism. For as she sees it, realism is an idle wheel in current science: science with a mind-dependent construal of the universe, call it *science*^{MD}, is not scientifically inferior to *science*^{MI} (science with a mind-independent gloss). Of course, if scientific norms do not favour science^{MD} over science^{MI}, adopting TN does not resolve the realism-idealism debate.

[22] An example of authors who seem to adopt such a view of science is Ladyman and Ross (2007, p. 28): '…there is no such thing as "scientific method", by which we mean: a particular set of positive rules for reasoning that all and only scientists do or should follow. There are of course many observed prohibitions (for example, "Do not induct on samples known to be selected in unrepresentative ways" and "Do not invent data"), but these apply to all sound reasoning, not to distinctively "scientific" reasoning'.

[23] With the usual caveat about the specifically mental constituents of the universe—our thoughts, feelings, etc.

We should be clear that this is really an objection to how TN is used rather than an objection to TN per se. Realists who wish to invoke TN to argue for realism must show that realism contributes to science's success more than idealism ever could. They must examine whether various scientific theories are compatible with an idealist interpretation, or more generally whether idealism is in tension with any parts of science. Although not exactly the same as traditional philosophical debates regarding realism and idealism, reflections of this sort—which one finds in contemporary philosophy of science or scientifically-minded metaphysics—have many affinities with them. The moral is that philosophical reflection on, and interpretation of, science is required in order to clarify what exactly it is that TN enjoins us to accept.[24]

Some idle wheels are easy to spot. When conservation of energy was posited as a distinct principle in the mid-19th century, some scientists (notably Lord Kelvin) propped it up with a theological explanation: energy is God's gift to the world, which is why it can neither be increased nor decreased by non-divine activity. The scientific principle has survived into the 21st century, though Kelvin's theological scholium has not. Similar scholia—e.g. that laws of nature are God's handiwork—were prevalent in older science (Boyle and Bacon come to mind), and still occasionally surface. Today, however, almost everyone thinks of such claims as detachable from the science—hence 'scholia'.

The idleness or industry of a particular wheel may not, however, always be easily detectable. Take Lewis's invocation of TN in *Parts of Classes*, for simplicity adapting his example from set theory to arithmetic. Contrast two interpretations of the statement that $1 + 2 = 3$. On an eliminative structuralist reading, '$1 + 2 = 3$' is interpreted as (roughly) 'in any structure instantiating the Dedekind-Peano axioms, the function representing addition takes the second element and the third element to the fourth element'. (The first element represents zero.) On a platonist reading, '$1 + 2 = 3$' is instead interpreted as 'the abstract entities 1 and 2 stand in the abstract addition relation to the abstract entity 3'. Now suppose that the platonist interpretation of '$1 + 2 = 3$' in fact correctly reflects how mathematicians understand the sentence. In combination with TN, does this entail that platonism is true? Not necessarily. Perhaps mathematical norms do not endorse the platonist interpretation of arithmetic any more than they endorse its structuralist interpretation. Careful scrutiny might even show that they endorse the latter. In other words, platonism *might* be an idle wheel in mathematics. There are no short cuts here of the type Lewis was hoping for.[25] What is clear is that espousing TN does not in itself give us much purchase

---

[24] Compare Daly and Liggins (2011, p. 323): 'Resolving those [philosophical] debates involves weighing up many kinds of data and many methodological considerations by means of a protracted and often difficult cost-benefit analysis'. The debate might also lead to the broader question of what constitute the aims of science.

[25] The Idle Wheel Objection is related to what in Paseau (2005) and Paseau (2008). I called the reinterpretation objection. I argued there that philosophy does not have a comparatively worse track record than science when it comes to questions of interpretation. I had in mind questions such as whether science^{MI} is scientifically preferable to science^{MD} or whether the platonist interpretation of arithmetic is mathematically preferable to its structuralist interpretation, or whether realism or instrumentalism is the correct conception of a particular scientific theory. If the interpretation inherits

on such philosophical questions. The hard work lies in showing which of various philosophically rival views is scientifically privileged.

It should be fairly clear, then, that the Track Record Argument does not offer a shortcut in the philosophical debate. In particular, the scientific merits and demerits of realism on the one hand and those of idealism on the other must be carefully weighed before TN can do the work asked of it. As stressed earlier, this is not an objection to TN itself. It is a criticism of TN's ability to move the dial philosophically.

21.8 Sixth Objection: No Dogmatism

The next objection is based on the fact that the Track Record Argument is not deductively valid. Past failures cannot logically compel one to reject a belief-acquisition method: one can insist with perfect logical consistency that a predictive method with a past success rate of 0% will get it right next time, or that a method with a past success rate of 100% will get it wrong next time. It is not even a symptom of illogicality per se to be 100% convinced that a method with a 0% past success rate will get it right next time, past evidence to the contrary. Nevertheless, its proponents will say that, despite its logical invalidity, the Track Record Argument's conclusion is warranted, since rationality goes beyond mere logical consistency. A comparatively poor track record should decrease one's confidence in the method if one is rational; in fact, other things being equal, it should lead one to bestow no confidence in the method if pitted against a historically superior one.

The Trumping Naturalist's response assumes that it is rational to assign a historically unreliable method no probability of success. Yet must there always come a point at which one's confidence diminishes to 0? Consider cautious Kath, who never assigns credence 0 or 1 to any proposition that is not a logical truth or falsehood.[26] Kath starts off with some initial credences (distinct from either 0 or 1) about the soundness of the same-as-today method and the roulette method for predicting the next day's temperature. Time goes by and Kath notices that the same-as-today method has a good track record whereas the roulette method has a poor one. Her confidence in the former's soundness goes up to 0.7 say and her confidence in the latter's soundness goes down to 0.2 say. The trend continues and Kath adjusts her credences accordingly, repeatedly raising her credence in the reliability of the same-as-today method and repeatedly lowering her credence in the reliability of the roulette method. Yet Kath, being a cautious sort by name and nature, never lets these credences actually

---

the scientific confirmation accrued by the theory, it is beyond dispute according to TN. But if the interpretation is an idle wheel, it is not scientifically endorsed and may therefore be overturned (by philosophy say) compatibly with TN. Thus the reinterpretation objection to TN in Paseau (2005) succeeds if and only if the interpretation of a domain is integral to its associated scientific theory. In other words, the reinterpretation objection is a special case of the Idle Wheel Objection.

[26] Though there does not appear to be anything special about logical truths in this regard. If you agree, replace cautious Kath by supercautious Susan, who never assigns credence 0 or 1 to *any* proposition.

reach 1 or 0, however much evidence comes in. She always leaves room for the possibility that the roulette method might get it right and that the same-as-today method might fail. There is always a sliver of doubt in her mind about extending past trends into the future.

Kath is clearly not irrational. In fact, many would laud her as a paragon of rationality. Surely her response is at least rationally permissible, if not rationally compulsory or optimally rational. A credential principle congenial to Kath, if she is naturalistically minded, might take the form:

> For any p, give p credence $s \times d_1 + (1 - s) \times d_2$ in p if scientific norms endorse p to degree d_1 and non-scientific norms endorse p to degree d_2,

where s might be 0.9 say or some other number close—but not equal—to 1.

A principle along these lines, in which $s < 1$, cannot be an expression of TN on pain of abandoning science's trumping role. Although a threshold credence account of belief—according to which a subject believes p iff only her credence in p is greater than some threshold—is not generally correct, it does reflect the difference between the subject believing p and not believing p in some cases. Suppose Kath is in such a case with threshold t, and that science's endorsement of p is to degree d_1 with $d_1 > t$. She might nevertheless disbelieve p if her non-scientific norms endorse p to a sufficiently low degree d_2. The simple arithmetical point is that $s \times d_1 + (1 - s) \times d_2$ may be lower than t even if d_1 is greater than t as long as s is smaller than 1. In such cases, philosophy may overturn science's recommendation regarding p. Kath's rule therefore captures a form of naturalism that falls short of giving science a trumping role, even for high values of s, so long as $s < 1$.

There is a twist in the tale. Cautious Kath is in fact behaving just as scientific norms recommend. For it is part of science that no proposition should be believed with utter certainty and that no method should be trusted entirely. Indeed, ask a scientist for the principal characteristic of science, and the chances are that this is what she will mention first: science enjoins that no proposition is beyond doubt. From an anthropological point of view, modern science differs from previous 'knowledge traditions' precisely in its willingness to admit our ignorance and fallibility.[27] The Track Record Argument's conclusion only follows on the basis of a dogmatism that science itself rejects. Cautious Kath's stance is an alternative naturalism more in line with science's anti-dogmatism than TN is.

In reaching this conclusion, we have followed a strict line about what qualifies as an expression of TN. Any argument that succeeds in establishing some version of our opening platitude—e.g. something along the lines of 'In assessing whether p, give most emphasis to what science says about p'—but falls short of underwriting a trumping role for science is not an argument for TN. This high standard, it should be stressed, is *self-imposed* by Trumping Naturalists.[28] In the next section, we develop an objection that threatens a looser version of naturalism.

---

[27] Harari (2011, p. 279).

[28] See the references in endnote 3. This point is also stressed by other critics of (what I have called) Trumping Naturalism, such as Daly and Liggins (2011, p. 322).

You may ask why so many good philosophers have endorsed a thesis susceptible to relatively straightforward refutation. One explanation is that TN lies at the end of a road along which philosophers are generally willing to travel very far. Overstatement can then be hard to resist. Perhaps another reason is that, as a philosopher, there is a certain transgressive thrill to be had in doing down your own discipline. In any case, as we have seen, TN is too strong to be plausible.

21.9 Seventh Objection: Schools

This section focuses on the clash between philosophy and science. In part, this is because it is the main case discussed in the literature, and in part because a single counterexample is enough to falsify a generalisation.

The Track Record Argument is less compelling to philosophers who break with the past. If you spearhead a philosophical enterprise radically different from anything that has come before, you stand outside the Track Record Argument's range. Most philosophers, however, are part of a historical tradition, whether they like it or not. As Trotsky is said to have remarked: though you may repudiate history, history will not repudiate you. There are of course many traditions in philosophy (most obviously Western and Eastern), many sub-traditions within them (e.g. empiricism or rationalism), species of these (e.g. logical empiricism), subspecies, and so on. Notoriously, consensus on major philosophical questions is hard to achieve and philosophers tend to divide into different camps or schools when it comes to their answers. Bourget and Chalmers's (2014) survey of c. 2,000 philosophers in 2009 revealed wide disagreement among them. Such sociological data opens up the possibility that, for many philosophers, past and present norms do not significantly diverge on whether *p* is true, for any given *p*. The thought, then, is that from the perspective of any given philosopher, there is broad historical agreement over what philosophical norms endorse.

The existence of philosophical schools—groups of thinkers with different views on different questions, without the usual implication that the differences are particularly systematic—weakens the Track Record Argument. The reason is that failure within any given school, *from the perspective of that very school*, is historically limited. Contrast this with science, in which today, broadly speaking, there are no schools.[29] Within each school, a philosopher regards her take on the question as roughly consistent with her particular tradition's history. Thus philosophers of any given school tend *not* to see the history of philosophy of their school as a history of

[29] I'm well aware that scientists disagree. My point is that these disagreements are more local than in the case of philosophy. Some scientific traditions reject theories for entire generations; but eventually there is convergence. The importance of scientific 'schools' is also relatively diminished today. These days, differences in 'schools' tend to reflect differences of interest, focus and approach more than *major* doctrinal disagreement, of which there is a relative paucity. These differences tend to be exaggerated by scientists themselves for professional gain, e.g. to enlarge one's circle of influence, increase grant capture, and the like.

failure—at least following some historical watershed, often the school's formation or the moment it reached maturity. The data of historical failure upon which the Track Record Argument is supposedly founded are therefore missing.

To illustrate the Schools Objection, let's run it on a few of Lewis's examples of alleged philosophical failure. These, you will recall, were that: (i) motion is impossible; (ii) the ontological argument is sound; (iii) nothing can be thought to exist outside the mind; (iv) time is unreal; (v) no theory has been made at all probable by evidence; (vi) an empirically ideal theory cannot possibly be false; (vii) it is a wide-open scientific question whether anyone has ever believed anything. The response to (i) is that any philosopher who thinks that Zeno's little piece of reasoning is not sound by today's standards is unlikely to think it was sound by the standards of Zeno's time. In fact, Zeno's argument was at the time—and always has been—understood by the great majority of philosophers as issuing in a paradox, not a true conclusion. Whatever Zeno himself may have thought, philosophy itself never included the impossibility of motion as one of its official theorems. In fact, it is not even clear whether in Zeno's time (the 5th century BC), his argument would have been considered philosophical, mathematical, or scientific, or some combination of the three.[30]

In response to (ii), any philosopher who thinks the ontological argument is unsound by today's standards is unlikely to think it was sound by earlier ones, e.g. those of St Anselm's time. Indeed, many of the classic replies to the argument appeared not long after Anselm's 11th-century formulation in the *Proslogion*.[31] Examples (iii)–(vii) are similar. To take the first two, any philosopher who thinks that idealism is not endorsed by today's standards is unlikely to think it was endorsed by the norms of Berkeley's time. And anyone who thinks that the unreality of time is not endorsed by today's philosophical norms is likely to think that things were ever so, or at least that they were so at the start of the 20th century, when McTaggart published his argument.

I have spoken of 'schools' because this vividly brings to mind differences in attitude between individual philosophers. But my suggestion is not that the differences between, say, a realist and an eliminativist about folk psychology have the same kind of depth and systematicity as those between empiricism and rationalism of the early modern period. The 'Schools' label has strong connotations, which I realise are entirely apt only in a few cases (e.g. Neo-Platonism, Hegelianism, Thomism). Still, with its force appropriately understood, it conveys the right idea.

To be clear: my claim is *not* that for all broadly philosophical *p*, philosophers fall into different schools regarding *p*. Although there is a lot of disagreement in philosophy, especially over its 'central problems', there is also a good deal of agreement, in particular over the validity of various arguments or conditional claims such as,

[30] White (1992, p. 168ff.) and Knorr (1982) comment on ancient discussions of Zeno's argument. Knorr remarks that contemporary appraisal of how Zeno's argument was classified at the time can only be speculative, based as it is on relatively little documentary evidence.

[31] Whether or not Anselm himself regarded it as an argument, a vexed issue.

say, 'If scepticism is true then internalism about evidence is also true'.[32] Indeed, one can argue that when it comes to these kinds of claims, philosophy has a good track record. Real and stable knowledge has been created over the centuries, in philosophy just as in science.[33] The focus here on cases involving disagreement simply reflects the fact that these are the ones adduced as evidence for science and philosophy's differential track records.

We should also distinguish between what philosophical norms (held by some philosophers) at some time recommend and what most philosophers at that time believed. Consider a contemporary Popperian who believes that although evidence can falsify a theory, no amount of evidence can ever make it probable.[34] Now this thesis is framed using probabilistic language that would have made no sense to pre-17th-century philosophers and little sense to most 17th- and 18th-century ones. The period relevant to a Track Record Argument against Popper's contention must therefore begin no earlier than the late 18th century. Most philosophers from this point on have *not* been Popperians: they have believed that evidence can and does make theories more probable. Yet Popperians will tend to think that Popper's arguments were in principle available in 1785, 1835 and 1885 just as much as in 1935, when Popper deployed them. The logical point that universal quantifications over all space and time are falsifiable but not verifiable could have been appreciated centuries ago. Hume had laid the foundations for Popper's argument in his formulation of the problem of induction in the middle of the 18th century. Bayes's theorem could have been appreciated in the late 18th century thanks to Bayes and Laplace's work. As Popperians see it, then, it would have been rational to accept Popper's arguments had they been produced anytime from the late 18th century onwards.[35] This illustrates the point that, from any given philosopher's perspective, the majority verdict at some point in the past should not be equated with what philosophical norms endorsed back then. Philosophers belonging to a school of thought regarding *p* will look back at the history of philosophy and typically read its philosophical norms as aligned with current ones regarding *p*, whatever the majority opinion, now or then.

Does the Schools Objection make it too easy to block the pessimistic induction by allowing philosophers to read their views back into the past? No: philosophers are no more illicitly reading their views back into the past than they are illicitly reading their views into the present. Popperians are simply being faithful to their philosophical

---

[32] Internalism about evidence being the view, roughly, that a subject's evidence supervenes on her internal mental states.

[33] Gutting (2009), esp. Chaps. 4 and 10.

[34] Popper (1935/1959). A representative quotation: 'I think that we shall have to get accustomed to the idea that we must not look upon science as a "body of knowledge", but rather as a system of hypotheses; that is to say, as a system of guesses or anticipations which in principle cannot be justified, but with which we work as long as they stand up to tests, and of which are never justified in saying that we know that they are "true" or "more or less certain" or even "probable"' (Popper, 1959, p. 317).

[35] In an appropriate formulation for the time. For the sake of argument, we have assumed that Popperians can be moved by inductive considerations such as the Track Record Argument. But if they are to be faithful to Popper, they have a much quicker response to the Track Record Argument.

perspective in thinking that philosophical norms have, from the late 18th century on, never endorsed the fact that evidence makes theory probable. Non-Popperians are likewise being faithful to their perspective in maintaining the opposite. One cannot help but use one's own philosophical convictions to judge what propositions past norms endorsed. Indeed, one reason why it is useful to speak of philosophical norms endorsing *p* rather than philosophy endorsing *p* is to emphasise the point that different philosophers take different views about what these norms endorse.

One might dismiss this whole way of proceeding as historically insensitive, because past philosophers framed problems and theories in terms not directly comparable to today's. As is often observed, strong Kuhnian claims about the incomparability of present and past theories are more plausible about philosophy than science. Yet such Kuhnian reservations sit ill with the Track Record Argument. Lewis's discussion assumes that, by contemporary lights, previous philosophical conclusions have been mistaken. If no such direct comparisons are possible, the Track Record Argument collapses. Since we are trying to give that argument the best run for its money, we cannot but ignore Kuhnian worries about incommensurability. Or to put it another way: to the extent that you are worried by Kuhn, you should to at least that degree have reservations about the Track Record Argument.

One might also object that those who come to different philosophical conclusions share the same norms but have different views as to what their norms endorse. One reason behind the different verdicts might be the possession of different data. In the temperature prediction case, for example, the fact that an observer predicts that tomorrow's temperature will be 25° and another 18° does not in itself imply that they are using different methods. They could both be using the roulette method but have spun different roulette wheels; or they might have spun the same wheel but read it differently; or they could both be using the same-as-today method but have a different record of today's temperature. In general, one can be hamstrung by one's data just as much as by one's standards/norms. Thus, it might be said that the Schools Objection falsely infers a difference in philosophical norms from a difference in philosophical views.

A full evaluation of this point would call for a detailed analysis of each case. No such analysis is required here, however, since the force of the Schools Objection is simply that from any given philosopher's perspective, there is a nice fit between what past philosophical norms have recommended and what current norms recommend. The Cartesian sceptic believes that the philosophical standards of 2024 endorse scepticism just as much as those of 1641 did, the year in which Descartes' *Meditations* was published; and the anti-sceptic thinks that neither set of standards endorses scepticism. Whether the sceptic and the anti-sceptic have different norms, or share the same norms but construe them differently is ultimately irrelevant. The essential point is that a given philosopher's verdict about past cases closely matches her verdict on present ones. It is this agreement that undermines the Track Record Argument, whatever its source.

Another accusation, only partly pre-empted earlier, is that the Schools Objection seems to be committed to past and present philosophical standards agreeing on any question whatsoever. As the general claim is false, it is fortunate that the Schools

Objection is not committed to it. Even from the perspective of a given philosophical school, philosophical standards of the past may return different verdicts from present ones. For example, consider a traditional version of the design argument for the existence of God, which claims that the best direct explanation for the biological complexity of organisms on Earth is that some powerful, purposeful being created them. At the beginning of the 19th century, this argument was overwhelmingly strong, strong enough to have impressed the young Charles Darwin. The argument was severely weakened, if not demolished, by the alternative explanation for biological complexity provided by Darwin's own theory of evolution. From our perspective, other versions of the argument from design, e.g. from the fine-tuning of physical constants, are much stronger. As a result of scientific progress, then, support for the (direct) creationist explanation of biological complexity has fallen sharply in the past two centuries.

A plausible example of a changed scientific and philosophical norm is the emphasis placed on common sense. Copernicus's theory that the Earth moves round the Sun offended against common sense when it was first propounded. One reason was that rapid motion at that time was associated with riding on horseback, or in a carriage over rutted roads, or with sailing on generally rough seas.[36] As no such sensation results from standing still on the Earth's surface, it simply did not *feel* as if the Earth moved. Nowadays, the importance afforded this type of consideration is much diminished, since we are aware that motion may be smooth. More generally, we have learnt that the universe is much stranger than common experience on the surface of the Earth suggests. Prior to the 20th century, philosophy and especially science respected common sense more than they do now.

The Schools Objection therefore does not overshoot since it concedes that philosophers of various stripes—schools—accept that there are past cases of philosophical failure. Philosophy's past success rate is higher than the Trumping Naturalist reckons when estimated from the perspective of various schools; but it is not 100%. The objection accordingly does not absolve philosophy of all past historical failure. But it does dramatically reduce its scope.

Finally, suppose that the Schools Objection does succeed in establishing that, as a result of the considerable disagreement in philosophy, there is no single perspective from which philosophy has a poor track record. Is this lack of agreement not itself damning? How can the deliverances of philosophy be trusted given the extent of our disagreement over what exactly these deliverances are? An equally anti-philosophical moral seems to be that, unlike science, convergence of opinion on matters philosophical is unlikely or even impossible.

These are certainly important criticisms of philosophy's credibility. But they do not undermine the Schools Objection. The argument that infers from the comparative lack of agreement over philosophical questions (perhaps compared to scientific ones) that philosophy is not credible (or at least less credible than science) is not a historical argument, since it depends on how things stand today rather than on the disciplines' track records. This synchronic Disagreement Argument is distinct from

[36] Gribbin (2003, pp. 11–12).

the diachronic Track Record Argument. The differences between the three arguments can be helpfully illustrated by means of our temperature prediction analogy. Suppose you are part of a scientific community that is now in its 50th year. Initially, half your peers used the same-as-today method and the other half used the roulette method to determine the next day's temperature. Every year, one percent of inquirers switched from the same-as-today method to the roulette method. Looking back in time from year 50 (the present), you observe perfect convergence in people's temperature prediction methods over time: in year 0 the split was 50–50%, now it is 100–0%. At present, then, there are no differences in opinion, since everyone employs the roulette method. And yet, for all that, assuming that our imaginary community's world is in other relevant ways like ours, the roulette method still has a poor track record compared to the same-as-today method. The community accordingly remains vulnerable to an inductive argument—the temperature analogue of the Track Record Argument—against the roulette method and in favour of the same-as-today method. Yet it is *not* susceptible to the temperature analogue of the Disagreement Argument, since by hypothesis there is no disagreement in the present. Nor is it susceptible to an analogue of the Lack of Convergence argument, since by hypothesis there is convergence over the past 50 years to the roulette method. Moral: the Track Record Argument is distinct from the Disagreement Argument and the Lack of Convergence Argument.

Now, my aim in this chapter is not to assess all possible pro-naturalism and anti-philosophy arguments. It is the more modest one of assessing the Track Record Argument. The Disagreement Argument and the Lack of Convergence Argument deserve extended, but separate, consideration.[37]

We conclude with methodological advice for philosophers flowing from the Schools Objection. Philosophers should be sensitive to whether *from their philosophical perspective* philosophy has a poor track record. If so, that should tend to undermine their philosophical norms. Other things being equal, one should adopt a philosophy that minimises philosophy's history of failure.

[37] Beebee (2018) sees what I have called the Disagreement Argument (which she calls the 'disagreement challenge') as well as the 'methodological challenge' (that our methods in philosophy, especially metaphysics, can stake no claim to truth-conduciveness) as undermining the view that philosophy, especially metaphysics, aims for knowledge.

21.10 Summary

In bullet point form, here are the seven objections to the Track Record Argument and an (unnuanced) summary assessment:

- No Clash Objection (§3): mistaken;
- Definitional Failure Objection (§4): mistaken;
- The Failure of Science Objection (§5): resistible;
- Good Philosophy is Scientific Objection (§6): does not affect a suitably reformulated conclusion;
- Idle Wheel Objection (§7): shows TN must be supplemented to be effective;
- No Dogmatism Objection (§8): undermines TN as stated;
- Schools Objection (§9): undermines a looser version of TN.

Roughly speaking, the first four objections do not undermine the Track Record Argument or its use, whereas the last three do. The Idle Wheel Objection (§6) shows that even if TN is true, much work of a broadly philosophical nature is needed before reading philosophically decisive claims from scientific practice. The No Dogmatism Objection (§7) succeeds in defeating the Track Record Argument and TN itself. The Schools Objection (§9) threatens a version of the argument that motivates less stringent naturalisms than TN.

Our opening platitude—that science is, generally speaking, more credible than non-science—is not in doubt. We *should* generally give science more weight in our deliberations than non-science. But the naturalist tendency that converts this sound advice into a trumping thesis goes too far, at least if its main basis is the Track Record Argument. A closer look at how philosophical debates evolve over time suggests that little more than the platitude can be squeezed out of respective track records. The Track Record Argument is not that strong.[38]

References

Beebee, H. (2018). Philosophical scepticism and the aims of philosophy. *Proceedings of the Aristotelian Society, 118*, 1–24.

Blome-Tillmann, M. (2009). Moral non-cognitivism and the grammar of morality. *Proceedings of the Aristotelian Society, 109*, 279–309.

Bourget, D., & Chalmers, D. J. (2014). What do philosophers believe? *Philosophical Studies, 170*, 465–500.

Burgess, J. (2004). Mathematics and bleak house. *Philosophia Mathematica, 12*, 18–36.

Burgess, J., & Rosen, G. (1997). *A subject with no object*. Oxford University Press.

Colyvan, M. (2001). *The indispensability of mathematics*. Oxford University Press.

Daly, C., & Liggins, D. (2011). Deferentialism. *Philosophical Studies, 156*, 321–337.

Feyerabend, P. (1975). *Against method*. Verso.

Feynman, R. (1985). *QED: The strange theory of light and matter*. Princeton University Press.

[38] Thanks to Tushar Menon for comments and to Pen Maddy for letting me see her draft reply to a draft version of this chapter, which led to a small number of changes.

Fine, K. (2009). The question of ontology. In D. Chalmers, D. Manley, & R. Wasserman (Eds.), *Metametaphysics: New essays on the foundations of ontology* (pp. 157–177). Oxford University Press.
Goodman, N., & Quine, W. V. O. (1947). Steps toward a constructive nominalism. *Journal of Symbolic Logic, 12*, 105–122.
Gosse, P. H. (1857). *Omphalos: An attempt to untie the geological knot*. R. Clay Printer.
Gould, S. J. (2001). *Rocks of ages*. Jonathan Cape.
Gribbin, H. (2003). *Science: A history*. Penguin.
Gutting, G. (2009). *What philosophers know*. Cambridge University Press.
Harari, Y. N. (2011). *Sapiens: A brief history of humankind*. Vintage.
Holton, G. (1988). *Thematic origins of scientific thought: Kepler to Einstein* (Rev. ed.), Harvard University Press.
Hofweber, T. (2009). Ambitious, yet modest, metaphysics. In D. Chalmers, D. Manley, & R. Wasserman (Eds.), *Metametaphysics: New essays on the foundations of ontology* (pp. 260–289). Oxford University Press.
Knorr, W. (1982). Infinity and continuity: The interaction of mathematics and philosophy in antiquity. In N. Kretzmann (Ed.), *Infinity and continuity in ancient and medieval thought* (pp. 112–45). Cornell University Press.
Kuhn, T. (1977). Objectivity, value judgment, and theory choice. In *The essential tension* (pp. 320–339). University of Chicago Press.
Ladyman, J., & Ross, D. (2007). *Every thing must go*. Oxford University Press.
Lakatos, I. (1978). Introduction: Science and pseudoscience. In *The methodology of scientific research programmes*. Cambridge University Press.
Laudan, L. (1981). A confutation of convergent realism. *Philosophy of Science, 48*, 19–48.
Leiter, B. (2004). *The future for philosophy*. Oxford University Press.
Lewis, D. (1991). *Parts of classes*. Blackwell.
Maddy, P. (1997). *Naturalism in mathematics*. Oxford University Press.
Maddy, P. (2007). *Second philosophy*. Oxford University Press.
Paseau, A. (2005). Naturalism in mathematics and the authority of philosophy. *British Journal for the Philosophy of Science, 56*, 399–418.
Paseau, A. (2007). Scientific platonism. In M. Leng, A. Paseau, & M. Potter (Eds.), *Mathematical knowledge* (pp. 123–149). Oxford University Press.
Paseau, A. (2008). Naturalism in the philosophy of mathematics. In E. Zalta (Ed.), *Stanford encyclopedia of philosophy*. http://plato.stanford.edu/entries/naturalism-mathematics
Popper, K. R. (1935/1959). *Logik der forschung*. (Springer, Trans. as *The logic of scientific discovery*). Hutchinson.
Putnam, H. (1971). Philosophy of logic. Repr. in his *Mathematics, matter and method: Philosophical papers vol. 1* (pp. 323–357). Cambridge University Press.
Putnam, H. (1983). Why reason can't be naturalized. Repr. in *Realism and reason, philosophical papers vol. 3* (pp. 229–247). Cambridge University Press.
Quine, W. V. O. (1981). Things and their places in theories. In his *Theories and things* (pp. 1–23). Harvard University Press.
Rosenberg, A. (2012). *Philosophy of science* (3rd ed.). Routledge.
Rovelli, C. (2011). *Anaximander* (Trans. by M. Lignana Rosenberg). Westholme.
Russo, L. (2004). *The forgotten revolution: How science was born in 300 BC and why it had to be reborn* (Trans. by S. Levy). Springer.
Schaffer, J. (2009). On what grounds what. In D. Chalmers, D. Manley, & R. Wasserman (Eds.), *Metametaphysics: New essays on the foundations of ontology* (pp. 347–383). Oxford University Press.
Shapin, S. (1996). *The scientific revolution*. University of Chicago Press.
van Fraassen, B. (1980). *The scientific image*. Oxford University Press.
Vargas, M. (2009). Revisionism. In J. Martin Fischer, R. Kane, D. Pereboom, & M. Vargas (Eds.), *Four views on free will* (pp. 126–165). Blackwell.

White, M. (1992). *The continuous and the discrete*. Oxford University Press.
Wigner, E. (1970). *Symmetries and reflections: Scientific essays of Eugene P Wigner*. MIT Press.
Wittgenstein, L. (1980). *Culture and value* (2nd ed., Trans. of *Vermischte Bemerkungen*, by P. Winch, G.H. von Wright (Eds.)). Basil Blackwell.

Chapter 22
Reply to Paseau

Penelope Maddy

Keywords Naturalism · Set theory · Method · Truth and existence · Second philosophy · Demarcation criterion · Barry Stroud

Professor Paseau raises the important question of how 'naturalism' is to be defended. Of course, these days, the term is applied so widely and variously as to be nearly useless, so Paseau identifies a specific target that he calls Trumping Naturalism[1]: 'If science endorses *p* then one should believe *p*' (p. 268). One might recoil from this characterization, but if the shoe fits, one must bear it. This isn't the first time I've felt some well-earned embarrassment about the overly dramatic passage Paseau cites from *Naturalism in Mathematics*: if mathematics and philosophy come into conflict, 'it is the philosophy that must give' (Maddy 1997, p. 161).[2] As he himself observes, this kind of declaration.

> lies at the end of a road along which philosophers are generally willing to travel very far. Overstatement can then be hard to resist. (p. 282)

In my own feeble defense, I might note that this bald declaration is actually described as a 'hint' derived from the previous discussions of Wittgenstein, Gödel, and Quine, a hint I hope to 'develop'. But however that may be, I'd like to point out that the position actually developed in the book doesn't fit with this rhetorical flourish.

---

[1] I applaud Paseau's refusal to treat 'trump' and its variants as skunked words. 'Trumping' perfectly captures what he wants to convey.

[2] See his footnote 2.

P. Maddy (✉)
Department of Logic and Philosophy of Science, University of California, Irvine, CA, USA
e-mail: pjmaddy@uci.edu

S. Arbeiter and J. Kennedy (eds.), *The Philosophy of Penelope Maddy*,
Outstanding Contributions to Logic 31, https://doi.org/10.1007/978-3-031-58425-1_22

Part of what makes the line of thought in this section of *Naturalism* so baroque is a prolonged attempt to separate the 'philosophy' from the 'mathematics' in historical debates over set theoretic method.[3] This was misguided, as I later realized[4]—all that's actually needed is a separation between kinds of considerations that influenced a particular methodological decision and those that didn't. Viewed from this angle, it's easy to see that though disagreements over ontology, semantics, and epistemology undeniably appear, they remain unresolved while the methodological questions at issue are settled in terms of the particular mathematical goals in play. Given that questions of mathematical existence, truth, and knowledge seem irrelevant distractions to the rational core of these historical debates, the proposal is to try to remove them from current, ongoing debates and see what's left. If this remainder takes the shape of preferring effective means toward legitimate ends, then the methodological decision in question should count as fully rational for that reason alone. Granted, the considerations set aside are those often labeled 'philosophical', but this is entirely beside the point: they aren't set aside because they're so labeled; they're set aside on the counsel of history; and that counsel proves wise when doing so reveals an underlying rationality. So, despite the unfortunately formulated 'hint', this isn't Trumping Naturalism.

It's important to note that nothing like a principled distinction between 'philosophy' and 'mathematics' emerges from this historical exercise, just the particulars of the individual cases at hand. In fact, more broadly, the characterization of Second Philosophy doesn't rely on any principled distinction between 'science' and 'non-science'. I take it as given that the methods of science evolve in unforeseeable ways, that it would be hamstrung by any decision in advance about what is and isn't allowable.[5] So rather than endorse any doctrine of the form 'trust science', I introduce the character of the Second Philosopher and show her beginning with observation and experimentation, progressing from there to theory formation and testing, correcting and improving as she goes—counting on the reader to get the hang of it.[6] Eventually, her all-encompassing inquiry brings her to questions that we would classify as 'philosophical', but that she regards as just more of the same: can we gain reliable information about the external world?, what grounds logical truth?, what is the nature of mathematics? Second Philosophy consists in the answers (I conjecture that) she would give to those questions. The key point here is that, along the way, she's open to considering any method, including astrology or Descartes' Method of Doubt, without condemning anything in advance as 'unscientific'. Those methods just don't measure up in her eyes.

[3] I call this the 'boundary problem' (Maddy 1997, pp. 188–193).

[4] See Maddy (2007), p. 349, footnote 12, and (2011), p. 53, footnote 33.

[5] E.g., before Perrin, many held that it was impossible in principle to establish the existence of entities too small to perceive, that they were inaccessible to the only allowable scientific methods. Then his innovative experimental designs and outcomes established a way of 'detecting' things too small to see. (For discussion, see Maddy (1997), pp. 135–143, (2007), pp. 402–407.).

[6] See Maddy (2007), pp. 1–3.

The point of all this is that the debate Paseau enters here depends on assuming there's a principled way of separating 'science' and 'philosophy', 'even if it's fuzzy', as he remarks (p. 270). So the Second Philosopher can't so much as state, let alone embrace, Trumping Naturalism.[7] Still, there remains the broader question of how any form of naturalism can be defended. For a view like Second Philosophy—more a stance, in van Fraassen's sense, than a set of beliefs[8]—the challenge would be to justify adoption of the Second Philosopher's overall approach to inquiry. This I've explicitly declined to attempt because I've taken it to be a fool's errand.[9]

The challenge, as I've understood it, is to ascend to a neutral arena and confront someone whose entire system of thought is dramatically different. The Second Philosopher can ably defend her methods piecemeal, but faced with the demand that she defend them wholesale, that is, that she set aside all her hard-won strategies for finding out about the world, and that, so-deprived, she defend ... well, it doesn't much matter what she's being asked to defend at that point, since she's lost access to all her means of defending! She sees herself as being asked to produce an altogether new kind of evidence, not the kind of thing she could reach by building on the beliefs and approaches she already has, but an extra-ordinary kind of evidence that would serve to tip the debate her way in this rarified, neutral arena. She readily admits that she can't do it.

But this whole line of thought is irrelevant to Paseau's discussion, because he's taking the Track Record Argument to be straightforwardly empirical.[10] Well, then, if the Second Philosopher is allowed her ordinary methods, she figures she has

[7] [Added later: since this was written, Professor Paseau has modified his characterization of the Trumping Naturalist to include this proviso: 'If in practice you always side with the scientific consensus against any ... appeal to non-science ... when the two clash, you are *ipso facto* a Trumping Naturalist—irrespective of whether you embrace or disavow the label, accept or reject the science/non-science distinction' (p. 271). Assuming this is intended to bring the Second Philosopher into the fold, the thought must be that there is a distinction between science and non-science; that the Second Philosopher accepts some methods/beliefs and rejects others; and that those she accepts happen to fall in the science bin and those she rejects in the non-science bin. At the same time, if you ask her to give a global characterization of the methods/beliefs that she accepts now or will accept in the future, she declines, on the grounds that she can't foresee how her pattern of acceptance will evolve (recall that in Maddy (2011), she eventually ends up allowing for the existence of abstracta, which involves accepting methods quite different from those she's accepted before). I have three observations: (1) from Paseau's perspective, the Second Philosopher is making a mistake; there is a way to characterize the methods/beliefs she has accepted or will accept in the future, i.e., as 'science', and the methods/beliefs she rejects or will reject in the future, i.e., as 'non-science', (2) on that characterization, her answers to questions traditionally classified as 'philosophical' arise from methods/beliefs she accepts, and she counts as a Trumping Naturalist because everything she accepts falls on the 'science' side of Paseau's distinction, so all her accounts of mathematics, logic, truth and reference, external world skepticism, and even about the most effective methods for approaching questions like these—all this must count as 'science', an odd outcome, and (3) in the end, I'm not sure Paseau maintains this liberal version of Trumping Naturalism (see footnote 15).].

[8] See Maddy (2022f).

[9] See Maddy (2007), p. 3, (2017a).

[10] [Added later: Paseau now characterizes Trumping Naturalism as 'meta-epistemological' rather than 'epistemological' (p. 273), but I'm not sure what difference this makes.].

much better arguments than that! Paseau's picture has 'science' on one side and 'philosophy' on the other, but the Second Philosopher approaches questions of both sorts in the same ecumenical way, and, as we've seen, at least some of those typically labeled as 'philosophical' strike her as deserving answers. So, for example, suppose a professional philosopher[11] argues that her inability to win the rarified debate, that is, her inability to defend her methods 'from scratch', implies that she has no better reason to believe what she does than to disbelieve it. In my little book on skepticism (Maddy 2017a), I give what I take to be a second-philosophical response on her behalf: her inability to produce extra-ordinary evidence doesn't undercut the efficacy of ordinary evidence she can produce. This isn't a Track Record argument against 'philosophy' in general; it's a more specific and articulated argument again one particular line of thought generally described as 'philosophical' (radical skepticism about our knowledge of the external world, based on the Dream Argument or the Argument from Illusion). Elsewhere, I give what I take to be second-philosophical responses to professional philosophers who claim that logical truth is necessary or that mathematics is the study of a Platonic realm that's epistemically problematic.[12]

In sum, then, we've seen that the Second Philosopher, lacking a principled distinction between 'science' and 'non-science', can't so much as state the Trumping Naturalist's credo.[13] (Neither can I.[14]) But, as we've seen, she can and does engage with questions that are typically classified as 'philosophical'. When she disagrees with what professional philosophers say in such cases, she appeals to empirical considerations more individualized than a simple Track Record Argument. But, assuming answers to these 'philosophical' questions would also be typically classified as 'philosophical', she produces 'philosophical' claims of her own, defended in her own terms. (Some professional philosophers may wish to dismiss these answers as not the right sort of thing in some way, so the Second Philosopher is also issuing an implicit challenge to explain precisely what's wrong with them.) For that matter, she certainly isn't prepared to forgo the insights that can be gleaned from the writings many historical greats—Descartes, Locke, Berkeley, Reid, Kant, Frege, Wittgenstein, Austin, Quine—so it's not clear she's anywhere near the Trumping Naturalist even in spirit. For her, it isn't 'science' versus 'philosophy'; it's just good ideas, good reasons, good evidence versus bad.

Paseau might feel, in fact, does feel, that this is unsatisfying:

> [If] Second Philosophy is characterised very vaguely, so as to avoid assimilation to a clear-cut form of naturalism … [it] might be acceptable, but at the cost of not saying very much. Subscribing to Second Philosophy would not in practice help resolve any methodological dilemmas. (private communication 10/20/20, quoted with permission)

[11] I use this term for a mainstream employee of a university Department of Philosophy—like the Second Philosopher, I have no principled criteria to deploy—or maybe better, for the kind of person Paseau has in mind when he speaks of a 'philosopher'.

[12] Maddy (1997), (2007), (2011), and (2014a).

[13] Recall footnote 7.

[14] The difference between me and the highly idealized Second Philosopher isn't that I use any different methods, but that she's so much wiser and more knowledgeable than I am. (van Fraassen presses this point in his (2015). See Maddy (2022f).

Resolving methodological dilemmas would mean

> [taking] sides in disputes between religious believers who repudiate parts of science and upholders of the scientific consensus [and] in debates about revising mathematics on the basis of philosophical ideas (e.g. à la Brouwer). (Ibid.)

Now in one sense, of course, the Second Philosopher does take sides in those debates—she examines the evidence of the religious believer or the motivations for curtailing mathematics and finds them wanting—but Paseau wants something else; he wants her to do this as part of a general counsel against the methods of 'religion' or the motivations of 'philosophers'. This she doesn't do.[15]

In an intriguing paper called 'The transparency of "naturalism"', Barry Stroud puts the point eloquently:

> Conclusions reached in Second Philosophy are not to be defended on the authority of something called 'science', or even something called 'Second Philosophy' or 'naturalism' either, but only on the basis of the evidence actually available or discoverable. Second Philosophy is accordingly not a set of doctrines or a specified method. It is at best a kind of inquiry 'without strict disciplinary allegiances'. We can understand it only by understanding in each case the questions it considers and the ways it tries to answer them. (Stroud (2009), p. 254)

This describes the approach I take in *Second Philosophy* very nicely—to understand who the Second Philosopher is, look to what she does—but, in a way, this is to agree with Paseau that Second Philosophy doesn't say very much. This is what Stroud has in mind when he speaks of its 'transparency':

> This shows what I call the transparency of 'naturalism'. ... you always have to look right through the term to see ... what it is meant to stand for. It has some determinate but still very general meaning as opposition to certain *a priori* philosophical doctrines. But otherwise we are left with nothing much more determinate than the idea of careful, informed investigation, or perhaps just responsible inquiry. That is not nothing to be left with. It is all we've got. But there is no need to decide in advance how inquiry into whatever interests us must, or must not, proceed. (Stroud (2009), p. 255)

The Second Philosophy as I describe it is where the Second Philosopher ends up simply by following her nose. My focus is on her answers to the various questions posed, which I commend to the reader's consideration. Insofar as Second Philosophy has any characterization, it is the approach that leads us here, by this route; insofar as it has any defense, it lies in what I hope the reader will see as the attractions of those outcomes. I hope this is not nothing![16]

---

[15] [Added later: recalling footnote 7, this is where I think Paseau is asking the second-philosophical Trumping Naturalist, not merely to act in accord with the science/non-science distinction, but to recognize and deploy that distinction in the course of these debates. In the end, it seems we disagree over the status of the liberal Trumping Naturalist allowed in footnote 7: is her epistemic state compromised by her inability to see his science/non-science distinction?; does she owe a global characterization of her methods/beliefs? As a Second Philosopher myself, I say no, my epistemic state is in order despite the fact that I don't think there's a way to characterize the methods/beliefs I now accept or will accept in the future. Paseau disagrees.]

[16] Thanks to Professor Paseau for his contribution and for a productive exchange on the topics it raises.

Reference

Maddy, P. (1997). *Naturalism in mathematics*. Oxford University Press.
Maddy, P. (2007). *Second philosophy*. Oxford University Press.
Maddy, P. (2011). *Defending the axioms*. Oxford University Press.
Maddy, P. (2014a). *The logical must*. Oxford University Press.
Maddy, P. (2017a). What do philosophers do? Oxford University Press.
Maddy, P. (2022f). On the question of realism. In [2022h] (pp. 49–90).
Maddy, P. (2022h). *A plea for natural philosophy and other essays*. Oxford University Press.
Stroud B [2009] 'The transparency of naturalism', reprinted in his Philosophers Past and Present, (Oxford: Oxford University Press, 2011), pp 240–255
van Fraassen, B. (2015). Naturalism in epistemology. In R. Williams & D. Robinson (Eds.) (pp. 63–95). Bloomsbury Academic Publishers.

Part III
History of Philosophy

Chapter 23
What Do Philosophers Do? Maddy, Moore (and Wittgenstein) II

Annalisa Coliva

Abstract The chapter takes issue with Professor Maddy's reply (2018) to my commentary (Coliva in International Journal for the Study of Skepticism 8:198–207, 2018) on her account of Moore's proof in What Philosophers do. Skepticism and the Practice of Philosophy (2017). In particular, I argue that transcendental idealism was the target of Moore's proof, rather than skepticism; that he did not dismiss Cartesian skepticism as illegitimate or even nonsensical; yet thought that since knowledge of the premises of his proof, such as "Here is one hand", and "Here is another", does not depend on being able to prove that one is not dreaming, contra Cartesian skepticism, he did have knowledge of them. Moore's defence of the good standing of his proof was thus based on embracing a proto-externalist account of knowledge, according to which knowledge (K) does not entail knowing how one knows, in the specific sense of being able to prove that one knows (KK).

I want to start by thanking the editors, Juliette Kennedy and Sophia Arbeiter, for inviting me to contribute to this volume in honor of Professor Penelope Maddy. This gives me an opportunity to engage with the views of a philosopher and a colleague I greatly admire. As my title suggests, however, it also gives me an opportunity to pursue our conversation further. For we first engaged on the occasion of the *Symposium* on Maddy's latest book *What do Philosophers Do? Skepticism and the Practice of Philosophy* (2017), held at the Pacific Division Meeting of the American Philosophical Association (San Diego 2018), published in the same year in the *International Journal for the Study of Skepticism*. Professor Maddy kindly responded to my comments, particularly on G. E. Moore and his celebrated "Proof of an external world" (1939), by putting forth an extremely interesting reading of some key passages

A. Coliva (✉)
University of California, Irvine, CA, USA
e-mail: a.coliva@uci.edu

S. Arbeiter and J. Kennedy (eds.), *The Philosophy of Penelope Maddy*,
Outstanding Contributions to Logic 31, https://doi.org/10.1007/978-3-031-58425-1_23

in that and other works by Moore. While appreciating the subtlety of her analysis, I remain unconvinced and I welcome this opportunity to explain why.[1]

23.1 Coliva on Maddy's Moore

In my commentary (Coliva, 2018) on Chap. 3, "The cure and beyond", of Maddy's (2017) book, I took issue with the following reasoning:

(i) that Moore, in the "Proof", is engaging with a skeptic;
(ii) that he is doing so by arguing that he has perfectly good evidence that he is not dreaming;
(iii) that he is aware that this evidence will not assuage a skeptic's doubts, because they depend on "extraordinary dreaming" (Maddy, 2017, p. 31) rather than ordinary dreaming;
(iv) and yet he is convinced that one should not be moved by this kind of challenge and should rather dismiss it as inappropriate or misguided. For no challenge based on extraordinary dreaming should be taken seriously.

As a result, and *contra* a skeptic, according to Moore

(v) we can know perfectly well that there is a hand here, where we see it, even if we cannot prove we are not dreaming *in the extraordinary sense in which a skeptic would require us to prove that we are not*—that is, in the sense that would require us to be able to prove it independently of any sensory evidence we take ourselves to have for thinking that we are awake.

The passages in Maddy's reconstruction I disagreed with are (i), (iv) and partly (v). Regarding (i)—that Moore is engaging with a skeptic—, by appealing to his "Reply to my critics" (1942), I claimed he was rather taking issue with an idealist. Thus, I claimed that his target was someone who denies the existence of material objects, rather than someone who denies that we can know of their existence. The key passage reads as follows:

> I have sometimes distinguished between two different propositions each of which has been made by some philosophers, namely (1) the proposition "There are no material things" and (2) the proposition "Nobody knows for certain that there are any material things". And in my last published writing, my British Academy lecture called "Proof of an external world" ... I implied with regard to the first of these propositions that it could be *proved* to be false in such a way as this; namely, by holding up one of your hands and saying "*This* hand is a material thing; therefore there is at least one material thing". But with regard to the second of these two propositions, which has, I think, been far more commonly asserted than the first, I do not think I ever implied that *it* could be *proved* to be false in any such simple way; e.g., by holding up one of your hands and saying "I know that this hand is a material thing; therefore at least one person knows that there is at least one material thing". (Moore, 1942, p. 668)

[1] Since Professor Maddy did not respond to my comments on her reading of Wittgenstein's *On Certainty* in writing, I will not pursue that part of our conversation here.

I did not take issue with (ii)—that Moore takes himself to have good sensory grounds for holding that there is hand where he sees it. Indeed, I myself (see Coliva, 2010, Chap. 1) have repeatedly insisted on that, even though I did not insist on that in my commentary.[2] Nor have I objected to (iii)—that he is aware that his evidence will not cut any ice with a skeptic, since the latter is basing her doubts on extraordinary dreaming. Indeed, in my commentary I have exploited this very idea to insist that Moore was aware of the peculiarities of the skeptical challenge and did not think of himself as engaging with, let alone responding to it, in "Proof of an external world".

In my commentary, however, I did insist, *contra* (iv) that, for Moore, that challenge is not simply to be dismissed as inappropriate or misguided. It would be so, according to Maddy's Moore, because it would be based on the conflation between ordinary dreaming—whose obtaining (or not) can be ascertained based on various empirical criteria—and extraordinary dreaming—whose obtaining (or not) could not be ascertained thus. On Maddy's reading, Moore objected to the legitimacy of the Cartesian skeptical challenge for it would ultimately require something impossible of us—that is, prove that we are not dreaming independently of *any* sensory (and possibly other "ordinary") evidence we take ourselves to have for thinking that we are awake. On the contrary, I have maintained that, in "Proof", Moore did not take himself to be engaging with a Cartesian skeptic and that the very little he says about Cartesian skepticism does not seem to support such a strong reading.[3] In particular, I have maintained that far from deeming the Cartesian challenge somewhat illegitimate, he did try to meet it, albeit unsuccessfully in my opinion, in other writings he was notoriously dissatisfied with—namely, "Four forms of skepticism" (1959a) and "Certainty" (1959b).[4] He did so by proposing his famous "gambit". That is, the argument whereby since his grounds for holding that he is awake and seeing a hand seem stronger to him than his skeptical opponent's for saying that he might be asleep and dreaming of seeing a hand, his claim that he knows that there his hand there is on more secure a ground than his opponent's denial of it.

Thus, I think that Moore, in the "Proof", did not object to the legitimacy of the skeptical challenge. If he ever did so, it was only in later writings, which, as I pointed out, he was not satisfied with anyway. Yet, on closer inspection, even then did Moore not dismiss the skeptical challenge as illegitimate due to its being based on the conflation between ordinary and extraordinary dreaming. For, no matter how unpersuasive his gambit was, it was based on the conviction—surely objectionable by the lights of a Cartesian skeptic—that his sensory evidence as of a hand in front of him would be more solid a ground than anything a skeptic would be using to insinuate that he might just be dreaming of it and on the commonsensical belief that he was more certain of being awake and holding a hand in front of him than of anything his opponent could use to claim the opposite.

---

[2] There will be more on this in §2.

[3] There will be more on this in §2.

[4] These papers were composed between 1940 and 1944 and were published only posthumously in 1959.

In fact, already in "Proof", did Moore state:

> How absurd it would be to suggest that I did not know it [that there was a hand when he asserted each of the premises of his proof], but only believed it, and perhaps it was not the case! You might as well suggest that I do not know that I am now standing up and talking – that perhaps after all I'm not, and that it's not quite certain that I am! (Moore, 1939, pp. 146–147)

I take this passage to express the characteristically commonsensical reaction of dismay vis-à-vis philosophical pretensions.

In "Four forms of scepticism", Moore writes:

> I don't see any reason to abandon my view that I do know for certain ... that I am not dreaming now. And the mere proposition, which I admit, that percepts of the same kind *in certain respects* do sometimes occur in dreams, is, I am quite certain, no good reason for saying: this percept *may* be one which is occurring in a dream" (1959a, pp. 222–223).

He also writes:

> It seems to me *more* certain that I *do* know that this is a pencil and that you are conscious, than that any single one of these four assumptions [from which it would follow that I did not know that] is true, let alone all four (Moore, 1959a, p. 226).

And in "Certainty" he wrote:

> I agree, therefore, with that part of this argument which asserts that if I don't know now that I'm not dreaming, it follows that I don't *know* that I am standing up, even if I both actually am and think that I am. But this first part of the argument is a consideration which cuts both ways. For, if it is true, it follows that it is also true that if I *do* know that I am standing up, then I do know that I am not dreaming. I can therefore just as well argue: since I do know that I'm standing up, it follows that I do know that I'm not dreaming; as my opponent can argue: since you don't know that you're not dreaming, it follows that you don't know that you're standing up. The one argument is just as good as the other, unless my opponent can give better reasons for asserting that I don't know that I'm not dreaming, than I can give for asserting that I do know that I am standing up (Moore, 1959b, p. 247).

Given the conclusion—that the two arguments are on a par—it is normally agreed that Moore himself acknowledged the impotence of his "gambit" and was therefore dissatisfied with the paper, to the point of never licensing it for publication.

Nevertheless, the reason why Moore's reasoning would be objectionable by the lights of a Cartesian skeptic is that the latter too would use the same kind of evidence Moore would appeal to but point out that it would be entirely compatible with the possibility that it might be dreamt of. Should Moore then appeal to the coherence between that evidence and further evidence made available to him by his senses, a skeptic would simply insist that that evidence too could be merely be dreamt of. One might of course object to metaphysical possibilities *tout court*, or, more mildly, to the fact that a metaphysical possibility would show anything relevant regarding what we know and take ourselves to know, or with respect to what is in fact the case. To the best of my knowledge, however, Moore never explicitly said anything along these lines, which sound much more attuned to some of Wittgenstein's "therapeutic" ideas.

Hence, I do agree with that part of (v) that says that, for Moore, we can know perfectly well that there is a hand here, where we see it, even if we cannot prove we are not dreaming. Yet, I don't agree with the reason Maddy proposes—that is, Moore's dismissal of the Cartesian skeptical challenge as based on the conflation between ordinary and extraordinary dreaming (and between the kind of ordinary and extraordinary evidence which dismissing each of these hypotheses would in fact require). Rather, I maintained that Moore considered that challenge legitimate and yet endorsed, in his 1939 paper, a kind of proto-externalist account of knowledge such that one may know that P (e.g. "Here is my hand") even though one is unable to prove how one knows that P. By rejecting the KK principle, according to which knowledge of how one knows that P is necessary for knowledge of P, Moore then maintained that he did know that there was a hand where he seemed to see it, even if he was unable to prove how. And, in particular, even though he was unable to prove it to a Cartesian skeptic's satisfaction, because he would not have been able to prove that he was not dreaming. Indeed, it is worth noticing that, towards the end of "Proof", Moore writes:

> How am I to prove that "Here's one hand and here's another?" I do not believe I can do it. In order to do it, I should need to prove for one thing, as Descartes pointed out, that I am not now dreaming. But how can I prove that I am not? I have, no doubt, conclusive reasons for asserting that I am not now dreaming; I have conclusive evidence that I am awake: but that is a very different thing from being able to prove it. I could not tell you what all my evidence is; and I should require to do this at least, in order to give you a proof. (Moore, 1939, p. 149).

Thus, he was perfectly aware that there was a Cartesian challenge he was not capable of meeting. Not so much, in my opinion, because, as he says, he was unable to list all his evidence for believing to be awake, but, rather, because all that evidence would have been compatible with the hypothesis that he was merely dreaming of it.

Thus, surely Maddy and I give an overall different assessment of Moore's performance. For she thinks he was ultimately right about his claim to knowledge regarding (at least) the premises of his proof, and even winning against his opponent, as he correctly exposed the skeptical challenge as illegitimate, because of its being based on the conflation between ordinary and extraordinary dreaming. In contrast, while I agree that Moore was right in his claim to knowledge regarding his premises, I do not think he produced a satisfactory proof, for reasons I have elaborated elsewhere (see Coliva, 2015, chapter 3). Furthermore, I do not think he had an adequate strategy against Cartesian skepticism. For neither the externalist maneuver just rehearsed, nor his famous gambit or the appeal to his current conclusive but incomplete evidence could have cut any ice with such a skeptic.

23.2 Maddy on Coliva's Moore

In her "Reply" to my commentary, Maddy (2018) mainly takes issue with my denial of (i)—that in "Proof" Moore was engaging with a skeptic. She recognizes that in Moore (1942) he himself claimed that he was refuting (or trying to refute) idealism rather than skepticism. However, she mentions three reasons that give her pause.

The first one is that in the discussion of Kant's idea of "things to be met with in space" Moore notices that, for Kant, they are not "things external to our minds". Maddy then contends that this is a brand of idealism and that, if that "were the target of the paper, we should expect some serious argumentation at this point" (Maddy, 2018, p. 232). However, she continues, we only find a quick dismissal of Kant's views consisting in noticing that Kant never managed to explain clearly in what sense things to be met with in space would not be external to our minds (cf. Moore, 1939, p. 139). Maddy thinks that such a cursory remark displays no real engagement with idealism.

I beg to disagree. First, let me notice that Moore's remark may be cursory, but the direct engagement with Kant's ideas and terminology is the main bulk of Moore's paper and that it occupies about eighteen pages out of twenty-three. By contrast, there is only a quick mention of Cartesian skepticism in the whole "Proof", consisting of less than ten lines. If numbers matter, and typically they do when we are doing history of philosophy, it is odd to suggest that Moore was not really engaging with Kant's idealism.

Second, the fact that what Moore says is not powerful an argument or a response, by the lights of his opponent, does not mean that he is not engaging with idealism. Indeed, he is, and he is quite bluntly dismissing it (or at least Kant's version of it) by accusing Kant (and other philosophers who followed him in this) of having been unable to clarify what they meant. In fact, Moore is contesting the intelligibility of Kant's distinction between a transcendental and an empirical sense of "things to be met with in space", such that, on the former account, plants and other objects would turn out *not* to be external to our minds. For Kant "things to be met with in space" in the transcendental sense are not external to our minds because they are phenomena which, as such, constitutively depend on the forms of pure sensibility—space and time—and on categories, which are all *a parte subjecti*. For Moore, in contrast, they are external to our minds because he thinks, as he explains later in "Proof", that they do not depend on anyone's experiencing them. Hence, the only legitimate sense of "things to be met with in space" is the empirical one, for Moore, which makes it analytical that they are external to our minds, as they can exist independently of anyone's mind. Personally, I think this is the key move in the paper and, if it were correct, it would be dialectically very powerful. For if it is true that there is only one intelligible sense of "things to be met with in space"—namely, the empirical one—and that the transcendental one is spurious, then clearly (transcendental) idealism would be a dead-end and Moore's further proof would just be teasing out the consequences of this conceptual truth.

Third, and connectedly, it should be kept in mind that, at this stage in the paper, Moore is trying to make precise what *he* means by "things outside of us", which he eventually considers to be equivalent to "things to be met with in space" in the non-transcendental and merely empirical sense that he thought was the only clear and admissible one. It is only by assigning a clear meaning to "things outside of us" that, according to Moore, a rigorous proof of their existence can be given.

Maddy's second reason consists in noticing that, by saying that a thing to be met with in space is something that could exist without being perceived and without anyone's having any experience of it (cf. Moore, 1939, p. 144), Moore is merely presupposing the falsity of idealism, while offering no real argument against it (cf. Maddy, 2018, p. 233). I agree with Maddy's judgement, but again, I am more inclined to conclude that Moore was not successfully engaging with Kant's transcendental idealism rather than not engaging with it at all. As a partial excuse for Moore's attitude, as we have just seen, one can insist on the fact that Moore was mainly trying to make clear what it meant, for him, to talk of things outside of us, and external to our minds. Hence, he was not considering each and every argument idealists could use to support their views (let them be transcendental idealist ones or phenomenalist ones). Furthermore, he thought of himself as having to provide that clarification to *then* be able to offer a proof of the existence of an external world. That proof famously proceeded via exhibition—"here's a hand, and here is another"—and exemplification. For hands are instances of things to be met with in space and are thus outside of us and external to our minds (in the empirical sense that Moore takes himself to have clarified and somewhat defended previously in the paper). Moore than took the foregoing to establish the conclusion—"an external world exists". Of course, an idealist would have not analyzed the premises and the phrases "external to our minds" and "to be met with in space" as Moore proposed, and would not have counted the exhibition of Moore's hands as an exemplification capable of establishing that the category of physical objects, understood *à la* Moore, is not empty. Hence, the proof is entirely question-begging. That Moore did not see that is the startling aspect of his proof any Moore scholar is going to be wrestling with. But, again, the fact that the proof is unsuccessful—albeit for reasons difficult to pin down, as the long history of its exegeses shows—is something not many interpreters have seriously questioned.

Maddy's third and last reason comes from Moore's "Reply" to his critics, where, once confronted with the phenomenalists' analysis of "material object", according to which no such object could be external to our minds, he claims that "with this meaning of 'there are no material thing'... it is really impossible to prove that that statement ["There are no material things"] is false in the way I gave" (Moore, 1942, p. 670). Maddy then takes Moore to be showing little interest in idealism (of a phenomenalist brand). Once again, I beg to disagree. For to say that he could not prove wrong the phenomenalist analysis of the meaning of "material object" (and of "hand" as well) by appealing to common sense is not to say that he was not engaging with his opponent's view. Rather, it is merely to recognize that common sense cannot prove idealism wrong. Yet, for Moore, this is consistent with the retention of the idea that common sense is right and idealism wrong. For, after all, to be a philosopher of common sense means, at least in part, to remain convinced that common sense would always have the upper hand vis-à-vis philosophical pretensions and analyses, even when one cannot prove the latter wrong. That is why Moore often dismissed them quite off hand.

In short, I think that Maddy cannot reconcile the sense of philosophical hopelessness Moore's writings elicit with her admiration for his overall stance, and that she is constantly trying to make him come out better in philosophical argument than he did. I confess to be less of a sympathetic reader. While I agree that Moore pointed out many interesting things and was surely right in claiming to have knowledge of the premises of his proof, he never succeeded in giving either a satisfactory proof of the existence of an external world or a credible refutation of Cartesian skepticism, as he himself acknowledged and never claimed to have provided.[5] As he wrote:

> How am I to prove now that 'Here's one hand, and here's another'? I do not believe I can do it. I should need to prove for one thing, as Descartes pointed out, that I am not now dreaming... I have conclusive reasons for asserting that I am not now dreaming...but that is very different from being able to prove it. (Moore, 1939, p. 149).

In partial defense of Maddy, there is a passage in "Proof" that might give the impression that Moore thought that the Cartesian skeptical challenge is illegitimate, but which, on closer inspection, does not show that, in my opinion. It is worth quoting it in full:

> But I am perfectly well aware that, in spite of all that I have said, many philosophers will still feel that I have not given any satisfactory proof. And I want briefly, in conclusion, to say something as to why this dissatisfaction with my proofs should be felt. One reason why, is, I think, this. Some people understand 'proof of an external world' as including a proof of things which I haven't attempted to prove and haven't proved...If I had proved the propositions which I used as *premises* in my two proofs, they would perhaps admit that I had proved the existence of external things, but, in the absence of such a proof (which, of course, I have neither given nor attempted to give), they will say that I have not given what they mean by a proof of the existence of external things... In other words, they want a proof of what I assert now when I hold up my hands and say 'Here's one hand and here's another'; and, in the other case, they want a proof of what I assert now when I say 'I did hold up two hands above this desk just now'. Of course, what they really want is not merely a proof of these two propositions, but something like *a general statement as to how any propositions of this sort may be proved. This, of course, I haven't given; and I do not believe it can be given: if this is what is meant by proof of the existence of external things, I do not believe that any proof of the existence of external things is possible* (emphasis added). Of course, in some cases what might be called a proof of propositions which seem like these can be got. If one of you suspected that one of my hands was artificial he might be said to get a proof of my proposition 'Here's one hand, and here's another', by coming up and examining the suspected hand close up, perhaps touching and pressing it, and so establishing that it really was a human hand. But I do not believe that any proof is possible in nearly all cases. (Moore, 1939, pp. 148–149)

What Moore is denying here is the possibility of giving something like a general proof of the existence of material objects. What can be done, in his opinion, is merely

[5] As we saw in §1, in his "Reply" he clearly claimed he was not responding to Cartesian skepticism in "Proof" and he was dissatisfied with his attempts to respond to skepticism in "Four forms of scepticism" and in "Certainty".

to exhibit particular physical objects, like hands and pencils, from which it follows that the category of physical objects is not empty, once it is admitted that these objects are exemplifications of that category, understood in the way proposed by Moore. Sometimes we can do a little bit more than that, by ascertaining by means of further sensory evidence whether a given object is a real, rather than an artificial hand, but only when a doubt of an entirely empirical kind has been raised. The fact that right after this passage Moore goes on to discuss the possibility of proving that he is not dreaming and admits he cannot do it may suggest that he was deeming the latter impossible too as it would be based on conflating ordinary and extraordinary evidence. But this would be a hasty conclusion based on mistaking the request of proving that one is not dreaming of seeing a hand, in a sense which would satisfy a Cartesian skeptic, which Moore is *not* deeming impossible, even though *he* says he cannot meet it, with the impossibility of giving a general proof of the existence of physical objects prior to exhibiting one of them and taking it follow from that that the external world exists.

Or else, and in further partial defense of Maddy's interpretation, and based on very thin, almost vanishing textual evidence, one might maintain that Moore, in the passage just quoted, was indeed claiming that a proof of the fact that he was not dreaming could not be given *at all*. Still, in that passage there is nothing that suggests the further "therapeutic" idea that that very request should be dismissed as illegitimate, precisely because it could not be met. Rather, and at most, the passage should be taken to show that even if the request is impossible to satisfy for him and for anyone else, one can make do. For, due to Moore's proto-externalist leanings, one can know that one is not dreaming and that there is a hand where one seems to see it, even if no one is—or indeed, could be—in a position to satisfy that skeptical request.

Finally, Maddy expresses dissatisfaction precisely with my attribution of a proto-externalist strategy to Moore. She writes: "So unlike the Plain Man with his externalist way of knowing, Moore thinks he has good arguments, good grounds [for holding that there are material things, such as hands]—it's just that they don't add up to proof" (Maddy, 2018, p. 235). And she continues "he thinks he knows he has hands, but he can't prove it, if proving requires to proceed 'from scratch'" (ibid.). First, I do agree that Moore admits to his inability to prove that he is not dreaming. However, in no way does this amount to finding the Cartesian skeptical challenge illegitimate, as it would be based on the conflation between ordinary and extraordinary dreaming and evidence. Second, I do agree, as remarked in §1, that Moore thought he had grounds for his claim that he knew that there were his hands where he seemed to see them. Yet, he also insisted that he was unable to list all his evidence and that he should have done at least that, in order to start providing a proof (see Moore, 1939, p. 149). Being unable to provide all his evidence and being unable to prove that he was not dreaming, thus failing to know how he knew the premises of his proof, were not, however, impediments to his knowledge of them. And *this*, I submit, is quite in harmony with an externalist way of conceiving of knowledge, by way of rejecting at least two possible interpretations of what the KK principle would require—that is, either being able to mention all of one's evidence for P, or being able to prove that a condition whose obtaining would be incompatible with knowledge of P, and yet would be compatible with all of one's available evidence, does not in fact obtain. As Moore himself wrote:

> But another reason why some people would feel dissatisfied with my proofs is, I think, not merely that they want a proof of something which I haven't proved [that is, the truth of the premises by ruling out that he was merely dreaming of his hands], but that they think that, if I cannot give such extra proofs, then the proofs that I have given are not conclusive proofs at all. And this, I think, is a definite mistake…I can know things, which I cannot prove; and among things which I certainly did know, even if (as I think) I could not prove them, were the premises of my two proofs. (Moore, 1939, pp. 149–150).

So Moore thought that he could know that there were hands where he seemed to see them even if he was unable to prove that he did, because he was unable to provide all his evidence for such a claim, and, connectedly, to prove that he was not dreaming. Unlike a mathematician and a certain kind of philosopher who would take it to be their responsibility to show how they know and to prove that certain countervailing scenarios do not obtain, if their views have a right to be considered knowledge and not mere hypotheses, Moore thought he did have knowledge of the premises of his proof nonetheless. He thus initiated, probably *malgré lui*, a certain line of response to skepticism subsequent externalist theorists would capitalize on. Namely, a line of response that would insist on the gap between the conditions for knowledge, on the one hand, and on the conditions for knowing how one knows, on the other, and on the non-necessity of meeting the latter to satisfy the former. (In fact, Susan Stebbing (1932) initiated this style of response to skeptical worries, as I claim in Coliva (2021) and she also insisted on the fact that we cannot doubt all of our empirical beliefs at once, but only piecemeal).

Notice, furthermore, that, ironically, Moore's proto-externalism would actually be reinforced and it would turn out to be better motivated, if one were to follow Maddy in thinking that he also deemed the Cartesian skeptical challenge illegitimate. For, in that case, it would be entirely obvious and non-question-begging against a Cartesian skeptic why Moore could have knowledge of the premises of his proof even if he was unable to prove, to a Cartesian skeptic's satisfaction, that he was not dreaming of seeing his hands. For, if the request of providing such a proof were illegitimate, and not merely impossible to meet (just for Moore or even for anyone else)—a fact, this, that, by itself, could only bring grist to a Cartesian skeptic's mill—then surely one could not be expected to satisfy it in order to have knowledge of "here is a hand". Conversely, externalism could nicely be combined with Maddy's "therapeutic" reading of Moore to show that, once the Cartesian skeptical challenge has been exposed as illegitimate, we can know the premises of Moore's proof and do so even if we cannot produce all our evidence in favor of them. For having knowledge would be independent of being able to produce the grounds—let alone, all the grounds—one's knowledge is based on. Alas, even if the combination of Maddy's "therapeutic" reading of Moore and of my externalist interpretation of him would actually make it for an interesting and powerful anti-skeptical account, I still think no such anti-skeptical argument is to be found in "Proof", and for the very good reason that, as Moore himself clearly stated in 1942, and as he hinted at already in the "Proof", Cartesian skepticism was not that paper's target.

As the conversation between Maddy and me should have amply shown by now, that paper was a direct engagement with Kant's transcendental idealism, which Moore dismissed as based on a (by his lights) unintelligible transcendental notion of "things to be met with in space". Once the ground had been cleared from such a notion, and the empirical, more commonsensical notion of "things which would exist without being perceived (by anyone ever)" had been reinstated, Moore thought he could easily remedy the "scandal to philosophy" denounced by Kant. Namely, the scandal consisting in the fact that "the existence of things outside of us ... must be accepted merely on faith, and that, if anyone thinks good to doubt their existence, we are unable to counter his doubts by any satisfactory proof" (Kant, 1787, B, xxxix).[6] For, once it is accepted that the only intelligible notion of "things outside of us" is the empirical one Moore is defending, and once it is admitted that instances of it can be exhibited, it then simply follows that the existence of an external world has been proved. It is only an irony that nothing of what Moore rested his confidence on would have been conceded by his opponent, as we have seen. Yet the fact that Moore's argument would not have persuaded a (transcendental) idealist is no reason to think that such an idealist was not the real target of his celebrated "Proof". What Moore's argument showed for sure, even if it turned out to be powerless against its opponent, is the force of our commonsensical picture of the world, which we take to be populated by mind-independent objects, no matter what idealists of whatever brand would have us believe. That, I submit, is Moore's main legacy, the one Wittgenstein greatly admired and tried to vindicate by devising arguments that, by showing the illegitimacy of Cartesian and other forms of skepticism, aimed to cure us from the temptation of giving in to them.[7] But that was Wittgenstein, not Moore; and certainly not the Moore of "Proof of an external world", at any rate.

References

Coliva, A. (2010). *Moore and Wittgenstein. Scepticism, certainty and common sense*. Palgrave.

Coliva, A. (2015). *Extended rationality. A hinge epistemology*. Palgrave.

Coliva, A. (2018). What do philosophers do? Maddy, Moore and Wittgenstein. *International Journal for the Study of Skepticism, 8*, 198–207.

Coliva, A. (2021). Stebbing Moore (and Wittgenstein) on common sense and metaphysical analysis. *British Journal for the History of Philosophy, 29*, 914–934.

Kant, I. (1787). *Critique of pure reason* (Trans. by N. Kemp Smith). Macmillan, 1929

Maddy, P. (2017). *What do philosophers do?* Oxford University Press.

Maddy, P. (2018). Replies to coliva, leite, and stroud. *International Journal for the Study of Skepticism, 8*, 231–244.

[6] Not by chance did "Proof" start with this quote from Kant.

[7] For a discussion of Wittgenstein's several anti-skeptical strategies in *On Certainty* (Wittgenstein 1969), see Coliva (2010, Chap. 3).

Moore, G. E. (1939). Proof of an external world. In his *Collected papers* (pp. 127–150). Allen & Unwin.
Moore, G. E. (1942). A reply to my critics. In P. A. Schilpp (Ed.) *The philosophy of G. E. Moore* (pp. 535–677).
Moore, G. E. (1959a). Four forms of skepticism. In his *Collected papers* (pp. 196–226). Allen & Unwin.
Moore, G. E. (1959b). Certainty. In his *Collected papers* (pp. 227–251). Allen & Unwin.
Stebbing, S. (1932). The method of analysis in metaphysics. *Proceedings of the Aristotelian Society, 33*, 65–94.
Wittgenstein, L. (1969). *On certainty*. Blackwell.

Chapter 24
Reply to Coliva

Penelope Maddy

Keywords G. E. Moore · Skepticism · Idealism · Externalism · Comparative anti-skepticism · Second philosophy · Dream argument · Common sense

It's always a pleasure to engage with Professor Coliva on the mysteries of G. E. Moore. We share a fascination with his writings, though, as her paper makes clear, in the end I'm the more sympathetic reader. One conspicuous point on which we disagree is whether Moore in his famous 'Proof of an external world' is addressing the skeptic, or exclusively the idealist.[1] I think our differences on this point have been well-aired by now, so I'll set it aside here with the simple observation that anyone who can prove there are 'things outside of us' must be epistemically better off than the skeptic would allow.[2] Instead, I'd like to focus on three other topics: the

[1] Referring to my (Coliva, 2018), Coliva maintains that I acknowledge that Moore himself, in his (1942) 'claimed that he was refuting (or trying to refute) idealism rather than skepticism' (p. 304). I didn't intend anything so strong when I wrote that this reading 'may well be right, but let me point to three passages that give me pause' (2018, p. 232)—I meant that those passages make me question the view that he is addressing idealism. For more on this topic, see Leitz [n.d.].

[2] Cf. Stroud (1984), pp. 136–137. One point on which I've so far offered no comment is the very passage from Moore (1942), p. 668–669, on which Coliva anchors her reading, so perhaps I owe a word on that. There Moore says that though he can prove the falsity of 'There are no material things' in the manner of 'Proof', he cannot prove the falsity of 'Nobody knows for certain that there are any material things' by an analogous argument. He does remark that the analogous argument would be '*a* good argument' for its conclusion, but apparently it wouldn't constitute a proof. Why not? Because 'the fact that an immensely greater number of philosophers have held' the latter than have held the former shows that there's 'an important difference between the two cases'. Why is this important? Moore doesn't say, but perhaps the intense scrutiny of these many philosophers shows that what's being required isn't just a proof in the sense of 'Proof', that what's being required is also a proof of the premises, which again Moore would admit he can't provide.

P. Maddy (✉)
Department of Logic and Philosophy of Science, University of California, Irvine, CA, USA
e-mail: pjmaddy@uci.edu

S. Arbeiter and J. Kennedy (eds.), *The Philosophy of Penelope Maddy*,
Outstanding Contributions to Logic 31, https://doi.org/10.1007/978-3-031-58425-1_24

coherence of the skeptical challenge, the role (if any) of the KK principal, and the anti-skeptical 'comparative' strategy.[3]

24.1 The Nature of the Skeptical Challenge

Coliva holds that 'On Maddy's reading, Moore objected to the legitimacy of the Cartesian skeptical challenge' on the grounds that 'it would be based on the conflation between ordinary … and extraordinary dreaming' (p. 301). In fact, I don't think I ever said this, and if I did, it was a slip. Let me try to explain.

Ordinary dreaming is the kind we're all familiar with, the kind that happens when we're asleep. Ordinary dreams are fragmented and illogical; we can't think straight or move about normally; things come and go, scenes changed dramatically without rhyme or reason. Waking experience is nothing like this. It's continuous and coherent; we can think and move normally, remember the past and plan for the future. By these ordinary means, any one of us can tell when we're awake, and some of us, those subject to lucid dreams, can tell when we're dreaming. None of this is enough to generate a serious skeptical challenge. The kind of dreaming that appears in skeptical argumentation is an 'in principle' kind of dreaming, a kind completely indistinguishable from waking experience. This is what I call 'extraordinary dreaming'. Ordinary evidence is enough to establish that I'm not ordinary dreaming, but to establish that I'm not extraordinary dreaming would require a kind of evidence different from all my commonsense and even scientific ways of finding out about the world. This is what I call 'extraordinary evidence'. It has its epistemic force without appeal to anything else I may believe about the way the world is.

So the skeptical challenge is: show you're not extraordinary dreaming that, for example, 'here's a hand'. This is functionally equivalent to the challenge to show you're not being deceived by an Evil Demon into believing 'here's a hand' or manipulated by a mad neuro-scientist who has your brain in a vat into believing 'here's a hand'. My Moore, along with my Plain Inquirer [4] or my Second Philosopher,[5] takes this to be the challenge to present extraordinary evidence. There's nothing incoherent about it—who wouldn't want such powerful evidence if it could be had? For that matter, Descartes even claimed to have a way of producing it, and both Moore and the Plain Inquirer/Second Philosopher are inclined to hear him out before declaring his approach of failure him.[6]

---

[3] Coliva calls this 'Moore's gambit' (see, e.g., Coliva (2018)).

[4] See Maddy (2017a).

[5] See Maddy (2007; 2011; 2014a; 2022h)

[6] See Moore (1933/4), p. 194: 'Here a method has been used by Descartes … I do not think this is successful … But I suppose this sort of argument *ought* to be investigated'. See also Maddy (2007), p. 18, and (2017), pp. 17–18, where the Second Philosopher and the Plain Inquirer, respectively, are willing to try Descartes's Method of Doubt with an open mind, always eager to develop new ways of finding out about the world. The trouble for all three comes when they find fault with his subsequent argumentation.

Granting, then, that my Moore takes the skeptical challenge to be coherent, despite his inability to answer it as it's intended, he can't be arguing for its incoherence via a purported conflation of ordinary dreaming with extraordinary dreaming. Such a conflation—or really, an elicit, often unnoticed, slide from ordinary dreaming to extraordinary dreaming—turns up not in my discussion of Moore, but in a separate attempt to explain the enduring attraction of the dream argument[7]: at the beginning, we take it as obvious that we can't know 'here's a hand' if we can't rule out the possibility that we're only ordinary dreaming that we see it and feel it and so on (which we can do); at the end, we're defeated by the challenge to rule out the possibility that I'm extraordinary dreaming these things (which we can't). In a philosophical mood, we don't concern ourselves with the move from what we actually dream to what we can dream 'in principle'. We take all too easily to entertaining a kind of dreaming that's entirely indistinguishable from ordinary life.

This observation can be seen as part of the therapeutic effort to diagnose the hidden mechanisms that lead us astray. Perhaps this is what prompts Colihim va to classify my reading of Moore as therapeutic, but in fact it plays no role there. In allying Moore with the Plain Inquirer and the Second Philosopher, I'm describing him as broadly naturalistic, as defending common sense, yes, but science as well, as an extension of common sense. The only evidence of this in 'Proof' is the brief passage about after-images, where he cites scientific textbooks and even reports a few modest experiments of his own, but his endorsement of scientific method takes center stage in a series of lectures delivered just a few years before. For example, on the existence of bodies, he writes:

> The question arises whether the method used by the sciences isn't the best and only proper method for settling such questions. (Moore (1933/4), p. 178)

And on evidence:

> Isn't the fact that the scientists say: Such-and-such is good evidence for so-and-so, a reason for saying: It is good evidence? (Ibid., p. 189)

I've reported on these important lectures and their implications in more detail elsewhere,[8] but I hope this is enough to suggest Moore's naturalistic inclinations.

24.2 The Role of the KK Principle

Despite her denial that Moore is addressing the skeptic in 'Proof', Coliva does find a 'proto-externalist' strain in the paper that would grant him knowledge that 'here's a hand' without his being able to prove it: as she sees it, he would know, but not know that he knows, rejecting the KK principle that Kp implies KKp.[9] Against this,

[7] See Maddy (2017), pp. 39–40.

[8] See Maddy (2017), pp. 169–175, (2022a).

[9] See pp. 302–303, 306–308, also Coliva (2018), §3.

I've noted that Moore claims to have 'conclusive evidence' that he's not dreaming, not the sort of thing an externalist would say (Moore, 1939, p. 149).[10] Closer to the mark, it seems to me, is her suggestion that Moore thinks 'one can make do' (p. 307), but not in her externalist sense.

Picking up from the previous section, the skeptic's challenge to Moore is to show that he's not extraordinary dreaming that 'here's a hand', to provide extraordinary evidence for his claim. This Moore admits he cannot do, despite having conclusive evidence that he's not ordinary dreaming. The question, always, is what follows from the fact that he can't prove his premises to the skeptic's satisfaction? Moore insists perhaps even more than that he can 'make do':

> I can know things that I cannot prove [i.e., for which I don't have extraordinary evidence]; and among things which I certainly did know, even if (as I think), I could not prove them, were the premisses of my two proofs [i.e., here's a hand and here's another]. I should say, therefore, that those ... who are dissatisfied with these proofs merely on the ground that I did not know the premisses, have no good reason for their dissatisfaction. (Moore (1939), p. 150)

Here I take Moore to side with the Plain Inquirer and the Second Philosopher in holding that, while extraordinary evidence would be a great thing to have, the lack of it doesn't undercut the ordinary evidence we do have. In the lectures, where ordinary scientific evidence is his focus, he writes:

> I have insisted the sciences give indirect proofs [the 'matter is real'] by proving that particular material objects have existed: I instanced the proof of the existence of Plesiosauri, Ichthyosauri, etc.; and of millions of stars. (Moore (1933/ 4), p. 194)

By the time of 'Proof', he's realized that he doesn't need dinosaurs or astronomical bodies, that his hands will do, but the sentiment is the same. I don't know that this amounts to 'winning against his opponent' (p. 303)—the skeptical challenge as intended hasn't been met—but it is to deny that our failure to meet that challenge has the dire consequences it's commonly thought to have for our everyday knowledge of the world.

24.3 The Comparative Strategy

Though she takes 'Proof' to be addressed to idealism, not skepticism, Coliva does locate an anti-skeptical argument in two of Moore's writings of the early 1940s, 'Certainty' (1941) and 'Four forms of scepticism' (1944). She quotes this passage from a former:

> I agree, therefore, with that part of this argument which asserts that if I don't know now that I'm not dreaming, it follows that I don't *know* that I am standing up ... But this ... cuts both ways. For, if it is true, it follows that it is also true that if I *do* know that I am standing up, then I do know that I am not dreaming. I can therefore just as well argue: since I do know

[10] See Maddy (2018), pp. 234–235.

> that I'm standing up, it follows that I do know that I'm not dreaming; as my opponent can argue: since you don't know that you're not dreaming, it follows that you don't know that you're standing up. (Moore (1941), p. 247)

She concludes that since he acknowledges that the two arguments are 'on a par', Moore recognizes the 'impotence' of this strategy (p. 302). But this is to miss Moore's important caveat immediately after the above-quoted remarks:

> The one argument is just as good as the other, unless my opponent can give better reasons for asserting that I don't know that I'm not dreaming, than I can give for asserting that I do know that I am standing up. (Ibid.)

The key to Moore's comparative anti-skeptical strategy is that he thinks it's more rational to believe the latter than to believe the former.

This comes out most clearly at the end of 'Four forms'. After laying out Russell's skeptical arguments in detail, isolating four of his premises, Moore concludes:

> What I can't help asking myself is this: Is it, in fact, as certain that all these four assumptions are true, as that I *do* know that this is a pencil ... ? I cannot help answering: It seems to me *more* certain that I *do* know that this is a pencil ... than that any single one of these four assumptions is true, let alone all four ... Nay more: I do not think it is *rational* to be as certain of any one of these four propositions, as of the proposition that I do know that this is a pencil. (Moore (1944), p. 226)

'Four forms' plays a leading role in Lycan's forceful reconstruction of the comparative strategy (Lycan 2007) and Weatherall's scrupulous elaboration (Weatherall, 2015),[11] though Weatherall argues that the comparative theme can also be found in Moore's writings going back to Moore (1909, 1910/11, or 1918). So it's not clear that Moore regards this approach as a failure.[12]

In any case, those of us who take Moore to be addressing skepticism in 'Proof' (at least in part) face the puzzle of how what goes on there relates to the comparative strategy. Lycan thinks the argument can be construed as comparative, but that this comes 'at a cost': 'either the proof is superfluous'—because we've used the comparative strategy to establish that he knows the premises—'or Moore must ignore the requirement that a proof's premises be known'—which is to give up its status as a proof in Moore's terms (Lycan (2007), p. 95). Weatherall's assessment is more generous. He argues that since the comparative strategy requires a particular skeptical argument to get started, Moore uses 'Proof', which was first presented as a lecture, to provoke the skeptic into providing one. Ingenious and attractive as this reading is, it's not open to me because I've taken 'Proof' to present a substantive,

[11] Weatherall's ultimate goal is to explain the role of 'Proof' in the context of Moore's overall comparative case against skepticism (see below).

[12] Coliva supports her negative assessment by noting Moore's dissatisfaction with 'Certainty' in the preface to Moore (1959) (p. 13). Given that the comparative strategy is best realized in 'Four forms', it's not clear that his worries about 'Certainty' are conclusive. Weatherall (2015), pp. 11–12, offers a persuasive explanation of how 'Certainty' goes astray: Moore fails to address a particular skeptical argument, with particular premises, which leaves him without a concrete target for his rationality comparison. By focusing on Russell's argumentation in 'Four forms', he doesn't make this mistake.

naturalistic case that the skeptical challenge, as he understands it, doesn't require a direct response.[13]

My take on the puzzle rests on the observation that in 'Proof', Moore has effectively turned the tables on the skeptic. Most discussions begin with skeptical argumentation leading to the radical skeptical challenge; the anti-skeptic is then on the defensive, required to meet that challenge and unseat the skeptic. In contrast, 'Proof' begins with Moore's long preamble of naturalistic observations on the nature of 'things outside of us'. Having established that perspective, he presents his proof, effectively challenging the skeptic to unseat him, instead—and that turns out to be more difficult than one might have thought. So one possibility is that the comparative strategy is a defensive tactic, applicable when the skeptic has the first move, and the argument in 'Proof' is an offensive maneuver, designed to claim the first move and put the skeptic on the defensive. Which to employ then depends on the dialectical situation.

References

Coliva, A. (2018). What do philosophers do? Maddy, Moore and Wittgenstein. *International Journal for the Study of Skepticism, 8*, 198–207.
Lycan, W. (2007). Moore's anti-skeptical strategies. In S. Nuccetelli & G. Seay (Eds.) *Themes from G. E. Moore* (pp. 84–99). Oxford University Press.
Maddy, P. (2007). Second philosophy, (Oxford: Oxford University Press).
Maddy, P. (2011). Defending the axioms, (Oxford: Oxford University Press).
Maddy, P. (2014a). The logical must, (Oxford: Oxford University Press).
Maddy, P. (2017a). What do philosophers do? (New York: Oxford University Press).
Maddy, P. (2017b). 'Set-theoretic foundations' In A. Caicedo., J. Cummings., P. Koellner., & P. Larson (Eds.), *Foundations of Mathematics: Essays in Honor of W. Hugh Woodin's 60th birthday,* (Providence, RI: American Mathematical Society), pp. 289–322.
Maddy, P. (2022h). *A plea for natural philosophy and other essays*, (Oxford: Oxford University Press).
Moore, G. E. (1959). *Philosophical papers*. George Allen and Unwin.
Moore, G. E. (1909). *Hume's philosophy*, in his (1922) (pp. 147–167).
Moore, G. E. (1910/11). *Hume's theory examined*, in his (1953) (pp. 108–126).
Moore, G. E. (1918). *Some judgments of perception*, in his (1922) (pp. 220–252).

[13] Charles Leitz has suggested that Weatherall's reading and my reading might be compatible: why shouldn't Moore be doing two things at once in 'Proof', both presenting an anti-skeptical argument and provoking the skeptic to produce an argument of his own (to which Moore will then apply the comparative strategy)? I think this reasonable question misses an important difference between the challenges Weatherall's Moore and mine present to the skeptic. Weatherall takes Moore's proof to challenge the skeptic to attack either his conclusion or his premises, either of which would amount to an anti-skeptical argument (Weatherall (2015), p. 18). But Moore has cheerfully admitted that his proof won't satisfy the skeptic; his key claim is that it doesn't need to, that the demand for extraordinary evidence doesn't need to be met. So the challenge my Moore presents to the skeptic is to explain why his lack of extraordinary evidence undermines his ordinary evidence. If that's his main project in 'Proof', then the skeptic's responding with some skeptical argument would be a distraction—what Moore wants is to get the skeptic's attention focused on this new challenge.

Moore, G. E. (1933/4). Selections from a course of lectures given in 1933–34. In C. Lewy (Ed.), *Lectures on philosophy*, in his (1966) (pp. 151–196). George Allen and Unwin
Moore, G. E. (1939). *Proof of an external world*, in his (1959) (pp. 127–150).
Moore, G. E. (1941). *Certainty*, in his (1959) (pp. 227–251).
Moore, G. E. (1944). *Four forms of skepticism*, in his (1959) (pp. 196–226).
Stroud, B. (1984). *The significance of philosophical skepticism*. Oxford University Press.
Weatherall, J. O. (2015). 'On G. E. Moore's "Proof of an external world"', *Pacific Philosophical Quarterly*, 98 (pp. 219–250).

Chapter 25
A Carnapian Solution to the Puzzle of Extrinsic Justifications

Zeynep Soysal

Abstract Set theorists commonly give *extrinsic justifications* for axioms, i.e., they present the fruitfulness of axioms with respect to certain mathematical goals as evidence for their truth. The puzzle of extrinsic justifications is the question of how the fruitfulness of an axiom can be evidence for its truth. In this paper, I first discuss some problems for Penelope Maddy's solution to the puzzle of extrinsic justifications. I then propose an alternative, Carnapian solution according to which the fruitfulness of an axiom can be evidence for its truth because there are analytic truths to the effect that sets are fruitful with respect to certain mathematical goals. I argue that the Carnapian solution is independently plausible, that it avoids the problems faced by Maddy's solution, and that it neatly aligns with Maddy's epistemological and metaphysical views concerning set theory, as well as with her methodological commitments.

Keywords Extrinsic justifications · Intrinsic justifications · Thin realism · Arealism · Analyticity in Mathematics · Descriptive Theories of Reference

25.1 Introduction

According to terminology that has become standard in the philosophy of set theory, giving an *extrinsic justification* for a mathematical axiom means, roughly, presenting the "fruitfulness" or "effectiveness" of the axiom as evidence for its truth. As Penelope Maddy has documented in many of her works, set theorists give extrinsic justifications for all types of axioms, from those as basic as the axioms of Zermelo–Fraenkel set theory with the Axiom of Choice (henceforth 'ZFC') to new axioms proposed at the cutting edge of research in set theory today.

I am grateful to Sophia Arbeiter, Juliet Floyd, Jens Kipper, Juliette Kennedy, and John Komdat for very helpful comments and discussions.

Z. Soysal (✉)
University of Rochester, Rochester, NY, USA
e-mail: zeynep.soysal@rochester.edu

S. Arbeiter and J. Kennedy (eds.), *The Philosophy of Penelope Maddy*,
Outstanding Contributions to Logic 31, https://doi.org/10.1007/978-3-031-58425-1_25

Set theorists' practice of giving extrinsic justifications for their axioms raises a puzzle: How could the mere fruitfulness or effectiveness of an axiom be evidence for its truth? Alternatively: How could the method of accepting set-theoretic axioms on the basis of extrinsic justifications (henceforth simply 'the method of extrinsic justifications') be *reliable*, i.e., mostly lead to truths?

For philosophers like Maddy who would like to maintain that the method of extrinsic justifications is reliable, the challenge is to explain why it is reliable. According to Maddy, this challenge is met by adopting either one of two metaphysical views about sets: Thin Realism or Arealism.

In this paper, I first explain the puzzle of extrinsic justifications in more detail (Sect. 25.2). I then discuss some problems for Maddy's proposed solution to the puzzle of extrinsic justifications (Sect. 25.3). Finally, I suggest an alternative, Carnapian solution (Sect. 25.4). In outline, the Carnapian solution involves the claim that it is analytically true that sets are in the relevant sense "fruitful" or "effective" for certain mathematical goals. On this view, the method of extrinsic justifications is reliable because the fruitfulness of an axiom is evidence that the axiom follows from analytic truths. I argue that the Carnapian solution is independently plausible, that it avoids the problems for Maddy's solution, and that it is in line with many of Maddy's central commitments.

25.2 The Puzzle of Extrinsic Justifications

The notion of extrinsic justifications for axioms is usually discussed with reference to Kurt Gödel's (1947).[1] There, Gödel proposes that set theorists should search for new axioms that would resolve questions left undecided by ZFC, such as whether the Continuum Hypothesis is true. Turning to the question of how one can come to know new axioms, Gödel contrasts what he calls the potential "intrinsic necessity" of an axiom with another kind of possible evidence for its truth:

> [E]ven disregarding the intrinsic necessity of some new axiom, and even in case it had no intrinsic necessity at all, a decision about its truth is possible also in another way, namely, inductively by studying its "success," that is, its fruitfulness in consequences and in particular in "verifiable" consequences [...]. There might exist axioms so abundant in their verifiable consequences, shedding so much light upon a whole discipline, and furnishing such powerful methods for solving given problems [...] that quite irrespectively of their intrinsic necessity they would have to be assumed at least in the same sense as any well-established physical theory. (Gödel, 1947, p. 182f.)

It is standard to interpret Gödel here as claiming that there are two legitimate ways to argue for the truth of an axiom. One can provide *intrinsic justification* for the axiom, i.e., one can present the fact that the axiom follows from the analysis of the concepts of set theory as evidence for its truth. One can also provide *extrinsic justification* for the axiom, i.e., one can present the "fruitfulness" or "success" of the axiom

---

[1] Earlier, Zermelo (1908) and Gödel (1944) also discuss extrinsic justifications.

with respect to certain mathematical goals as evidence for its truth. Gödel seems to consider intrinsic justifications in general to be stronger than extrinsic justifications,[2] but he considers them both to be legitimate ways to justify new axioms beyond ZFC.

Maddy argues that extrinsic justifications in this broad Gödelian sense play a central role in the practice of set theory, and that they have done so since the very early days of the axiomatization of set theory. For instance, set theorists such as Dmitry Mirimanoff (1917), John von Neumann (1925), Ernst Zermelo (1930), Abraham A. Fraenkel, Yehoshua Bar-Hillel, and Azriel Levy (1973), as well as philosophers such as George Boolos (1971) and Michael Hallett (1984), give extrinsic justifications for some of the most basic axioms of ZFC, such as Extensionality, Foundation, Empty Set, Power Set, and Replacement (Maddy, 1988, p. 483f., 486f., 489f.; 1997, Sect. I.3). However, extrinsic justifications become particularly puzzling when they justify axioms that aren't (or aren't strongly) intrinsically justified as well.[3] As Maddy (2011, p. 126, 135) notes, the earliest such justification is usually taken to be the one Zermelo (1908) gives for the Axiom of Choice (henceforth 'AC'). More recent examples include the argument against the Axiom of Constructibility (henceforth '$V = L$'),[4] and the argument for the Axiom of Determinacy for $L(\mathbb{R})$ (henceforth '$\mathrm{AD}^{L(\mathbb{R})}$').[5] To have an example to work with throughout the paper, let me outline some elements of the justification that set theorists give for $V \neq L$, specifically as described by Maddy (1993, 1997).

$V = L$, introduced by Gödel (1938), states that every set occurs at some stage of the Constructible Universe, L, i.e., every set is a member of some L_α for some ordinal α, where L_0 is the empty set, successor stages $L_{\alpha+1}$ are all the subsets of L_α definable from L_α,[6] and limit stages are unions of the previous stages. $V = L$ is independent from ZFC (Gödel, 1938; Cohen, 1963). Gödel initially thought that $V = L$ could be *absolutely undecidable*, i.e., that it could fail to be settled by any justified axiom.[7] He soon came to hold instead that every independent sentence might be decided by new axioms of infinity (Gödel, 1946), and seemed inclined against $V = L$.[8] It indeed turned out, as Dana Scott (1961) proved, that ZFC extended with the axiom stating the existence of a measurable cardinal (henceforth 'MC') implies $V \neq L$. Scott's result strongly swayed set theorists' opinions against $V = L$. For many, Scott's result shows that $V = L$ is "restrictive" or "limitative," and therefore false. Hugh Woodin states, for instance:

---

[2] As, for instance, Parsons (2014) explains, analyticity has been central to Gödel's views on justification in mathematics.

[3] It is generally accepted that there is strong intrinsic justification for many of the ZF axioms. For a survey of intrinsic justifications for ZF, see (Maddy, 1988).

[4] See, for instance, (Maddy, 1993, 1997, 2007; Arrigoni, 2011).

[5] See, for instance, (Maddy, 2011; Koellner, 2013).

[6] Given sets X and A, X is *definable* from A if and only if there is a formula $\varphi(x_1, ..., x_n, x)$ of the (first-order) language of set theory, and elements, $a_1, ..., a_n$ of A, such that $X = \{a \in A | <A, \in> \models \varphi(a_1, ..., a_n, a)\}$.

[7] For discussion of Gödel's views on absolute undecidability, see, for instance, (Koellner, 2006).

[8] For discussion of Gödel's views on $V = L$, see, for instance, (Tait, 2005; Kennedy, 2014).

> [T]he axiom $V = L$ limits the large cardinal axioms which can hold and so the axiom is *false*. (Woodin, 2010, p. 1)

John Steel puts it slightly differently; echoing Maddy (1997), he states that $V = L$ is "restrictive" specifically given set theorists' goal to "maximize":

> [W]hat we are attempting to maximize here is the *interpretative power* of our set theory. [...] In this light we can see why most set theorists reject $V = L$ as restrictive: adopting it restricts the interpretative power of the language of set theory. (Steel, 2000, p. 423)

Scott's result is thus taken to show that $V = L$ is restrictive in the sense that accepting it would limit the interpretive power of set theory, as well as in the more straightforward sense that it would entail that there are no measurables or other *large* large cardinals (i.e., large cardinals incompatible with $V = L$). To unpack the latter construal of restrictiveness, one should say more about the justification for large large cardinals, which is also usually taken to be extrinsic. As Maddy states, for instance, MC has "identifiable benefits" as it settles some open questions in descriptive set theory, such as Π^1_1-determinacy (Martin, 1970; Maddy, 1997, 79f., 231). I won't discuss the argument for large large cardinals further here, as it is extensive.[9] It suffices to note that the restrictiveness of $V = L$ given its incompatibility with MC and other large large cardinals is standardly presented as evidence against it.

According to Maddy (1993), $V = L$ is also restrictive in a sense independent of Scott's theorem. Maddy argues that even before Scott's result, $V = L$ was deemed a "potentially limitative assumption" by set theorists (Maddy, 1993, p. 38), because it fits with a discredited "definabilist" maxim according to which "mathematics is the study of the definable, the representable, the expressible" (Maddy, 1993, p. 20). Maddy explains that this maxim started to be seen with suspicion as a number of anomalies emerged, and the consensus among mathematicians became that there is no good reason to think that every mathematical object is in some sense "definable." $V = L$, then, "harkens back to a now discredited methodological maxim," the successor "combinatorialist" maxim doesn't give reason to think that it is true, and the failures of the definabilist maxim give some reason to expect it to be false (Maddy, 1993, p. 38). See, for instance:

> The key argument against accepting $V = L$ [...] is that [it] [...] appears to restrict unduly the notion of *arbitrary* set of integers; there is no a priori reason why every subset of ω should be definable [...]. (Moschovakis, 1980, p. 610)

In sum, the fact that $V = L$ is restrictive in one of the senses outlined above, and thus that $V \neq L$ is fruitful given the set-theoretic goal to maximize, is standardly given as evidence for the truth of $V \neq L$. Let me flag a couple of complications. First, even though the majority of set theorists reject $V = L$, there are some notable exceptions. Ronald Jensen, for instance, states that $V = L$ is a "very attractive" and "mathematically fruitful" axiom (Jensen, 1995, p. 398), and seems to hold that there isn't a unique right answer to whether $V = L$ is true—there may be two equally good ways to extend the axioms of set theory, depending on whether one takes the

[9] For discussion, see, for instance (Koellner, 2013).

"arithmetical" or the "geometrical point of view" on set theory (Jensen, 1995, p. 400f.). There is thus at least some disagreement among set theorists over whether $V \neq L$ is justified.[10] Second, among those who believe that $V \neq L$ is justified, there is some disagreement over how it is justified. In particular, some set theorists describe the justification for $V \neq L$ as intrinsic. According to Tatiana Arrigoni (2011), for instance, the main consideration standardly presented in favor of $V \neq L$ is that it follows from the "maximum iterative concept of set" (Arrigoni, 2011, p. 337). See also, for instance:

> [$V = L$] is not generally accepted as true. Indeed, it is likely to be false according to the iterative concept of set. [...] By intention we view sets as arbitrary multiplicities regardless of how or if they can be defined. (Wang, 1974, p. 196)

Even some elements of the justification Maddy (1993) discusses can plausibly be interpreted as intrinsic: One could take it that it is conceptually false that mathematical objects must be definable, and thus there is no "a priori reason," as Yiannis Moschovakis says, to think that $V = L$.

This second complication is important. In my view, it reveals that just as Maddy (1988; 2011) argued that no axiom is justified *purely intrinsically*, it may also be that no axiom is justified *purely extrinsically*.[11] A philosophical account of justification in set theory should explain this seeming lack of purely intrinsic and purely extrinsic justifications. At the very end of (2011), Maddy points towards a view that could provide such an explanation: She suggests that the "fundamental justificatory force" in set theory is extrinsic, in that even intrinsic considerations are subsumed under more general extrinsic considerations (Maddy, 2011, p. 136). In Sect. 25.4, I will develop the opposite picture. In my view, the fundamental justificatory force in set theory is intrinsic, in that all extrinsic and intrinsic considerations given in support of axioms are considerations about what follows from the concept of set. My view appeals to a distinction between different kinds of analytic truths: An extrinsic consideration for an axiom is evidence that it follows from what I will call 'opaque' analytic truths, while an intrinsic consideration is evidence that it follows from 'transparent' analytic truths.

Setting aside these complications for now, let us turn to the puzzle of extrinsic justifications. The puzzle of extrinsic justifications is the question of how the fruitfulness of an axiom—given certain set-theoretic goals, like the goal to maximize—could be evidence for its truth. In other words, set theorists present extrinsic considerations as evidence for axioms, and accept axioms on the basis of these considerations, but how can these purported pieces of evidence actually be evidence for the truth of these axioms? As Maddy puts it, specifically concerning the argument for $V \neq L$:

> The argument based on our maximizing goal shows why adding $V = L$ leads to a theory of the sort we dislike, a restrictive theory, but how does its being a theory we dislike provide

---

[10] Jensen's views are discussed further by Arrigoni (2011). Others often interpreted to accept $V = L$ include Devlin (1977) and Quine (1992).

[11] AC is often given intrinsic justification (see, for instance, (Shoenfield, 1977; Hallett, 1984; Ferreirós, 2011)). Plausibly, the case for $AD^{L(\mathbb{R})}$ also includes some intrinsic justifications.

compelling evidence that it's false? To put the problem more generally: granting that selecting for maximization generates theories we like, what reason do we have to think it likely to generate theories that are true? (Maddy, 2007, p. 366)

The puzzle of extrinsic justifications has been raised by many philosophers.[12] And it isn't very easy to solve. Notice, for instance, that the parallel Gödel (1947) draws with justifications in physics doesn't immediately help: When a physical hypothesis yields true predictions, i.e., predictions that are observed or otherwise verified to be correct, this is evidence for the truth of the hypothesis (let us assume).[13] But, at least for all that we have said so far, the kinds of consequences of an axiom that can be presented as extrinsic evidence for it needn't be verified by standardly accepted (and non-puzzling) methods, like observation or proof. As we saw, settling open questions, illuminating a theory, and maximizing, are all consequences of an axiom that can be used to justify it extrinsically. But what is the independent reason for why these open questions should have these particular solutions, or have any solution at all? Or again, as Maddy asks, what is the independent verification for the correctness of maximal theories? The puzzle of extrinsic justifications thus isn't trivially solved by analogy with justifications in the natural sciences. If we want to maintain that the method of extrinsic justifications is generally reliable, we need an account of the practice or subject matter of set theory that explains the method's reliability.

25.3 Maddy's Solution

To understand Maddy's solution to the puzzle of extrinsic justifications, we must first examine her philosophical methodology, which is a form of methodological naturalism she calls 'Second Philosophy'. The adherent of this method, the Second Philosopher, accepts broadly scientific methods of inquiry, including those of social sciences like anthropology and linguistics (Maddy, 2007, p. 2). The Second Philosopher doesn't merely defer to the scientists, but is an active participant in the sciences (Maddy, 2011, p. 39). Moreover, she accepts and pursues pure mathematics, including set theory, because of its value in empirical theorizing (Maddy, 2011, p. 39) and because it might have unexpected applications in her empirical investigations (Maddy, 2014, p. 230).

Maddy claims that, from a Second Philosopher's perspective, there is no fact of the matter whether set theory is an activity "in the business of uncovering truths" like the empirical sciences are (Maddy, 2014, p. 230). She argues that if one wants to say that set theory is a body of truths, then one must accept *Thin Realism*, i.e., the view that "sets just are the sort of thing set theory describes; this is all there is to them" (Maddy, 2011, p. 61). And if one wants to say that set theory isn't a body of truths, then one must accept *Arealism*, i.e., the view that set theory is just "the

---

[12] See, for instance, Tiles (1989, p. 208), Feferman (2000), Tait (2005, 283f.), Kennedy and Väänänen (2015, p. 84). See also Maddy (2011, p. 58, 130; 2014).

[13] Even this is contested; see, for instance (Chakravartty, 2017, Sect. 2.1).

activity of developing a theory of sets that will effectively serve a concrete and ever-evolving range of mathematical purposes" (Maddy, 2011, p. 89). Maddy argues that the Second Philosopher's data support Thin Realism and Arealism equally well, and thus the choice between them "is an inconsequential matter of preference" (Maddy, 2014, p. 230).

Both Thin Realism and Arealism, then, provide a solution to (or, in Arealism's case, a dissolution of) the puzzle of extrinsic justifications. In outline, according to Thin Realism, the method of extrinsic justifications is reliable "because sets just are the sort of thing that can be learned about by application of such set theoretic methods" (Maddy, 2007, p. 374). Arealism doesn't face the same kind of puzzle of extrinsic justifications: Set-theoretic claims aren't true, so there is no question of whether the method of extrinsic justifications is reliable. Arealists can nonetheless explain why this method is reasonable, because on their view set theorists are only in the business of "devising effective ways of doing various pressing mathematical jobs" (Maddy, 2014, p. 229), and accepting axioms on the basis of their fruitfulness with respect to such jobs is reasonable (Maddy, 2007, p. 388).

In this paper, I will assume that set theorists are in the business of uncovering truths, and thus set aside Arealism. There are a couple of reasons for this. Firstly, Arealism is revisionary given the ordinary practice in set theory of calling set-theoretic axioms and theorems 'true'. I think it would be good to have a theory that preserves set theorists' presupposition that set theory is a body of truths. This may be a matter of preference, but this may also be a requirement on any philosophical account of set theory, because, for instance, one arguably can't account for the applicability of pure mathematics without truth (Liston, 2007).[14] Secondly, the *default position* for many philosophers of set theory, i.e., the intuitive (albeit defeasible) starting point for philosophical theorizing, is that set theorists make truth-evaluable statements, and that the commonly accepted axioms and their consequences are true in the ordinary sense of 'true'.[15] On this default position, the puzzle of extrinsic justifications really is puzzling. One of my main goals in this paper is to solve the puzzle of extrinsic justifications from this default position.

Let us then turn to Thin Realism. Maddy introduces Thin Realism in contrast with the view she calls 'Robust Realism', which is the view that sets form an objective reality, are abstracta, and exist independently of and are causally isolated from us (Maddy, 2011, p. 57f.). According to Maddy, Robust Realism can't satisfactorily answer Benacerraf's challenge to explain how we can have mathematical knowledge (Maddy, 2011, p. 57). Moreover, Maddy argues that it is very hard to see how the method of extrinsic justifications can be reliable if Robust Realism is true (Maddy, 2007, p. 366; 2011, p. 57–59). From a Second Philosopher's perspective, these are reasons to think that Robust Realism is false (Maddy, 2007, p. 366; 2011, p. 59).

Thin Realism, then, is supposed to be the alternative to Robust Realism on which Benacerraf's challenge is answered, and from which it follows that the method of

---

[14] For a reply to Liston, see (Maddy, 2011, pp. 89–96).

[15] I take this talk of "default position" from Koellner (2009, p. 91ff., footnote 38), and say more about it in Sect. 25.4.

extrinsic justifications is reliable. Once again, Thin Realism is the thesis that sets just are the things that set theory describes (Maddy, 2011, p. 61, 71, 75) and that can be learned about by application of set-theoretic methods (Maddy, 2007, p. 374; 2011, p. 63, 71, 75). As we saw in Sect. 25.2, one example of a set-theoretic method is the method of accepting axioms on the basis of their fruitfulness for the goal to maximize. Thin Realism thus tells us that sets are things that can be learned about by accepting principles that are fruitful for maximizing, and thus, in outline, that sets are things that are correctly described by principles that are fruitful for maximizing (henceforth simply 'sets are fruitful for maximizing'). More generally:

> [T]he fact of measurable cardinals being mathematically fruitful in ways x, y, z (and these advantages not being outweighed by accompanying disadvantages) is evidence for their existence. Why? Because of what sets are: repositories of mathematical depth [i.e., mathematical fruitfulness]. (Maddy, 2011, p. 82f.)

On Maddy's view, then, the reliability of the method of extrinsic justifications simply follows from the metaphysical nature of sets: If it is part of the nature of sets that anything that is derived by using set-theoretic methods is true of sets, then these methods have to be reliable.

My own view is that Thin Realism is basically correct as a metaphysical thesis about the nature of sets; the Carnapian solution I propose in Sect. 25.4 will closely align with Maddy concerning Thin Realism. However, Maddy doesn't provide an *explanation* of why Thin Realism is the correct metaphysical account of sets, i.e., an explanation of why it is that sets just are the sort of thing that can be learned about by application of set-theoretic methods. And this, in my view, is the biggest problem facing Maddy's solution to the puzzle of extrinsic justifications. Without an account of why sets have a metaphysical nature from which it follows that the method of extrinsic justifications is reliable, we have made little if any progress in addressing the question of why the method of extrinsic justifications is reliable.[16] As I will explain, Maddy resists giving an explanation of why Thin Realism is correct because, she argues, any attempt to do so inevitably goes against the methodology of Second Philosophy (Maddy, 2011, p. 71, 77). She settles for treating Thin Realism as a "brute fact" (Maddy 2011, p. 64), and argues that the most Thin Realists can do to relieve nagging doubts is to explain why the methods of set theory are fully objective, and thus—according to Thin Realism—why truths about sets are fully objective (Maddy, 2011, pp. 77–83). In my view, in order to solve the puzzle of extrinsic justifications, one needs to explain why there is this "close connection between sets and set theory, between sets and set-theoretic method" (Maddy, 2011, p. 64), and one can do so while remaining in line with Maddy's methodological commitments. This will be my task in Sect. 25.4: I will propose an account of how the meanings of set-theoretic expressions are determined (i.e., a *metasemantic* account), which will explain why 'set' applies to things that are known by set-theoretic methods, and thus, in turn, will explain why the method of extrinsic justifications is reliable.

Maddy (2011, pp. 64–71) considers three candidate explanations for why Thin Realism is true: a Kantian, a Carnapian, and a Deflationist explanation. She rejects all

[16] Linnebo (2012, p. 135) also raises this concern.

three on the grounds that they either go against Second-Philosophical methodology, or are incompatible with Thin Realism. For instance, the Kantian explanation of Thin Realism is that sets are "constituted" by our set-theoretic practices (Maddy, 2011, p. 65), but, Maddy argues, this is incompatible with Thin Realism since it entails that "there is considerably more to [sets] than what set theory tells us" (Maddy, 2011, p. 66). On the Carnapian explanation, "the connection between sets and set-theoretic methods is 'analytic' or 'conceptual'" (Maddy, 2011, p. 67). I address Maddy's criticisms of the Carnapian explanation in Sect. 25.4, where I develop precisely this kind of view. More generally, Maddy seems to think that there is a fundamental obstacle facing any candidate explanation of Thin Realism. As I understand it, there are two lines of thought that lead Maddy to this conclusion. First, Maddy seems to think that any account of why sets are the way Thin Realism says they are inevitably goes beyond what set theory says about sets, and, therefore, conflicts with Thin Realism—just like the Kantian account:

> [The Thin Realist is] [...] left with the claim that sets simply are the things that set theory describes, and it's hard to see how any further explanation could be given without introducing an external metaphysics that goes beyond what set theory tells us. (Maddy, 2011, p. 71)[17]

But this line of thought is mistaken. For instance, the Carnapian explanation I develop in Sect. 25.4 involves metasemantic claims about set-theoretic language and doesn't add anything about the metaphysical nature of sets to Thin Realism. Metasemantic claims about set-theoretic language aren't claims about the metaphysical nature of sets, and even Thin Realists would accept that there must be metasemantic facts about set-theoretic language, even though set theory doesn't tell us about them.

Maddy's second line of thought against explanations of Thin Realism seems to be that asking for such an explanation is asking to "supplement" the epistemology of mathematics; it is asking for more evidence, and a different kind of evidence, than what set theorists give, and thus inevitably involves putting into question the reliability of the methods of set theory. For instance, Maddy contrasts her approach with that of philosophers who question

> how what mathematicians actually do could have any connection to the subject matter they're attempting to describe. [...] [T]hey begin with the metaphysics and are led to confusion about the methods. In contrast, the Second Philosopher begins with the methods, finds them good, then devises a minimal metaphysics to suit the case. (Maddy, 2011, p. 86)

However, being confused about the methods of set theory or asking for an account of how they are reliable is compatible with granting that they are reliable. As Linnebo puts this point:

> Our conviction *that* certain methods are uniquely reliable in a given domain does not make it misplaced to ask *why* they are so. (Linnebo, 2012, p. 135)

We can take it as a desideratum for a metaphysical account of sets that it is consistent with the reliability of the method of extrinsic justifications. But we should still be able to ask for an explanation of why the method of extrinsic justifications is reliable, or,

[17] See also (Maddy, 2011, p. 64, 77).

equivalently, of why a metaphysical view that one "devises" to entail that the method of extrinsic justifications is reliable is actually correct about the nature of sets.

Without an account of why Thin Realism is correct, then, we are no closer to understanding why the method of extrinsic justifications is reliable. Moreover, even independently of one's interest in the puzzle of extrinsic justifications, and simply as a proponent of Thin Realism, it would be better to have an explanation of why Thin Realism is the correct metaphysical view about sets, which is a point Maddy herself acknowledges (Maddy, 2011, p. 64, 77). I thus turn to the task of providing such an explanation.

25.4 A Carnapian Solution

I will assume the following to be part of our default position. The standardly accepted axioms for set theory and their consequences are *true*—in whatever is the ordinary sense of 'true'. Expressions of the ordinary (often informal) language set theorists use are *meaningful*. Moreover, and as is standardly accepted as a platitude, the truth-value of a sentence is determined by three factors: the meanings of the expressions in the sentence, the way they are combined together in the sentence, and the world. The first two factors determine what specific character the sentence says the world has. For the sentence to be true, the world needs to have that character. In particular, then, the truth of axioms of set theory is in part determined by the meanings of the expressions of the language used by set theorists.

There are many questions to ask about the default position. For instance, we should ask what kinds of things *meanings* of mathematical expressions are—and, in particular, to what extent they can be modeled using standard notions like extensions (sets of objects), intensions (functions from possible worlds to extensions), etc. We should also ask what exactly the "mathematical world" is such that it partly determines the truth-values of mathematical sentences.

In my view, however, there are important advantages to starting with a metasemantic question: Assuming that mathematical expressions have meanings (whatever they are), what could *determine* that they have those meanings in the first place? One advantage of starting here is that we have a pretty good sense of the kinds of answers available for metasemantic questions of this kind, and some of them are quickly ruled out in the case of mathematical expressions. For instance, *causal theories* state that what determines the meaning (and thereby the reference) of an expression is some sort of causal relation between one's use of the expression and the referents of the expression. Since, on a standard view, abstract mathematical objects like sets can't bear causal relations, causal theories arguably can't answer our metasemantic question.[18] There are other standardly discussed *externalist* metasemantic theories (i.e., theories on which meaning is determined at least in part by features external to the speaker),

[18] Interestingly, Maddy (1980) herself famously proposed a causal theory of reference for mathematical expressions, but gave it up by (1997).

but they won't be particularly attractive to Second Philosophers. For instance, one could hold, along the lines proposed by David Lewis (1984), that the *metaphysical naturalness* of sets make them "reference magnets," and therefore partly determines the reference of 'set'—but the notion of metaphysical naturalness wouldn't be acceptable to Second Philosophers. Among the available traditional metasemantic theories, then, there is only one left to try, viz. (*metasemantic*) *descriptivism*. The general idea of descriptivism applied to the language of set theory is that what determines the meanings of set-theoretic expressions, such as 'set', is set theorists' pattern of usage of 'set'. This general idea is very plausible, as Maddy herself seems to agree:

> To understand how we came to use 'set' to refer to sets, we need only examine how the word came to be used as it is. (Maddy, 2007, p. 372)

In my view, descriptivism—specifically a kind inspired by Rudolf Carnap (1955), and developed along the lines of Lewis (1970), Frank Jackson (1998a; 1998b), and David Chalmers (2002; 2012)[19]—is the correct kind of metasemantic account for mathematical languages.[20] Elsewhere, I argue for descriptivism (Soysal, 2021; Kipper & Soysal 2022), and defend it against arguments purported to show that descriptivism in mathematics yields radical indeterminacy of meaning (Soysal, 2020). I won't defend descriptivism further here. What I will show is that if one accepts descriptivism for the language of set theory—which, as we saw, is a plausible and a natural answer to our metasemantic question—then one has a straightforward explanation of the reliability of the method of extrinsic justifications. This discussion will thus further serve to show that starting at the level of metasemantics can have very fruitful consequences for the philosophy of mathematics.

On the version of descriptivism that I favor, what determines what a speaker means with an expression is that the speaker *associates* a property with the expression—the expression refers to whatever has the associated property. On Jackson's (1998b) construal, the property a speaker associates with an expression is the property that underlies her *pattern of usage* of the expression, i.e., her dispositions to apply the expression when she is rationally unimpaired. The picture is familiar from Carnap (1955): A speaker has dispositions to apply an expression—given an expression, the speaker will be disposed to apply it to some thing(s), if presented with an actual or hypothetical case. These linguistic dispositions determine the *meaning* of the expression in the sense that they determine what Carnap calls the 'intension' of the expression, which is the general condition that underlies this pattern of application, i.e., the condition an object satisfies if and only if the speaker is disposed to apply the expression to it (Carnap, 1955, p. 41f.).[21]

Given that the property a speaker associates with an expression is what underlies the speaker's pattern of application of the expression, and in particular what under-

---

[19] Although Chalmers doesn't use the label 'descriptivist' for his view.

[20] More specifically, I hold the slightly weaker view that at least part of what determines what set theorists express with 'set' is that they associate a property with 'set', but the weaker view has the same consequences I explain next.

[21] Carnap later (1956) defines 'intension' in the standard sense above. See also (Chalmers, 2002; 2011) for a development of this kind of view.

lies her dispositions to apply the expression given descriptions of hypothetical cases, the associated property can itself be expressed in words by an *associated description*, which involves or entails sentences involving the expression that the speaker is disposed to accept no matter what she learns, and no matter what hypothetical case she considers (when she is rationally unimpaired).[22] In general, there might be a relatively short, definition-like, associated description that captures the associated property, but there needn't be.[23] For instance, for a classic example like 'game', it might be that the associated description consists of as many sentences of the form 'If something is [add complete description of chess], then it is a game' as we have dispositions to apply 'game' in hypothetical cases, and that no more compact associated description can be given (though this is unlikely).[24]

The sentences that are part of an associated description are standardly taken to be knowable a priori (since they can be known by reflection on one's judgments concerning hypothetical cases), necessary,[25] and "true in virtue of meaning" in at least some sense. For example, because one of the properties we associate with 'natural number' is, say, the property of having a successor, a necessary condition for being in the extension of 'natural number' is to have this property, and therefore the sentence 'If something is a natural number, then it has a successor' is true. Following Carnap (1955), such sentences can thus plausibly be called 'analytic' or 'conceptually true'.[26]

Assuming that descriptivism is the correct account of what determines what set theorists express with expressions like 'set', the key question is: What is in the description set theorists associate with 'set'? By definition, this associated description involves or entails sentences involving 'set' that set theorists are disposed to accept no matter what. Answering this question thus requires examining the practice and "proper methods" of set theorists: What set theorists explicitly or implicitly accept concerning sets and seem to be disposed to accept no matter what is *evidence* for which sentences concerning sets they are disposed to accept no matter what.

Methodologically, the descriptivist thus aligns with Maddy (2007, 2011). Recall from Sect. 25.3 that, according to Thin Realism, sets just are the things that set theory describes and that are known by set-theoretic methods. Presumably, knowing what is part of set theory and its methods requires examining what set theorists

---

22 For simplicity, I henceforth omit the qualification of being rationally unimpaired.

23 For discussion, see, for instance, (Jackson, 1998b, p. 203f.; Jackson et al., 2000; Chalmers, 2002, p. 143f.; 2012, pp. 16–19, 381–385).

24 It might also be that the description associated with an expression inevitably involves other uninterpreted expressions that get their meanings by being associated with the same description, as in (Lewis, 1970; Jackson, 1998a). In (Soysal, 2020, 2021), I argue that the expressions 'set' and 'membership' will in this way get their meanings determined jointly by being associated with a description involving them both.

25 At least if the associated description isn't de jure rigid, but this issue doesn't arise for mathematical expressions. For discussion, see (Jackson, 1998a, p. 143f.; Lewis, 1970, pp. 435–438).

26 Jackson (1998a, p. 146) also calls such sentences 'analytic' if they are necessary (see footnote 25 above). I argue further for a metasemantic notion of analyticity like this in (Soysal, 2021).

explicitly or implicitly accept concerning sets.[27] So, if we assume Thin Realism, we have to examine what set theorists explicitly or implicitly accept concerning sets in order to come to know truths about sets. This kind of evidence for what is true about sets when assuming Thin Realism also counts as evidence for which sentences concerning sets set theorists are disposed to accept no matter what, which in turn—assuming descriptivism—is evidence for what is analytically true about sets. So, if descriptivism is true, evidence from the practice of set theory becomes evidence for what determines meaning, and thus for what is analytically true of sets. As promised, descriptivism adds an explanatory layer between "sets and set-theoretic methods," and a metasemantic explanation for why Thin Realism is correct.

Using this kind of evidence, we can conclude that at least some of the basic axioms of ZFC are part of the description set theorists associate with 'set'.[28] Moreover—and here is the key point—we can also conclude that the description that set theorists associate with 'set' includes sentences that express *functional* properties of sets, i.e., sentences that express roles that sets play in mathematical theorizing, such as 'Sets are fruitful for maximizing'. This conclusion is warranted by the very same reasons for which Maddy concludes that "sets just are the sort of thing that can be learned about by application of such set theoretic methods [such as the maximization argument for $V \neq L$]" (Maddy, 2007, p. 374), or that sets are "repositories of mathematical depth" (Maddy, 2011, p. 83): All these conclusions are based on what set theorists actually accept and seem to be disposed to accept no matter what. As Maddy argues, and as we saw in Sect. 25.2, the method of extrinsic justifications is part and parcel of set-theoretic practice; since the very introduction of sets by Georg Cantor and Richard Dedekind, the idea that sets are fruitful for certain mathematical purposes guides the development of set theory (Maddy, 2011, pp. 41–45). Thus, if we accept descriptivism about the language of set theory, some sentences expressing functional properties of sets, such as 'Sets are fruitful for maximizing', are plausibly analytic.

We can now return to the puzzle of extrinsic justifications, and provide the Carnapian solution to it—I call it 'Carnapian' because it emerges from a form of descriptivism inspired by Carnap. If sentences such as 'Sets are fruitful for maximizing' are analytic, then, given a candidate axiom, if the truth of the axiom entails that sets can't play the functional role of being fruitful for maximizing (for instance, if the axiom is "restrictive" like $V = L$), then this is evidence against the axiom, because it is evidence that the axiom is incompatible with an analytic truth. (This schema captures the structure of the argument against $V = L$ from Sect. 25.2.) Similarly, if the axiom's truth entails that sets play the functional role of being fruitful for maximizing, then this is evidence for the axiom. To compare: Thin Realism's explanation for the reliability of extrinsic justifications proceeds from the truth of sentences like 'Sets are fruitful for maximizing', without explaining why they are true (except to

---

[27] More precisely, we should examine what set theorists accept concerning sets *when they are rationally unimpaired*. As Maddy (2011, p. 55, 74f.; 2014, p. 228) acknowledges, some idealization is required in the statement of Thin Realism, since set theorists can of course make mistakes due to inattention, inebriation, etc.

[28] I argue for this in more detail in (Soysal, 2021).

say that this is what set theory tells us). The Carnapian solution fills the gap by adding that these sentences are *analytically* true, and by providing a metasemantic framework from which this follows.

Let me note some interesting consequences of the framework I am proposing with respect to the possibility of pluralism in set theory, which some of Maddy's reviewers have argued Thin Realism can't adequately address (Incurvati & Smith, 2012; Leng, 2016). First, some analytic truths about sets might be vague, i.e., it might be that no precise description captures the property set theorists associate with 'set'. If, for instance, 'maximizing' is vague (which is highly plausible), then it is possible that there are jointly incompatible, but equally well justified, candidate axioms about sets, because what best fulfills the role of maximizing is underdetermined. A second way pluralism might emerge is if different set theorists associate slightly different properties with 'set', because, for instance, they associate slightly different functional roles with sets.[29] As we saw in §25.2, there are disagreements over whether some axioms of set theory are justified. Descriptivists here have strategies to explain such disagreements. For instance, disagreements might be verbal if disagreeing set theorists associate different properties—and thus strictly speaking express different meanings—with 'set'. Disagreements might be in some sense "faultless" if there is more than one way to fulfill some functional role associated with sets. Disagreements might also be substantial (i.e., not verbal) when they concern which mathematical objects best fulfill certain functional roles. In general, then, descriptivism opens up interesting avenues of investigation for the possibility of pluralism in mathematics, and these can be explored by Thin Realists as well.

Lastly, let me turn to addressing the criticisms I mentioned in §25.3 that Maddy (2011, pp. 66–69) raises specifically against Carnapian explanations. Maddy states that if Carnapians want the method of extrinsic justifications to come out as reliable, then they have to claim that both intrinsic and extrinsic considerations are analytic. Her first criticism of the Carnapian explanation, then, is that this would be "obscuring a distinction important to our understanding of the practice" (Maddy, 2011, p. 69). As I just explained, I indeed think extrinsic considerations are analytic in the sense that sentences expressing functional properties of sets are included in the description set theorists associate with 'set'. This is because sentences expressing functional properties of sets are plausibly among the sentences set theorists are disposed to accept no matter what, and thus, in at least that sense, they are on a par with sentences that have traditionally been categorized as intrinsically justified, like the basic axioms of ZFC. However, this doesn't entail that we lose the intrinsic–extrinsic distinction—we can recapture it within the descriptivist framework. We can define *transparent* analytic truths about sets to be analytic truths that reveal essential properties of sets without much or any further reasoning, and *opaque* analytic truths about sets to be ones that reveal essential properties of sets only with further a priori (or even perhaps a posteriori) reasoning.[30] The idea is that we can come to know something about the

[29] Incurvati and Smith (2012) consider this kind of scenario.

[30] This comes close to Chalmers's (2012, pp. 385–392) distinction between analytic and a priori truths within his scrutability framework.

nature of sets on the basis of knowing one of their functional properties only by first figuring out what fulfills the functional role. For instance, we might be disposed to accept both 'Sets are fruitful for maximizing' and 'Sets are extensional', but only the latter reveals something about the nature of sets without further reasoning. Thus, sentences traditionally categorized as "extrinsically" justified, like $V \neq L$, are ones justified because they follow from opaque analytic truths, while sentences categorized as "intrinsically" justified, like Extensionality, are ones justified because they follow from transparent analytic truths.

Some questions concerning the transparent–opaque distinction remain to be worked out. First, what an "essential property" of sets is might be vague, in which case some vagueness is built into the distinction; more should thus be said concerning whether the distinction can be precisified. Second, the distinction comes in degrees—the more "further reasoning" is required to come to know essential properties of sets, the more opaque the analytic truth is. Thus, more should be said about how to quantify amounts of, or characterize types of, further reasoning required. In any case, the intrinsic–extrinsic distinction, too, is arguably vague and gradable, and we can already see that the transparent–opaque distinction comes close enough to it that it can equally serve our understanding of the practice of set theory.

Maddy's second criticism of the Carnapian explanation is that it is committed to set theorists' adoption of the concept of set being a "purely pragmatic matter of linguistic convention," whereas the Thin Realist "takes Cantor and Dedekind to have discovered the existence of these mathematical objects, not to have come to a convenient decision" (Maddy, 2011, p. 68). Let me explain why the Carnapian view isn't committed to this claim by making some clarifications. First, whoever accepts the default position accepts that the truth of a sentence in general (not just in set theory) depends in part on the meanings of the expressions in it. For instance, if 'chess' had referred to pigeons, the sentence 'Chess is a game' would have been false. However, once facts about meanings and their determinants are in place—even on a Carnapian metasemantics—the truths expressed by the sentences aren't true by convention or decision. For instance, if the property we associate with 'bachelor' is the property of being unmarried, male, and adult, then it isn't up to us whether 'All bachelors are unmarried' is true, just like it isn't up to us whether 'Rudolf Carnap is a bachelor' is true—whether these sentences are true depends on the way the meanings are combined in the sentences and on the relations holding between these meanings.[31]

Moreover, it also isn't a matter of convention or decision whether the properties we associate with our expressions are actually instantiated. It might be, for instance, that the properties we associate with an expression are mutually inconsistent, which might mean that this expression doesn't have an extension (or an intension). So, more generally, the fact that we associate certain properties with certain expressions doesn't imply that the truths we express—or *whether* we express truths—are merely matters of convention or decision on the Carnapian view.

[31] Carnap (1963, p. 915) himself makes this point.

Finally, the reason why we associate a particular property with an expression needn't be a matter of convention or decision either. Of course, it can be—sometimes, the reason why we are disposed to apply an expression to certain things given hypothetical and actual cases is that we have decided to use, or stipulated the use of, the word in that way. But the Carnapian view isn't committed to this always being the case. It allows that other kinds of reasons explain our linguistic dispositions. In particular, Carnapians can agree with Maddy concerning why we express certain concepts with our mathematical expressions:

> So what guides our concept-formation, beyond the logical requirement of consistency, is the way some logically possible concepts track deep mathematical strains the others miss. (Maddy, 2011, p. 79)

Tracking strains of mathematical fruitfulness and other set-theoretic goals can certainly be the reasons why we have the linguistic dispositions that we have, and therefore, according to descriptivism, why we express what we express. The Carnapian view thus nicely fits with Maddy's commitments yet again.

25.5 Conclusion

The puzzle of extrinsic justifications is the question of how set theorists' method of extrinsic justifications is reliable. Maddy's solution involves saying that sets just are what set theory says they are, and set theory includes claims about sets being fruitful for certain purposes. The main problem with Maddy's solution is that we aren't given an explanation of why sets just are what set theory says they are—we lack the "mechanisms" that connect what set theory says about sets to what is true about sets. The Carnapian solution provides this kind of mechanism. I first explained why a Carnapian descriptivism is the most plausible metasemantics for the language of set theory. I then explained that, based on observations of the practice of set theory, we can conclude that sentences expressing functional properties of sets are part of the description associated with 'set', and that they are thus analytic. This entails that the method of extrinsic justifications is reliable; for instance, the restrictiveness of an axiom is evidence against that axiom because it is evidence that it is incompatible with an analytic truth.

The Carnapian metasemantic approach I took in this paper not only solves the puzzle of extrinsic justifications, but it also supplements the needed explanations for Maddy's Thin Realism. Moreover, the Carnapian view fits with Maddy's methodological principles, and with her central commitments concerning the nature of sets and the purposes and origins of set-theoretic concepts. There are, of course, numerous further questions to be answered about this Carnapian descriptivism. But, even if only for the reasons given in this paper, I take it to be a view worth considering seriously for all Second Philosophers.

References

Arrigoni, T. (2011). $V = L$ and intuitive plausibility in set theory: A case study. *The Bulletin of Symbolic Logic, 17*(3), 337–360.
Boolos, G. (1971). The iterative conception of set. *Journal of Philosophy, 68*(8), 215–231.
Carnap, R. (1955). Meaning and synonymy in natural languages. *Philosophical Studies, 6*(3), 33–47.
Carnap, R. (1956). *Meaning and necessity: A study in semantics and modal logic*. University of Chicago Press.
Carnap, R. (1963). W. V. Quine on logical truth. In P. Arthur Schilpp (Ed.), *The Philosophy of Rudolf Carnap, Library of Living Philosophers Vol. 11*, 915–921. Open Court.
Chakravartty, A. (2017). Scientific Realism. In E. N. Zalta (Ed.), *The Stanford Encyclopedia of Philosophy (Summer 2017 Edition)*.
Chalmers, D. J. (2002). On sense and intension. *Philosophical Perspectives, 16*, 135–182.
Chalmers, D. (2011). Revisability and conceptual change in "Two Dogmas of Empiricism". *The Journal of Philosophy, 108*(8), 387–415.
Chalmers, D. (2012). *Constructing the world*. Oxford University Press.
Cohen, P. (1963). The independence of the continuum hypothesis. *PNAS, 50*(6), 1143–1148.
Devlin, K. J. (1977). *The axiom of constructibility: A guide for the mathematician*. Springer.
Feferman, S. (2000). Why the Programs for New Axioms Need to Be Questioned. In (Feferman et al. 2000), 401–413.
Feferman, S., Friedman, H., Maddy, P., & Steel, J. (2000). Does mathematics need new axioms? *Bulletin of Symbolic Logic, 6*, 401–446.
Ferreirós, J. (2011). On arbitrary sets and ZFC. *The Bulletin of Symbolic Logic, 17*(3), 361–393.
Fraenkel, A. A., Bar-Hillel, Y., & Levy, A. (1973). *Foundations of set theory*. North-Holland.
Gödel, K. (1938). The consistency of the axiom of choice and of the generalized continuum-hypothesis. In (Gödel 1990), 26–27.
Gödel, K. (1944). Russell's mathematical logic. In (Gödel 1990), 119–141.
Gödel, K. (1946). Remarks before the princeton bicentennial conference on problems in mathematics. In (Gödel 1990), 150–153.
Gödel, K. (1947). What Is Cantor's Continuum Problem? In (Gödel 1990), 176–187.
Gödel, K. (1990). *Collected Works, Volume II: Publications 1938–1974* (Edited by Solomon Feferman, John W. Dawson Jr., Stephen C. Kleene, Gregory H. Moore, Robert M. Solovay, and Jean van Heijenoort). Oxford University Press.
Hallett, M. (1984). *Cantorian set theory and limitation of size*. Oxford University Press.
Incurvati, L., & Smith, P. (2012). Review of (Maddy 2011). *Mind, 121*(481), 195–200.
Jackson, F. (1998a). *From metaphysics to ethics: A defence of conceptual analysis*. Oxford University Press.
Jackson, F. (1998). Reference and description revisited. *Philosophical Perspectives, 12*, 201–218.
Jackson, F., Pettit, P., & Smith, M. (2000). Ethical particularism and patterns. In B. Hooker & M. O. Little (Eds.), *Moral Particularism*, 79–99. Oxford University Press.
Jensen, R. (1995). Inner models and large cardinals. *The Bulletin of Symbolic Logic, 1*, 393–407.
Kennedy, J. (Ed.). (2014). *Interpreting Gödel: Critical Essays*. Cambridge University Press.
Kennedy, J. (2014). Gödel's 1946 Princeton Bicentennial Lecture: An Appreciation. In (Kennedy 2014), 109–130.
Kennedy, J., & Väänänen, J. (2015). Aesthetics and the dream of objectivity: Notes from set theory. *Inquiry: An Interdisciplinary Journal of Philosophy, 58*(1), 83–98.
Kipper, J., & Soysal, Z. (2022). A Kripkean argument for descriptivism. *Noûs, 56*(3), 654–669.
Koellner, P. (2006). On the question of absolute undecidability. *Philosophia Mathematica, 14*(2), 153–188.
Koellner, P. (2009). Truth in Mathematics: The Question of Pluralism. In O. Bueno & Ø. Linnebo (Eds.), *New Waves in philosophy of mathematics* (pp. 80–116). Palgrave Macmillan.
Koellner, P. (2013). Large cardinals and determinacy. In E. N. Zalta (Ed.), *The Stanford Encyclopedia of Philosophy (Spring 2014 Edition)*.

Leng, M. (2016). Critical review of (Maddy 2011). *The Philosophical Quarterly, 66*(265), 823–832.
Lewis, D. (1970). How to define theoretical terms. *The Journal of Philosophy, 67*(13), 427–446.
Lewis, D. (1984). Putnam's Paradox. *Australasian Journal of Philosophy, 62*(3), 221–236.
Liston, M. (2007). Review of (Maddy 2007). *Notre Dame Philosophy Reviews*.
Linnebo, Ø. (2012). Review of (Maddy 2011). *Philosophy, 87*(339), 113–137.
Maddy, P. (1980). Perception and mathematical intuition. *The Philosophical Review, 89*(2), 163–196.
Maddy, P. (1988). Believing the axioms I. *The Journal of Symbolic Logic, 53*(2), 481–511.
Maddy, P. (1993). Does V Equal L? *The Journal of Symbolic Logic, 58*(1), 15–41.
Maddy, P. (1997). *Naturalism in mathematics*. Clarendon Press.
Maddy, P. (2007). *Second philosophy: A naturalistic method*. Oxford University Press.
Maddy, P. (2011). *Defending the axioms: On the philosophical foundations of set theory*. Oxford University Press.
Maddy, P. (2014). A second philosophy of arithmetic. *The Review of Symbolic Logic, 7*(2), 222–249.
Martin, D. A. (1970). Measurable Cardinals and analytic games. *Fundamenta Mathematicae, 66*, 287–291.
Mirimanoff, D. (1917). Les Antinomies de Russell et de Burali-Forti et le Problème Fondamental de la Théorie des Ensembles. *L'Enseignement Mathématique, 19*, 37–52.
Moschovakis, Y. (1980). *Descriptive set theory*. North-Holland.
Parsons, C. (2014). Analyticity for realists. In (Kennedy 2014), 131–150.
Quine, W. V. O. (1992). *Pursuit of truth: Revised edition*. Harvard University Press.
Scott, D. (1961). Measurable cardinals and constructible sets. *Bulletin de l'Académie Polonaise des Sciences (Série des Sciences Mathématiques, Astronomiques et Physiques)*(9), 521–524.
Shoenfield, J. R. (1977). Axioms of Set Theory. In J. Barwise (Ed.), *Handbook of mathematical logic* (pp. 321–344). North-Holland.
Soysal, Z. (2020). Descriptivism about the reference of set-theoretic expressions: revisiting Putnam's model-theoretic arguments. *The Monist, 103*(4), 442–454.
Soysal, Z. (2021). From metasemantics to analyticity. *Philosophy and Phenomenological Research, 103*(1), 57–76.
Steel, J. (2000). Mathematics needs new axioms. In (Feferman et al. 2000), 422–433.
Tait, W. (2005). Gödel's Unpublished papers on foundations of mathematics. In *The Provenance of Pure Reason*. Oxford University Press, 276–313.
Tiles, M. (1989). *The philosophy of set theory*. Basil Blackwell.
van Heijenoort, J. (Ed.). (1967). *From Frege to Gödel: A source-book in mathematical logic, 1879–1931*. Harvard University Press.
von Neumann, J. (1925). An axiomatization of set theory. In (van Heijenoort 1967), 393–413.
Wang, H. (1974). *From mathematics to philosophy*. Routledge and Kegan Paul.
Woodin, H. (2010). Strong axioms of infinity and the search for V. *Proceedings of the International Congress of Mathematicians*, Hyderabad, India.
Zermelo, E. (1908). A new proof of the possibility of a well-ordering. In (van Heijenoort 1967), 139–141.
Zermelo, E. (1930). Über Grenzzahlen und Mengenbereiche: Neue Untersuchungen über die Grundlagen der Mengenlehre. *Fundamenta Mathematicae, 14*, 29–47.

Chapter 26
Reply to Soysal

Penelope Maddy

Keywords Thin Realism/Arealism · Intrinsic/extrinsic · Mathematical depth · Analyticity · Metasemantic descriptivism · Carnap

Professor Soysal explores a particularly delicate point in the scaffolding of Thin Realism—its distinguishing feature, in fact[1]—namely its trivial epistemology. She believes this point is so delicate as to be problematic, and she generously offers a fix. I'm grateful for her efforts to save me from myself, but perhaps predictably, I fear that I'm not in a position to benefit from them. In fact, I suspect that Thin Realism as I understand it is actually too meager a realism to satisfy a thinker like Soysal, who embraces what she calls the 'default position': the axioms of set theory and their consequences are true.[2] This is just what a Thin Realist would say. The trouble is that I take Thin Realism to be a mere cosmetic variant of Arealism,[3] so while it's perfectly fine to speak in the Thin Realist idiom, it's not perfectly fine to do so while denying the Arealist the right to speak in hers.[4] Given that this is precisely what Soysal does, her Thin Realism can't be quite the same as my Thin Realism—hers isn't a mere

[1] That is, what distinguishes it from Robust Realism, which requires a non-trivial epistemology (see Maddy (2011), p. 63).

[2] p. 325. It later comes out that the default position includes more than this, perhaps a commitment to meanings and a compositional theory of thereof (p. 328). This may well be connected with her meta-semantic theorizing, which I don't fully understand and hope to skirt as much as possible (cf. footnote15).

[3] For the record, this isn't a merely epistemic lack—we can't tell which—it's a metaphysical lack—there's no fact of the matter at issue. They're two different ways of describing the same thing.

[4] See Maddy (2011), p. 113.

P. Maddy (✉)
Department of Logic and Philosophy of Science, University of California, Irvine, CA, USA
e-mail: pjmaddy@uci.edu

S. Arbeiter and J. Kennedy (eds.), *The Philosophy of Penelope Maddy*,
Outstanding Contributions to Logic 31, https://doi.org/10.1007/978-3-031-58425-1_26

cosmetic variant of Arealism. In the end, I hope to isolate another symptom of the mismatch between her figure and mine.

But first, before engaging with Soysal's project of improving Thin Realism, I'd like to say a few words in defense of Arealism. Soysal offers two objections in passing. The first is that Arealism violates the default position now characterized more fully:

> the intuitive (albeit defeasible) starting point for philosophical theorizing is that set theorists make truth-evaluable statements, and that the commonly accepted axioms and their consequences are true in the ordinary sense of 'true'. (p. 325).

This may indeed be the starting point—it's where I myself started many years ago[5]—but there are (it seems to me) sound reasons to reject Robust forms of Realism,[6] and once Thin Realism is isolated, I find no fact of the matter to distinguish it from Arealism. (Perhaps it's worth noting that the Arealist is quite at home with the notion of 'truth in a model' and may even speak more loosely of 'truth in V'.) So I take the real weight of Soysal's case to rest on her second objection, namely that the Arealist 'arguably can't account for the applicability of pure mathematics' (ibid.). This is a familiar charge, beginning with the well-known indispensability arguments, but again, I've argued at length that the case for mathematical ontology based on the role of mathematics in natural science is ineffective and offered an alternative picture of how mathematics works in application.[7] If I'm right on these points, the only mistake the Arealist makes it to deny that the Thin Realist's idiom is equally legitimate.[8]

Speaking with the Arealist is straightforward: she offers extrinsic reasons for adopting a new concept or method or axiom, viewing them as effective means toward important mathematical ends—an approach that's clearly rational since the overarching goal of the practice is the development of deep mathematics. The Thin Realist needs a defter touch. She begins sounding the same as a Robust Realist, declaring sets to be objective, non-spatiotemporal, acausal, and so on,[9] but when it comes to epistemology, she offers her thin metaphysics: sets just are the sort of thing that we can find out about using precisely these methods (the same methods as the Arealist). Where the Arealist sees good grounds for adopting an axiom, the Thin Realist sees evidence for its truth.

---

[5] See Maddy (1990).

[6] Most observers take some version of the famous Benacerraf challenge to be the most telling objection to Robust Realism, but my fundamental concern comes before that: even if a given claim has all the mathematical support a Thin Realist could ask for, a Robust Realist would say that this isn't enough, that we have to check whatever conditions the non-trivial epistemology imposes (see Maddy (2011), pp. 57–59).

[7] See Maddy (1997), §II.6, Maddy (2007), IV.2, Maddy (2011), I, IV.2 (which includes a discussion of Liston's twist on indispensability, cited by Soysal).

[8] For the record, I personally find the Arealist idiom more congenial. It was only the gentle but persistent prodding of John Steel and John Burgess that first gave me a glimpse of how a non-Robust Realism might go (see Maddy (2011), pp. 60–64).

[9] Soysal's description of Robust Realism—'the view that sets form an objective reality, are abstracta, and exist independently of and are causally isolated from us' (p. 325)—applies equally well to Thin Realism.

Soysal's concern about my presentation of Thin Realism is that it doesn't explain why sets are like this is, why they're such that these methods deliver reliable information about them (see pp. 326–327). In fact, I think I do offer something by way of explanation somewhere in this vicinity, but it depends essentially on the assumption of that mathematical depth is objective.[10] I've tried to convey the intuitive idea here with the concept of 'group': I take it as objective fact that 'group' generates good/important/useful/fruitful mathematics, that it's a 'deep' concept, that the mathematics it generates is 'deep'.[11] The mathematical goals in my means/ends analyses, then, are legitimate only insofar as they're deep.

Right from the beginning, sets have opened the way to deep advances: Cantor furthered his work on trigonometric series; Dedekind found the key to his abstract algebra; Zermelo and those who followed unlocked an ideally flexible framework for certain important foundational jobs.[12] I claim, then, that sets are markers for deep mathematics. Given that our set-theoretic methods are designed to find effective means to various mathematical ends, and that those mathematical ends are legitimate only insofar as they foster deep mathematics, it follows that those methods are designed to track depth, to track sets. In other words, what sets are—repositories of depth—explains why our methods—intended depth trackers—give us access to them, why they are things that can be found out about in these ways.

This explanation doesn't satisfy Soysal—not that I entirely blame her[13]—so she offers an alternative, an alternative that rests what she terms 'metasemantic descriptivism'. I confess that this opens philosophical terrain unfamiliar to me.[14] As I dimly understand it, for a given term, there's a property that underlies our speech dispositions (the term refers to things with that property), this property is expressible in words, and what follows from that expression is what a speaker among us 'is disposed to accept no matter what she learns, and no matter what hypothetical case she considers' (p. 330). So, for example, the property of having a successor (partly) underlies our use of the term 'natural number', so 'if something is a natural number, then it has a successor' is analytically true (p. 330). For the case of set theory, then, if we observe that set theorists are inclined to accept something 'no matter what', we have evidence that it is analytic (pp. 330–331). These analyticities will include

[10] See Maddy (2011), III.4. Soysal (e.g., pp. 326–327, 328) cites various passages where I acknowledge the difficulty of explaining a trivial epistemology without tipping over into something nontrivial, but these occur in the course of the discussion and aren't intended as the final word. So, for example, when I say the epistemic connection 'has so far been treated as a brute fact' (ibid., p. 64, cited by Soysal on p. 326), the 'so far' is intended to signal that this isn't the end of the story.

[11] The examples in Maddy (2011), II.2, are intended to strike the reader similarly. I once hoped to get beyond this reliance on intuitive reactions to examples when explicating depth, but I haven't succeeded (see Ernst et al. (2015)).

[12] See Maddy (2017b), (2019a).

[13] In Ernst et al. (2015), pp. 245–248, I admit that if depth isn't objective, I don't see that Thin Realism can survive. Arealism would carry on as before—it's just that the worthiness of our mathematical goals would presumably trace to something more inter-subjective.

[14] Perhaps due to my deflationary tendencies (see Maddy (2007), part II, (2022g)). The semantic story summarized in the following sentence raises a number of questions that trouble me, but this isn't the occasion to pursue them.

'at least some of the basic axioms ZFC', but also—and here's where it gets interesting—such sentences as 'sets are fruitful for maximizing' (p. 331). Applying this to the case of $V = L$[15]: if it's restrictive, 'this is evidence that [it] is incompatible with an analytic truth' (ibid.). To put the case the other way around, extrinsic evidence for a claim 'is evidence that [it] follows from analytic truths' (p. 320), so we have extrinsic evidence that $V \neq L$ (p. 322).[16]

We have here an explanation of why one kind of extrinsic consideration—the restrictiveness of $V = L$—is evidence for $V \neq L$, an explanation of why sets are such that this counts as evidence for that. The explanation is that the connection between this evidence and that claim is analytic, or more generally, that 'sets are fruitful for maximizing' is analytic. Evidence for this is to come from 'examining what set theorists explicitly or implicitly accept concerning sets' because 'which sentences concerning sets set theorists are disposed to accept no matter what … is evidence for what is analytically true about sets' (p. 331). So our reason for thinking that 'sets are fruitful for maximizing' is analytic is that we think set theorists are probably disposed to accept it come what may.

Now the iterative conception itself involves a combinatorial, anti-definabilist element that naturally aligns with some version of maximization,[17] so 'sets are fruitful for maximizing' is the sort of thing that is sometimes regarded as following from or implicit in the concept of set. This can make Soysal's analyticity claim feel plausible, intuitive. But consider Cantor introducing sets so as to facilitate the generalization of his theorems on trigonometric series. If his success here is to count as extrinsic evidence for positing sets, is 'sets facilitate theorems on trigonometric series' to be regarded as analytic? Is it rooted in the speech dispositions of set theorists, as the meta-semantic analysis requires? What about Dedekind's goal of representation-free definitions? Was this goal something set theorists were disposed to assent to when he first undertook to pursue it? Was 'providing a useful measure of consistency strength' analytically linked to sets or was it a welcome benefit of large cardinal axioms? Perhaps a typical conceptualist would balk at the thought that these evidential connections are derived from the concept, but of course Soysal isn't answerable to the typical conceptualist![18]

---

[15] The central case study in Maddy (1997).

[16] I have a little trouble getting Soysal's argument into deductive form. Perhaps the analytic truth is 'restrictive theories are (likely to be?) false', so when $V = L$ is shown to be restrictive, it follows that $V = L$ is (likely to be?) false. (Incidentally, this whole discussion is using 'maximizing' and 'restrictive' loosely, not in the more precise senses developed in Maddy (1997) and elsewhere. See Schatz (2019) for a comparative study.).

[17] As Soysal points out (see p. 323). I think this gets things backwards: not maximization is good because it's part of the iterative picture, but the iterative picture is a good heuristic because it encourages maximization, which furthers the deep foundational goal.

[18] These questions raise other concerns that lead inevitably to particulars of meta-semantic descriptivism that I hope to avoid. E.g., we might ask whether the property Cantor's usage assigned to 'set' is that same as the property we now assign, whether what was analytic for Cantor is also analytic for us. Here Soysal suggests that the two properties may not be the same, but that they might be co-extensive (cf. p. 332).

There remains the matter of 'come what may'. Even at the outset, when the default position was described as accepting that the 'axioms for set theory and their consequences are *true*' (p. 328), I felt inclined to whisper 'probably true?', but this slight discomfort intensified when extrinsic epistemic connections were described as being held true 'no matter what'. To take an easy example, if my Thin Realist were presented with sound mathematical reasons to prefer a new multiverse theory to her current universe theory (roughly ZFC plus large cardinals), she'd conclude that she'd been wrong to think set theory was the study of a single universe and happily change her mind about a whole array of items that might be thought to be analytic.[19] For that matter, didn't an upheaval of this magnitude happen once before, when the conviction that sets are extensions of concepts gave way to the combinatorial iterative conception? Looking forward, can we be so absolutely sure that 'sets are fruitful for maximizing' mightn't one day give way to overriding mathematical imperatives in another direction?

These questions miss an important feature of Soysal's account, namely, the epistemic uncertainty of the meta-semantic descriptivist's judgments of analyticity. A claim is only analytic if the set-theoretic community is inclined to accept it in all possible situations, but all we have is our current sample. So, for example, back in 1990, I may have thought I would endorse 'set theory is the study of a single universe of sets' no matter what, I may have thought this to be analytic, embedded in the very language of ZFC, but now, with serious work on multiverse theories, there's at least the chance that a theory in a language of sets and worlds will prove mathematically advantageous, and I'm no longer so sure I'm inclined to endorse that universe claim 'no matter what'. What's happened here, according to meta-semantic descriptivism, isn't that I may come to reject an analytic truth, but that it may turn out not to have been analytic after all.

This move brings out just how far Soysal's appeal to Carnap and analyticity departs from what's typically found in the philosophy of mathematics. The familiar syntax-era Carnap is often seen as meeting the epistemological challenge: what grounds do you have for your logical and mathematical beliefs? Answer: they're analytic, part of the language I've pragmatically chosen; I won't give them up short of revolutionary linguistic change. This move faced the familiar Quinean objection—what distinguishes a change of language from an ordinary change of theory?—but my point here is that Soysal's Carnapian analyticity isn't designed to play this role. According to meta-semantic descriptivism, all a set theorist can currently say about a particular set-theoretic belief is that she's tentatively committed to it, that she can't imagine a situation in which she'd give it up, but this hardly makes for a satisfying answer to the original epistemological question—not to mention that it aggravates the Quinean problem; what the set theorist just said hardly distinguishes her set-theoretic belief from other beliefs she thinks likely to be true. But Soysal isn't out to explain our knowledge of set theory; her question isn't—how do we know that

[19] Cf. Maddy (2011), p. 63, footnote 7. Maddy (2017b), and 'On multiversism', this volume, explore the multiverse possibility in light of more recent developments. See also Maddy and Meadows (2020). So I think there's no obstacle to my Thin Realist embracing pluralism (Soysal agrees, p. 332).

'sets are effective for maximizing'?—but—what makes it the case that sets are this way? She takes her inspiration from the semantic-era Carnap and hopes appeal to a descendent of his notion of analyticity will defeat a skeptic who questions Thin Realism as a metaphysical account.

This brings us back, finally, to my worry that the Thin Realist she's defending isn't my Thin Realist. Consider, for example, how our respective thinkers would respond to the question 'why is avoiding restrictive theories a good way to find out about sets?' Both hold that sets are the sort of thing that we can find out about by employing proper set-theoretic methods, but it turns out there are two ways of reading this. For Soysal, the way to go about answering the question just posed is to figure out whether avoiding restrictive theories is a proper set-theoretic method and this

> requires examining what set theorists explicitly or implicitly accept concerning sets ... which sentences concerning sets set theorists are disposed to accept no matter what. (p. 334)

(And, by the way, according to meta-semantic descriptivism, this links up with what is 'analytically true about sets' (ibid.).) So her Thin Realist's answer to my question is that avoiding restrictive theories is a good way to find out about sets because it is one of the things that set theorists except come what may.

But this isn't the answer my Thin Realist would give. Why is avoiding restrictive theories a good way to find out about sets? Because the foundational goal of set theory demands that set theory provide as generous an arena for classical mathematics as possible, and restrictive theories run counter to that goal. What's happened here is that I take 'proper set theoretic methods' to be those whose rationality is established by detailed means-ends analyses particular to each case, while Soysal takes them to be the ones whose rationality rests on the set-theoretic community's inclination to endorse them come what may. For what it's worth, this seems wrong to me: it isn't our tendency to endorse a particular method that ultimately makes it good; it's that it's an effective means toward a deep mathematical goal. So in the end, at least for now, I'm inclined to stick with my slightly delicate—perhaps even precarious, but I hope ultimately workable—version of Thin Realism. Nevertheless, I very much appreciate Professor Soysal's sophisticated attempt to do better![20]

References

Ernst, M. et al. (2015). 'Mathematical depth', special issue of *Philosophia Mathematica, 23*(2).
Maddy, P. (1990). *Realism in mathematics*. Oxford University Press.
Maddy, P. (1997). *Naturalism in mathematics*. Oxford University Press.
Maddy, P. (2011). *Defending the axioms*. Oxford University Press.
Maddy, P. (2007). *Second philosophy*. Oxford University Press.
Maddy, P. (2017b). Set-theoretic foundations. In A. Caicedo, J. Cummings, P. Koellner, & P. Larson (Eds.), *Foundations of mathematics: Essays in Honor of W. Hugh Woodin's 60th birthday* (pp. 289–322). American Mathematical Society.

[20] My thanks to Professor Soysal for her contribution and for correcting the misunderstandings in an early draft.

Maddy, P. (2019a). What do we want a foundation to do? In S. Centrone, D. Kant, & D. Sarikaya (Eds.), *Reflections on the foundations of mathematics* (pp. 293–311). Springer.
Maddy, P., & Meadows, T. (2020). A reconstruction of Steel's multiverse project. *Bulletin of Symbolic Logic, 26*, 118–169.
Schatz, J. (2019). *Axiom selection by maximization: V=Ultimate L vs. forcing axioms*, Ph.D. dissertation, UC Irvine.

Chapter 27
Wittgenstein's Wayward Student: The Unauthorized Autobiography

Curtis Franks

Annotated by Curtis Franks

Abstract Our hero has been confined until now to the space allotted him by other writers who only want to bounce their own ideas off of him because of their interest in the resulting trajectories. Quite reasonably he feels that he's never been given a fair shake. His efforts to understand his critics turn up points where their arguments strike him as unconvincing or altogether absent. He delights in these discoveries, but his readers will see them instead as evidence of his own confusion and as clues to the sort of positions his critics adopt. When he meets another idealized character in Penelope Maddy's Second Philosopher, he is intrigued by the difference in the treatment he receives. She raises many of the same questions that haunt him, the very considerations that others dismissed. But her reaction only further alienates him. Struggling to fathom how she can carry on without the answers he seeks, he unwittingly reveals what sets the Second Philosopher's position apart from related naturalistic accounts by chronicling how it leads him to confront the idleness of his demands.

The device of an unauthorized autobiography has been used at least since Larry Rivers's *What did I do?* and recurs in work by Ray Davies, Tommy Chong, Lemony Snicket, and others—the present work differing from these by actually being unauthorized. Wittgenstein's wayward student is the persona found in Wittgenstein's *Philosophical Investigations* and *Remarks on the Foundations of Mathematics*, but the particular label is due to Maddy. Earlier (1985) Goldfarb had used "wayward child," which the author obviously finds misleading.

Curtis Franks—With boundless gratitude to Penelope Maddy, whose vision and imagination are exceeded only by the impact of her patience with her own wayward student.

C. Franks (✉)
University of Notre Dame, South Bend, IN, USA
e-mail: cfranks@nd.edu

S. Arbeiter and J. Kennedy (eds.), *The Philosophy of Penelope Maddy*,
Outstanding Contributions to Logic 31, https://doi.org/10.1007/978-3-031-58425-1_27

Keywords Naturalism · Wittgenstein · Bolzano · Brouwer · Carroll · Maddy · Strawson · Logic

27.1 First Things

I think the first thing you ought to know about me, if you're going to appreciate much of what I have to say later on about my main scientific and philosophical agenda—the nature of logic and mathematics as it happens to be—is that I'm a lot older than I'm often made out to be. You know me as Wittgenstein's wayward student—not because I ever was a student of the man, and I suppose if I had been I would by now at least be on a ways in years—but because in his *gutes Buch*[1] I play the role of a student who displays, systematically and with impunity, certain "deviant" reactions to instruction, training, and rules, and in particular to the rule "add 2," no matter how many correctives his teacher issues.

When you hear about such a student, it seems inappropriate just to describe him as poor or lagging. I strike you, no doubt, as fantastically confused, as displaying the sort of behavior that just couldn't arise in any ordinary course of instruction. And the point Wittgenstein wants you to think about, the point to which we will return later, is *why* I strike you this way. What you *don't* think much about, I trust, what you just take for granted, is that I am a child, and a pretty young one at that. Who but a child, even in such a preposterous scene as Wittgenstein's, is being taught basic arithmetical operations, much less failing to catch on right away?

The situation reminds me of those Renaissance paintings of Abraham poised to sacrifice his young son. The rabbinical commentators tell us that the "lad" was a full thirty-seven years old, notwithstanding the Bible's narrative depiction of a passive figure lying unawares atop the altar, naively asking his father if he'd forgotten the sacrificial ram. Similarly, you should know, I found my way into Wittgenstein's book after more years of writers conjuring me into their work than I care to think about—ordinarily at least, although I find myself now compelled to recount the whole sorry history of these cameo appearances. Suffice it to say, for now, that my role in the *Investigations* was hardly my first, and, far from being a child, I was by that time already quite old.

The first thing you need to know about Wittgenstein, meanwhile, is that he talks to himself. I mean *throughout* the *Investigations* one finds extended passages comprised of records of voices debating some delicate topic or another—one voice defending a philosophical position about meaning, another trying to debunk it, sometimes even what seems to be a third voice butting in to clear up a misunderstanding between the first two or to tell the debunker to cool it off—and it's tempting, even natural, to think that somewhere in that stew is the voice of Wittgenstein himself, pitted against his foes.

---

[1] *Philosophische Untersuchungen*, cited in text by the author as *Investigations*.

But then one comes to a remark like, "How does the philosophical problem about mental processes and states and about behaviorism arise?" (§308) and it becomes clear that the whole chorus of voices that preceded it was a display of how philosophical thinking creeps up on us and leads us to think that we have to fit ourselves into one side of a debate: "Our thinking plays us a queer trick. We want ...to say 'Either such an image is in his mind, or it is not; there is no third possibility!'" (§352). Well, Wittgenstein proposes, there may be nowhere to run in this arena other than into one or the other corner, but don't mistake the arena for reality. And pay attention to how you found your way in!

"For this," Wittgenstein declares, "is what disputes between [for example] Idealists, Solipsists, and Realists look like" (§402). So arises a second temptation: to take most of the voices in the *Investigations* as caricatures of the various philosophical schools that Wittgenstein is mocking. But this can't be right, either. Because, you see, this litany of voices just won't stop. Just when you think he's smothered them for good, Wittgenstein turns his head, asks another innocent question, and before you know it gets drawn into another arena, or sometimes, through a different entrance, back into the same one he just left, or thought he'd left.

And the point of all this repetitiveness and lack of resolution is supposed to be that Wittgenstein can't quite close the deal. The voices are his own. Or perhaps they are the personification of the spirit of the times in which we live. And he wants to show us how insidious they are, how tricky it is to escape from them, and how there is no argumentative strategy that allows you to shut them down and move on. Instead, he says, "the results" he is aiming for are "the dents that the understanding has got by running its head up against the limits of language" (§119).[2] He records himself talking to himself in order to show us how easily, perhaps inevitably, and unwittingly we slip into philosophical messes, and in order to bang up his own habits of thinking in his attempts to find his way out of them. Again, these dents aren't the cost of what Wittgenstein is up to. They are the point.

Anyway, it's within *that* project that you find me, summoned from the annals of history, seated in some surreal classroom, unable to take to the instruction that some exasperated teacher is offering. And I tell you, it's more than a little uncomfortable to star in someone else's identity crisis.

27.2 No Laughing Matter

Now, one obvious consequence of the way I am situated in Wittgenstein's discussion is that I don't get much of a say about anything. And I don't just mean that I don't get many turns to speak, though in fact I do get but a few—I get no chance at all to carry on with anyone other than this teacher. Some other voices opine about my lot. One suggests that the instruction I've been issued is ambiguous, equally well interpreted

[2] I do not know why the author renders "*Beulen*" here as "dents" rather than "bumps." Elsewhere he follows Anscombe's translation.

as I have done and as the teacher intended—that some other more explicit instruction is needed. Another disagrees, claiming that no amount of instruction will do, that no statement or formula or even belabored training session could ever capture the teacher's full intention, which has infinite applications. And so on and so on.

But what you can't find anywhere is Wittgenstein just saying outright what I'm doing wrong, or where the teacher has gone wrong, or whatever. And this, it seems, is because he's more interested in showcasing the assembly of voices and exposing the frustrating cacophony that ensues than in putting forward his own theory of meaning or account of what it is to understand a rule. He is, in a phrase, more interested in playing around at my expense than in any serious philosophical work.

It's a point I can't overemphasize. In Lewis Carroll's treatment of the issue, he casts me as a tortoise—*a tortoise!*—and, instead of speaking directly to me, has me in conversation with a particularly feeble Achilles. I'm sure it's a rollicking good time for the right audience, but you'll excuse me if I can't appreciate such joking at my own expense. Fair to say that if there is a serious topic being discussed at all, it is being treated lightly, rather than with the care and sobriety that real philosophical work calls for.

To give you a full sense of the predicament, consider Norman Malcolm's report that "Wittgenstein once said that a serious and good philosophical work could be written that would consist entirely of jokes (without being facetious)" (2001, pp. 27–28). Now try to imagine making headway on, say, the problem of reference with a joke. I have my suspicions about this, but let us even grant that a comedic genius like Shakespeare could pull it off. Is there any evidence that Wittgenstein could?

Speaking of Shakespeare, let me take this opportunity to weigh in on the great authorship controversy (you will recall that I have been around for quite a long time) and report first-hand that the man did not, after all, write all those wonderful plays. In fact, someone else also named William Shakespeare did.

But to return to the issue at hand, think about the elaborate layers separating Wittgenstein, or Carroll, or any of my other detainers from me, to say nothing of the distance between me and *their* readers. Apparently direct confrontation would spoil the joke, so these writers keep their distance—but at what cost to philosophy? Ordinarily one thinks that the task of philosophers is to interpret such phenomenon as they are writing about, whereas what these writers do is better described as passing the buck. If anything, they *complicate* the matter, so that their readers are left to debate just what point they themselves are trying to make.

For this reason, it seems to me, one cannot assume without further ado from the similarity of my role in Carroll's and in Wittgenstein's writing, or in anyone else's for that matter, that one and the same idea is being put forward time and time again. It's often unclear to me whether any idea about anything at all is on offer.

27.3 Naturalism

When I appear in the *Investigations*, I've already been taught various things about basic arithmetic. To my teacher's satisfaction, I've demonstrated reasonable mastery of the material. Then I'm asked to perform some skip-counting with big numbers, 882, 884, 886, 888, 890, and this all goes pretty well. When I get to 998, I continue in the same way, 1000, 1004, but the teacher objects that I should instead have followed 1000 up with 1002.

Naturally, I ask why. He replies that I was meant to add 2. I insist that I had done just that. The teacher says that I'd added 4, whereas I was supposed to continue on in the same way. But so far as I have been able to understand, I had added 2, just as I had up to that point.

Here's where those voices chime in. One remarks that I've somehow mistaken the instruction "add 2" as "add 2 up until the number 1000 and add 4 thereafter"—something that all the training and testing I'd been through couldn't reveal until this point. To clear this up, another suggests, I ought to be told that adding 2 is no different among 4-digit numbers as it is among 1, 2, and 3-digit numbers. Someone else, rightly, observes that I don't take myself to have done anything different. I didn't think that "add 2" meant "add 2 up until the number 1000 and add 4 thereafter," but "add 2 no matter what number you're starting with." And that's what I thought I was doing.

This is when my favorite voice calls out: "But isn't the same at least the same?" (§215).

And I suppose there are a number of different things you could ask about this exchange, but the thing I think we ought to focus on is what exactly a proper understanding of a rule like "add 2" amounts to. For apparently I have misunderstood that rule. My teacher expected me to know, based on my ability to perform in accordance with it as expected for so long, that when starting with 1000 I was supposed to proceed in a way that strikes me as entirely different. So even though it is expected that this amount of training would suffice for me to see that 1002, and not 1004, is the right way to continue, it didn't. What does?

Wittgenstein's readers have suggested three different types of answer to this question. The first type of response seeks the thing—an algebraic expression that describes an entire infinite sequence, a computing machine, a mental image, the teacher's intention—that in addition to the training I received would have prevented any misunderstanding. The *Investigations* are pretty widely read as emphatic about the inadequacy of this sort of answer.

So a second type of answer is to grant that no amount of training, no mental image or intention, and no formula or machine of any sort, could ever suffice to guarantee understanding. Because there are always multiple ways to move consistently beyond any given data, there's also always a question about what makes one of those ways correct. And—this is the rub—nothing does. According to this answer, we have to get by with community agreement of some sort—i.e., grounds for taking our responses to be right and others' wrong are lacking, but this is of no practical concern in ordinary circumstances because beings like us have other ways of reaching consensus. This "skeptical solution," as it has been called by Saul Kripke (1982) and others, following Hume, sees me in a soft light. I'm not mistaken, just different.

Then there's the naturalistic answer. According to the naturalist, the ambitious seeker, chasing down the elemental thing that understanding consists of, and the skeptic, denying that there is any such thing, share in a common confusion. Wittgenstein called it a bewitchment. These philosophers both start out looking for an answer to the question, "What makes 1002 the correct response to the instruction 'add 2' when one is starting at 1000?" One of them is sure that something, perhaps a self-interpreting fundamental rule or state of mind, will be found that fills this role. The other denies that any such thing could, because 1002 is in fact no more correct than 1004. These attitudes seem to be poles apart. What could their commonality be?

Our seeker and our skeptic both demand that something outside of our ordinary practice of issuing and following instructions somehow *ground* that practice in order for there to be any sense in saying that some applications of a rule are correct. Well, if you think this demand is a fool's errand, the person who keeps searching doesn't look an awful lot sillier than the person who throws up his hands and says "Oh well, I guess we'll have to get by without grounds, somehow or another."

And that's the naturalist's read of the terrain. The seeker's various theories about what makes 1002 the correct response all fail, but this doesn't mean that we lack grounds for how we respond. "The …naturalist's point," in P. F. Strawson's words, "is that there can only be a *lack* where there is a *need*" (1985, p. 41). What makes 1002 the correct response to the instruction "add 2" when one is starting at 1000 is just that you are starting at 1000 and asked to add 2, and "there is, philosophically speaking, nothing behind all this, and no good need for anything beyond or behind it all to constitute a philosophical explanation of it" (pp. 77–78). To look beyond is to lose your grip on what following a rule is.

Warren Goldfarb memorialized this attitude. The skeptic, he wrote, concludes that "nothing determines how to go on with the series" from our inability to point to any ground for the correctness of an application of a rule. "Yet the scenario doesn't show anything like that," he replied. "[H]ow to go on is certainly determined, in the sense that if [the wayward student] is to continue the series we asked him to continue, then he must go on to write 1002, 1004, 1006." Instead, Goldfarb elaborated, "the conclusion …is that he is incapable of understanding what we mean" (1985, p. 485). Recall my description of the skeptic's line of thought above. I said that there is always a question about which among multiple ways to go on is correct. According to Goldfarb, that very idea is a step away from reality into the philosophical arena, "for in ordinary life there isn't always a question" (p. 486).

Well, where does that leave *me*? If you tell me there are grounds for the teacher's expectations, I'll ask what they are. If you tell me there are no grounds, that I'm just taking to training differently than other students, because I'm not part of their community or something, I can't be too surprised. I didn't exactly wind up in this classroom in the most ordinary way to begin with. But how am I supposed to take the news that I'm *just wrong* and that there's nothing more to say about why I'm wrong beyond just repeating the same instruction? Isn't that just the description of the difficulty that we wanted to resolve? "OK, now, you're up to 1000. Add 2 again. Oops, you've got 1004 there, are you sure?"—"Yes, that's what I meant to write."—"Well, you're wrong. The answer is 1002."—"1002? Why is that correct?" "Because you were to add 2, and you were up to 1000." ...

27.4 Turtles All The Way Down

I believe that Peter Winch was the first person to identify me as the tortoise in Lewis Carroll's (1895) essay about logical inference. I mean, I'd already made a big splash in this role, and as Winch himself noted (1958, p. 57), the "moral" of the parable that I'd been drawn into there, "that the actual process of drawing an inference, which is after all at the heart of logic, is something which cannot be represented as a logical formula," is a little hard to miss. Winch apologized for "boring" his readers by pointing it out. But I'm not aware of anyone before Winch seeing through my disguise enough to connect this essay with Wittgenstein's thought.

So what is the connection?

In Carroll's essay, I'm pondering with Achilles a few propositions labeled A, B, and Z, allegedly so related that Z follows logically from A and B.[3] Achilles and I agree that even granting the truth of A and B, it would be rash to infer Z without further ado, that a perfectly rational person could assent to A and B and, for all that, deny Z. So I inform Achilles that I'll play the part of a perfectly rational person for a change, and I challenge him to get me to see my way to the truth of Z by logic alone.

Achilles begins by asking me to consider the conditional expression, "If A and B, then Z." This strikes me as obviously true, quite independently of my commitment to A and B, so I happily assent. Achilles expects this to prepare me to see my way to the truth of Z, by inferring it from A, B, and "If A and B, then Z." But this, too, strikes me as too hasty. Pressed about my reluctance, I point out that Z doesn't seem to follow straightaway from A, B, and "If A and B, then Z" any more than it does from A and B alone. Surely, though, Achilles insists, I would agree that if A, B, and "If A and B, then Z" are all true, so too must Z be. And I assure him that, now that he has drawn my attention to it, I do indeed. Somehow Achilles expects me at that point to be prepared to infer Z form A, B, and "If A and B, then Z," but I remind him that we had just agreed that the inference isn't licensed without the extra premise, "If A, B, and 'If A and B, then Z,' then Z."

The narrator of the story—the person who lets this discussion play out without ever weighing in himself, and who *I never get to talk to*—departs at this point and leaves the reader to sort out on her own what to think.

---

[3] Carroll's sentences are geometric and related according to the inference pattern *modus ponens*, but as the author does not dwell on their contents and later considers in this vein the inference pattern known as disjunctive syllogism, it is appropriate to think of A as "It's either red or it's green," B as "It's not red," and Z as "It's green."

Now, the first thing to think is that the regress that arises in Carroll's story is very similar to one that can arise in Wittgenstein's. Achilles's search for the missing item that will carry the inference home only leads to a situation like the one he began in. Barry Stroud saw a perfect analogy in the search for the thing that will secure proper understanding. He focussed on the proposal that mental representations will fill the role:

> Even if "representations" were little plastic flash-cards and "the mind" was a special little pocket they fit into, and [understanding] was defined as a card's being in that pocket, we could still not explain someone's thinking that *p* as just a matter of there being a card with "*p*" on it in that person's special pocket. ..."Representations" alone are not enough because, at the very least, the person also has to understand what is on his card, he has to "grasp" the "content" that is represented there. ...But then those graspings or understandings are themselves presumably psychological phenomena [...and] to say that they too consist of nothing more than the presence before the mind of "objects" or "representations" would lead to a regress. (1991, p. 131)

But what interest me more are the *dis*analogies between my roles as the tortoise and as the student. Just consider the available reactions to the tortoise regress. Is a skeptical response even possible?

As a wayward student I'm supposed to elicit the thought that, if there's nothing that can be said or given to me to ensure that I will fasten onto the right understanding of a rule, then there's nothing you or my teacher or anyone else can point to as a reason to know that they've understood it correctly, either. The skeptic concludes that there just is no fact of the matter about whether I or my teacher or anyone else understand correctly—even denies that there is something to understand correctly—and, in the accounts by Dummett, Kripke, Bloor, and others, follows this up with a theory about how one (i.e., everyone but me) gets by in the absence of such facts of understanding. The naturalistic response is supposed, then, to be hard-won against the pull of this sort of skepticism.

Does this line of thought make sense in the case of logical inference? It doesn't seem at all right to conclude from my conversation with Achilles that there just is no fact of the matter about whether Z follows from A and B, moreover to suggest that it is a matter of contingent social practice that people infer Z from A and B, that they could just as well reason otherwise. And in fact, no one suggests anything like this.

Instead, the naturalistic response just seems immediate. Winch spent no time considering whether A and B might not fully justify Z in the absence of some conditional statement or another, because he thought that Carroll's joke is more about me raising the question than with the defect of any of Achilles's particular attempts to answer it: "To insist on any further justification is not to be extra cautious; it is to display a misunderstanding of what inference is" (p. 57). Similarly, Stroud observed that "it is difficult to see how the tortoise could understand and accept A and B without accepting Z" (1979, p. 44). One is reminded of Goldfarb's idea that "the correct conclusion is" not that no one means anything at all determinate, but simply "that *he* is incapable of understanding what *we* mean."

Why does one of these scenarios invite a skeptical response and the other not do so? Interestingly, commentators like Winch and Stroud who see the connection between them do not explore this disanalogy. In "Inference, Belief, and Understanding," for example, Stroud summed up his account with comments that would apply quite broadly to the general point about understanding and applying a rule, with no mention of any extra wrinkle added by the fact that his route to this conclusion began with considerations specifically about a rule of logical inference: "it cannot be a condition of explaining how a person comes to understand a word or sentence in one way rather than another that there should always be something he goes by or relies on, in the sense of something he can be said to know or believe, to determine that the expression is to be applied this way rather than that" (1979, p. 33) .

Similarly, in "Wittgenstein and Logical Necessity," despite the title's suggestion to the contrary, Stroud's analysis does not appear to rely on anything particular about logic. He began just by observing that

> Part of human behaviour consists of calculating sums, distances, quantities, of making inferences, drawing conclusions, and so forth. …There are various facts which make [such behavior] possible …. For example, our memories are generally good enough for us not to take numbers twice in counting up to 12, and not to leave any out …and so on. The inhabitants of the earth might have lacked these and other simple abilities, and if so, there would be no such thing as calculating at all. In that way the possibility of calculating depends on such contingent facts. (1965, p. 12)

This sets up his naturalistic strategy to dehumanize me: throw the human tendency to "take '1002, 1004, …' to be going on in the same way as putting down '996, 998, 1000, …'" into the same pile of contingent facts responsible for how they calculate. He elaborated:

> It is a fact that we naturally go on in this way, but people might not have done so. Since they might naturally have followed the rule in a different way, our rules alone do not …guarantee that that they will not be taken or understood in deviant ways […because] a rule is not something which stands apart from our understanding of it …. How we naturally understand and follow the rule determines which applications of it are correct, and the way the rule is followed will depend in part on what we take to be "going on in the same way." (p. 12)

That's the naturalistic response, all right. But doesn't it seem extraordinary to treat "making inferences" and "drawing conclusions" as just more cases of applying a rule? Stroud said that a being such as me "would not be fully intelligible" to humans, that I "would ultimately be unfathomable" (1965, p. 13). My inhumanity as a student boils down to the fact that I can't follow the rule correctly. If humans were different in just the right way in how they took the instruction "add 2," the idea goes, then I could understand them, and they me. But isn't my error as a tortoise somehow deeper than this? Could humans have naturally followed the rules of logic in a different way and still be reasoning correctly? If not, then it is plain why as a tortoise I do not elicit a skeptical response, hence why the naturalistic response is not hard-won.

Penelope Maddy seems to be the first person to observe this elementary distinction. She located a discussion in Wittgenstein's *Remarks on the Foundations of Mathematics* where, she says "we find the interlocutor lodging a line of protest that …doesn't appear in the *Investigations*" (2014, p. 77). The discussion there involves someone drawing inferences other than the ones humans customarily draw rather than someone just finding himself, like me, lacking sufficient grounds to draw any inference at all, but the point is the same: "But isn't there a truth corresponding to logical inference? Isn't it true that this follows from that?" Wittgenstein asked. "What should happen if we made a different inference—how should we get in conflict with the truth?" (1933–44, §5).

To illustrate, Maddy considers someone adopting the inference patterns associated with "tonk," a connective introduced just like the disjunction and eliminated just like the conjunction. This person and the wayward student both "come into conflict with the truth," Maddy says, but in different ways. The student continues skip counting to 1004 "in the context of" the human practice of following the rule, "where it's incorrect." The deviant logician, on the other hand, is "following the tonk rule correctly; it's just a bad rule" (p. 78).

We've seen how the naturalist tries to avoid dealing with me. When I demand an explanation for what I've done wrong by writing down 1004, this figure just points back to the fact that I was at 1000 and instructed to add 2, refusing to acknowledge any deeper reason why 1002 rather than 1004 is correct (insisting, in fact, that any such acknowledgement is a step into the philosophical arena leading to regress or skepticism). My demand for an explanation for what I've done wrong by not finding my way to 'It's green" from "Either it's red or it's green, and it's not red" could be met with a like-minded response: just repeat the premises, maybe with some emphasis or exasperation. This appears to be both Stroud's and Winch's attitude, but it is even less satisfying in this case than in the case of arbitrary rule-following.

Maddy offers a more direct engagement. Calling some rules good and others bad acknowledges something making them so. Couldn't an explanation of my alleged error point to this something?

27.5 Science

Decades before I appeared in Carroll's essay, I snuck my way into the margins of Bernard Bolzano's *Wissenschaftslehre*, "tempting" him to count the rule according to which a valid inference is drawn among the logical grounds of that inference's conclusion (1837). I have always marveled at Bolzano's attitude about this. Rather than claim that my suggestion is inhuman or that it somehow reveals that I harbor a deep misunderstanding of what logical inference is, he wrote that the ensuing regress (together with what I can only imagine is a theological assumption that the grounds for any truth have to be finite in number) "forces" him to accept that A and B (returning to our example) are the complete grounds of the truth of Z (§199).[4]

---

[4] Neither Carroll's original geometrical example nor our substitute instance of disjunctive syllogism accurately illustrate Bolzano's program. According to Bolzano, the ground/consequent relation is anti-symmetric and analytic, so the grounds of compound truths are always simpler than those truths. The author is apparently unaware of this central feature of Bolzano's logical theory.

I think the reason behind Bolzano's reluctance to acknowledge that nothing can be said in favor of Z over and above A and B is that doing so seems to leave unanswered any question about what makes this particular inference correct. That's why it was a surprise to him to learn that A and B are the entire ground for Z. Of course, I didn't mean to tip Bolzano off to any such discovery. I meant for him to realize that even if "It's green" follows from "It's either red or it's green" and "It's not red," the reason it does must have something to do with the fact that in our world, when a disjunction is true and one of its disjuncts is false, the other disjunct must be true. That's what Bolzano called the "temptation" to think that "It's either red or it's green" and "It's not red" cannot be the full grounds of "It's green." It's what I call plain sense.

How can a card-carrying naturalist deal with this question?

After rehearsing what he called "perhaps the best skepticism-rebutting argument in favor of the existence of body," namely, that the hypothesis of an external world is part of the most plausible available explanation of our memory and experience, Strawson wrote:

> We accept or believe the scientific theories (when we do) just because we believe they supply the best available explanations of the phenomena they deal with. That is our reason for accepting them. But no one accepts the existence of the physical world *because* it supplies the best available explanation [of anything]. That is no one's reason for accepting it. Anyone who claimed it was his reason would be pretending. ...Similarly, the best argument against other-minds skepticism is, probably, that, given the non-uniqueness of one's physical constitution ...it is in the highest degree improbable that .... But, again, this is no one's reason for believing in the existence of other minds, of other people .... We simply react to others as to other ...people. (pp. 20–21)

A naturalistic defense of ordinary logical inference in this vein could go like this: Perhaps we infer "It's green" from "It's either red or it's green" and "It's not red" because our habits of reasoning have been shaped by features of the world in which our species evolved and in which we as individuals matured. So the reason it's green, whenever it's either red or green and it's not red, is that the world we navigate is one with certain regularities and the like. Those features are, perhaps, both the reason the inference is correct and the reason why it's the inference we draw. But it is no one's reason for inferring as they do. The reason each of us infer that it's green is just that it's either red or green and it's not red.

Bolzano, though, was not after agents' reasons for inferring as they do, but ontological grounds for those inferences' correctness. Of course an agent need not be consciously aware of and appeal to such ontological grounds to be justified in her inferences. But Bolzano wasn't interested in an agent's justification. He was interested in the actual reasons for a claim's truth.[5] His retreat to what looks like the naturalist's position—the conviction that "It's either red or it's green" and "It's not red" are all the reasons there are that "It's green"—is based on a realization that what seems like an additional essential fact, the validity of the conditional "If A and B, then Z," cannot play any role.

[5] This distinction between agents' justification and actual reasons underlies Bolzano's emphatic distinction between the grounding relationship under discussion here and the relationship of mere derivability.

Maddy's arrival on this scene could hardly be more eventful. But to make sense of her position, I first need to return to the ditch she dug between the general problem of rule-following and the special problem of the correctness of logical rules. Her first unexpected move is to fill it back in.

Taking up the question, "what makes the tonk rule bad?" that she so carefully teased apart from the question, "what makes '1004' the wrong answer?" Maddy says, "I think Wittgenstein's answer to this question marshals the same considerations as his answer to the [other] question—which is why the two tend to blend together." We already noted Maddy's observation that, whereas it makes sense to say that 1004 is the wrong answer to the instruction "add 2" when you're starting at 1000, it makes no sense to say that "add 2" is the correct rule. Still, she says, what makes 1004 the wrong response to the instruction is that the instruction "is embedded in a practice that requires 1002," and it is perfectly legitimate to ask what makes "add 2" a good rule in that practice:

> The practice of elementary arithmetic is as it is for a range of reasons, among them that we have certain interests and goals in which counting and the arithmetical operations are important and that we find "+2" more natural than a rule that changes to what we call "+4" after 1000. ...There might be nothing particularly disadvantageous about employing an arithmetical rule that changes from our "+2" to our "+4" after 1000, but a rule that calls for random choices or coin flips at each point would not serve the purposes of the practice of arithmetic. (2014, p. 78)

"Add 2" is not the correct rule in the sense that it is the only good one, but it is a good one all the same, one of the rules that facilitates one of our important human activities.

The same sort of considerations are what make the tonk rule bad: "There would be nothing particularly disadvantageous about using an exclusive 'or' in place of our inclusive 'or,' but using 'tonk' defeats the purpose of inferring." That purpose: "to know about the world around us" and to have "a reliable method of moving from truths to truths." Summing up: "*Both* correct application of existing rules *and* the choice between possible rules rest on ...our interests, our nature, and the world's regularities" (2014, p. 78).

How far we've gone while standing in one place! Maddy's inquiry began with her observation that logical inference presents a set of considerations absent from the general case of rule-following. Commentators like Winch and Stroud who missed this distinction suggested that giving in to the temptation to offer an explanation of the correctness of Z that cuts deeper than referral to A and B reveals a misunderstanding of the nature of inference. Spotting it led Maddy, like Bolzano, to find sensible an investigation of what makes some inferential practices good. But that investigation revealed that rules of inference do not in fact differ substantially from other rules. "Our interests, our nature, and the world's regularities" together license some inferences and not others, just as they, in different proportion, explain what sorts of rules comprise our counting and calculating practices. The key distinction prompting Maddy's investigation is exposed by that very investigation as an illusion.

What has not proved illusory, according to Maddy, is the fact that inquiry into the nature of logical truth is both coherent and possible:

> In the logic case, if someone asks why I'm justified in concluding that it's green, given that it's either red or green and it's not red, there are things I can say—"look here, there are only two possibilities, don't you see? and one of them is ruled out"—but in the end, if my conversation partner doesn't get the point, I have nothing more to offer. (2014, p. 102)

So far Maddy is perfectly in line with the naturalist we've become familiar with. She agrees with Strawson that providing reasons outside these ordinary ones would be both ineffective and insincere. It would be ineffective because no one who balks at the inference in the first place can be expected to come around after hearing an account of how or why we came to adopt it. It would be insincere because they're not, after all, our reasons for concluding that it's green. "But the point at issue now," she continues, "is what's open to the philosopher, to the observer of, the commentator on, these various behaviors," and here she feels herself parting ways with the naturalist we've depicted:

> Questions like ["if it's red or green, and it's not red, why must it be green?" or more generally "what is the nature of logical truth?"], traditionally understood as "philosophical," would remain open after Wittgenstein's therapeutic efforts were complete, and ...attempts to answer them would remain legitimate. (2014, p. 110)

But it's wrong to construe this parting of ways as a flight from naturalism. Following Strawson, it seems instead appropriate to describe Maddy as "an exponent of a more thoroughgoing naturalism," someone who "could accept the question, *What causes induce us to believe in the existence of body?*" or *What causes induce us to infer according to the pattern known as disjunctive syllogism?* "as one we may well ask, as one that can be referred to empirical psychology, to the study of infantile development" (1985, p. 12).

Here is a naturalist who keeps old philosophical questions alive by transforming them, sometimes, it seems to me, mutilating them nearly beyond the point of recognition. Someone asks, "Why do you suppose bread will nourish you?" You answer, "Because it has always done so in the past," prompting the obvious follow-up, "But why suppose what has worked in the past will do so again?" The skeptical response is to concede that you have no grounds for this supposition. A Strawsonian response, by contrast, maintains that "the correct way with the professional skeptical doubt" of this sort "is not to attempt to rebut it with argument, but to point out that it is idle, unreal, a pretense; and then the rebutting arguments will appear as equally idle; the reasons produced in those arguments to justify induction ...are not, and do not become, *our* reasons for these beliefs; there is no such thing as *the reasons for which we hold* these beliefs" (pp. 19–20).

Our "thoroughgoing naturalist," though, gleefully and deliberately *misunderstands* the question: "Why do I suppose that the future will be like the past? Why, because my species evolved in and I matured in an environment exhibiting key regularities, so my cognitive apparatus is attuned both to pick up those features and to anticipate like regularity. There is some question still at what stage of my development this sort of anticipation became fully formed, how much and in what ways it is determined by my neurological structure as opposed to experience ...." It won't help to point out to these types that they aren't answering the question you're asking. I know. I've tried.

27.6 Second Philosophy

Let's not get too down about Maddy, though. Here finally is someone who finds it perfectly legitimate to understand the question "Why must it be green?" so that the coy "Because it's either red or green, and it's not red" won't do for an answer.

Maddy takes her lead from Robert Fogelin, who balked at Wittgenstein's suggestion that "the desire to find physiological explanations of psychological phenomena" was some sort of prejudice of the age of science. Fogelin said such phenomena "just seem to be of the wrong *order* to be simply brute or inexplicable." He pointed out how absurd it would be to adopt a similar attitude about rain: "Imagine someone saying 'it just rains, that's all; explanation has to stop somewhere'" (1987, p. 207). Similarly, according to Maddy, someone might ask for a justification of Z—be prepared to take A and B as sufficient grounds or not—but in any case to see them as the right sort of candidate grounds. Someone else, though, can ask for a justification of Z *knowing* that A and B are its alleged grounds (perhaps even convinced that they are), and instead be soliciting an explanation of what makes them so, of the validity of the inference from A and B to Z. This is the sort of insistence that Carroll mocked but Bolzano took seriously. Maddy doesn't just side with Bolzano. She's even more resilient in her inquiry.

So how does she do it? Not directly, of course, but with a "device," a "character," a "sort of idealized inquirer" she dubs the Second Philosopher (2007, p. 1). Such is my fate. When I'm not cast in some such role myself, I'm represented in negative by someone who is. David Pears, incidentally, has attributed common misreadings of Wittgenstein to a tendency to "mistake the play within the play for the main drama" (1983, p. 12). Never having been welcome in the main drama, it's hard for me to see these as misreadings. Nor can I much appreciate the hand-wringing over questions like whether Wittgenstein's disdain for science is really directed at science per-se or merely at the idea that science will fill, rather than supplant, the role that philosophy used to play. The *Investigations* are chock full of appeals to science ("What is the relationship between name and thing named?"—"Well, what *is* it? Look at [this example] or at another one. There you will see ...." §37) in deliberate, even comical misunderstandings, clinging to the idea that empirical answers are the right sort to offer to philosophical questions. If the play-bill instructs you about when to take these comments seriously, I suppose part of the joke is my naivety.

Still, I can't help thinking that my very distance from the audience, and from Maddy for that matter, draws me especially close to the Second Philosopher. Even with no opinion about whether what she has to offer ought really be called philosophy, or whether Wittgenstein or anyone else would accept it as such, I think I can appreciate what this character has to say about logic.

Which brings us to Maddy's second big surprise: "Why must it be green?" Second Philosopher: "Actually, it needn't be."

Now, this is just what I've been saying all along! But before we get ahead of ourselves, we need to understand what sort of reply this is. The Second Philosopher still faults me for failing to infer that it's green from it being either red or green and not red. She just observes that on her account, Z follows from A and B only contingently, so that there's no sense in which it "must" so follow: "the force of the logical must is compromised" (2014, p. 31).

Maddy describes several routes to this view of logic, but her principal reference point is the framework of Kant's transcendental idealism. On Kant's view, she says, "the truths of logic can be understood empirically," as depending on very general structural features of the world, "but the force of the logical must [can be] explained only transcendentally" (2014, p. 13). That explanation: Logic is true of the world because of its most general structural features. Humans believe logical truths because no other sort of discursive, conceptual thinking is possible. But the world we experience is a projection of such constraints on our thinking. No wonder, then, that the world would be structured so as to support logical inference (2007, p. 223).

The Second Philosopher wants to preserve a kernel of this Kantian picture, the ideas that (1) the truths of logic depend on general structural features of the world and that (2) the human belief in logic can be accounted for in terms of "primitive cognitive mechanisms." But she's not at all satisfied with Kant's "transcendental" move of explaining (1) with (2). There is, she claims, just no evidence that the world is structured as it is *because* we think the way we do, compared with considerable evidence of a converse explanation: (3) "human beings are so configured cognitively because they live in a world that is so structured physically" (p. 226).

The casualties to logic of this view are legion. To begin, as already noted, the "force of the logical must" gives way immediately. Kant's transcendental argument supplied us with an explanation of logical truths' necessity. Its second-philosophical converse tells us why "we tend to overlook the contingency of rudimentary logic" (p. 273). In place of a picture of why logic is necessary, we get a picture of why we erroneously think it is. Z follows from A and B, though it needn't do so—despite the fact that humans cannot readily conceive of how it could fail.

Worse, a little second-philosophical poking around uncovers more cracks in the Kantian scheme. The weirdness of the quantum world, for example, makes it plain that necessity gives way, not to "always, but only contingently so," but to "in many but not all cases," for the truth of logical principles depends on the world exhibiting certain structures, which it does "in many but not all respects" (p. 244). The link between (1) and (2) has to be modified to (3'): Humans' primitive cognitive mechanisms are the way they are because they live in a world by and large appropriately structured for that way of thinking to come out right and they interact almost exclusively with its structures of that sort (p. 245). Z needn't follow from A and B, and in fact *doesn't* when "it" is a photon and terms like "red" and "green" are replaced with "horizontally polarized" and "vertically polarized."

Finally, some logical principles that the Second Philosopher urges us to accept fail fairly consistently, even in the ordinary world that we habitually navigate. Maddy cites the principle of excluded middle and the material conditional analysis of if-then claims. These are "idealizations" no different than the friction-less planes evoked by physicists.[6] They are "admittedly false," but still effective "in a range of applications" and worth adopting for that reason alone (p. 289). "Just as in the rest of science," the Second Philosopher maintains, "we must be sensitive to the benefits and dangers of idealization, and satisfy ourselves that the idealization is appropriate to the case at hand" (p. 287).

The putative benefits recommending these idealizations are that they lend simplicity to the overall framework of reasoning (via truth-functionality at the propositional level) (pp. 285, 292), that they are familiar to us (p. 288), that they appreciably strengthen our inferential resources (p. 288), and that they usher in meta-theoretic niceties like the Deduction Theorem (p. 287). It's possible to contest these individually: Much less than classical logic is needed for the Deduction Theorem to hold; the material conditional strikes very few people as familiar from their days of untutored reasoning; etc. But let that not distract us from the view itself. The propriety of laws of logic lies not in their necessity, nor even in their universal applicability. Few, if any, are applicable in all contexts, and some are more widely apt than others. Even when a logical law is applicable, though, this need not be due to it being literally true. Maddy cites approvingly H. G. Wells's observation of the mismatch between logical validity and objective truth: "In my way of thinking, relentless logic is only another name for stupidity—for a sort of intellectual pigheadedness. If you push ...enquiry through a series of valid syllogisms—never committing any generally recognized fallacy—you nevertheless leave behind you at each step a ...marginal loss of objective truth and you get deflections that are difficult to trace, at each phase in the process (1908)."

The combined force of these points suggests that even as the Second Philosopher distinguishes "rudimentary logic" from classical logic, they differ from one another only in the degree to which, and the frequency in which, their principles are true and, what's not the same thing, in the range of their application. So what exactly is my error if I hesitate to infer Z from A and B? Other than the unhelpful reminder that I've accepted there being only two possibilities and ruled one of them out, the Second Philosopher has no all-purpose answer. She will look at the details of the case. Some worldly structures may be in place to underwrite the inference. Should they be absent, the inference may yet be effective. In any case, everyone draws it as a matter of course. Usually, that's a good sign, quantum weirdness aside. Just don't get carried away!

---

[6] Maddy does not point out that adding the law of excluded middle to the framework of basic inferences presumably licensed by her rudimentary logic makes "if A, then B" and "either B or not A" interderivable, so that really only one "idealization" separates classical logic from rudimentary logic. Instead she wages a war on two fronts, contending with issues of vagueness in her defense of LEM (pp. 285–91) and navigating the "paradoxes of the material conditional" separately (pp. 291–96). Possibly this is due to one direction of the interderivability depending on the *ex falso quodlibet* rule, which figures prominently in her discussion of the material conditional. It is unclear whether *ex falso quodlibet* is part of rudimentary logic, whose description Maddy deliberately leaves vague (p. 290). On the other hand, as she points out herself in 2012 (p. 484), Maddy's flagship example of rudimentary inference, disjunctive syllogism, is not derivable (though it is admissible) in the bare inferentialist framework of the the minimal calculus—its validity rests on *ex falso quodlibet*.

I think it's time to learn a little more about the Second Philosopher and how she arrived at such iconoclasm. Her name suggests that she's a philosopher who has caught a second wind. Perhaps after some initial exhaustion from the repeated fizzling out of her ambitions at Cartesian styled first-philosophy, she has, rather than simply thrown in the towel, re-tooled with methods continuous with science.

Maddy emphasizes that this is not the case—and that, although her position resists "outright definition," we should get to know her by her ways (2007, p. 1). So far we've seen her reverse a key argument of Kant's in light of the lack of evidence for his "transcendental idealism" and the extensive evidence for its converse. It turns out that several features of this one move are indicative of the Second Philosopher's style: Although she has no working definition of what counts as science or active bias towards science over other topics or pursuits, she's looking for evidence for claims of all sorts—the same sort of evidence, in fact—for what would commonly be recognized as a scientific claim as for what is thought of as philosophical or logical. Moreover, she finds obvious attractions in any number of schemes, as exemplified by the "considerable appeal" she sees in the central facet of Kant's account of logic. So she willingly sets out to see how much of those schemes can be salvaged under scrutiny, even if their origins are, by her lights, arcane. The Second Philosopher takes up questions of meaning, truth, and all the rest "under the low and leaden sky," not as a consolation prize, but because nothing has ever made "transcendental," "ultimate," or *a priori* perspectives appealing to her.

What really brings out her true colors, though, are her reactions to attempts to lure her out of these tried and true ways. Recall Strawson's description of the thorough-going naturalist, the inquirer who "could accept the question" *What causes induce us to infer according to the pattern known as disjunctive syllogism?* "as one we may well ask." In the remainder of that passage, paraphrased to fit our logical example, Strawson said this inquirer would remand that question to psychology and the study of child development, "but would do so in the justified expectation that answers to it would in fact take for granted" the validity of disjunctive syllogism (1985, p. 12).

This is precisely the Second Philosopher's tack: "She acknowledges …that she can't justify anything without using the methods of inquiry she's developed for that purpose; the very suggestion that she attempt this seems wrong-headed to her" (2007, p. 32). But given her provisional attitude towards the validity of such logical principles, one might expect a distinctive response to allegations that her inquiry is fueled by the very principles that it's meant to account for. After all, the outcome of her inquiry is that these principles are, despite all appearances, only contingently true and even fallible.

But the Second Philosopher is unmoved. "Our Second Philosopher freely acknowledges one poignant aspect of the human condition: we can't step outside our system of beliefs and methods and justify them from an external perspective; the only perspective we can occupy is our own" (2007, p. 35). Never mind that her appraisal of logic falls short of what ordinarily is thought of as justification! "Her study of various methods of reasoning …are all [conducted] relentlessly 'from the inside'" (2007, p. 32).

Maddy lets us see up close how this relentlessness colors the second philosophy of logic. The first vignette occurs when the Second Philosopher "rests with her conclusion that the world has, at least, a distinctive structure of individual objects" of the sort that rudimentary logic depends on:

> Here some will object on grounds of circularity—we use our beliefs about bounded, cohesive, spatiotemporally continuous objects to confirm that the world is populated with such objects .... We noted [before][7] that this objection often betrays the objector's underlying wish for a "higher," extra-scientific justification of science, so that science can be preferred to pseudoscience ...on some neutral grounds. But the Second Philosopher finds no footing outside her own methods; she simply argues that astrology and creationism are wrong, on her own terms, unimpressed by the reply that they could run a parallel argument against her. She knows what's wrong with that argument, too! (2007, p. 235)

But, one wants to object, what else could you expect to discover, having taken for granted the very thing you've verified! "The Second Philosopher's justifications are not circular in this sense," Maddy replies. The discoveries about the quantum world are enough to illustrate "that their success is not a foregone conclusion" (2007, p. 235).

Strawson, again, anticipated the ensuing dynamic:

> Our inescapable natural commitment is to a general frame of belief and to a general style ...of belief-formation. But *within* that frame and style, the requirement ...that our beliefs should form a consistent and coherent system, may be given full play. Thus, for example, though Hume did not think that a rational justification of induction in general was either necessary or possible, he could quite consistently proceed to frame "rules for judging cause and effect." Though it is Nature which commits us to inductive belief-formation in general, it is Reason which leads us to refine and elaborate our inductive canons and procedures and, in their light, to criticize, and sometimes to reject, what in detail we find ourselves naturally inclined to believe. (1985, p. 14)

An obvious concern, though, in the case of logic, is that in our attempts to distinguish cases where idealization is innocent from cases where it isn't we will be susceptible to the very deceptions of idealization that we are trying to harness. This sets the stage for a second vignette. After reminding us that "we must come to terms with the fact that even this rudimentary logic cannot be assumed to apply effectively in every situation," Maddy presents the Second Philosopher's project of validating disjunctive syllogism in a specific situation: "The last card in her hand is either a club or a heart; it can't be a heart (because Joe played the last one); so she's holding a club." How do we know that this inference is valid, that nothing akin to quantum phenomena are at play? "The cards and bridge players here are objective individuals; each card has a rank and suit; each card either has been played or remains in one and only one player's hand; and so on. The world of the bridge game is [the sort in which] disjunctive syllogism is valid" (2007, p. 272).

But doesn't this "defense" of disjunctive syllogism make use of disjunctive syllogism? In reply to this question, Maddy just suggests we observe that "the seemingly transparent soundness proof can be carried out only because the necessary

[7] See 2007, p. 108.

objects, properties, relations, dependencies, are available in this context, ...there to be exploited." But isn't it an assumption that they are there?—Not at all, it's a discovery!—But isn't that discovery aided by disjunctive syllogism as well, to say nothing of other fallible principles and idealizations that might or might not mislead in this task?—Indeed, and the kitchen sink!—Then how will you ever know that you haven't been misled?—"All this can be disheartening to the philosopher in search of certainty, but [I] never undertook to philosophize from a point of view more secure than that of science" (2007, p. 299).

Maddy sizes the Second Philosopher up pretty well: "Her equanimity in the face of this admission may frustrate us" (2007, p. 35).

27.7 Mathematics

Wittgenstein had abandoned philosophy after writing the *Logisch-Philosophische Abhandlung* and for years resisted others' efforts to rekindle his interest when Friedrich Waismann and Herbert Feigl "managed to coax Wittgenstein, after much resistance, to join [them] in attending the lecture" of the Dutch topologist Luitzen Egbertus Jan Brouwer. The lecture, "Mathematics, Science, and Language," was delivered in Vienna in Brouwer (1928). According the Feigl, "a great event took place" just afterwards:

> Suddenly and very volubly Wittgenstein began talking philosophy—at great length. Perhaps this was the turning point, for ever since that time, 1929, when he moved to Cambridge University Wittgenstein was a philosopher again. (*Feigl* 1981, p. 64)

Feigl later wrote in a letter to George Pitcher that the evening "marked the return of Wittgenstein to strong philosophical interests and activities" as he immediately "began sketching ideas that were the beginnings of his later writings" (*Pitcher* 1964, p. 8n).

Also attending Brouwer's lecture were such luminaries as Kurt Gödel and Rudolf Carnap. And although there is little doubt about Lewis Carroll's considerable direct influence on Wittgenstein's late thought, I'm convinced that my appearance that evening is what led to my role in the *Investigations*.

Brouwer described language as a natural development in the social history of the human species, organized in order to facilitate increasingly complex tasks. To set the frame for a genetic analysis of language, he contemplated a simple social order organized with the help of the first forms of linguistic behavior: "At primitive levels of culture and in primitive relationships, the transmission of the will is achieved by a simple gesture; and here the cry is particularly effective." This scene, of course, would later be echoed in the opening sections of the *Investigations*. So too would Brouwer's continuation: "But in matters belonging to the organization of a higher human community the tasks to be imposed are too various and too complicated to be brought about by simple cries," leading to more elaborate forms of communication (§I.3).

As an outgrowth of human social behavior, language by its nature does not simply or straightforwardly encode concepts. Uprooted from its context, an utterance or inscription has no meaning. It's only in its social role, among humans who share a natural history and set of purposes, that language functions.

Naturally, this leaves me in a predicament of being unable to interpret occasional instructions thrown my way, when these instructions' natural homes are contexts foreign to my own experience and training. "For transmission of the will mediated by speech," Brouwer declared, "there exists neither exactness nor certainty" (§II). Therefore such bits of language will be open to multiple interpretations, with nothing inherent in the language favoring one among them.[8] And crucially, "for pure mathematics as well there exists no certain language, i.e. no language that excludes misunderstandings in conversation and, when it is being used to prop up the memory, protects against errors ...." So even simple instructions like "add 2" will be subject to misunderstanding.

By now we are familiar with the wide range of potential correctives one might propose for this situation: an algebraic formula, a mental image, etc. Brouwer focussed on one, which he associated with a prominent research program in the foundations of mathematics: "This circumstance cannot be remedied, as the formalist school attempts, by submitting mathematical language itself to a mathematical contemplation, revising it ...and then coming to an understanding about it in a language of the second order, a metalanguage." After all, if the new-fangled metalanguage is to function at all, it must to that extent figure into the social practice of a human community, so though it may "indeed guard against misunderstandings and errors with great probability ..., in keeping with the essence of language it cannot do so with absolute certainty" (§II).

Far from leading Brouwer to skeptical conclusions or despair, this realization prompted a familiar naturalistic pronouncement: "On the other hand stability and exactness of language is not necessary in practice because people are drilled by a common will to an automatism of understanding incomplete sentences." There can only be a lack where there is a need, and humans, sharing as they do an evolutionary and social history together with like training, habituation, and community, need no such exactness, or "completion," in their language. Their every utterance is made possible by, and therefore is informed by, this elaborate backdrop.

The right conclusion to draw from my inability to catch on, therefore, is not that something is defective with language, but, alas, that something is defective with me—specifically, that I am not part of the same world as those for whom the instruction is meaningful. The idea that something could be supplied that would guide me to the right interpretation is the same as the idea that there is something guiding ordinary humans to the right interpretation. Brouwer called it "the false belief in a magic reach of language ..., a reach that exceeds its character as a means for transmission of the will" (§II).

---

[8] I am unsure just what the author has in mind, but Pitcher presents some evidence of influence in (1965).

If the image coming into focus here evokes Wittgenstein's dismissal of his early view of language—as a "picture linked with, reaching up to reality" (1921, §2.1511), as one in which "naming appears as a queer connection of a word with an object" and "as an occult process" (§34)—in favor of considering language as "part of the natural history of human beings" (§415) and "seeing that nothing out of the ordinary is involved" (§94), consider that Brouwer even fielded an objection based on the possibility of a "private language":

> Language is therefore absolutely a function of the activity of social man. Even if the individual man in total isolation uses language to prop up his memory, this is only because he must take the sciences and the organization of the community into consideration. (§I.3)[9]

Pressing further, Brouwer diagnosed the misunderstanding of language just described as "the natural consequence of a much older, more primal, more consequential and more deeply rooted error, namely, the reckless trust in classical logic" (§II). To explain both what he meant by trusting logic and in what sense doing so is careless, Brouwer turned to etiology:

> Already in antiquity man possessed a very perfect language (i.e. one that practically excluded misunderstandings) for the mathematical contemplation of finite groups of things of the objective spatio-temporal world. For this language there are certain forms of transition from correct statements to other correct statements; these forms of transition ...were gathered together under the name of logical principles. (§II)

These "principles proved themselves," Brouwer said, as the statements derived from them turned up time and time again to be independently verified. "And then one came to trust statements that had been derived by means of the logical principles even when they were not susceptible to direct check" (§II).

The suggestion here is that our "trust" is out of line because such statements might in fact be false. Brouwer did caution against this, but in fact it is just a consequence of a more general, and insidious, error. Our "trust" in logical principles leads us to see logic as something sublime when in fact it is mundane. "The practical reliability of the logical principles," at least when they are reliable, "rests on the fact that much of the world of perception, in its finite organization" exhibits a certain "loyalty and contentedness" (§II).

Of course we do not tend to think that its being green because it's either red or green and not red depends in any way on contingent features of the world, much less the fact that we are considering it only in finite swaths. We think the inference is inevitable, conceptual, necessarily secure in a way that empirical principles cannot be. Brouwer has an explanation for this fact about us, one that hooks into his more general account of language: "Man has for ages been blind to this sober interpretation because he did not recognize that words are nothing but a means of transmission of the will, and regarded them [instead] as a means of indicating fetish-like concepts." As they are "supposed to represent *a priori* laws governing the concepts and their

---

[9] This sentence appears only in the 1933 Dutch version of the lecture, translated by Heyting in *Brouwer* 1975.

linkages," logical principles do not strike us as devices whose reliability could depend on features of the world, however ubiquitous (§II).

Now, the Second Philosopher disagrees with some aspects of this account. Humans feel the force of the logical must, not because they are basically prone to superstition, but because evolutionary pressure has attuned them to those features of the world where logic applies and because their cognitive development occurred almost exclusively in this environment. Finitude is not obviously the relevant feature for the applicability of rudimentary or even classical logical principles. And it's not at all clear that general misconceptions about language figure into the illusion of logic as somehow special, even if on this point Wittgenstein might agree with Brouwer.

But Brouwer's core account, his "sober interpretation," is the Second Philosopher's own. The truths of logic depend on the world exhibiting certain very general regularities, which by and large it does, though not always. Our belief in logical truths, even our mistaken view that they are true universally and independently of any facts about the world, is due to our habits of thinking, themselves contingent facts about us. And these habits of thinking are ours because in our natural history of navigating the world we interacted nearly exclusively with those structures that allow our logical principles to "prove themselves."

The Second Philosopher agrees with Brouwer, too, that the characteristic laws of classical logic fare worse in this scheme than a set of more rudimentary principles. Sometimes Brouwer is mistaken for an advocate of a traditional view about "intuitionistic" logical principles—as holding that just these are *a priori* truths, constitutive of reasoning, whereas the validity of LEM is a contingent matter. We've seen that he actually held that no principles of logic are true *a priori*, that their appearance as necessities of thought is an illusion, and that at most such principles can be "practically reliable." Still, an unmistakable thread in Brouwer's polemic is the insistence that LEM's practical reliability is extremely limited compared to that of the intuitionistic laws—precisely the Second Philosopher's assessment.

Then these two part ways. Faced with the realization that LEM isn't literally true beyond a limited range of applications, Brouwer wondered whether it still functions, "in the mathematics of the infinite, as a practically reliable means of transition between constructions." Famously, he concluded that it cannot. Because logic, as language, has no inherent meaning and serves only to transmit a mathematician's will, the logical derivation of a statement is a promissory note that a direct mental construction can be performed. LEM, "although it is not *correct*, nevertheless, if one presupposes it exclusively for finite species of properties, is *consistent*." Statements derivable with LEM from verifiable truths about infinite assemblages, on the other hand, are among those that are not susceptible to direct verification of this sort. As a consequence, Brouwer advocated for the "collapse" of "considerable portions of the previous mathematical edifice" (§III).

Faced with the same realization, the Second Philosopher goes to the other extreme. After all, LEM is an idealization. "The debate isn't over [its] truth or falsity ...but over its effectiveness ...in a range of applications" (2007, p. 289). Brouwer offers one proposal for what it would be for a logical principle to be effective in mathematical reasoning. But his focus on constructive verification and mathematical contemplation strikes the Second Philosopher as dogmatic. As every application has its own purpose, no one criterion could determine whether a tool is effective in carrying it out. We need to find out what the study of infinite aggregation is good for—something unlikely to be vividly present outside the details of set theory itself—and determine from there whether LEM or other principles facilitate or hinder that goal. She concludes that classical logic and much more, even if misleading and inappropriate in other applications, are particularly appropriate in this arena.

And this is what I'm referring to as my presence at Brouwer's lecture.[10] Fielding my challenges about simple arithmetical operations like "add 2" and basic logical inferences like disjunctive syllogism, Brouwer sounds at first just like the Second Philosopher: "Why must these operations yield these conclusions?"—"They needn't. The 'must' is illusory. But this is still the right way of calculating, the right way of inferring, for these rickety, creaky, *contingent* reasons." Brouwer managed to close the door—even to board up the windows—to keep my worries about getting past 1000 out of his laboratory. But he used some ceiling planks to do so, leaving a space for me to sneak back in just above ω and $\aleph_0$. For his account of the laws of logic and mathematics is quickly blown up into a semantic theory, underpinning those laws' applicability and ready to be deployed against their more extravagant uses.

With the Second Philosopher, meanwhile, we have a naturalist who sticks to her guns. No more than the schoolmarm owes me any special account of my calculation errors beyond an appeal to the very practice that I'm failing to grasp, the set-theorist owes nothing to one who would see her edifice collapse because of some antecedently drawn-up account of what makes her methods legitimate.

Now we saw the Second Philosopher flag the illegitimacy of classical logic, just in quantum physics rather than in set theory. But notice that her rejection of LEM has the opposite lilt of Brouwer's: Rather than use theories about the role of logic to evaluate the special sciences as they are practiced, she turns to those special sciences as testing grounds for theoretical accounts of logic, happy to watch those accounts deflate and fail. She doesn't expect the same phenomena that underwrite the legitimacy of LEM in the course of ordinary reasoning to play the same role in the study of infinite aggregates or to figure into an account of what goes wrong in applications of LEM to photon polarization. Her critical evaluation of LEM is shot through with set theory and quantum physics, as the occasion demands. Criticizing these disciplines based on a preconception about what role LEM could ever play in them is exactly backwards!

27.8 The Last Word

I fear I've now reached the time at which we are most gifted with prophecy, and what I have collected over all these years are questions, not answers. They will be

[10] This misunderstanding seems to lie behind the popular, but historically strained, opinion that Wittgenstein's return to philosophy was prompted, not by an affinity with anything Brouwer was saying, but by an aversion to it. Details about this and a related opinion can be found in (2003).

recognizable from my most recent encounter, but each carries the weight of history, which is why I've prolonged my tale. If logic isn't necessarily true, why pursue the development of its theory? Why even call it logic, or, for that matter, anything at all, if it differs from ordinary physical laws only in its breadth of applicability and generality? Why assume, unlike Brouwer, that LEM applies to infinite sets, if you agree with him that it's a mere idealization?

For some time I suspected that the Second Philosopher could never satisfactorily answer these and other questions. Her recorded attempts all seem circular and her response to being called out for that inscrutable.

But it's hard to sustain suspicion about what someone just owns up to. Here's Maddy wrapping up her introduction of the Second Philosopher: "Finally, of course, none of this amounts to an argument that we should all strive to conduct ourselves as Second Philosophers. My hope is that the appeal of the approach will be obvious to the susceptible" (2007, p. 3).

"Where does that leave the unsusceptible?" I often wondered. Now I'm beginning to understand. The Talmud (Pesachim 110b) says that demons prey on whoever takes measures to guard themselves from them, making those provisions the more urgent. People who don't give them a thought are left alone. And so it is with me. I torment just those who take precautions against me. For everyone else, I simply don't exist.

I can't enroll in your class on my own. But how easily, until now, I've always obtained instructors' permission! Don't you see that your provisional account can never explain how logic, "viewed all at once," hooks up with the world, "as it is in itself?" That your defense of disjunctive syllogism rests on that very principle? That your use of LEM to settle questions about sets already presumes their answers? That your explanation of *this* feature of logic and your justification of *that* logical inference are rough-and-ready and for that reason not getting to the heart of the matter?

—"Granted, it isn't the sort of explanation/justification that would persuade in all possible situations ...and thus not the sort of thing to satisfy the traditional philosopher, but that's as it should be" (*Maddy* 2014, p. 119).

References

Bolzano, B. (1837). *Wissenschaftslehre. Versuch einer ausführlichen und grössenteils neuen Darstellung der Logik mit steter Rücksicht auf deren bisherigen Bearbeiter.* Sulzbach: Seidel. Translated by B. Terrel in B. Bolzano *Theory of Science*, J. Berg (Ed.) 1973. Boston: D. Reidel Publishing Company.

Brouwer, L. E. J. (1928). Mathematics, science, and language. *Ewald, 1996*, 1175–85.

Brouwer, L. E. J. (1975). *Collected works 1. Philosophy and foundations of Mathematics*. A. Heyting (Ed.). Amsterdam: North-Holland.

Carroll, L. (1895). What the tortoise said to Achilles. *Mind*, pp. 278–80.

Ewald, W. (1996). *From Kant to Hilbert: A sourcebook in the foundations of Mathematics* (Vol. II). New York: Oxford University Press.

Feigl, H. (1981). The Wiener Kreis in America. In R. S. Cohen (Ed.), *Inquiries and provocations: selected writings 1929–1974* (pp. 57–94). Dordrecht: D. Reidel.

Fogelin, R. (1987). *Wittgenstein* (2nd ed.). London: Routledge.
Goldfarb, W. (1985). Kripke on Wittgenstein on rules. *The Journal of Philosophy, 82*(9), 471–88.
Kripke, S. (1982). *Wittgenstein on rules and private language*. Cambridge: Harvard University Press.
Maddy, P. (2007). *Second philosophy: A naturalistic method*. New York: Oxford university Press.
Maddy, P. (2012). The philosophy of logic. *The Bulletin of Symbolic Logic, 18*(4), 481–504.
Maddy, P. (2014). *The logical must: Wittgenstein on logic*. New York: Oxford University Press.
Marion, M. (2003). Wittgenstein and Brouwer. *Synthese, 137*, 103–27.
Malcolm, N. (2001). *Wittgenstein: A memoir* (2nd ed.). New York: Oxford University Press.
Pears, D. (1983). Man as the measure (review of Kripke 1982). *The London Review of Books, 5*(15), 12–13.
Pitcher, G. (1964). *The philosophy of Wittgenstein*. Englewood Cliffs, New Jersey: Prentice-Hall.
Pitcher, G. (1965). Wittgenstein, nonsense, and Lewis Carroll. *The Massachusetts Review, 6*(3), 591–611.
Strawson, P. F. (1985). *Skepticism and naturalism: Some varieties*. New York: Columbia University Press.
Stroud, B. (1965). Wittgenstein and logical necessity. *Philosophical Review, 74*. Reprinted in *Stroud 2000*.
Stroud, B. (1979). Inference, belief, and understanding. *Mind, 88*. Reprinted in *Stroud 2000*.
Stroud, B. (1991). The background of thought. In E. LePore and R. Van Guleck (Eds.) *Searle and his critics*. Blackwell, Oxford. Reprinted in *Stroud 2000*.
Stroud, B. (2000). *Meaning, understanding, and practice*. New York: Oxford University Press.
Winch, P. (1958). *The idea of a social sicence and its relation to philosophy*. 2nd Edition published in 1990. Routledge.
Wittgenstein, L. (1921). *Logisch-Philosophische Abhandlung. Annalen der Naturphilosophie, 14*.
Wittgenstein, L. (1933–44). *Remarks on the foundations of Mathematics*, revised edition. Translated by G. E. M. Anscombe. G. H. von Wright, R. Rhees, and G. E. M. Anscombe (Eds.). Cambridge, MA: MIT Press. 1978.
Wittgenstein, L. (1953). *Philosophische Untersuchungen*. Translated by G. E. M. Anscombe as *Philosophical investigations*. Second Edition (1958). Walden, Massachussetts: Blackwell.
Wells, H. G. (1908). *First and last things: A confession of faith and a rule of life*. New York: G. B. Putnam's Sons.

Chapter 28
Reply to Franks

Penelope Maddy

Keywords Naturalism · Second philosophy · Wittgenstein · Skepticism · Talmud

Professor Franks's Wayward Student grasps the second-philosophical approach so sensitively and so delightfully that I'm left with nothing to do but express my appreciation. Though his philosophizing doesn't consist entirely of jokes, as Wittgenstein proposed, it certainly does generate a few good ones. Dragged as he is, against his will, into a series of high-stakes debates, the Wayward Student's very being poses its own series of stark challenges: meanings that don't rely on any contingencies, logic that holds in all possible worlds, justification of LEM securely grounded in a prior account of the metaphysical subject matter of mathematics—to which list we might add the radical skeptic's demand for a form of evidence whose effectiveness doesn't depend on what the actual world is like. In response, the Second Philosopher persists in offering ordinary observations of what works in practice, in subordinating the metaphysics of mathematics to its particular methods and goals, in appealing to empirical theories of the world and our cognition. Can't she see that these ephemera aren't even answers of the right sort?! Then, suddenly, at the very end, the Wayward Student comes to see the whole matter from entirely the other way 'round. Drawing inspiration from the Talmud, he wonders: perhaps insisting on the Right Sort of Answer is the Big Mistake that sets us out on the wrong path in the first place! To which I can only respond: Amen.

P. Maddy (✉)
Department of Logic and Philosophy of Science, University of California, Irvine, CA, USA
e-mail: pjmaddy@uci.edu

S. Arbeiter and J. Kennedy (eds.), *The Philosophy of Penelope Maddy*,
Outstanding Contributions to Logic 31, https://doi.org/10.1007/978-3-031-58425-1_28

Part IV
Mathematical Practice

Chapter 29
Realizing Infinity

Maryanthe Malliaris and Assaf Peretz

Abstract What happens when mathematics realizes infinity. When are mathematical definitions actually useful?

Keywords Logic · Set theory · Model theory · Infinity and beyond · Definitions in mathematics · Creativity in mathematics

I

Many people find reading fiction beneficial to their life. This is depicted in "The Lion, The Witch And The Wardrobe" where we are shown how visiting a possibly fictitious land behind the wardrobe tremendously improves the visitors' lives. One can ask why are these things useful in people's lives. One can also hypothesize whether the very distinction between reality and fiction is closer to a fictitious one than to a real one; more than likely a question of scale rather than any absolute distinction.

II

Mathematics has a unique place in our lives. The place of complete precision—at least as far as we can see. This precision is powerful. In mathematics we don't describe some triangle in the sand, but an idealized triangle, with lines which have no width and can go on infinitely long, and where the meeting place of two lines is a single point. We completely define what "a line" is and we (at least mathematicians) believe that these idealized ideas are useful to us in our everyday lives. We know

---

Date: Version of February 2, 2021; minor edits July 5, 2022; minor edits June 11, 2023.

M. Malliaris (✉)
University of Chicago, Chicago, USA
e-mail: mem@math.uchicago.edu

A. Peretz
New York City, NY, USA
e-mail: assafperetz@proton.me

S. Arbeiter and J. Kennedy (eds.), *The Philosophy of Penelope Maddy*,
Outstanding Contributions to Logic 31, https://doi.org/10.1007/978-3-031-58425-1_29

there is no **real** line in the world. Yet introducing the concept, the definition, of a line has been extremely useful in our history.

Similarly introducing the concept/definition of a variable x has been extremely powerful in our history. The variable x does not exist per se, it is not real, and yet, its method has been powerful. The introduction of the 'variable' is a methodology, but this methodology necessitates a unique creation—that of the mysterious 'variable'.

Already we can see that there is a difference between the introduction of a line and that of a variable. Both may be "fictitious"—in the sense of not existing in the 'real world'—but they are so in quite a different manner.

Another important "fictitious," or should we say imaginary element was defined as "$i = \sqrt{-1}$." This too proved to be quite useful, though it took thousands of years to come to fruition, from its first appearance to the Greeks, to being useful to Euler and Gauss.

Mathematical definitions give us the power of creation. The moment we define something it exists. We breathe life into it. Contrary to most works of fiction though, mathematics, demanding complete precision, has provided itself with the quality of dis-creation, the power to annihilate from existence. If Montaigne can say that he may contradict himself but never contradict the truth, mathematics can't do that. It does not allow contradiction, and if a definition leads to contradiction it is dis-created, and annihilated from existence. As a metaphor, we could say that math constantly creates matter, and then anti-matter, and then uses the energy created by their meeting in order to get to where it wants to get to—the goal, the Proof.

III

Infinity as a methodology existed in different manners in mathematics from early on. A major creation by Leibniz and Newton was that of the methodology of the limit which goes to infinity, a clear mathematical definition of a progression to infinity. The invention of calculus, we can say, has been useful.[1]

At the end of the 19th century mathematics took another step and created infinity itself. It gave a definition for, and thus created and brought to life, "infinity." Today, more than a century later, this creation is still ignored and barely understood by all but a few mathematicians (and by most scientists, who have a presumption of the finitude of the physical world).

[1] Another example of a version of realization of infinity is the definition of the point of compactification. This too has proved quite useful.

IIII

A leap of faith:

For religious people, we are finitary beings and to reach God, to reach the infinite, we need to actually cross infinity. It is an impossibility, and thus what it needs is a leap of faith. Until the mathematical invention of the infinite, infinity was a goal: a line can go on and on and on, numbers can strive to get there in their limit. Through this limit we could integrate and produce many useful things, but it was always an impossible goal. *Its realization by the use of a mere definition is what is so staggering about mathematics and its power.*

The moment infinity is defined, it is created and starts to have skeleton structure we can investigate. Its investigation yielded some surprising results, the first of which is perhaps Cantor realizing at the end of the 19th century that there are different sizes of infinity.

IIIII

The power of creation, of realization in mathematics, is a great power, but like works of fiction, some realizations are more impactful than others. Every mathematician can invent/create a unique world with the snap of their fingers, a world which they may extensively work to describe, through proofs. There is an area however where the creation-power of mathematics is far greater than that of fiction, and that is in its power of annihilation; of dis-creation. Unlike fiction (which has the power to harm but not to dis-create[2]) years of work can be annihilated in a minute by another mathematician's disproof of a created world.

IIIIII

Just so there isn't the risk of being misunderstood we add the following:

We are not saying that mathematics is fiction, nor that certain mathematical definitions are more real than others. In fact, Kant saw mathematics as (synthetic) a priori. Descartes saw math as the hidden language for the book of the world. Our point is about the wondrous out-worldly power of creation by definition in mathematics.

In the history of philosophy from Plato to Kant, in different ways, mathematics had a meta-physical status, that is, the status of being beyond the physical world, however we might understand this beyond. Mathematics, in a way, is distinguished

[2] Even God, per the bible, only destroys, not dis-creates.

from the real, and yet we can use it and learn from it in the real world. That, in the end, is what we are doing in life in the real world.

IIIIII

We give a mundane example of what we discussed above.

> John walked near his favorite cafe when he noticed a poolah on the side of the road. We define a Poolah as an animal with three legs, no hands, three eyes, a single beak, and two tails. John hadn't seen a poolah in a while, so he quickly proceeded towards it to take a closer look, when all of a sudden the poolah jumped and started running away.

Fiction can create a poolah. That is what it does. But in-real-life we can't just create a poolah by defining it. Life doesn't work that way. But mathematics does. Mathematics possesses the power of creation by definition. The moment of a mathematical definition is a moment of creation. Defining $i = \sqrt{-1}$ gives i a life of its own. For example, we can derive that $i^2 = -1$. The moment of definition lets the creation live its life—or be disproved as impossible and dis-created.

IIIIIII

0, x, i, infinity, limit structures, transcendentals, π, triangles, and any other mathematical definition—are some more real than others? Is $1 + 1 = 2$ more real than $10^{20} + 10^{20} = 2 \times 10^{20}$ or $x + x = 2 \times x$? While some people want to cling to a supposed real and fictitious mathematics, we see it as a matter of what is actually useful (which, as mentioned above, may take centuries to know).

Before delving more in-depth into these questions, and the power of the definition of the infinite, through concrete examples, we propose the following formal definition of the usefulness of definitions:

Definition of Usefulness

(1) We say a definition is *somewhat useful* if it is an integral part of a correct proof of a theorem.
(2) We say a definition is *actually useful* if it is an integral part of a correct proof of a theorem which doesn't include the definition either in its assumptions or in its result.

IIIIIIII

Paul Cohen, in lecture notes explaining his celebrated work on the continuum hypothesis (Cohen, 2008), writes that "Cantor's work in set theory was the subject of much criticism to the effect that it dealt with fictions."

Why was this felt to be different from the use of infinity in classical mathematics? Is such a feeling of difference justified?

It may seem at first glance that in the earlier cases, infinity's role is mainly as an organizing principle, helping us to see in a clearer and clearer way how finite objects behave. Consider statements and expressions like: *There is no largest prime. The sequence* $1/2^n$ *converges to 0 as* $n \to \infty$. $\int_3^5 3x^2 dx$. *There is a nondecreasing function* $f : \mathbb{N} \to \mathbb{N}$ *such that for every n, and every graph G of size at least* $f(n)$, *G contains a homogeneous set of size* n. Indeed, for many practical applications of calculus, once one knows the limit theory works, it is sufficient to get close enough.

Still, this calm feeling of a steady progression to the very large, with steadily better information all the way up, can already be challenged in interesting cases. As an example from combinatorics, Szemerédi's regularity lemma says roughly that *huge* finite graphs can be well-approximated by much smaller random graphs. That is, given $\epsilon > 0$, one defines a pair of vertex sets (A, B) in a graph to be "ϵ-regular" if, informally, the edges between A and B are uniformly distributed (up to error ϵ), similarly to a random bipartite graph with edge probability equal to the density $d(A, B)$. Then the lemma says for every $\epsilon > 0$ there is $N = N(\epsilon)$ such the vertices of any sufficiently large graph can be partitioned into no more than N pieces of approximately equal size so that most pairs of pieces are ϵ-regular. In principle, this can give us a reasonable approximate picture of a truly huge finite graph. However, by a theorem of Gowers (1997), N necessarily has tower-type dependence on ϵ, so the scale at which the picture appears is already well beyond our own.

Such a lemma may grant us a kind of second sight, showing how certain mathematical objects behave at huge, but still finite, scales, which were otherwise not within our power to observe. Yet the behavior of graphs at the intermediate scales may remain quite mysterious.

Already with infinity as a limit, when the vastness of the infinite takes up a certain residence among finite objects, the path from the small to the huge is not necessarily experienced as an unbroken path.

IIIIIIIII

To isolate what is surprising about the definition of infinity, we may try to separate what is *generally* or *often* surprising about mathematical definitions, from what may be specifically surprising in this case.

"*Call the set H of vertices of the graph G homogeneous if either no two vertices in H are connected by an edge, or every two distinct vertices in H are connected by an edge.*" "*Let p be the largest prime.*" "*Consider a geometry satisfying Euclid's first four postulates.*" "*Call a real number transcendental if it is not the root of any polynomial with rational coefficients.*"

Recall that is meaningful to give a definition before we have evidence of anything satisfying it, and indeed to prove examples exist on general grounds without having any in hand. If a definition eventually leads to a contradiction it is thrown out—as we have said, dis-created ("let p be the largest prime")—but such a definition was not necessarily useless; proof by contradiction has a long history, and its inherent mysteries need not reflect a mystification coming from the infinite. Recall that a definition may appear to generalize when what it really does is simplify.[3] Finally, a definition which encompasses many obvious or natural examples—even, seemingly, "minimally so"—doesn't a priori rule out less familiar examples of the same definition.[4]

IIIIIIIIII

Let us consider[5] how deeply the definition of the infinite draws from the finite, beginning with Cantor's precise ideas which unify and organize our understanding of finite sets, their size, their extensions, and their arithmetic. We might say that a correct understanding of the finite opens a door by which the infinite also enters.

First is an organization of the idea of size. According to Cantor's definition, two sets A, B have the same size, in symbols $|A| = |B|$, if they can be put in bijection (one-to-one correspondence). Say $|A| \leq |B|$ if A has the same size as a subset of B, and $|A| < |B|$ if $|A| \leq |B|$ but there is no bijection. These definitions make sense even if we do not know the sizes of A or B, and clearly are meaningful for finite sets. According to Dedekind's definition, a set is infinite if it can be put in bijection with a strict subset of itself.

Notice this tells us we cannot increase the size of an infinite set by adding a single element.

Cantor noticed that there is another operation which increases the size of any set, regardless of its size: the power set operation (recall that the power set of X, is the set of subsets of X). If $X = \{0, 1, 2\}$, then $\mathcal{P}(X) = \{\{\}, \{0\}, \{1\}, \{2\}, \{0, 1\}, \{0, 2\}, \{1, 2\}, \{0, 1, 2\}\}$ and so $3 = |X| < |\mathcal{P}(X)| = 8$. According to Cantor's famous theorem, $|X| < |\mathcal{P}(X)|$ for any set X, thus, there are different infinite sizes, and what's more, no largest infinite size.

Meanwhile, suppose we write $0 = \emptyset$, $1 = \{0\}$, $2 = \{0, 1\}$ and in general $n = \{0, \ldots, n-1\}$. Then in von Neumann's language, each natural number n is a set which is *transitive* and *linearly ordered by* $\in$; call such a set an ordinal. They can be thought of as order-types of well ordered sets. The class of ordinals are linearly ordered, and closed under successor (if α is an ordinal, so is $\alpha \cup \{\alpha\}$, called "$\alpha + 1$") and under union. So the natural numbers are not the only ordinals: we have

[3] See also Malliaris and Peretz (2017).

[4] See Peretz (2006).

[5] Looking, we should emphasize, only as logicians of our own time, not historians of mathematics.

their union $\omega = \{0, 1, 2, \ldots\}$, then $\omega + 1$, and so on. Define the cardinals to be ordinals which cannot be put in bijection with any smaller ordinal. Then any ordinal α has a unique size, or *cardinality*, $|\alpha|$—the cardinal of the same size. The cardinals inherit a well-ordering from the ordinals. (Assuming the well-ordering principle, any set can be well-ordered, then is order-isomorphic to a unique ordinal, and so has a well-defined size or "cardinality" too.)

The event of defining infinite sizes/cardinals makes the infinite literal and useful. Cantor's cardinals are linearly ordered:

$$0, 1, 2, 3, \ldots \aleph_0, \aleph_1, \aleph_2, \ldots$$

There are arithmetic operations which specialize to the usual operations when the input is finite. However, once one of the cardinals involved is infinite, new phenomena appear. For instance, the fundamental theorem of cardinal arithmetic says:

> Let κ, λ be nonzero cardinals, at least one of them infinite.
> Then $\kappa + \lambda = \kappa \times \lambda = \max\{\kappa, \lambda\}$.

IIIIIIIIIII

We may test Cantor's discovery against our Definition of Usefulness taking as evidence one of his own proofs.

As above, call a real number algebraic if it is the root of some polynomial with rational coefficients, and transcendental if it is not. It had been an open question to demonstrate the existence of transcendental numbers. Using his new understanding of infinite sizes, Cantor solved this problem by the following proof. Call a set *countable* if it can be put in bijection with the natural numbers. Each polynomial in x with rational coefficients is determined by a finite sequence of rational numbers, whose length depends on its degree. The rationals are countable. For fixed n, the set of length n sequences of elements of a countable set is countable. The union of countably many at most countable sets is countable. So the set of finite sequences of rationals is countable. Each of our countably many polynomials has at most finitely many roots. And thus the number of algebraic numbers is countable. The reals have the same size as the power set of $\mathbb{N}$, and so are uncountable. Thus, not only do transcendental numbers exist, there are many more of them than algebraic numbers. Note that this proof does not produce any examples of transcendental numbers.

This proof allegedly caused a sensation.

Moreover, Cantor's cardinals revealed "hidden in plain sight" hard questions about objects people thought they knew well. For example, $|X| < |P(X)|$ begs the question: how much larger? Taking $X = \{0, 1, 2\}$, 3 is quite a bit smaller than 8. Since the reals have the same size as the power set of the natural numbers, this begs the question of whether there is an infinite set of reals "of intermediate size," too large to be put in bijection with $\mathbb{N}$ and not large enough to be put in bijection with $\mathbb{R}$. This led, of course, to the continuum hypothesis, Hilbert's first problem (on his list of open questions presented at the 1900 International Congress of Mathematicians). Addition and multiplication of infinite cardinals worked so simply, it's interesting that exponentiation appears more difficult.

Thus already from here we can see that, according to our Definition of Usefulness Cantor's definition is *actually useful.*

IIIIIIIIIIII

Insofar as infinite cardinals and ordinals come "out from under the overcoat" of the finite, we might expect that they would carry certain useful features of the natural numbers into new areas and new domains. An example of this is that proof by induction can now be extended to transfinite induction, and definition by recursion to transfinite recursion.

For example, a theorem in the wonderful book of Komjath and Totik (2006) says that it is possible to decompose Euclidean space, $\mathbb{R}^3$, as the disjoint union of circles of radius 1. The proof begins by enumerating the points of $\mathbb{R}^3$ as $\langle p_\alpha : \alpha < \kappa \rangle$, where $\kappa = |\mathbb{R}| = |\mathbb{R}^3|$, and constructs by transfinite recursion a sequence of circles C_α, each either empty or of radius 1, such that each C_α is disjoint from all previous circles, and if p_α is not contained in a previous circle, it is contained in C_α. Let "large" mean size κ, and "small" mean size strictly less than κ. The proof turns on the fact that if α is an ordinal $< \kappa$, then $|\alpha|$ is small, and if μ, λ are small cardinals, then $\mu \cdot \lambda$ is small (note this includes the case where both μ, λ are finite).[6]

Induction and recursion on the natural numbers rely on two things: that the natural numbers are well ordered, and that as we walk increasingly through them, there are always few things behind us compared to what is ahead of us. Cardinals make these properties available to us in walks through a much larger landscape.

[6] More precisely, a circle is a locus of points, all lying in a plane and equidistant from some point called the center. The proof builds by transfinite recursion a sequence of circles C_α, each either empty or of radius 1, such that each C_α is disjoint from all previous circles and such that p_α is contained in one of the circles up to and including C_α. Suppose we arrive to stage α. If p_α is already covered, let C_α be empty and go to the next stage. Otherwise, we first choose a plane P_α containing p_α such that no previous nontrivial circle C_β lies in P_α. We can do this because there are a large number of planes through p_α, and the number of circles chosen so far, $|\alpha|$, is small. Of course, some of the previous circles may intersect our plane. So let us call a point of P_α "bad" if it intersects any earlier C_β. Each C_β can be responsible for at most two bad points, so the number of bad points in P_α is small. It would suffice to choose a circle of radius 1 lying in P_α, going through p_α and avoiding all bad points. Consider a fixed bad point b. The number of circles of radius 1 in P_α which can go through both p_α and b is at most two. So each bad point b rules out at most two circles. There are a large number of circles of radius 1 through p_α in our plane P_α, and a small number which are ruled out, and so we must succeed.

IIIIIIIIIIIII

As model theorists, we see a picture perhaps currently less visible in other areas of mathematics: mathematical information observed across the hierarchy of cardinals can be distilled into powerful definitions and theorems which themselves, after the fact, may make no reference to these origins. (As written elsewhere (Malliaris, 2018), perhaps model theory may have a special ability to make translations of this kind.)

We now sketch such an example from the monumental work of Shelah (1990). We will give just a few details, chosen to illustrate how, along the way, certain properties of cardinals are being used.

A (first-order) *theory* is a set of axioms, which we can identify with the class of structures satisfying it, for example: algebraically closed fields of characteristic zero, real closed fields, or dense linear orders without endpoints.

We begin with *counting types*. Roughly speaking, a type in model theory tells us what can be said about one element of a structure in terms of another set of elements. Given a particular infinite λ, we say a theory is λ-stable if over any set of parameters of size λ in any model of the theory, there are no more than λ types; otherwise, it is λ-unstable.

If K is any algebraically closed, hence infinite, field of characterisic p and size λ, then an element can interact with K either by being a root of some minimal polynomial with coefficients in K (there are $\lambda = |K|$ such polynomials), or by being transcendental over K. So the number of types over K is precisely λ, and we say the theory of algebraically closed fields of fixed characteristic is λ-stable *for every infinite* λ. On the other hand, for every infinite λ there exists a dense linear order without endpoints of size λ, call it M, over which there are strictly more than λ cuts; thus, the number of types over M is $> \lambda$. The theory of dense linear orders without endpoints is λ-*unstable for every infinite* λ. A priori, there is perhaps little reason to expect such a pure dichotomy across all theories.

One of Shelah's discoveries was that if a theory T is unstable in some λ with $\lambda = \lambda^{|T|}$, then there must be a single formula causing the number of types to be large. The proof goes by writing the type as a product of its restrictions to each formula, of which there are $|T|$-many, and noticing one of the factors must have size $> \lambda$, as otherwise the $|T|$-fold product of λ, i.e. $\lambda^{|T|}$, would have size λ. Thus, instability is a local phenomenon, caused by unstable formulas, which we discover by using the arithmetic of a particular cardinal.

Shelah proved a dichotomy theorem:

Theorem: Any theory T is either

(a) stable, i.e. λ-stable for all infinite lambda such that $\lambda = \lambda^{|T|}$.

(b) unstable, i.e. λ-unstable for all infinite lambda.

This dichotomy, which had a foundational impact on modern model theory, is powerful in part because Shelah was able to show that there are many other characterizations of this partition. For example, it may be re-framed in terms of ranks, dimensions, definability of types, bounds on the orbits of certain automorphism groups, and also completely "locally" in terms of unstable formulas. Let us state the last of these. Say a formula $\varphi(\bar{x}, \bar{y})$ has the order propery if in models of T, for arbitrarily large finite n, there exist $\langle \bar{a}_i : i < n \rangle$ and $\langle \bar{b}_j : j < n \rangle$ such that $\varphi[\bar{a}_i, \bar{b}_j]$ holds if and only if $i < j$. Shelah proved:

Theorem: A theory T is unstable if and only if it contains a formula with the *order property*.

By the time we arrive to this theorem, the definition of the order property is simply a kind of long asymmetry, and makes no mention of infinite cardinals. Algebraically closed fields do not have asymmetries of this kind; neither do free groups on fixed finite numbers of generators. Infinite random graphs and real closed fields do.

IIIIIIIIIIIII

What can the infinite tell us about the finite?

One can answer this question in many ways; we will give a short example which is recent and close to our hearts.

Stability has been an extremely productive and powerful influence within model theory, explaining surprising similarities and differences among classes of infinite structures (such as algebraically closed versus real closed fields). It long seemed to be about infinite structures in a deep way. Recently, the picture has shifted.

Consider an a priori different setting. The finite Ramsey theorem tells us that for any finite n there is $M = M(n)$ such that any finite graph G of size at least M contains a homogeneous set of size n. Approximately, a graph of size n will have a homogeneous set of size at least $\log n$. Precisely determining the bounds or "Ramsey numbers" for all graphs or for certain classes of graphs leads to much challenging combinatorics.

Homogeneous sets may be thought of as special cases of the following model theoretic definition, which uses all formulas, not only the formula expressing whether an edge holds. (This connection was already made in Ramsey's original paper (Ramsey, 1929).)

Definition. In any mathematical structure, call a sequence of elements indiscernible if for any finite k, any two strictly increasing subsequences of length k satisfy the same first order formulas.

Part of the work on stability in Shelah (1990) is the following surprising theorem about infinite models of stable theories: *If T is stable then in any infinite model M of T of size κ^+ there is a set of indiscernibles of size κ^+*. Informally, in infinite models of stable theories of reasonable sizes, one can extract indiscernible sequences (indeed, sets) of the same size. As an example, in any algebraically closed field of fixed characteristic and size $\aleph_1$, one can find a transcendence basis of size $\aleph_1$ (whose elements in some sense interact completely generically with each other).

It turns out that this theorem about infinite stable theories makes a precise and useful prediction about the finite world. A 2014 theorem of Malliaris and Shelah [9] shows that one can in some sense "finitize" stability in the case of graphs by forbidding a certain family of configurations, the half-graphs, which are finite versions of the order property for the graph edge relation. It becomes possible to build an approximation to the structural understanding of stability within these "stable-like" graphs, and thus to prove, mirroring the infinite case, that in so-called finite stable graphs there are much larger homogeneous sets than predicted by Ramsey theory. A simple version of the resulting theorem says: *There is a constant $c = c(k)$ such that if G is a finite graph which does not contain any k-half graph then G homogeneous set of size $|G|^c$.*

In this simple example, and in analogous refinements of Szemerédi regularity (Malliaris & Shelah, 2014, 2021), one uses an understanding of infinite stability phenomena (themselves revealed in the infinite world through an understanding of the hierarchy of cardinals) as a precise guide to prove theorems about the finite world.

IIIIIIIIIIIIII

If in the course of this paper we have invoked our Definition of Usefulness to start to see how, in their first century and half, Cantor's infinite cardinals show themselves to be delicately and deeply interwoven with the fabric of our mathematical world and our mathematical thought, the Definition of Usefulness shouldn't however be used *against* cardinals, as a way of silencing them or banishing them from hypotheses of theorems.

There is a greatness in certain mathematical ideas—as also in certain works of fiction—which challenges us by enlarging the boundaries of what we thought possible without ever directly breaking with the world we knew. Emerson, writing on character, says of certain great figures that "the largest part of their power was latent." Mathematics, having as it does a longer history, may take centuries or even millenia to understand and to assimilate a new definition, but this doesn't mean its power isn't felt already in our time.

References

1. Cohen, P. (2008). *Set theory and the continuum hypothesis* (Dover ed., page 2).
2. Gowers, W. T. (1997). Lower bounds of tower type for Szemerédi's uniformity lemma. *Geometric and Functional Analysis, 7*, 322–337.
3. Komjath, P., Totik, V. (2006). *Problems and theorems in classical set theory.* Problem books in mathematics. Springer. Section 13.
4. Malliaris, M. (2018). Model theory and ultraproducts. In: *Proceedings of the International Congress of Mathematicians-Rio de Janeiro 2018* (Vol. II). Invited lectures (pp. 83–97). World Scientific Publishing

5. Malliaris, M., & Peretz, A. (2017) What simplicity is not. *Simplicity: Ideals of practice in mathematics and the arts*. Mathematics in Culture and the Arts (pp. 51–58). Springer.
6. Malliaris, M., & Shelah, S. (2014). Regularity lemmas for stable graphs. *Transactions of the American Mathematical Society, 366*, 1551–1585.
7. Malliaris, M., & Shelah, S. (2021). Notes on stable regularity. In: *Bulletin of Symbolic Logic* (Vol. 27, No. 4, pp. 415–425).
8. Peretz, A. (2006). Geometry of forking in simple theories. *The Journal of Symbolic Logic, 71*(1), 347–359.
9. Ramsey, F. P. (1929). On a problem of formal logic. *Proceedings of the London Mathematical Society, 30*, 338–384.
10. Shelah, S. (1990). *Classification theory and the number of non-isomorphic models* (1st ed., 1978. (rev. ed. 1990)). North-Holland.

Chapter 30
Second Philosophy and the Depth Metaphor

Juliette Kennedy

Abstract We examine Maddy's concept of mathematical depth against the background of her naturalism.

30.1 Introduction

What is the place of depth in Maddy's Second Philosophy? The answer to this question is that the concept of mathematical depth lies at the very center of it:

> A mathematician may blanch and stammer, unsure of himself, when confronted with questions of truth and existence, but on judgments of mathematical importance and depth he brims with conviction. For this reason alone, a philosophical position that puts this notion center stage should be worthy of our attention.[1]

This feels slightly anomalous. In her writings on Second Philosophy Maddy has mainly dealt with the standard concerns of philosophers of mathematics working in the post-war period—that is to say, epistemological and metaphysical concerns. Along with a substantial body of work on realism, she has written about modality, necessity and the nature of the logical; she has written on Kant's, Frege's and Wittgenstein's philosophy of mathematics; most of all she has written on naturalism relativized to the problem of objectivity in set theory, in particular the question, how

---

[1] (Maddy 2011, p. 117). Maddy mentions a collection of apparently different virtues in *Defending the Axioms*. But these are all used to pick out the same general phenomenon:

> A generous variety of expressions is typically used to pick out the phenomenon I'm after here: mathematical depth, mathematical fruitfulness, mathematical effectiveness, mathematical importance, mathematical productivity, and so on. (Maddy 2011, p. 80).

Since depth is mentioned most prominently in that work and in the proceedings volume of a conference devoted to the concept of depth organised at UC Irvine in April 2014, we will focus on depth in this essay.

J. Kennedy (✉)
University of Helsinki, Helsinki, Finland
e-mail: juliette.kennedy@helsinki.fi

S. Arbeiter and J. Kennedy (eds.), *The Philosophy of Penelope Maddy*,
Outstanding Contributions to Logic 31, https://doi.org/10.1007/978-3-031-58425-1_30

to justify the introduction of new axioms for set theory, failing a robust conception of mathematical existence.

One may think that the concept of depth is slightly out of place here, entangled as it is with aesthetics—"depth" being a term, one might think, of mere aesthetic appraisal. But Maddy unites these two apparently opposing philosophical imperatives—that is, to take both mathematical objectivity and aesthetic value seriously, philosophically—under a single banner: mathematical depth is objective, and judgements of mathematical depth are *objective* judgements:

> ...the topography of mathematical depth...stands over and above the merely logical connections between statements, and furthermore, it is entirely objective...[2]
>
> It also bears repeating that judgments of mathematical depth are not subjective...[3]

In fact, not only is depth objective, but the objectivity of mathematics *tout court* is underwritten, in her view—or more precisely in the view of the thin realist/strong arealist—by the objectivity of depth.[4] As Blanchette has written in her essay for this volume, "Mathematics is objective, on the Thin Realist's account, because it's about the contours of mathematical depth, which contours hold independently of us."[5]

Thus a mathematician's appraisal of depth is not simply an expression of his or her subjective aesthetic pleasure—or at least, it is not only that; such an appraisal is responding to, on some level, objective structure.

Maddy's centralization of depth is carried out in a specific context, namely the second philosophical program of extrinsic justification for new axioms of set theory; but there is of course a great deal more to say about Maddy's general standpoint. This may be described variously as vestigially or thinly realist within the Second Philosophy point of view she introduced in Maddy (2009), her own position within the graded set of philosophical commitments she introduced in that work—robust realism, thin realism and strong arealism—amounting to a rejection of robust realism while embracing both thin realism and strong arealism as equally good *idioms*, in the sense of being "equally well-supported by precisely the same objective reality: those facts of mathematical depth."[6]

---

[2] (Maddy 2011, p. 81).

[3] (Maddy 2011, p. 81).

[4] See below for the discussion of the terms "thin realist" and "strong arealist."

[5] Maddy has even written, bravely, that the purpose of doing (pure) mathematics, its ultimate goal, is to discover deep mathematics:

> "...the thin realist begins with an assessment of the effectiveness of set-theoretic methods for achieving goals of the practice (ultimately...the discovery of deep mathematics)."

(Ernst et al. 2015, p. 247).

[6] (Maddy 2011, p. 112). Maddy continues:

> These facts are what matter, what make pure mathematics the distinctive discipline that it is, and that discipline is equally well described as the Thin Realist does or as the Arealist does. Once we see this, we can feel free to employ either mode of expression, as we choose—even to move back and forth between them at will.

It may seem counterintuitive to embrace either as a mere idiom; and moreover to embrace both idioms as equally well-supported. For the thin realist "[mathematics: JK] is a body of truths about objectively existing things...,"[7] while the arealist (either strong or weak) altogether *eschews* talk of truth and existence. The point is contested, also by contributors to this volume.[8] And while the distinction matters in this paper we do not pursue the debate over their equivalence here, beyond saying what we said in Kennedy (2014), that for Maddy that ship sailed long ago. The point is that *objectivity* secures the validity of both the thin realist and strong arealist idioms—"those facts of mathematical depth" which track the objectivity of mathematics. This is what interests the Second Philosopher, namely depth, and the idea of its objectivity.

Thin realism has a central place in Maddy (2011), so we now take a moment to describe that view. Relativized to set theory, thin realism is the view that:

> ...sets just are the sort of thing set theory describes; this is all there is to them; for questions about sets, set theory is the only relevant authority.[9]

A priori epistemology is off the table, as are any external justificatory strategies, and the space between metholodology and object is gap free:

> ...the Thin Realist holds the set-theoretic methods are the reliable avenue to the facts about sets, that no external guarantee is necessary or possible. So the fundamental diagnostic is this: the Robust Realist requires a non-trivial account of the reliability of set-theoretic methods, an account that goes beyond what set theory tells us; for the Thin Realist, set theory itself gives the whole story; the reliability of its methods is a plain fact about what sets are.
>
> The key here is that the second-philosophical Thin Realist begins from her confidence in the authority of set-theoretic methods when it comes to determining what's true and false about sets, and from the observation that her more familiar methods appear irrelevant; she concludes that the call for a non-trivial, external epistemology is misplaced.[10]

Thin Realism does tolerate existential commitments, of a kind:

> Set-theoretic concepts are formed in response to set-theoretic goals just as the concept 'group' was formed in response to algebraic goals....
>
> Of course the set-theoretic cases we've been concerned with involve not definitions but existence assumptions—like the introduction of sets in the first place or the addition of large cardinals—and new hypotheses—like determinacy—but in these cases, too, far more than consistency is at stake: these favored candidates differ from alternatives and near-neighbors

---

> The proposal, then, comes to this: Thin Realism and Arealism are equally accurate, second-philosophical descriptions of the nature of pure mathematics. They are alternative ways of expressing the very same account of the objective facts that underlie mathematical practice.

In correspondence with the author Maddy drew a useful distinction within arealism. As she wrote: "The Thin Realism and Arealism of *Defending the Axioms* are anchored in an objective notion of depth. If depth turns out not to be objective, my thought is that Thin Realism collapses, and we're left with a 'weak' form of Arealism that tracks the goals and values our interests and abilities lead us to favor. The Arealism of DA would then be 'strong', in contrast."

[7] (Maddy 2011, p. 103).

[8] See Burgess's and Blanchette's essays, this volume.

[9] (Maddy 2011, p. 61).

[10] (Maddy 2011, p. 63).

> in that they track what we might call the topography of mathematical depth. This topography stands over and above the merely logical connections between statements, and furthermore, it is entirely objective: just as it's not up to us which bits of pure mathematics best serve the needs of natural science, just as it's not up to us that it would be counterproductive to insist that all 'groups' be commutative, it's also not up to us that appealing to sets and transfinite ordinals allows us to capture facts about the uniqueness of trigonometric representations, that the Axiom of Choice takes an amazing range of different forms and plays a fundamental role in many different areas, that large cardinals arrange themselves into a hierarchy that serves as an effective measure of consistency strength, that determinacy is the root regularity property for projective sets and interrelates with large cardinals, and so on. These are the facts that play a role analogous to Kant's Euclidean space, the facts that constrain our set-theoretic methods, and these facts, unlike Kant's, are not traceable to ourselves as subjects.[11]

Sets, then, are just what set theoretic methodology tells us they are, and to say anything about their existence beyond that is to stray beyond the parameters of the position. The thin realist's notion of objectivity is similarly thin, involving mind independence—"it's not up to us"—but not necessarily objects in the "thick" sense, or truthmakers or whatnot, though a thin notion of object is still included.

Metaphysically speaking, we are deep in deflationary territory—territory also explored by Putnam in *Ethics without Ontology* Putnam (2004), albeit arguing for the objectivity of *ethical* judgements. Putnam's is also an objectivity without objects and in that sense his view is reminiscent of Maddy's strong arealism.

A word about the other two "realisms." As one would expect from the terminology, robust realism is a straightforwardly platonistic conception of mathematical existence, while arealism, whether strong or weak, eschews truth and existence: "the Arealist has no grounds on which to regard...[mathematical] objects as real or its claims as truths."[12]

But...*depth.* As we saw, for the thin realist, the objectivity of mathematics, however thin it is, is overlaid on what seems to be a prior objectivity, namely that of the contours of depth. One may think that this concept of mathematical objectivity is overly thin, as with aesthetic judgements generally, assessments of depth, though *seemingly* univocal among mathematicians, can also be unstable, ephemeral, and above all *personal.*

Maddy is aware that pinning an "uncomfortably metaphorical"[13] concept such as depth to objectivity is precarious, rhetorically, in the sense that if the one goes, the other goes too. Thin realism and mathematical depth, are, in other words, symbiotic, and the same presumably goes for strong arealism:

> But what if depth should turn out not to be 'in the math', but only 'in us'? I say this without pretending to a defense, but my inclination is to think that in the imagined eventuality, the ground beneath thin realism would collapse. We would be left with a version of arealism as the only viable option remaining, and a version that grasps the first horn of Arana's dilemma: the strains we choose to follow among the vast array of logical interconnections reflect not

[11] (Maddy 2011, p. 80).

[12] (Maddy 2011 p. 96).

[13] See Ernst et al. (2015).

> any independent mathematical virtues, but the goals and values our interests and abilities prompt us to pursue.[14]

The prospect of depth tracking *subjectivity* rather than objectivity seems to force Maddy, then, away from thin realism and toward weak arealism. This would complete the circle on a philosophical evolution that begins with perceptual realism and arrives, now, at its complete denial—albeit a development taking place over a lifetime.

In this chapter I consider Maddy's use of the depth metaphor as a tool for philosophical progress. I will contextualise her view by means of Rorty's and Wittgenstein's writings on metaphor in general and the work metaphors do in mathematics; and from the Ordinary Language Philosophy point of view that (in this case) assessments of depth are incorrigibly context- or occasion-sensitive—a possibility Maddy herself entertains in Ernst et al. (2015), when she suggests that the concept of mathematical depth may be "irreducibly specific."[15] I will fold in Putnam's problematization of ethical judgement from his *Ethics without Ontology* Putnam (2004), transposed to the question whether it is feasible to regard aesthetic appraisal, here assessments of mathematical depth, as untethered from questions of objective existence—Putnam's claim, transposed from the domain of ethics to that of aesthetics, being that aesthetics has 'nothing to do with ontology'.

30.2 Rorty and the Literalization of Metaphor

Rorty's elaborate theory of metaphor as a device for individual and social transformation is tied to his anti-representationalist view of language, representationalism being the view of language that it is, simply put, a medium through which human beings represent the world to themselves. Rorty's is an unMaddian view of language, it would seem, but his theory of metaphor has one aspect that we find useful here, and this is the idea of the literalized metaphor.

To understand what Rorty means by the phrase "literalized metaphor" we must begin with the recognition that metaphors are, for Rorty, pieces of private language

---

[14] Maddy is referring here to Arana's "On the Depth of Szemerédi's Theorem" Arana (2015). This was one of the papers presented at the above-mentioned conference about mathematical depth organized at UC Irvine in April 2014. (Ernst et al. 2015, p. 248).

[15] The full quote:

> So, this line of thought continues, perhaps the features that make any individual mathematical item deep are irreducibly specific to the context and texture of that particular item, not to any identifiable subclass of that context and texture. We might notice, for example, that analytic number theory ties elementary number theory with analysis, that large cardinals tie analysis with set theory, but the general fact of 'tying together' would not be what makes either of these deep; what makes them deep would lie in the details of how the tying is accomplished and what it achieves. There are, after all, many ways of tying things together that are superficial or mere trickery.
>
> (Ernst et al. 2015, p. 246).

originating "outside" of language—"outside" in the sense of unintelligibility. "Private acts of redescription"[16] irrupting at the edge of meaning, a (good) metaphor is a piece of brilliant nonsense. Nonsense, that is, lacking in meaning, because, as Rorty writes, "To have a meaning is to have a place in a language game. Metaphors, by definition, do not."[17] Unlike Wittgenstein's notion of *Unsinn*, perhaps, Rorty's nonsense is just straightforwardly so. When the (possibly apocryphal) mathematician says "Theorems aren't deep. Oceans are deep." s/he is simply saying that it is nonsense to speak of a theorem as being deep.

In Rorty's terminology, the mathematician who insists on oceans being deep but not theorems is saying so in the face of a *live* metaphor, a piece of language acting, in turn, on language, by calling for the transformation of nonsense into sensibility—calling, in short, for (the creation of) new meaning. At this moment of nonsense breaking into sensibility and meaning, the metaphor enters the social realm now as a literalized or dead metaphor—that is, as a public, shared piece of language:

> Between...[between living and dead metaphor] we cross the fuzzy and fluctuating line between natural and non-natural meaning, between stimulus and cognition, between a noise having a place in a pattern of justification of belief. Or, more precisely, we begin to cross this line if and when these unfamiliar noises acquire familiarity and lose vitality through being not just mentioned...but used: used in arguments, cited to justify beliefs, treated as counters within a social practice, employed correctly or incorrectly.[18]

And indeed the depth metaphor became literalized in the sense that by now every mathematician understands what it means to say that a theorem is deep, our apocryphal mathematician notwithstanding.

Rorty sees this process of the creation and death of metaphors as the "fuel of liberalism," with metaphors acting as engines for self-renewal, individually—because transformation "starts at home, with redescription"[19]—and writ large, because a human society can only be transformed on the level of the individual—one by one, as it were. For Rorty, the vitality, even the viability of a shared social practice—indeed the viability of human society itself—is contingent upon the ability of that practice to continually transform itself—that is, to produce new metaphors:

> Both perception and inference leave our language, our way of dividing up the realm of possibility, unchanged...A metaphor is, so to speak, a voice from outside logical space, rather than an empirical filling up of a portion of that space, or a logical-philosophical clarification of the structure of that space. It is a call to change one's language and one's life, rather than a proposal about how to systematize either.[20]

---

[16] (Arribas 2007, p. 328).

[17] (Rorty 1989, p. 18).

[18] (Rorty 1991, p. 171).

[19] (Arribas 2007, p. 315).

[20] (Rorty 1991, pp. 12–13).

As an aside, the poet Robert Frost, for whom metaphor was "all of thinking", alludes similarly to the fragility of human society sans metaphor, when he writes:

> What I am pointing out is that unless you are at home in the metaphor, unless you have had your proper poetical education in the metaphor, you are not safe anywhere. Because you are not at ease with figurative values: you don't know the metaphor in its strength and its weakness. You are not safe with science; you are not safe in history.[21]

From this point of view—"point of view" being an example of an intelligible/dead metaphor—because we know that we are safe in our mathematics, it must be the case that mathematics does not exist in a state sans metaphor, and indeed every mathematician knows that the process of incorporating new metaphors into mathematics is an endless one.

30.3 Wittgenstein's Treatment of the Mathematical Metaphor

Floyd and Mühlhölzer's *Wittgenstein's Annotations to Hardy's Course of Pure Mathematics: An Investigation of Wittgenstein's Non-Extensionalist Understanding of the Real Numbers* Floyd and Mühlhölzer (2020) is about Wittgenstein's non-extensionalism, but Wittgenstein's critique of the mathematician's metaphorical language is a running theme in it.

Extensionalism and metaphor are related to each other, for Wittgenstein, in the sense that extensions are, for him, *metaphor-like*. This is the basis of his critique of extensionalism: that it is an obscurantist phraseology enabled by metaphor. Floyd and Mühlhölzer describe Wittgenstein's nonextensionalism as follows:

> With the term "extension" Wittgenstein has two things in mind. First, he will strictly distinguish between sequences of numbers that the extensionalist considers to be, in Cantor's sense, "finished" [fertig] entities or sets—these are the "extensions"—from the techniques or rules by means of which such entities may be produced, assessed, or accessed. If there are such techniques, the extensionalist's interest is ultimately only in their results, the produced sequences, and not the possible processes or conceptual motifs or definitions leading to them. For the non-extensionalist, on the other hand, it is the processes and structured conceptual motifs, the grammar or logic of the notions, we should be concerned with. The most important cases discussed in Wittgenstein's texts are given by the conception of a real number as a rule-governed calculational procedure for which we can see that for any given n it will generate n digits: for example, a recipe for generating more and more successive digits of $\sqrt{2}$. About the number $\sqrt{2}$ Wittgenstein explicitly says that "we [as extensionalists] have a tendency to think that there is one result produced by $\sqrt{2}$, viz., an infinite decimal fraction." But on his view $\sqrt{2}$ produces a series of results, but no single result: "$\sqrt{2}$ is a rule for producing a fraction, not an extension."[22]

[21] Frost, "Education by Poetry," Amherst College address, *Amherst Graduates' Quarterly*, February 1931.

[22] (Floyd and Mühlhölzer 2020, p. 31).

The authors note that regarding infinite decimals as finished entities enables the metaphor of a gap free real line—to the mathematician, of course. As for the philosopher, to accept metaphors like gap-freeness is to give in to a wrong picture:

> Wittgenstein's investigation of the limitations of the extensionalist viewpoint is motivated by a diagnosis: that it all too easily leads to illusory thinking. It should therefore be refused as anything more than metaphor. In this sense he does not reject the extensionalist viewpoint altogether. Instead, he accepts it as a sort of specific phraseology, but one that should be assigned a relatively peripheral place, or rather, one that actually has a peripheral place already within our mathematical practice.[23]

Hardy's use of metaphor in *Course of Pure Mathematics* irritates Wittgenstein. When Hardy speaks of figures that 'look and behave like a real line,' Wittgenstein's marks such passages with, simply, *Unsinn*. Deny yourself the metaphor and then you'll see what's really going on with the figure, Wittgenstein seems to be saying. "See things as they are!"

One way of thinking about what Wittgenstein is doing in the annotations is that he is probing the breakdown of the mathematical metaphor—the touch and go of it, in Frost's terms:

> All metaphor breaks down somewhere. That is the beauty of it. It is touch and go with the metaphor, and until you have lived with it long enough you don't know when it is going. You don't know how much you can get out of it and when it will cease to yield. It is a very living thing. It is as life itself.[24]

Wittgenstein's rejection of the mathematical metaphor, that is to say his resistance to *literalization*, springs from his desire, on our behalf, that we keep *Unsinn* before our eyes; that we understand that the motley of conceptual paths delivering the real numbers, the techniques of diagonalization, continued fractions, and the like—given to us in the form of unique, space-filling entities—are given only metaphorically.

The inside/outside motif (or metaphor!) appears frequently in Rorty's writings on metaphor, for example nonsense in the form of live metaphor is said to be "outside" of logic and even of language. The metaphor is also present in Maddy's writings, when she mentions going "beyond what set theory tells us"[25] in connection with robust realism. Wittgenstein had a rather strong sense of "inside and outside," of what is interior or exterior to *mathematics*. Making an issue the validity of the law of excluded middle, for one example, is to give in to a *philosophical* distraction, for Wittgenstein, which leads us outside of mathematics[26]—away from the "processes and structured conceptual motifs, the grammar or logic of the notions,"[27] comprising the subject.[28]

---

[23] (Floyd and Mühlhölzer 2020, p. 34).

[24] Frost, "Education by Poetry."

[25] See above.

[26] See (Floyd and Mühlhölzer 2020, p. 74).

[27] See Floyd and Mühlhölzer's quote above.

[28] Floyd has commented on Maddy's concept of depth, thin realism and "the realistic spirit" in connection with aspect perception. See (Floyd 2021, p. 4).

30.4 Maddy's Use of the Depth Metaphor

Seen against the background of Rorty's and Wittgenstein's albeit very different view of metaphor, Maddy's appropriation of depth has an interesting profile. For Rorty, the banalization/literalization of metaphor is the way society generates new vocabularies (new meaning) for itself, as we saw. While for Maddy, a particular metaphor—that of mathematical depth—tracks the contours of objective mathematical terrain.

As for Rorty's idea that metaphor originates "outside" language and logic, it seems right to say that the concept of depth is exterior to the workings of mathematics, that is to say of mathematics as a set of "working wheels," to use Wittgenstein's terminology. Mathematics has a will of its own, in other words, and any commitment to the truth conduciveness of aesthetic value—for that is what is at issue here—will have to be set alongside the nonnegotiable existence, or not, of a proof. And in cases where mathematical justification becomes more complicated, philosophically, as with set theory and its many important independent statements, the Second Philosopher attempting to adjudicate among these statements will have to be guided by *internal* considerations, presumably.

The question then becomes, can the Second Philosopher tolerate this reliance on the depth metaphor? This voice, as Rorty puts it, from outside logical space? In other words, do assessments of mathematical depth play any part in set-theoretical methodology?

Maddy speaks with the authority of a philosopher with no small amount of mathematical experience, especially with set theory, and it is exactly in set theory that the walls between truth and proof, and that of aesthetic value, are beginning to dissolve.

We argued the point at length in Kennedy and Väänänen (2015), claiming there that the incorporation of aesthetic value into set theoretic practice *has* become part of set theoretic methodology in certain important cases; that in key cases the set theorist's aesthetic judgement can become entangled with their mathematical work in ways that make it difficult to distinguish aesthetic judgements from judgements of truth. Thus by our lights Maddy's appropriation of depth is doing important work; and Maddy is therefore completely correct in including aesthetic value (in the form of depth) in the second philosophical arsenal of justificatory strategies for new axioms for set theory. In short, considerations of depth are central to set-theoretic practice in important cases after all, rather than being exterior to mathematics in the either the Rortyan sense of being nonsense, or the Wittgensteinian sense of being a philosophical distraction.

We refer the reader to Maddy's Maddy (2011) for her extensive treatment of the continuum hypothesis, as well as her treatment of new axiom candidates in connection with maximality and other set-theoretic values, such as lawlikeness. In this connection, a recent example involves the HOD dichotomy, so-called, which divides set theoretic territory into lawlikeness on the one hand, and chaos on the other.[29]

---

[29] The HOD dichotomy, due to Woodin, states that assuming the existence of large cardinals, the set-theoretic universe is either very close to the class of hereditarily ordinal-definable sets or very far from it.

The situation, however, is precarious. Aesthetic value may turn out not be truth conducive after all; depth may *not* track objectivity. The HOD conjecture, for example, which is a central focus of set theorists nowadays, could be decided in the direction of "chaos," the less preferable outcome for the set theorist, presumably. Still, at the end of the day, set theorists, like all mathematicians, are beholden to the objective standards of the subject, namely the existence of a proof, and not aesthetic value.

The issue we are concerned with here is depth, and the claim that depth is objective. What kind of objectivity is at issue here, and how important is the objectivity of depth, or even objectivity simpliciter, for the Second Philosopher? Can we pose a different question than the one we asked at the beginning of this essay and ask, what is the place not of depth but of objectivity, both in Second Philosophy and in mathematics generally? Is objectivity *inside* or *outside*?

As for the place of objectivity in mathematics generally, of course mathematicians employ a vague and general notion of *existence* in their work—realism delivered by extensionalism, realism in the sense of the "realistic spirit," as Cora Diamond refers to it in Diamond (1995); realism couched in metaphorical language—a misleading phraseology, for Wittgenstein, and delivered by the extensional point of view, which he is critical of. As for objectivity in the quotidian sense, and the rather elusive idea that mathematical outcomes are "not up to us," this is also widespread in mathematics and may be regarded as part and parcel of the realistic spirit.

As for the Second Philosopher—whether speaking in the thin or in the strong arealist idiom—her position is clear: there is a strong notion of objectivity at issue in both, and it is metaphysically thin, though not objectless in the case of the thin realist idiom. According to Wittgenstein's strict conception of the inside and outside of mathematics then, the Second Philosopher has here acceded to a philosophical distraction; she has let herself be lead *outside* the practice.

Can objectivity can be pulled *inside*, as Maddy pulls depth inside? That is, is it possible that objectivity—either in either the thin or arealist sense—is part of set-theoretic methodology after all, as with depth?

Maddy notices that the word "depth" is woven into the mathematician's everyday vocabulary, and in fact mathematicians talk about depth all the time. She has also argued for the neccessity of the concept of mathematical depth for set-theoretic practice, if not for its objective existence; and we have argued that aesthetic value of various kinds is entangled with judgements of truth in set theory. We noted earlier Maddy's remark that if depth turns out to be only 'in us' and not in the math, the ground beneath thin realism would collapse. This seems right, by our lights, not because of the possible subjectivity of depth, but because mathematics, in particular set-theory, has no need to accommodate the philosophical concept of the abstract, mind independent object, however thin; nor does it have any need for the concept of objectivity. Objectivity, in other words, whether object-based (as in thin realism) or not (as in arealism), should be a non-starter for the Second Philosopher—however thin that objectivity is, under either thin or (strong) arealism. Objectivity cannot be pulled inside.

In sum, apart from coming up conversationally among mathematicians in their hesitant/philosophical mode, as Maddy observes in the opening quote of this paper, we cannot see objectivity at work in set-theoretic practice in the way that we can see depth at work. The concept of objectivity as applied to depth seems in any case to glue a term belonging to a kind of static metaphysical discourse, the discourse of "final vocabularies" and the like, to a living concept, the concept of mathematical outcomes being not up to us.

We do remain haunted by Maddy's question whether depth is incorrigibly subjective; whether depth is *only* in us—because after all, isn't the math is in us too? Not in the sense that the truth claims we want to make in mathematics have some degree of contingency—and indeed, as we are arguing here, shouldn't the Second Philosopher give up that project anyway, of grounding the subject in even very vestigial objectivity?

We mentioned Putnam at various points in this chapter. His argument in *Ethics without Ontology* for the objectivity, sans objects, of ethical judgement, mirrors, in crucial places, Maddy's arguments for the objectivity of mathematics (on the grounds of its tracking depth). When Putnam said that philosophical ethics construed as "merely a subspecies of metaphysical debate ...doesn't do us any good";[30]when he writes of ethical statements as being fully in the range of the notions of truth and falsity, good and bad argument, and the like—*bona fide* statements, in short[31]—he may as well have been speaking about judgements of depth in the character of Maddy's Second Philosopher.

Does the "irreducible specificity" of depth pull against thin realism, and presumably arealism? Putnam asks whether a general *ethical* theory is called for, or even possible; ethics, he worries, might be "situational in the extreme," and this may undermine the main thesis of *Ethics without Ontology*, he worries, namely that *ethical* reasoning is objective.

By the situationality of ethics Putnam is referring to the idea that a moral expression can only be read against the background of a form of life; that in ethics, the "general has no priority over the specific."[32] And indeed Putnam seems to be at his most profound in *Ethics without Ontology* when he urges on us a *personalized* Kantian ethics:

> Kantian ethics, I have argued (as Hegel already argued) is, in fact, empty and formal unless we supply it with content precisely from Aristotelian and Levinasian and yet other directions...Levinas is right to remind us that even if the ethical person acts in accordance with the Categorical Imperative, her focus is not on the Kantian principle as an abstract rule, but on the particular other person she is trying to help.[33]

---

[30] Bouveresse and Putnam (2020), cited in Laugier Laugier (2013).

[31] (Putnam 2004, p. 77).

[32] Diamond, *Fourth BWS Annual Conference: Wittgenstein and the Swansea School*, Gregynog, Wales 16–17 July 2011.

[33] (Putnam 2004, p. 26).

Maddy's is a Kantian aesthetics of depth, in the sense that she takes the objectivity of taste, in the form of judgements of mathematical depth, as a given. But Putnam's observation that we tend to act situationally in our ethical lives, in response to the suffering of a particular person in a particular situation, for example, rather than to an abstract principle like the Categorical Imperative, may be applicable also here. One could argue, in other words, that aesthetic judgement, here a judgement of depth, is also situational. Mathematics is rife with every possible variety of human being, after all, each bringing with them an individualized aesthetic sensibility.[34]

But then, if depth is irreducibly specific; if the expression "mathematical depth" doesn't refer to anything but an atomized—and moreover a seemingly infinitely atomizable—collection of occasion-sensitive concepts, doesn't this undermine the claim that depth is objective? Here again we have a conflict taking place at the border between semantics and pragmatics in the sense, say, of Baz (2012). As Baz puts it:

> Why *must* our capacity to competently employ each and every one of our words presuppose a capacity just to apply this word to cases, apart from any particular interest and without doing anything else with the word beyond sheer application? Why *must* there be a purely semantic component to the understanding of every general term—a component that may theoretically be separated from all of the rest of what this understanding involves and requires, and that may fully be cashed in terms of 'reference' and 'truth-conditions?[35]

This is of course a very longstanding debate in philosophy, between the Ordinary Language Philosophers and, it seems, everyone else. It is a debate between those who wish to treat theoretical philosophical terms such as "knowledge," or "existence" as referring expressions with an isolatable semantic component, analysable in terms of truth-conditions and causal reference,[36] or some other kind of reference, and so forth—and those who see all this as a useless errand. We do not take up the debate here, except to suggest that this line of thought might be less congenial to the Second Philosopher than the Ordinary Language Philosophy framework might be. The Second Philosopher stands alot to gain by tying Second Philosophy to objectivity, even if it is "only" the objectivity of depth; but she may also risk coherence, by making space for "outside" considerations.

Throughout this essay we have been circling around what we see as a tension in the Second Philosopher's position, between a full-blooded notion of objectivity—a notion, we have argued, that the Second Philosopher doesn't need—and the kind of objectivity in the quotidian sense of mathematical outcomes being "not up to us."

Maddy's consequentiality as a philosopher means that she is willing to see the ground beneath her thin realist collapse if it leads to a clearer picture. What of objectivity-based (i.e. strong) arealism? Could objectivity-based arealism go too? One senses that Maddy's Second Philosopher will always be drawn to the project of accounting for the legitimacy of mathematics based on objectivity in some form

---

[34] Of course one could empirically test whether mathematicians speak univocally on aesthetic issues, or even simply on depth and importance.

[35] (Baz 2012, p. 120).

[36] (Baz 2012, p. 119).

or other. But at the end of the day it seems that this being "not up to us" is all that matters:

> If Thin Realism and Arealism are equally accurate, second-philosophical descriptions of the nature of pure mathematics, just alternative ways of expressing the very same account of the objective facts that underlie mathematical practice, then we have here a form of objectivity in mathematics that doesn't depend on the existence of mathematical objects or the truth of mathematical statements, or even on the non-existence of mathematical objects or the rejection of mathematical claims. This form of objectivity is, as you might say, post-metaphysical. Though it doesn't involve truths about a mathematical ontology, it does involve an array of facts like the sort of thing we roughly express by saying that the concept of group opens up a lot of deep mathematics. Indeed, it seems fair to say that the objectivity of such facts is more robust than thin: an Evil Demon could deceive us wholesale about what follows from what or about where the deepest, most fruitful strains lie. Still, the coherence of the radical skeptical challenge isn't enough to revive Benacerraf-style worries: though it's hard to see why our set-theoretic methods should track the truth about the Robust Realist's ontology, they're clearly well-designed (the Demon aside) to track set-theoretic depth.

Maddy has pushed the philosophy of mathematics community into new territory by drawing their focus away from a priori metaphysics and toward the practice—and indeed she is one of the first to do so, and this is one of the many valuable contributions she has made to thought. And if all that we are left with is the recognition that mathematical facts seem to stand or fall apart from anything we do; if all we have left is the feeling that such facts, even as to their importance, or their depth, are not up to us, shouldn't that be enough?

There are undoubtedly philosophers for whom, unlike Maddy, this is *not* enough; philosophers who are hard at work in attempting to deliver a full blown, global theory of truth and existence in mathematics.

To them I would say: this is what solid ground looks like, and one takes a risk in one's thought, by leaving it.

References

Arana, A. (2015). On the depth of Szemerédi's theorem. *Philos. Math. (3)*, *23*(2), 163–176.

Arribas, S. (2007). The call of metaphor: Richard rorty's politics of language. *Logos. Anales del Seminario de Metaf?sica*, *40*(1), 305–328.

Baz, A. (2012). *When words are called for: A defense of ordinary language philosophy*. Harvard University Press.

Bouveresse, J., & Putnam, H. (2020). A conversation between Jacques Bouveresse and Hilary Putnam. *The Monist, 103*(4), 481–492.

Diamond, C. (1995). *The realistic spirit: Wittgenstein, philosophy, and the mind*. MIT Press.

Ernst, M., Heis, J., Maddy, P., Bennett McNulty, M., & Owen Weatherall, J. (2015). Afterword to special issue on mathematical depth. *Philos. Math. (3)*, *23*(2), 242–254.

Floyd, J. (2021). *Wittgenstein's philosophy of mathematics, Elements*. Cambridge University Press.

Floyd, J., & Mühlhölzer, F. (2020). *Wittgenstein's annotations to Hardy's course of pure mathematics—an investigation of Wittgenstein's non-extensionalist understanding of the real numbers*. Vol. 7 of *Nordic Wittgenstein Studies*. Cham: Springer.

Kennedy, J. (2014). Defending the axioms: on the philosophical foundations of set theory [book review of MR2779203]. *Bull. Symb. Log.*, *20*(1), 91–94.

Kennedy, J., & Väänänen, J. (2015). Aesthetics and the dream of objectivity: Notes from set theory. *Inquiry, 58*(1), 83–98.
Laugier, S. (2013). *Why we need ordinary language philosophy*. University of Chicago Press.
Maddy, P. (2009). *Second philosophy. A naturalistic method*. Oxford: Oxford University Press.
Maddy, P. (2011). *Defending the axioms: on the philosophical foundations of set theory*. Oxford: Oxford University Press.
Putnam, H. (2004). *Ethics without Ontology*. Harvard University Press.
Rorty, R. (1989). *Contingency. Irony and solidarity*. Cambridge University Press.
Rorty, R. (1991). Essays on heidegger and others. In *Philosophical Papers*, vol. 2. Cambridge University Press.
Rorty, R. (1991). Objectivity, relativism, and truth. In *Philosophical Papers*, vol. 1. Cambridge University Press.

Chapter 31
Reply to Kennedy

Penelope Maddy

Keywords Mathematical depth · Aesthetics · Set theory · Objectivity · Naturalism

Professor Kennedy[1] goes straight to the heart of the set-theoretic metaphysics of *Defending the Axioms*: mathematical depth and the question of its objectivity. She considers both the unabashed Objectivism of that book—the Thin Realism/Arealism described there—and the doubts raised in a workshop report a few years later (Ernst et al., 2015)—the possibility that failure of objectivity could undermine Thin Realism and leave only a weakened form of Arealism.[2] Along the way, she runs the notion of depth through a whirlwind of compare and contrast with thoughts of Wittgenstein, Rorty, and Putnam, ending with a quotation from a book by Avner Baz that raises the spirit of J. L. Austin (a great philosophical hero of mine[3]). There obviously isn't room here to react to all of this, but I do want to touch on two fundamental questions she raises, one about depth and one about objectivity.

[1] My heartfelt thanks to Professor Kennedy and Professor Arbeiter for their good efforts in preparing this volume.

[2] For the record, in Ernst et al. (2015), I take the objectivity of depth, the objectivity underpinning Thin Realism/(strong) Arealism, to be an open question; in other words, I'm not yet ready to give up on strong Arealism and settle conclusively for weak. (As Kennedy indicates, we use 'strong Arealism' for the position in Maddy (2011), backed by objective depth, and 'weak Arealism' for the position floated on p. 248 of Ernst et al. (2015), that is, what's left of Arealism if depth should turn out not to be objective.).

[3] See, e.g., Maddy (2017a), Maddy (2022g).

P. Maddy (✉)
Department of Logic and Philosophy of Science, University of California, Irvine, CA, USA
e-mail: pjmaddy@uci.edu

S. Arbeiter and J. Kennedy (eds.), *The Philosophy of Penelope Maddy*,
Outstanding Contributions to Logic 31, https://doi.org/10.1007/978-3-031-58425-1_31

At the outset, Kennedy is inclined to view 'depth' as a term of aesthetic appraisal.[4] Untutored as I am in aesthetics, this identification makes me uneasy. So, for example, a set theorist might offer the role of the large cardinal hierarchy in calibrating consistency strength as one component in an argument in favor of adding large cardinal axioms to our fundamental theory of sets. Is this an aesthetic judgment? It strikes me instead as a simple means/ends defense: it's mathematically useful to have a reliable measure of risk, and these axioms deliver it.[5] Or consider the definition of Woodin cardinal. I suspect few would regard it as beautiful, but the role of Woodins in the hypotheses of major theorems—beginning with the prize-winning Martin-Steel-Woodin result on determinacy—lead most to recognize it as a mathematically important concept, an especially significant milestone along the hierarchy of cardinals.[6] So at least at first blush, it's not obvious to me that depth—for axioms or for concepts—is a species of aesthetic virtue.

Still, at least for now, I doubt that much hangs on this. What's undeniably significant is Kennedy's concern about 'irreducible specificity',[7] a concern she derives from Putnam's writings on moral philosophy. In our examples, the mathematical virtues of the large cardinal hierarchy and the concept of Woodin cardinal are understood as good reasons to adopt the axioms and the concept, respectively. The question concerns the justificatory force those virtues provide for their respective adoptions. One possibility is that it derives from the specific features of the particular mathematical context, its logical interconnections and its local goals; another that it derives from more general, higher-level features that these particular cases share with many others. Putnam's worry is that if moral reasoning depends entirely on the particulars of each case, then it cannot be objective; Kennedy's corresponding concern is that if depth is 'irreducibly specific', then it cannot be objective.

To see what's at stake, consider the inference from the premises 'if I have the key, then I can open the door' and 'I have the key' to 'I can open the door'. The justification that the premises provide for the conclusion derives, not from the details of this key and that door, but from the form of the inference, that is, from modus ponens as a general rule. The corresponding question, then, is whether the status of various mathematical virtues as legitimate reasons derives, on the one hand, from local details or, on the other, from their being indicators of a more general feature of depth—where depth would then be subject to analysis or investigation at a higher level of generality, an investigation somehow comparable to that of logical validity.

I confess to considerable skepticism that it's the latter. Still, even if justificatory force is 'irreducibly specific', I resist the conclusion that depth can't be objective, that it can't be an objective matter of fact that a measure of consistency strength or the

---

[4] See also Kennedy and Väänänen (2015).

[5] This is called Risk Assessment in Maddy (2017b) and Maddy (2019a). Even adherents of so-called 'univalent foundations' turn to set theory to assess the riskiness of their type theories ((2019a), p. 309).

[6] Recent results suggest that extendible cardinals may eventually achieve this sort of distinction.

[7] Kennedy apparently draws the phrase from Ernst et al. (2015), p. 246.

definition of Woodin cardinal is, in and of itself, mathematically productive in objectively significant ways. In fact, I've proposed an analogous position on confirmation in natural science[8]: that the epistemic force of an observation or experimental result is a function of the details of that evidence and of its theoretical context, not of a general feature it shares with other somehow analogous cases (as in the case of logic inference). This isn't to say there might not be useful or illuminating higher-level investigations of empirical confirmation or mathematical depth in a broad range of cases, but the thought is that any general principles uncovered would be descriptive, not normative. If a compelling example came along that didn't fit the principal, the example wouldn't be rejected on that account—rather, the principal would be adjusted to fit.[9]

A second fundamental point pressed by Kennedy concerns not the objectivity of depth but the notion of objectivity itself; her charge is that objectivity, unlike depth, is 'outside' mathematical practice, 'a philosophical distraction' (p. 396) drawn from 'a … static metaphysical discourse' that the Second Philosopher would do well to 'give up' (p. 397). I resist this characterization. While the notion of objectivity at issue here is indeed 'outside' mathematics, the Second Philosopher's broad empirical purview includes studying mathematics from outside as a particular human practice among many. From this scientific perspective, she can weigh its significance (for example, its dramatic role in her most penetrating empirical inquiries) and ask what kind of discourse it is (as compared with, say, archeology, astrology, or literary fiction). Is it a referential or descriptive undertaking? Does it have a subject matter? If so, how do its methods manage to track that subject matter? If its assertions can be properly judged as correct or incorrect, are those judgments objective or simply reflections of human preferences or mores?[10]

These are all perfectly legitimate questions, the very sort of thing addressed in *Defending the Axioms*. There's a simple division of labor here: *Naturalism in Mathematics* aims to investigate and evaluate mathematical methods from inside, from a perspective no different from the practitioner; *Defending* then steps outside, into an ordinary scientific posture, and asks what sets must be like, how the epistemology must work, if these are the proper methods. In short, the one book is methodology, the other metaphysics—together, they illustrate what the philosophy of mathematics looks like from the Second Philosopher's naturalistic point of view.

[8] See Maddy (2007), pp. 401–403, (2022f), §iv.

[9] I suggest that this is what happened with 'detection' when Perrin's experimental results confirmed Einstein's predictions: this was taken to show that direct observation wasn't the only form of detection, after all (Maddy (2007), p. 402, Maddy (2022f), p. 86).

[10] I don't distinguish objectivity here from Kennedy's 'not up to us' (p. 397), but it is more than her 'feeling that such facts … are not up to us' (p. 399).

References

Ernst, M., Heis, J., Maddy, P., McNulty, M. B., & Weatherall, J. 0. (Eds.). (2015). *Mathematical depth*. Special issue of *Philosophia Mathematica* 23(2).

Kennedy, J., & Väänänen, J. (2015). Aesthetics dream objectivity: Notes from set theory. *Inquiry, 58*, 83–98.

Maddy, P. (2007). *Second philosophy*. Oxford University Press.

Maddy, P. (2011). *Defending the axioms*. Oxford University Press.

Maddy, P. (2017a). *What do philosophers do?* Oxford University Press.

Maddy, P. (2017b). Set-theoretic foundations. In A. Caicedo, J. Cummings, P. Koellner, & P. Larson (Eds.), *Foundations of mathematics: Essays in honor of W. Hugh Woodin's 60th birthday* (pp. 289–322). American Mathematical Society.

Maddy, P. (2019a). What do we want a foundation to do?. In S. Centrone, D. Kant, D. Sarikaya (Eds.), *Reflections on the foundations of mathematics* (pp. 293–311). Springer.

Maddy, P. (2022f). On the question of realism. In [2022h] (pp. 49–90).

Maddy, P. (2022g). A note on truth and reference. In [2022h] (pp. 173–187).

Maddy, P. (2022h). A plea for natural philosophy and other essays. Oxford University Press.

Chapter 32
The Flight from Application

Mark Wilson

Abstract Through Quinean appeal to their alleged "ontological needs," contemporary philosophical thinking continues to blend the working methodologies of contemporary physics and mathematics into a coarse amalgam that fails to capture the concrete considerations that inform how these entangled subjects enlarge and thrive within real life practice. In our individual manners, Maddy and I have each attempted to combat the ensuing methodological distortions by considering in closer detail the actual motivations that operate within the various sectors of "pure" and "applied" mathematical division. To this end, Maddy has occasionally cited traditional deflationist motifs as a convenient tool for isolating her specific localized projects. In this essay, I recommend against employing this particular mode of "divide and conquer" tactic.

Keywords Pure and applied mathematics · Quine · Motivations for mathematical research · Warehouse conception of prefabricated applications

> I think that it is a relatively good approximation to truth (which is much too complicated to allow anything but approximations) that mathematical ideas originate in empirics. But, once they are conceived, the subject begins to live a peculiar life of its own and is governed by almost entirely aesthetical motivations. In other words, at a great distance from its empirical source, or after much "abstract" inbreeding, a mathematical subject is in danger of degeneration. Whenever this stage is reached the only remedy seems to me to be the rejuvenating return to the source: the reinjection of more or less directly empirical ideas.—John von Neumann[1]

For several decades the philosophical world has been captivated by the discomforts of W.V. Quine's well-known "indispensability argument," which argues that mathematical objects must be ontologically tolerated on an equal footing with the quarks and the quail, insofar as the former continue to appear within the descriptive "postulations" of physical science. In response, no rescuing knight has more valiantly

M. Wilson (✉)
University of Pittsburgh, Pittsburgh, USA
e-mail: mawilson@pitt.edu

S. Arbeiter and J. Kennedy (eds.), *The Philosophy of Penelope Maddy*,
Outstanding Contributions to Logic 31, https://doi.org/10.1007/978-3-031-58425-1_32

labored to dispel these enchantments than Penelope Maddy, through a long series of enlightening books and papers.

Her central target is Quine's misguided form of "naturalism" which leads him to characterize "starry-eyed set theorists" in this disconcerting manner:

> What now of those parts of mathematics that share no empirical meaning, because of never getting applied in natural science? What of the higher reaches of set theory? We see them as meaningful because they are couched in the same grammar and vocabulary that generate the applied parts of mathematics. We are just sparing ourselves the unnatural gerrymandering of grammar that would be needed to exclude them... The main axioms of set theory are generalities operative already in the applicable part of the domain. Further sentences such as the continuum hypothesis and the axiom of choice, which are independent of those axioms, can still be submitted to the considerations of simplicity, economy, and naturalness that contribute to the molding of scientific theories generally. Such considerations support Gödel's axiom of constructibility ... It inactivates the more gratuitous flights of higher set theory.[2]

As Maddy correctly documents, such an attitude towards "V = L" is completely opposite to those prevailing amongst mathematicians who know their subject better.

Considered purely as a matter of developmental epistemology, Quine plainly has the dependencies of real life investigation largely reversed. As Lord Kelvin contemporaneously noted, most of descriptive successes of nineteenth century physics represent applications of Fourier analysis in some mode or other. But these physicists came to recognize the full power of the method only because they had antecedently trusted functional constructions that implicitly rely upon the Axiom of Choice; they did not postulate the latter "for the sake of the physics." Standard set theory as it became later codified grew in an organic manner from the patterns of limiting value construction that Sturm and Liouville utilized in justifying Fourier-like eigenfunction decomposition across a wide range of unexpected applications. To be sure, these investigations eventually strayed into abstract mathematical regions that Quine would designate as "starry-eyed," despite the fact that from the very beginning Sturm and Liouville had set their sights upon those very stars, rather than following some amorphous urging of "physics' needs."

Through these subsequent successes, modern set theory became firmly established as the court of final appeal with respect to the internal coherence of a purported mathematical arrangement, providing the basic platform upon which the vital existence theorems for partial differential equations rest (and many allied questions as well). And Maddy has made all of this clear.

However, in articulating her case for methodological liberation, she occasionally acquiesces in rhetorical motifs that can leave an undischarged residue of Quinean presumption in place with respect to the alleged "needs" of applied mathematics. This "undischarged residue" will be my topic in this essay. Insofar as I can see, these concerns do not affect her central project much, but they remain pertinent to the considerable swatch of philosophers who continue to accept Quinesan themes. And sometimes she employs phraseology that strikes me as unfortunate in these respects. For example, in her well-known article, "How Applied Mathematics Became Pure," she writes:

> What we have traced is a more or less simultaneous rise of pure mathematics and reevaluation of applied mathematics. Before all these, back in Newton's or Euler's day, the methods of mathematics and the methods of science were one and the same... [T]he needs of celestial mechanics (for Newton) or rational mechanics (for Euler) are the needs of mathematics. From this perspective, the correctness of a new mathematical method – say the infinitary methods of the calculus or the expanded notion of function– is established by its role in application. In contrast, after the developments we have been tracing, mathematics has been freed to pursue inquiries without application, it is encouraged to stock the warehouses with structures and leave the choices to the natural scientists... In this brave new world, where can the pure mathematician turn for guidance? How can we properly determine if a new sort of entity is acceptable or a new method of proof reliable? What constrains our methodological choices? I think there are answers to these questions ... – but that is a story for another day.[3]

This "story for another day" is then supplied within her reports on the motivations behind current research in higher set theory, which I have already praised.

Borrowing terminology from this very passage, Robert Batterman has criticized a popular presumption at the center of many contemporary attempts to "explain the utility of applied mathematics" in a simple manner. Here is a typical expression of what Bob calls the "warehouse conception of physical structure":

> [W]hy is it that mathematics is so useful in empirical science? One answer is simply that mathematics is a rich source of structures and therein lays its utility. Some mathematical structure is either designed to, or otherwise found to, accurately capture the important structural relations of an empirical set up, and we can thus read off important facts about the empirical set up from the mathematics.[4]

But exactly what such authors have in mind under the heading of "structures" or "models" is invariably left unclear, as we shall see when we examine a few examples.

Let me hasten to add that Maddy herself does not accept these simple assumptions of "structural isomorphism"—much of "How Applied Mathematics Became Pure" explicitly argues against them. But it will still be useful if we can articulate a better appreciation of how von Neumann's "reinjections of more or less directly empirical ideas" operate, so that mathematics' flight from applications isn't quite as severe as the "warehouse of structures" metaphor may suggest.

I will draw my sermon for today from a canny remark from the celebrated analyst Hans Lewy:

> [Mathematical analysis only supplies] kind of hesitant statements... In some ways analysis is more like life than certain other parts of mathematics... There are some fields... where the statements are very clear, the hypotheses are clear, the conclusions are clear and the whole drift of the subject is also quite clear. But not in analysis. To present analysis in this form does violence to the subject, in my opinion. The conclusions are or should be considered temporary, also the conclusions, sort of temporary. As soon as you try to lay down exact conditions, you artificially restrict the subject.[5]

Levy has in mind the fact that, within a concrete applicational setting, a variety of descriptive considerations must often be tweaked into harmonious adjustment with one another, through complex adjustments that belie uniformist labels such as "structure" or "model." More specifically, the proper "meanings" for the interior differential operators within a set of modeling equations must frequently become adjusted to suit the requirements of their attached boundary conditions, in a hard-to-anticipate manner that explains why Levy characterizes these "meanings" as "hesitant."

Before we consider a specific "hesitancy" of the type Lewy has in mind, let me first give a brisk overview of a modern modeling tactic that emerges as very central within my recent *Physics Avoidance*,[6] for the technique places robust computational flesh on the more skeletal suggestions of my earlier *Wandering Significance*.[7] Descriptive strategies of this type are generically designated as "multiscalar models," for they rely upon a self-correcting network of submodels associated with characteristic length ΔL and time Δt scales that communicate with each other through specialized messages called "homogenizations" (I'll call these networks "architectures" in the sequel). By distributing computing tasks amongst the submodels following a relatively simple form of "divide and conquer" policy, engineers have achieved truly astounding advances in the reliable simulation of complexly layered states of affairs. One can't walk through the science halls of a modern university and not encounter many posters advertising "A Multiscalar Approach to ...".

Here's a much simplified, two-level illustration of how these "architectures" operate (Fig. 32.1). Suppose we have a large mass of buried granite and wish to ascertain what might happen if a strong earthquake wave passes through. A very simple computational architecture for this situation might proceed as follows. (Step i) At a macroscopic scale ΔH, first model the rock as a continuous Hookean blob in the manner of standard mechanics. (Step ii) At any point where the predicted stresses become high, frame a lower submodel ΔL in which a localized region around p is instead modeled as a crystalline laminate made of component grains of quartz, feldspar, etc. (Step iii) Relying on the local environment induced around ΔL by the higher scale ΔH submodel, determine whether these crystals will (a) simply shear in place without alteration in the border region adhesions or (b) reattach their bonds in the layered manner of the metamorphic rock gneiss. (Step iv) If condition (iiia) holds, the ΔH modeling doesn't require correction, but if (iiib) obtains, a corrective instruction must be sent to ΔH to alter its descriptive parameters to anisotropic values. This step is represented by the feedback arrow in the diagram provided. (Step v) Perform these corrections over and over again until ones ΔH and ΔL assessments eventually fall into harmony with one another.

As such, our scheme clearly represents a sophisticated version of the venerable tactic of computing answers through "successive approximation." The task of computing the curvature of a loaded clothesline supplies a sterling exemplar (Fig. 32.2). It operates by simply starting out with an arbitrary guess G_0 with respect to the clothesline which it progressively tweaks in localized adjustments to determine if we can thereby lessen the bending energy assigned to the cord. We will thereby generate a sequence of improving estimates, $G_1, G_2, \ldots$. that will (hopefully[8]) converge to a fixed point limit G_∞ capturing the proper loaded curvature.

This same tactic of iterated corrected convergence is central to the descriptive advances obtainable through a multiscalar architecture. Our granite example works with just two submodels associated with the scale lengths ΔH and ΔL. These submodels progressively evaluate whether a proposed trial scenario satisfies the behavioral expectations that normally obtain at that scale length in question (PA

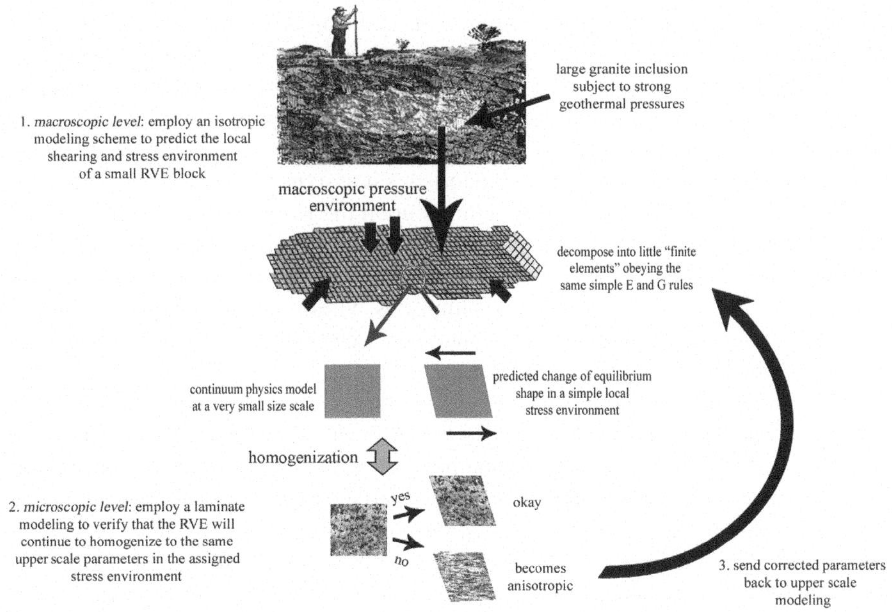

Fig. 32.1 Multiscale modeling of a granite block

Fig. 32.2 Relaxation method

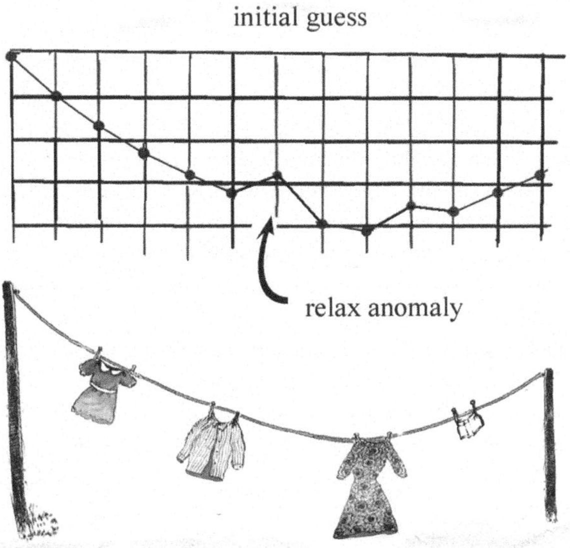

calls these "normally dominating behaviors"). Thus the ΔH submodel evaluates our rock as a continuous blob obeying some form of stress/strain relationship, whereas the ΔL submodel employs a laminate modeling to evaluate whether its inter-crystalline bonds will shift within a localized stress environment. Starting with an initial ΔH level guess G_0 about our rock's condition, the scheme progressively scrutinizes G_0 by determining whether the local characteristics it assigns to the rock are consistent with the behaviors that we normally witness upon the length scale ΔL. If the latter modeling instead decides, "No, the bonds will shift within a stress environment that intensive," our architecture then sends a feedback message back to ΔH demanding that it tweak its initial stress/strain assumptions within the relevant rock locality. By iterating these corrective tests over and over, we obtain a (hopefully) improving sequence of estimates G_1, G_2, ... just as before.

Prima facie, one might suppose that engaging in all of these repetitious improvements will take much longer than simply trying to calculate the rock's condition on a wholly ΔL basis. But this armchair supposition is wildly mistaken. The computational efficiencies obtainable through multiscalar successive approximation can be truly enormous.[9] On this theoretical basis, *Physics Avoidance* argues that many of our everyday modes of reasoning with respect to complex circumstances tacitly employ similar strategies in their operations, even if we fail to notice the backgrounded architecture that helps us achieve those skills. How could we do otherwise? Practicality demands a certain minimum of compliance.

Let's now observe that the mathematical arrangements that we loosely call "solutions" or "models" of a differential equation modeling frequently incorporate interactive subcomponents much like the varied ΔX levels in our multiscalar exemplar. Recognition of these unnoticed "architectures" began reshaping our understanding of standard modeling arrangements in the mid-twentieth century and illustrate some of the background pertinent to Lewy's "hesitancies."

To these purposes, let's employ Poisson's equation as a simple method for modeling a two-dimensional drumhead saddled with a pile of rocks. In this situation the relevant modeling equation is $\partial^2 h/\partial x^2 + \partial^2 h/\partial y^2 = W/k$, applicable to the interior region of the membrane and claiming that it statically maintains its loading W by bending in a manner proportional to k (this "bending" is encapsulated within in the second derivative expression "$\partial^2 h/\partial x^2 + \partial^2 h/\partial y^2$" where h is the membrane's displacement from the plane). As such, the situation represents a two-dimensional analog to the equation applicable to the loaded strings of Physics 101, viz., the one-dimensional static equation $\partial^2 h/\partial x^2 = W/k$. Considered naively, we expect that if a string is loaded by a unit weight at a single location P, the cord will bend into a V shape centered upon P as illustrated (Fig. 32.3). Simple considerations of symmetry further suggest that a comparable "loaded at P" outcome for our rectangular-shaped membrane will consist in two V-like patterns cross-multiplied against one another, again as illustrated. These patterns comprise the so-called are "influence" or "Green's functions for the string and membrane, for they capture their responses in the most elemental circumstances of being loaded at a single location P.

Fig. 32.3 Influence functions

In linear circumstances, the outcomes of more complicated loadings can be conveniently computed by simply adding together a parcel of influence functions centered at all of the relevant locations.

In making these determinations, we have tacitly assumed that the edges of our two examples are completely tacked down, with no wiggling permitted along these boundaries. Such an assumption is officially labeled a "Dirichlet boundary condition." If our membrane is instead held in place by springy clamps, the boundaries are governed by a so-called "Neumann condition" (or something more complicated). Comparable influence functions can be derived for situations in which the boundary is tugged by a constant load placed upon one of these boundary points. Once again in linear circumstances, we can compute the effects of wider patterns of boundary loading by adding up the appropriate influence functions. This is the natural manner of proceeding if we move to three dimensions and consider how a rectangular box will sag when we scatter a bunch of rocks on top.

All of these influence function decompositions may seem perfectly kosher at first glance, but from the very earliest days when equations of these types were utilized by applied mathematicians, it was realized that arrangements were somehow akilter[10]: the "influence functions" we have constructed can't properly qualify as "solutions" to their respective interior equations in a traditional sense, for these formulas demand second derivatives (e.g., $\partial^2 h/\partial x^2$) as their means of expressing what "bending by a load" requires. But these operations become ill-defined if our strings and membranes distort too sharply at P and other locales. For this reason, such constructions were traditionally labeled as "formal solutions," indicating that they really aren't "solutions" at all.

This "formal" label meant that our influence functions became treated for most of their subsequent career as mathematical outcasts of a sort, tolerated for the many practicalities they facilitate but shunned for their lack of rigorous placement (in the mode whereby ruthless bosses exploit their "courtesy employees" while denying them proper recognition and compensation for their assistance[11]). Gabriel Barton captures these traditional attitudes as follows:

> Of course, in practice the [point load procedure] is freely applied even to regions that do have corners and edges: for instance, we often deal with rectangular boundaries. While the

Fig. 32.4 Finite element decomposition

> mathematician might regard such boundary value problems as ill-defined at a corner, the physicist takes the explicit expression [derived formally by the point load method] as his solution; if he needs [a value for a quantity] ψ near a corner, then he simply consults the [formal expression for ψ] in order to see what the solution is. The justification is that, on the one hand, the corners of real physical boundaries, like metal boxes, are not mathematically sharp; while, on the other hand, only in regions negligibly small by normal standards is there any appreciable difference between the solution calculated for the true shape with rounded corners, and the idealized solution calculated for idealized rectangular shapes, albeit ignoring any mathematical ambiguities at the corners.[12]

For later purposes, let me interject the observation that, in a certain sense, a huge difference separates these two "solutions": the perfectly rectangular specimen can often be encapsulated within a discrete analytic formula (as in our example), whereas any rounded corner compatriot can only be captured within lengthy numerical routines. We'll later find that this divergence signals a tremendous disparity in the informational utilities of the two representations.

As often occurs with exploited labor of all kinds, abuses of rigor begin to proliferate as the utilities of the unsanctioned practices expanded more widely. Increasing swatches of applied mathematical inquiry became tolerated as "formal solutions" and eventually generated initiated crises with respect to trustworthy reasoning technique. A good example developed during World War II when aircraft designers dissected airplane wings into little squares (the official term is "tessellate") and utilized our P-loaded influence functions as a basis for computing the stresses to which the wing becomes subject under an exterior load (Fig. 32.4). These sallies into uncharted territory led to both successful and dreadful computational outcomes. Plainly these "officially unsanctioned" practices demanded a more rigorous, yet sympathetic, brand of mathematical scrutiny than their traditional dismissal to the unsanctioned sidelines.

As applied mathematicians pondered these issues, they gradually concluded that unacknowledged issues of *descriptive injustice* lay at the core of those old-fashioned "utilize but don't sanction" policies. After all, doesn't the gambit of describing a drumhead as "governed by the Poisson equation" embody every bit as much "idealization" as the characterization of its boundary being shaped like a perfect square? By accepting the continuum physics equation as an apt descriptive tool at all, we have inherently "idealized" (= smoothed over) a large amount of lower scale complexity with respect to the twisted polymer fibers that comprise a real life drumhead. In this sense, our membrane equation represents a natural mode of description on a largish scale of observation only, in the same way that our granite boulder is modeled as a continuous blob the higher (ΔH) submodel of our previous example (in contrast, the lower scale ΔL submodels utilize a somewhat different set of "laminate physics" tools). So why did the old-fashioned treatments presume that interiors were somehow less "idealized" than edges? One can't have a drumhead without both!

Such considerations, of course, raise the further question, "Why employ differential equations at all in addressing membrane behaviors that only become salient at a decidedly macroscopic scale?" We'll return to this issue later, but let us presently concentrate upon the comparative question of why prior tradition has treated the mathematical demands ("Give me two derivatives!") of the interior equations to qualify as less "idealized" than the comparable expectations we consign to boundaries upon the same macroscopic scale ("I can't see how cutting out an expanse of fabric with perfectly rounded corners will prove any easier than shaping it into a perfect rectangle").

Following this line of revisionist thinking, applied mathematicians came to two conclusions: perhaps our influence function woes merely reflect the facts that (1) the notion of "bending energy" has been improperly registered within our double derivative ($\partial^2 h/\partial x^2$) demands and (2) our two modes of macroscopic description (edges and interiors) don't immediately agree with one other in the simple manner we might have wanted. With respect to (1), an applied mathematician can assign a coherent amount of "bending energy" to the little areas **A** surrounding a problematic interior point P through computing the strain energies that the **A**'s would possess if their bumps at P were smoothed over a bit.[13] Mathematicians say that P's problematic behaviors have been "mollified" through these consultations with its tamer neighbors.

With respect to (2), we should imitate the divide-and-conquer strategies utilized within our multiscalar models and consign interiors and boundaries to separate spaces **δI** and **I.** We can then seek a manner of messaging that allows these two modes of description to "talk to one another" without fully agreeing upon the properties of a problematic point P. Such a tolerance of incompatible descriptions is crucial to the efficiencies of our granite example: its ΔH level characterization "sees" a smooth continuum around the point P while the ΔL level accounting instead detects a conglomeration of crystals in the same region. Nonetheless the overall scheme regards these level-sensitive evaluations as coexisting in descriptive harmony as long as the ΔL submodel doesn't predict a substantial degree of grain slippage around P. The secret for insuring this harmony lies in the special "homogenization messages" that the submodels exchange with one another. Just so with our **δI** and **I** pairings: the latter sends a so-called "trace deposit" to **δI**.

Under these arrangements, the interior region **I** may detect a singularity forming in a limit as we draw closer to P, but the mollified "trace deposit" message it sends to **δI** may only require a point load W there. The rationale for these **δI** and **I** subcategories don't reflect some ontological divide present within the external world itself, but simply the fact the descriptive tools that capture boundary and interior behaviors most effectively at the macroscopic level don't fit together as perfectly as neatly as we might like. We tame these disagreements by jerry-rigging the manner in which the two schemes communicate with one another. The necessity of doing so doesn't stem from the natural world itself; it merely reflects an awkwardness within the mathematical tools with which we naturally work.

So far so good, but a "worser problem" of the same ilk emerges when we turn our attention to boundaries that contain inset corners (like the notch pictured), because here the ill-matched blowups don't become adequately tamed under the simple mollifications we previously utilized. In real life, too, these situations are energetically problematic because sharp inset corners act as stress concentrators that can develop large cracks (such as the ones that destroyed the Liberty ships built during World War II, which were unwisely outfitted with rectangular cargo hatches) (Fig. 32.5). For this reason, learning that a particular form of **<δI,I>** matchup will generate Liberty ship-like stress concentrations is a valuable thing to learn, even if any attempt to implement those arrangements will immediately crack apart. So applied mathematicians have developed very clever schemes for tolerating wider notions of "solution" that can categorize and locate the locales where various forms of energetic problem are likely to arise. Such knowledge is vital within computer practice as well, for conventional numerical methods must skirt irregular regions such as these (numerical methods are so dumb that they can't distinguish a large stress from a genuine blowup). Engineering novices are often surprised to find that their advanced primers expect them to master a large quantity of rather subtle functional analysis. But these are the diagnostic tools required to anticipate the behavioral problems they are likely to encounter in the context of real life design.

Inherent within these considerations is a lesson established in from the nineteenth century: in terms of descriptive informativeness, singularities frequently serve as more effective markers of behavioral abnormalities than conventional, if unrealistic, high values. Thus applied mathematicians have learned to cherish poles, blowups, shock waves, wavefronts and a large menagerie of other signals that we might otherwise dismissive as descriptive failures.

To complete this litany of hesitancies, I'll briefly mention a further descriptive inequity within our **δI/I** pairings, viz., the **δI** data must generally be articulated in a lower dimensional manner than the interiors **I.** Cornelius Lanczos observes:

> And here we have first of all to record the fact that from the physical standpoint a "boundary condition" is always a simplified description of an unknown mechanism which acts upon our system from the outside... Imposed boundary conditions are merely circumscribed interventions from outside which express in simplified language the coupling which in fact exists between the system and the outside world.[14]

Indeed, the physical processes active along the outer surfaces of a real life metal parallelopipedon are considerably more complicated than those applicable within

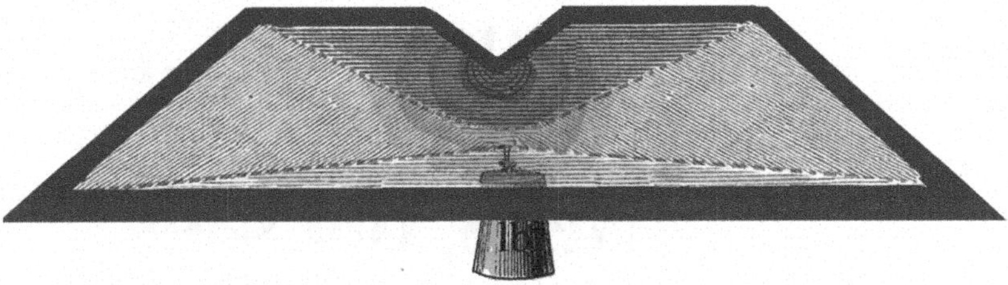

Fig. 32.5 Liberty ship failure

its interior. It's merely that the former are compact enough that we can usually condense their properties onto two-dimensional sheets and get away with it. But that is only "usually." If the outer layer hasn't been highly polished, it will contain innumerable microcracks that remain stable under normal conditions. But they will begin to enlarge if ambient conditions become extreme. The standard method for handling these matters (pioneered by A.A. Griffith in 1920) adopts the submodel tactics of a multiscalar scheme and shifts to a ΔL level investigation in which a local stretch of the smooth boundary appearing in the ΔH treatment of the box is replaced by an altered arrangement containing a trial crack. If no energetic advantage ensues if the little crack lengthens, the technique leaves the ΔH treatment as it was, but if not, a warning correction is sent to the higher level.[15]

Fig. 32.6 Helpful information

What can we learn from these ruminations? Popular magazines often advertise "information you can use" (Fig. 32.6).

Well, we've just observed that successfully encoding data about the external world within a linguistic format capable of guiding computation along profitable pathways often requires that "the information we can use" be found within the separated corners of interconnected networks that are more architecturally elaborate than we might anticipate while musing in our philosophical armchairs. This delicate interlacing of component parts explains why Lewy views the "proper interpretation" of a differential equation modeling as hostage to a shifting array of motivational considerations.

The various policies just examined for fitting together ill-matched pieces provide sterling exemplars of the ways in which the requirements of effective physical description offer what von Neumann calls a "rejuvenating return to the source: the reinjection of more or less directly empirical ideas." In fact, an apt choice of submodel levels within a multiscalar scheme usually attempts to mimic the natural structural layering objectively found in nature.[16]

The philosophical conceit that these jerry-rigged assemblies represent structural possibilities foreseen by the pure mathematician and which sit prefabricated upon the shelves of its grand warehouse can encourage the lack of attention to diagnostic detail that is characteristic of contemporary philosophy of science.[17] Even prior to this, many authors view the fictitious pretense that real life "theories" emerge as unified wholes brought into being through a concentrated blast of creative postulation as a harmless idealization–an innocuous species of epistemological simplification that allows a methodologist to survey "how science works in principle" without worrying about messy developmental details.[18] Likewise, an "already constructed within pure mathematics" presumption suppresses the vital architectural clues that only become apparent when the ad hoc assembly work characteristic of effective applied mathematics is inspected closely.

Certainly, vagaries as to what the terms "theory" and "model" connote in everyday usage play a significant role in perpetuating these reductive attitudes. We standardly characterize the multiscalar architecture **A** presented above as "a *theory* of morphological adjustment in granite" and speak of its stabilized adjustments as providing a "*model*" **M** of a particular seismic disturbance. But neither **A** nor **M** closely resemble the $<\mathbf{D}, \phi_1, \phi_2, \phi_3, \ldots>$ arrangements that logicians standardly study under these same headings. Indeed, underspecified schematisms of a $<\mathbf{D}, \phi_1, \phi_2, \phi_3, \ldots>$ class disguise or suppress the crucial clues that suggest that some form of piecemeal architectural assembly is active in the background of a descriptive practice, even if these arrangements have not been explicitly diagnosed as yet. Within the logicians' "models," no overt representations of interiors and boundaries, homogenization messages between submodels, lowered dimensionalities, etc. are found, despite the fact that these clues supply the surest symptomatology that some non-trivial controlling architecture provides vital structural assistance. And standard "theory T thinking" actively encourages these suppressions, in its attempts to speak about the 'scientific method" in the most sweeping terms possible. A large portion of *Physics Avoidance* is devoted to arguing that philosophers of science abandon their most useful diagnostic tools in doing so.

Let's turn to another significant advantage that intercommunicating architectures offer: a capacity for readily accepting profitable infusions of reliability from varied sources. In their justly celebrated *Treatise on Natural Philosophy*, William Thomson (Lord Kelvin) and P.G. Tait write:

> Enough, however, has been said to show, first, our utter ignorance as to the true and complete solution of any physical question by the only perfect method, that of the consideration of the circumstances which affect the motion of every portion, separately, of each body concerned; and, second, the practically sufficient manner in which practical questions may be attacked by limiting their generality, the limitations introduced being themselves deduced from experience, and being therefore Nature's own solution (to a less or greater degree of accuracy) of the infinite additional number of equations by which we should otherwise have been encumbered.[19]

In constructing a suitable modeling of nature, we shouldn't attempt to build everything from scratch, but instead exploit the reliable partial knowledge we currently possess as effectively as we can. Here's an illustration of what Thomson and Tait have in mind. Suppose that on a macroscopic scale, we encounter a wooden rod submerged within a fluid. Suppose further that we wish to develop an account of the system's interactions upon a molecular scale. But our macroscopic knowledge of the stick's rigidity allows us to conclude that, however else the forces between our molecules behave, some subgroup of them manages to keep the rod's own molecules (marked as A, B and C in the diagram (Fig. 32.7) in close proximity to one another. We'd like to exploit this robust, albeit partial, knowledge of molecular location effectively if we can. The standard technique of "Lagrange multipliers" for doing so transmits our upper scale "the stick is rigid" knowledge into the lower domain by inserting so-called "constraint forces" (symbolized by white arrows) within the molecular arena. As the regular (black arrow) molecular forces move the particles about, we continually compensate for their activities by providing the white arrow forces with exactly the strengths required to pull A, B and C back to where they're supposed to

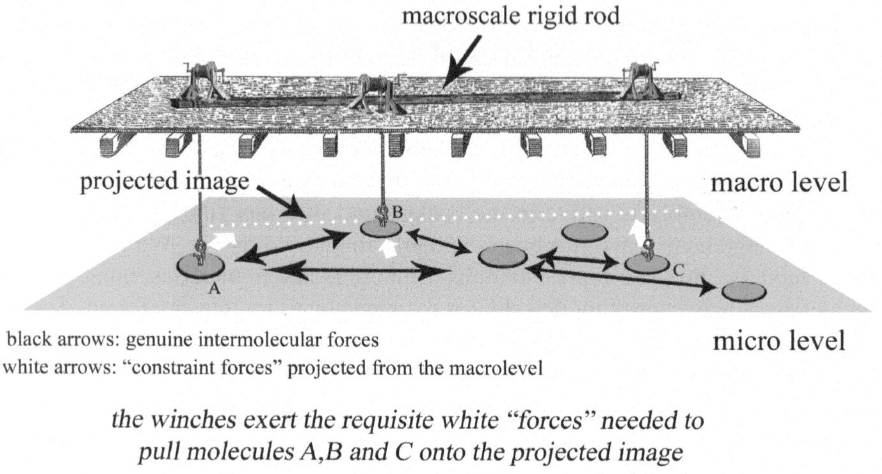

Fig. 32.7 Lagrangian multipliers

be (I've symbolized these adjustments) by little winches on the macroscopic scale; their exact values generally need to be computed through successive approximation).

Employing the strategic terminology that we applied with respect to the multiscalar models, the Lagrange multipliers supply the corrective feedback messages the macrolevel needs to send to the microlevel to ensure that its computations don't violate the partial data we already possess with respect to the total system. These upper scale projections assume the form of white arrow "forces" because talk of vector addition represents a language that the lower scale modeling "understands" (the basic source of multiscalar efficiency lies in the fact that its submodels manage to "talk to one another" without sharing a proper common language).

As Kelvin and Tait vigorously emphasize in their own discussion of these basic circumstances, exploiting our higher scale knowledge in this manner not only greatly simplifies the calculations we must consider, the tactic also injects a strong measure of *enhanced reliability* into our modeling effort. If we actually attempt to develop concrete hypotheses with respect to the forces that hold wooden rods together, we would surely come up with wildly speculative and unreliable answers. This is the manner in which mixed scale approaches to physical modeling inject strong doses of improved reliability into our proceedings.

But suppose that we fail to recognize the cross-scalar architecture inherent in this method and naively attempt to devise a single layer "theory" of how all the "forces" (white and black arrows) operate within the microscopic layer alone, without attending to their origins in the higher submodel above. Such attempts are likely to dead end in bewilderment because our two classes of "forces" must operate in subtly different manners from one other and will robustly resist any consistent attempt to regulate their behaviors in conventional axiomatics.

In fact, such attempts have a long history, beginning with Heinrich Hertz's 1892 endeavors and further enshrined within Hilbert's Sixth Problem on axiomatization.

And the resulting struggles produced a rather remarkable legacy.[20] Although it is debatable that any of these efforts have proved wholly successful in view of the multiscalar issues just raised, the logical empiricists became so entranced by the potential promise of the axiomatic ideal that they began to presume that pretending that such syntactic embodiments have actually been produced merely represents a convenient "idealization," providing an empty stage from which they can generalize about scientific methodology in a sweeping manner. Good examples can be found in loose remarks to the effect that "A Newtonian possible world represents a model of the axioms of Newtonian physics," I doubt that anyone can parse out what this is supposed to mean in concrete diagnostic terms.

In any case, the central purpose of this section is to emphasize the value of the increased reasoning reliability that can be obtained through judicious injections of data already known, where the "judiciousness" of the injection generally requires careful attention to the observation scale upon which the partial data has been obtained. The "harmless idealization" that real life "theories" emerge like Minerva from the brows of individual creators pays little attention to this increasing reliability: either the genius' "theory" gets everything right or a replacement must be found. But real science can't be distinguished from errant speculation if it is not steadily constructed with strong injections of established trustworthiness.

Allied considerations of reliability injection allow us to address another question already raised, but passed by, earlier. Why on earth should engineers fuss with these delicate technicalities with respect to meaningful derivatives, given that our membrane equation can only be regarded as a smoothed over representation, valid only to a finite observation scale ΔH pitched far above the molecular level? This is a question that Maddy herself raises,[21] but let us examine the manner in which Tim Poston and Ian Stewart articulate this common methodological puzzle:

> We have gone beyond the casual assumption of "configurations pretty much like smooth functions if we don't look too closely"...We are making very delicate hypotheses (like square integrable second distributional derivatives) about exactly what we would see if we could look arbitrarily close: hypotheses we know to be false. In what sense these "infinitesimal" hypotheses can be said to model the local average nature of elastic solids is a dark and gleaming mystery. The results of the model, in macroscopic predictions, are very successful indeed; but it is hard to say in what sense the theory is more exact than a computer treatment of a finite element model like that of section 4, but with many more rods and strings... Long calculations that you do not understand, whether by number crunching or Banach space methods, will land you with bridges that fall down. A theory is only as good as your understanding of it.[22]

In truth, this "mystery" isn't so "dark and gleaming," because it neglects the advisability of infusing our reasoning tactics with strong injections of *certifications of reliability* akin to those offered by ΔH scale partial information. Projecting a higher scale ΔL physical modeling down to the infinitesimal level of differential equations generally offers the best landscape for detecting the warning signals of potential computational failure at the higher scale ΔL where we start (say, if we employ finite element computations of width ΔL in the manner that our authors suggest). Here's a simple illustration of how these "warning signals at the infinitesimal level"

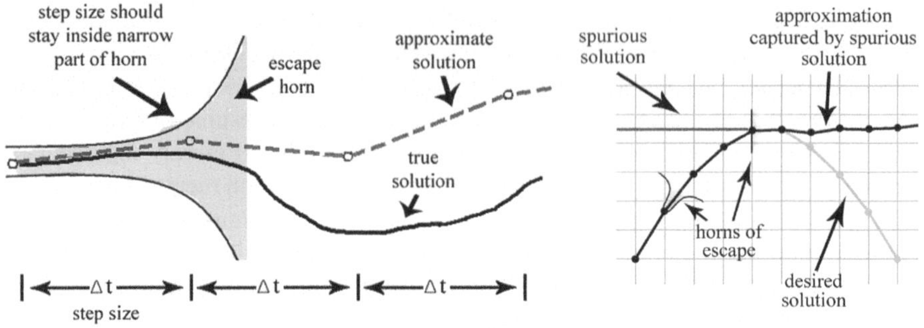

Fig. 32.8 Correctness of Euler's method

appear. A cannonball's frictionless flight can be captured by the simple equational model: $dx/dt = c$; $d^2y/dt^2 = -g$ (call these *). The simplest form of finite difference approximation (Euler's method) will supply a credible portrait of its flight on graph paper if we choose the grid size small enough. However, the total energy of our ball will be conserved over the course of the flight, which suggests that we might alternatively employ modeling equations that exploit this fact, $dx/dt = c$; ½ mdy/dt + gmy = E_0 (call these **). However, when we apply Euler's numerical method to this alternative equation set, we usually compute strange graphs that remain odd no matter how finely we render the grid size. In particular, our computations will show cannonballs that wobble mysteriously after they have reached their highest crest. What has gone wrong in our computations, given that we utilized the same reasoning method in both cases?

It turns out that we can articulate a very brisk answer if we project our cannonball's ΔH-appropriate behaviors down to the infinitesimal differential equation level. There we find that our (*) set of modeling equations everywhere obey a so-called Lipshitz condition locally. This proviso sets limits on how far an Eulerian approximation can depart from a true solution over the span of one time step unit Δt. (Fig. 32.8) This feature confines our approximation within a certain region that we might call the relevant equation's "escape horn." Its confining geometry guarantees as we adjust our step size Δt ever smaller, the graph drawn by our approximation technique will ever more closely resemble the proper flight path of our ball. If so, we can certify that an application of Euler's reasoning technique will prove reliable in the applicational circumstances before us (such a guarantee is called a "correctness proof").

In situation (**), however, the Lipshitz requirement fails as soon as the ball's path becomes horizontal. At this juncture, the formula's "escape horn" opens up completely no matter how finely we assign a mesh size Δt is set. This gap allows a host of spurious solutions to snake through the problematic horizontal in a manner that confuses our Eulerian tactics thoroughly.

In this manner, the "failure of a Lipschitz condition" on an infinitesimal scale provides a *palpable warning signal* with respect to the trustworthy application of a numerical method that only attempts to supply physical characterizations appropriate to the higher observation scale ΔL. And this is a familiar circumstance within the

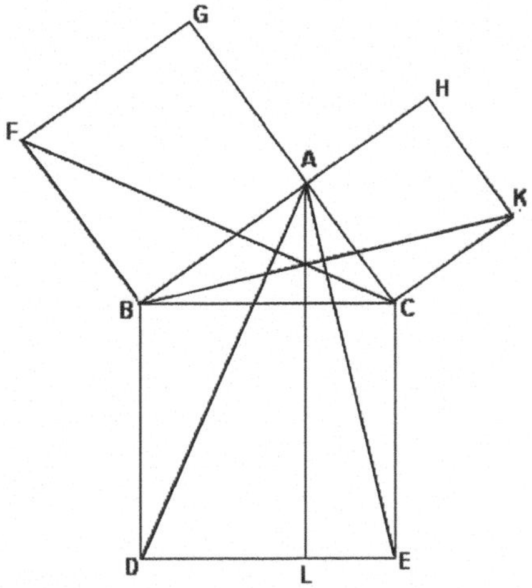

Fig. 32.9 Pythagorean embedding

realm of reliable reasoning: the most valuable warning signals with respect to a technique's potential failures lies within some mathematical landscape that appears superficially distant from our intended application. As Thomas Hill observed in 1857:

> The conception of the inconceivable [imaginary], this measurement of what not only does not, but cannot exist, is one of the finest achievements of the human intellect. No one can deny that such imaginings are indeed imaginary. But they lead to results grander than any which flow from the imagination of the poet. The imaginary calculus is one of the master keys to physical science. These realms of the inconceivable afford in many places our only mode of passage to the domains of positive knowledge. Light itself lay in darkness until this imaginary calculus threw light upon light. And in all modern researches into electricity, magnetism, and heat, and other subtle physical inquiries, these are the most powerful instruments.[23]

The *locus classicus* of this out-of-territory assistance lies in the manner in which singularities upon the complex plane alert us to potential failures of the series expansions we apply along the real line.

It's worth remarking that Kant's conception of the synthetic apriori nature of mathematical constructions traces to the remarkable fact that our assurance that the squares on a triangle's sides add up in a Pythagorean manner becomes significantly enhanced after we embed the figure in a network of auxiliary lines in Euclid's manner (Fig. 32.9). The fact that the reliability of our reasoning patterns can become greatly enhanced through the consideration of auxiliary embeddings such as this may strike us as rather mysterious, but the tactic represents a significant ingredient in the methods whereby we gradually tighten our descriptive hold on the world about us.

At this point we can attend to a significant error on Quine's part that has played a significant role in the manner in which he construes "physics needs." Relying upon the axiomatics-all-at-once pretenses of standard logical empiricism, he construes "the needs of physics" as reflecting the requirements of formalization within an axiomatic framework (viz., how far up in the analytic hierarchy must one reach to codify the required axioms properly). This same conception is responsible for his doubts (already cited) that set theory reaches too far beyond these limits to be regarded as anything other than "gratuitous flights of fancy." But physics' real "needs" are better conceptualized in terms of the reliable signposts of reasoning reliability and we just determined that these can potentially lie anywhere in mathematics' grand kingdom. Riemann applied the notorious "Dirichlet reasoning principle" within the realms of complex analysis whereas Lord Kelvin cited it in his reasoning about physical membranes. The fact that counterexamples could be found anywhere casts its reliability into disrepute within both applications, repaired only after Hilbert and others found clever ways to skirt these difficulties. The fact that a significant challenge to a tactics' presumed reliability first appears within the realms of higher set theory can prove just as damning as it emerges within some relatively tame class of real-valued functions.

Let's now consider another significant aspect of the "applications problem" as it is commonly conceived. Although Quine himself does not explicitly subscribe to the doctrine, a substantial aspect of the lingering methodological misapprehensions encouraged by his writings derive from the common presupposition that the supportive "semantics" of ordinary empirical statements proves relatively straightforward whereas a correspondingly attractive picture for mixed or purely mathematical claims is hard to come by. Paul Benacerraf articulates an influential expression of this widely accepted dichotomy within his well-known "Mathematical Truth":

> For Hermione to know that the black object she is holding is a truffle is for her (or at least requires her) to be in a certain (perhaps psychological) state. It also requires the cooperation of the rest of the world, at least to the extent of permitting the object she is holding to be a truffle. Further - and this is the part I would emphasize - in the normal case, that the black object she is holding is a truffle must figure in a suitable way in a causal explanation of her belief that the black object she is holding is a truffle. But what is a "suitable way"? I will not try to say (except that)... it must be possible to establish an appropriate sort of connection between the truth conditions of p (as given by an adequate truth definition for the language in which p is expressed) and the grounds on which p is said to be known, at least for propositions that one must come to know - that are not innate.... As an account of our knowledge about medium-sized objects, in the present, this is along the right lines. It will involve, causally, some direct reference to the facts known, and, through that, reference to these objects themselves. Furthermore, such knowledge (of houses, trees, truffles, dogs, and bread boxes) presents the clearest case and the easiest to deal with.[24]

Note the presumption that the exterior truth-conditions of "this truffle is black" most likely consists in some simple species of <object, property> combination that can be subsequently extended to longer sentential constructions on a sentence by sentence recursive basis (hence the essay's appeals to "Tarski's recursive treatment of truth-conditions"[25]).

I regard such a conception of "truth conditions" as entirely mythological, founded upon a rather useless policy of projecting raw linguistic structure onto external reality. And certainly a large number of other philosophers (including Quine himself) concur in this skepticism with respect to a "naïve semantics" of this character. But it is exactly along this rejectionist axis that my own philosophical predilections diverge most significantly from Maddy's. In *Second Philosophy*, she characterizes these differences as follows:

I think we can safely conclude that Wilson's correlational point of view and the stance of our second-philosophical disquotationalist resemble each other far more than either resembles more familiar forms of inflationism or deflationism. Both take the serious work to be the investigation of word—world connections and both reject the call for a single 'mechanism of reference'. Our disquotationalist is perhaps more skeptical than Wilson's correlationalist on a prediction: that further study will uncover unified supports for word—world correlations in local patches. But, if the evidence goes against her, our disquotationalist will happily convert, because both these theorists carry out their inquiries in an entirely second-philosophical spirit.[26]

But rejecting the simple picture of word/world alignment that Benacerraf favors doesn't mean that working out a more plausible account of these ties shouldn't remain a central philosophical goal. I began my career as a student of Hilary Putnam's and was captivated by the suggestion in his early writings that the sundry histories of improving scientific development can serve as an excellent proving ground upon which theses about effective word/world coordination could be aptly tested. And I never fully understood why he afterwards gravitated to Kripke-style rigidity with respect to language and advanced untested declarations such as "The terms in a mature science typically refer."[27] It struck me that we know quite a bit about the behaviors of large scale complex materials and can also accurately trace the developmental pathways that scientists have followed in uncovering useful strategies for reasoning about these behaviors in fruitful ways. As scientific students of linguistic process, shouldn't we directly investigate how language and the world have managed to reach a more practicable accord through a long sequence of clever strategizations and adaptative tweakings? Such concerns can be viewed as linguistically specialized forms of the questions central within biology (viz., how might an adaptive organism take effective advantage of the opportunities and dangers operative within the environmental niche before it?). As such, the questions I hoped to address comprise part of the natural history of the human species.

And the overriding lesson I have extricated from my studies is that the accommodationist strategies that function most effectively within real life application are far more intricate architecturally than we've been taught to anticipate. I believe that this lesson is vividly illustrated in the tremendous computational advantages obtainable through the relatively simple divide-and-conquer policies of the various multiscalar schemes I've studied.

I am essentially a pragmatist at heart and believe that language cannot successfully attach itself to the external world through good intentions alone ("I'd really like to attach this word 'black' to a simple external trait"), but through evolving operational entanglements that allow us to classify and calculate in manners that respond to our environmental circumstances in effective ways. Benacerraf is correct in presuming that some form of language/world entanglement makes it reasonable to categorize his contention that "this truffle is black" as true, but he fails to recognize these connections become forged within a controlling architecture of considerably subtler strategic nuance.[28]

In light of these animating concerns, Maddy is undoubtedly right to classify my investigations as those of a "second philosopher," although I might prefer "student of the natural history of well-adapted thinking."[29] What troubles me in the paragraph cited are the sympathies expressed with respect to deflationist views like those of Hartry Field. To be sure, such authors generally reject simplistic conceptions of "truth-conditions" of the sort favored by Benacerraf as illusory projections of raw linguistic structure onto empirical reality for the same reasons as I do. But they rarely attempt to replace these rude "semantic" doctrines with more informative studies of the descriptive strategies that interweave our linguistic behaviors with our environments in progressively improving manners. In this essay, I've attempted to sketch some of the tactics we follow in strengthening these bonds of practicality.

In these affairs, the word "true" does not function as the idle operation of syntactic projection that the deflationists portray but serves a wide range of tasks with respect to what PA calls "language management." Here is a simple illustration. Suppose we wish to introduce some novel form of multiscalar architecture into our descriptive toolkit. One of the tasks we confront is deciding how the homogenization messages between submodels should be formulated. When we discussed these concerns earlier, I utilized the metaphor of determining what events transpiring on a characteristic length scale ΔL *will look like* from the perspective of a different scale length ΔH? But these same concerns can be equally registered as questions about the behavior of "truth" upon the contrasted scales: which sentences X can be maintained as irreproachably true at the length scale ΔH in light of the fact that sentences Y hold true at the scale length ΔL? As such, these quasi—"semantical" judgements play a critical role in framing the new architectures that allow us to progressively improve our descriptive skills through Neurath's boat-style refurbishing. However, enlarging further on these themes would carry us too far afield.

For these reasons, I am reluctant to follow Maddy in her attempts to develop her deflationist propensities along the lines of the "thin realisms" and "arealisms" she favors. As she escapes the grubby coils of Quinean need, she underplays the strategic considerations that better reflect how the genuine "needs" of practical reasoning are addressed in real life. With respect to her central objectives (outlining the motives that currently direct set theoretic research), this neglect matters little. Nonetheless, some of her rhetoric unwittingly reinforces the popular picture of mathematical endeavor in which its "purist" activities remain hermetically housed within a grand mansion whose evolving interior adornments are determined solely by formal aesthetics and structural outreach. Although most of the inhabitants dwelling therein may not notice the resulting tremors, savvy workmen from the applied department regularly mount the old house on a truck bed and transport it down the road in search of firmer empirical grounding.

Notes

1. "The Mathematician", in R. B. Heywood, ed., *The Works of the Mind* (Chicago: University of Chicago Press, 1947).
2. *Pursuit of Truth* (Cambridge: Harvard University Press, 1990), p. 95.

3. *The Review of Symbolic Logic* **1**(1) (2008), p. 36.
4. Otávio Bueno and Mark Colyvan, "An Inferential Conception of the Application of Mathematics," *Noûs* **45**(2) (2011), p. 345. Bob offered this critique at an APA session involving Maddy, Michael Friedman and myself. I would to thank Bob and Michael Liston for many helpful discussions on issues such as this.
5. Constance Reid, "Hans Lewy.1904-1988" in P. Hilton, F. Hirzebruch and R. Remmert, ed., *Miscellanea Mathematica* (Berlin: Springer-Verlag, 1991), p. 264.
6. (Oxford: Oxford University Press, 2017). I have borrowed the present exposition from my contribution to a symposium on *Physics Avoidance,* forthcoming in *Philosophy and Phenomenological Research*, which discusses allied themes in a complementary manner.
7. (Oxford: Oxford University Press, 2006).
8. As we shall later see, the basic tactics of set theoretic thinking enter the scene in struggling with this "hopefully" proviso.
9. Robert Batterman, "The Tyranny of Scales" in Robert Batterman (ed), *The Oxford Handbook of Philosophy of Physics* (New York: Oxford University Press, 2017).
10. This is the celebrated dispute between d'Alembert and Euler over the equation for a vibrating string.
11. Jesper Lützen, *The Prehistory of the Theory of Distributi*ons (Amsterdam: Springer, 1982) surveys a variety of these early opinions.
12. *Elements of Green's Functions and Propagation* (Oxford: Clarendon Press, 1989), pp. 158–159.
13. This is a quick attempt to characterize the "descriptive philosophy" inherent in Laurent Schwartz' notion of a "distribution," which Robert S. Strichartz explicates as follows (*A Guide to Distribution Theory and Fourier Transforms* (Singapore: World Scientific, 1994), p. 2):

    [I]*t is often more meaningful physically to discuss quantities like* $\int f(x)\theta(x)dx$ *than the value of f at a particular point x. The secret of successful mathematics is to eliminate all unnecessary and irrelevant information ... Since we have decided that the value of f at x is essentially impossible to measure, let's stop requiring our function to have value at x. That means we are considering a wider class of objects. Call them* generalized functions. *What we will require of a generalized function is that something akin to exist for a suitable choice of averaging functions (call them test functions).*

    I elaborate upon this modeling strategy more fully in *Physics Avoidance* (pp. 324 - 61).
14. *Linear Differential Operators* (London: van Nostrand, 1961), p. 505.
15. Similar satellite submodels are needed when interior dimensions are lowered, as when a beam is modeled as a two-dimensional rectangle.
16. Many of the key factors responsible for a set of physical behaviors (such as the tone of a piano string) are not mentioned within their governing set of interior differential equations and are only reflected in the manner in which the surrounding reasoning architecture is pieced together. Frictional decay is often treated in this purely architectural manner.
17. Mark Steiner's mystically tinged *The Applicability of Mathematics as a Philosophical Problem* (Cambridge: Harvard University Press, 1998) has contributed significantly to the persistence of these "warehouse" misconceptions.
18. In other writings, I have called these propensities "the theory T syndrome.".
19. Vol. II (Cambridge: Cambridge University Press, 1912), p. 3.
20. In my forthcoming *Imitation of Rigor* (Oxford University Press), I survey this history in greater detail, including their motivating tensions within mechanical practice.
21. *Second Philosophy*, p. 317.
22. Tim Poston and Ian Stewart, *Catastrophe Theory* (London: Pitman, 1978), p. 300.
23. "The Imagination in Mathematics," *North American Review* **85** (1857), p. 285.
24. Paul Benacerraf, "Mathematical Truth" in Hilary Putnam and Paul Benacerraf, eds., *Philosophy of Mathematics: Selected Readings*, 2nd Edition (Cambridge: Cambridge University Press, 1984), pp. 412-4 (rearranged).

25. Benacerraf himself uses the phrase "Tarski's theory of truth," but I would restrict that phrase to Tarski's tactics in defining the "truth predicate" for a certain domain.
26. *Second Philosophy,* p. 192.
27. Hilary Putnam, *Meaning and the Moral Sciences* (London: Routledge and Kegan Paul, 1978), p. 21.
28. In *Wandering Significance*, I survey the interlocking manners in which we adjudicate whether two objects are properly described as being of "the same color"; the relevant criteria tacitly embody a quite sophisticated awareness of colorific variation across a wide array of environments. Through these contextual acknowledgements of physical complexity, our everyday color usage illustrates the discriminative complexities that J.L. Austin praises in "A Plea for Excuses":

    these surely are likely to be more numerous, more sound, since they have stood up to the long test of the survival of the fittest and more subtle, at least in all ordinary and reasonably practical matters, than any that you or I are likely to think up in our arm-chairs of an afternoon—the most favored alternative method. (*Philosophical Papers* (Oxford: Oxford University Press, 1961), p. 130).

    That being said, I lack a robust grip on the operational architecture responsible for these descriptive virtues comparable to sharper rationales I can supply for a typical multiscalar modeling.
29. Although Descartes is often regarded as a paragon of "first philosopher," he actually figures out what the external world is like by reverse engineering the environmental considerations that explicate why our sensory systems are fashioned as they are. In this wise, his tactics resemble my own predilections more closely than those of "furniture of the earth realists" such as L. Susan Stebbing or John MacDowell. Maddy agrees with me in this assessment, by the way.

Chapter 33
Reply to Wilson

Penelope Maddy

Keywords Pure and applied mathematics · Quine · Set theory · Truth and reference · Second Philosophy · von Neumann

It's heartening to find Professor Wilson allying himself with the Second Philosopher here—and also in his recent 'A second pilgrim's progress'. I've always thought of him this way myself, as a kind of unselfconscious exemplar of the approach. In fact, early drafts of the section of *Second Philosophy* devoted to his views (§II.6) were titled 'The Second Philosopher at work'. One of the goals of the book is to demonstrate that this austere form of naturalism doesn't mean the end of philosophy, as many might fear, that there's plenty of second-philosophical work to do; Wilson's work is then offered as another sample case, beside my own. Of course, it's perfectly possible for two Second Philosophers to disagree on what the evidence supports,[1] and Wilson has done his share of disagreeing! Despite this, I continue to think that any fundamental-sounding disagreements between us are only apparent, and I'm glad to have this opportunity to try to make that case.

Much of what troubles Wilson is the rhetoric, not the content of my second-philosophical views. For example, he begins with a quotation from a paper of mine that traces the evolution of our understanding of mathematics from a direct study of features of the world to a more self-directed inquiry that nevertheless continues to play a central role in the natural sciences. Near the end, I describe contemporary mathematics as 'stock[ing] the warehouses with structures and leav[ing] the choices

[1] See Maddy (2007), p. 3: 'one might sign on as a Second Philosopher while thinking I've gone astray in my pursuit of the particulars'.

P. Maddy (✉)
Department of Logic and Philosophy of Science, University of California, Irvine, CA, USA
e-mail: pjmaddy@uci.edu

S. Arbeiter and J. Kennedy (eds.), *The Philosophy of Penelope Maddy*,
Outstanding Contributions to Logic 31, https://doi.org/10.1007/978-3-031-58425-1_33

to the natural scientists' (Maddy 2008, p. 36). For the record, I now wince a bit at this formulation myself. At a minimum, the term 'structures' may invite the reader to think of the structures of model theory, correlated with theories in first-order logic, and perhaps even to imagine the scientist selecting from these vast stores the one perfectly isomorphic to the worldly situation in question. As Wilson grants, I never say the former and explicitly argue against the latter, but still, I think the world would be a better place if I'd used the term 'tools' instead of 'structures'.

What seems to me a more serious defect in this passage is the suggestion that the universe in which these tools live, presumably that of set theory, is so rich that for any given physical situation, it must include a ready-made tool to suit. A central theme of Wilson's opus is that this is emphatically not the case, that concocting a bit of mathematics that's both close enough to capture some aspect of the situation and workable enough to yield useful information is an extremely difficult and delicate undertaking.[2] I regret that I didn't emphasize this point in the paper in question, but elsewhere I've happily and explicitly helped myself to it in describing the Second Philosopher's perspective.[3]

Another area of at least apparent disagreement is the tricky matter of truth, in particular, my tentative embrace of a brand of disquotationalism.[4] Again, one aspect of Wilson's concern is clearly rhetorical: deflationary theorists are often motivated by a conviction that all study of word-world connections is objectionably 'metaphysical'; obviously, neither of us believes this.[5] (Just as obviously, Wilson joins me in rejecting the other extreme, a simple correspondence theory.[6]) Closer to home, the disquotationalist Hartry Field characterizes those word-world relations in terms of 'indication relations': for example, my belief that 'it's raining' is a good indicator of whether or not it's raining.[7] Understandably, Wilson views this as wildly oversimplified, and I follow him in this, explicitly appropriating his method of complex cases studies (involving 'directivities' and 'supports', both direct and distributed,

---

[2] Here again, 'tools' might have helped: after all, tools often need to be customized, and in a sufficiently demanding situation, newly constructed. Metaphors aside, presumably all the applied mathematician's creations are already present in the set theorist's universe, V; it's just that set-theoretic treatment doesn't call attention to them. Category theorists have long lodged an analogous objection to set-theoretic foundations, namely that the methods of set theory don't highlight the sets of interest to the algebraist. I think both the applied mathematician and the algebraist are right about this: set theory doesn't tell us which sets will be important to either of them. To do this would be to provide what I call 'Essential Guidance', which might be a good thing, but which isn't among the important foundational jobs set theory is designed to do and does so well. See Maddy (2017b), (2019a).

[3] See Maddy (2007), §II.6, also pp. 342–343.

[4] In Maddy (2007), Part II. In Maddy (2022g), I back away from this label, on the grounds that it suggests something like a theory of truth (as opposed to an ordinary language investigation of the role of the word 'true'). Wilson would probably still object.

[5] See Maddy (2007), pp. 176–177, (2022g).

[6] See Wilson (2006), p. 558.

[7] See Maddy (2007), pp. 136–138, for discussion and references.

simple and 'façade-like'[8]) into the Second Philosopher's perspective.[9] So far, then, Wilson's concern is that in entertaining disquotationalism, I might seem to endorse these mistaken views.

But beyond this, there are times when Wilson seems keen to insist on a substantive role for truth and reference. In *Wandering Significance*, this comes in times of semantic agnosticism—when a stretch of discourse is functioning effectively, but no direct correlational support is known.[10] Wilson is prepared to grant that the use of 'true' may 'conform docilely obedient to deflationist pattern' so long as language continues at 'its most unruffled' (Wilson, 2006, p. 633), but during these periods of uncertainty, another usage surfaces: we might predict that some particular sentence is likely to survive when the worldly supports are better understood and regularized, that its worldly support is more direct than the rest. Wilson observes that we might express this by saying that the preferred element is 'true', but it seems to me that his own system has subtler and more nuanced resources for expressing this sentiment (as the previous sentence demonstrates). In the current paper, he finds a robust role for truth when we say that sentences X are true at scale ΔH 'in light of the fact that sentences Y hold true at scale ΔL' (p. 424), but his own descriptions in his multiscalar case studies don't rely on any substantive notion of truth.

A potentially more significant motivation turns up in his final paragraph: he blames my disquotationalism for the line on mathematical ontology—what I call Thin Realism[11] – that's introduced toward the end of *Second Philosophy* (§IV.4) and developed in *Defending the Axioms*. As it happens, one of the reasons I included the second-philosophical discussion of truth and reference in Part II of Maddy (2007) was to provide the background for debunking this supposed connection late in Part IV (pp. 370–376). (Wilson isn't the only one who thinks this way.) The argumentation there is less perspicuous than I'd like, but the underlying thought is simple: there's no reason a Robust Realist can't be a disquotational theorist, and conversely, no reason a Thin Realist can't be a correspondence theorist. The keys here are that Robust Realism is a metaphysical thesis not a semantics thesis, and what's funny about Thin Realism is in the metaphysics, not the theory of truth.[12]

It's worth noting, by the way, that Wilson's apparent affection for robust truth and reference is somewhat puzzling on its face, because the fundamental question for him about a belief or pattern of reasoning isn't 'what makes this true?', but 'what makes this effective?' In many cases, the answer is simple—believing 'cats have whiskers'

---

[8] See Wilson (2006). §II.6 of my (2007) provides a Cliff Notes-like summary of that sometimes-overwhelming book. Maddy (2022g) makes heavy use of the machinery of directivities and supports.

[9] See Maddy (2007), p. 182, 192.

[10] See Wilson (2006), pp. 632–634.

[11] The distinction between my Thin Realism and what I call Robust Realism is epistemological: Robust Realism requires a nontrivial epistemology to access its abstract subject matter; Thin Realism gets by on a trivial epistemology due to its trivial subject matter.

[12] And of course truth is irrelevant to Arealism, where ontology is rejected entirely and replaced by method. Maddy (2011) aims to show that Thin Realism and Arealism are merely cosmetic variants, so the only 'realism' I countenance these days (that is, since the soon-disowned Maddy, 1990) is far distant from the Robust Realism of thinkers like G. H. Hardy or Kurt Gödel.

is effective because cats have whiskers—but in the cases that interest Wilson, it can involve something as complex as one of his elaborate semantic façades. If all he wants is to stop the glib disquotationalist from thinking that—'snow is white' is true if and only if snow is white—and—'cat' refers to cats—is all there is to say about word-world relations, he could do that without positing a fact of the matter about which of the myriad connections across the various patches of a given façade amount to reference and which don't, which worldly supports are referred-to and which are merely correlated-with.

In any case, setting aside his wish to avoid these various objectionable associations that tend to dog the term 'disquotationalism', the most explicit positive move I've found in the direction of an alternative is Wilson's suggestion 'that "refers" operates in a context sensitive manner very much like that of ... "hardness"' (Wilson, 2006, p. 430), that is to say, it operates on a complex semantic façade. Readers of *Wandering Significance* will well remember its extended analysis of hardness: unlike the case of temperature, where various measuring techniques were eventually shown to track a single quantity, the various ways of measuring hardness—squeezing, rapping, scratching, etc.—each turned out to apply only to a certain material and to measure its specific underlying physical property.[13] To provide such a thing for 'refers' would be at least as tall a task as the one mentioned earlier: not just identifying, in the patches of a given façade, which correlations are cases of reference and which aren't, but regimenting all cases of reference into patches, each of which is correlated with its own worldly reference relation, just as 'hardness' on the metal patch is correlated with yield strength and on the rubber patch with Young's modulus of elasticity. This wouldn't be the simple, unified account of reference once envisioned by causal theorists, but it seems to present some of the same challenges.

Fortunately, as far as I can tell, a robust reference relation isn't in any way essential to Wilson's rich account of how words relate to the world, and there's no obstacle to his using the words 'reference' and 'truth' in particular situations without committing himself to any unified worldly support, even on a patch, as long as the terms function as rough-and-ready stand-ins for the specific semantic machinery revealed in the course of his individual case studies.[14] Likewise, disquotationalism isn't essential to my overall philosophical project: none of my accounts of mathematical ontology or logic or second-philosophical epistemology or metaphysics or anything else depend on ruling out substantive theories of truth or reference.[15] To state what may be obvious by this point, what matters most to both of us—more than ontology, more than the nature of truth and reference—is method. And here Wilson's thinking sometimes appears to threaten the core of my project in the philosophy of set theory. In the end, I don't think this threat is real, so let me try to explain why not.

---

[13] See Wilson (2006), pp. 335–345.

[14] For that matter, there's no obstacle to using the terms more loosely in many contexts (see Maddy 2022g).

[15] The success of substantive theories would be a boon to the historicist arguments considered in (Maddy 2022f).

My uneasiness begins with Wilson's epigraph from von Neumann, where mathematics 'at a great distance from its empirical source' is said to 'live a peculiar life of its own ... governed by almost entirely aesthetical motivations' and to stand 'in danger of degeneration'. Perhaps this is my turn to worry about associations; this is just the sort of thing that so-called mainstream mathematicians sometimes say about developments in contemporary set theory, except they would claim that the danger of degeneracy has been realized. Much of my intellectual life has been devoted to arguing that higher set theory is governed by norms of rationality more rigorous and more specific than mere aesthetics, but there's no denying that it lies at a great distance from natural science and that it's governed by its own norms. This is also true of other branches of abstract pure mathematics: abstract algebra, for example, has its own range of methods and goals different from those of set theory.[16] For various reasons, those armed with Wittgensteinian pruners[17] tend to eye set theory first.

Wilson has obvious sympathies along these lines. In 'A Second Pilgrims Progress', he declares himself uncomfortable with what he describes as my account of

> an unfettered Republic of Pure Mathematics whose governing tenets are entirely determined by the 'practices' of its democratically nominated inhabitants – the mathematicians. (Wilson, 2017b, p. 363)

This he opposes, declaring:

> Like most would-be empiricists, I'd rather not allow any portion of language to escape so flagrantly from the tribunal of experience. (Ibid.)

von Neumann's danger of degeneracy becomes the fear of

> fall[ing] into the stagnating grasp of cults ... sustained within the practices laid down by some mutually reinforcing gaggle of gurus and disciples, supplying, perhaps, a psychologically fulfilling form of life but not one that displays an improving empirical grip on nature. (Ibid., pp. 363–364)

What makes these passages difficult to interpret is that Wilson goes on to declare that 'it is wrong to look upon set theoretic practice in this deprecatory manner' (ibid., p. 364).

Before trying to square this circle, let me insert a few clarifications on behalf of the Second Philosopher. First, I don't see the 'practices', or at least the correct practices, of the 'Republic of Pure Mathematics' as determined by balloting. My claim is that the various branches have their own various mathematical goals, and that their methods are rational in so far as they are selected for their efficacy in pursuing those goals. Of course, those goals might be illusory, subject to the dictates of fashion or coercion or human intellectual frailty, so I really mean that the methods are rational in so far

---

[16] Recall footnote 2.

[17] For those unfamiliar, see Wittgenstein (1933), p. 433: 'Philosophical clarity will have the same effect on the growth of mathematics as sunlight has on the growth of potato shoots. (In a dark cellar they grow several metres long)'. In his (2017), p. 364, Wilson quotes the corresponding passage from Wittgenstein (1937), p. 381.

as the goals they facilitate are themselves mathematically important, mathematically deep.[18] Second, I take it as obvious that one of the overarching goals of the Republic is to provide a wide range of tools for science. After all, the Second Philosopher wouldn't treat mathematics as she does—as a practice to engage in, not just to study in sociology or anthropology—if it weren't for its origins in and continuing relevance to natural science.[19]

Returning to the von Neumann/Wilson concern about degeneracy/culthood, then, should we try to prune mathematics back to its healthy core of direct applicability? Wittgenstein or Quine might say yes, but Wilson does not—he notes, for example, that the effectiveness of applied methods often rests on mathematics that was first undertaken for the mathematicians' own reasons (p. 406), on 'signposts ... that ... can potentially lie anywhere in mathematics' grand kingdom' (p. 422). As the Second Philosopher observes, the mathematics that has served natural science so well, often in such unexpected ways, is the mathematics that's been generated by the varied methods mathematicians have actually used in pursuit of their various mathematical goals – including, but not limited to, being of use in application—not by some other more limited principles. In other words, as we've seen John Burgess remarking, 'by producing so much applicable material mathematics earns ... a right to conduct its affairs autonomously' (quoted with permission in 'Reply to Burgess', this volume). Wilson apparently agrees.

So what about set theory proper? It arose in straightforwardly mathematical settings – Cantor's work in analysis and Dedekind's in abstract algebra—and retains close interconnections with both fields today,[20] but soon, with Zermelo's axiomatization, a new goal entered the story: set theory aims to provide a 'foundation' for classical mathematics.[21] Given the widespread confusion about what a 'foundation' is expected to do, I've attempted to isolate a range of mathematical jobs that would seem to merit the label 'foundational' and that set theory does well.[22] These include the mathematically fruitful provision of a Generous Arena in which all its objects[23] can be considered side-by-side and results freely imported and exported from one

[18] See (Maddy, 2011), Ernst et al. (2015). Though I believe depth to be an objective matter—e.g., that the mathematical superiority of the concept 'group' over other logically consistent nearby alternatives isn't just a matter of human preference—I can't pretend to have made a conclusive case (see Ernst et al., 2015, pp. 245–248). One indicator of depth that all seem to accept is surprising interconnections between various apparently unrelated branches of mathematics. The axiom of choice would pass this test, but one of the criticisms of contemporary higher set theory is that it hasn't, at least so far. (There's been push-back on this, of course). However, there have been surprising interconnections between apparently unrelated areas within set theory itself: e.g., between determinacy hypotheses and large cardinal axioms (see Koellner, 2014, §3). On the larger question of higher set theory's interconnections with broader mathematics, time will tell.

[19] See, e.g., Maddy (2007), pp. 344–351, (2011), pp. 53, 101.

[20] Not to mention that even in the esoteric reaches of the subject that have been my focus the leading open problem concerns the Continuum Hypothesis, a question about sets of real numbers. I'm sure Wilson would be the first to acknowledge that the continuum has its roots in application.

[21] See Maddy (2011), pp. 41–47, for a quick review of these developments.

[22] See Maddy (2017b), (2019a).

[23] Or better, set-theoretic surrogates for all mathematical objects.

branch to another.[24] Now if the Republic of Pure Mathematics is to be allowed its free development, then this foundational Arena has to be Generous enough not to interfere; in this way, I argue, the combination of mathematics' earned right and set theory's foundational goal generate methodological guidance for set theory.[25] Perhaps this much can be traced, however indirectly, to 'the tribunal of experience'.

But this isn't all there is to set theory: it isn't clear, for example, that the axiom of choice can be justified as an effective means toward the foundational end.[26] The reasons for its adoption are many and varied,[27] but there's no denying that among these are its mathematically beneficial affects within set theory itself,[28] for example, the order it brings to the theory of transfinite numbers. Wilson himself observes that Choice wasn't adopted 'for the sake of physics', but for abstract mathematical reasons that Quine would consider 'starry-eyed' (p. 406). These, too, Wilson apparently accepts, the same sort of thing as is now offered for more esoteric set-theoretic claims in the realm of large cardinals and determinacy.[29] Set theory, like any other branch of mathematics, is allowed to add some of its own goals and methods.

If any point of disagreement remains between us here, it may be that Wilson believes these apparently autonomous mathematical methods are ultimately justified by their eventual role in our empirical interactions with the world, but if so, 'eventual' here must be so open-ended that it's hard to see how this reservation has any practical force. Perhaps Wilson would follow Quine in claiming that the most carefully reasoned claims of contemporary set theorists in defense of their decisions are at best instruments to an end quite different from what they imagine, but however that may be, I lean toward the view that there are independent mathematical values.

In truth, though, I don't believe that anything so hopelessly 'in principle'[30]—is set-theoretic rationality *ultimately* instrumental or independent?—could be the real issue between the likes of Wilson and me.[31] In fact, I believe our apparent disagreement here is once again only rhetorical. To see this, consider that though both of us reject

---

[24] Wilson notes another job: 'modern set theory became firmly established as the court of final appeal with respect to the internal coherence of a purported mathematical arrangement' (p. 406). The idea that set theory acts as guarantor for the coherence of a mathematical notion is the beginning of what I call Risk Assessment (see Maddy (2017b), pp. 293–295, (2019a), pp. 297–298. Incidentally, see Maddy (2017b), pp. 296–298, (2019a), p. 299, for some cautionary examination of the 'court of final appeal'.

[25] E.g., the maxims MAXIMIZE and UNIFY from Maddy (1997), pp. 208–212.

[26] Is the axiom generous in choice sets or restrictive of disjoint families?

[27] See Moore (1982). Pages 54–57 of Maddy (1997) give a quick overview.

[28] E.g., see Zermelo initial defense in his (1908), pp. 187–189.

[29] E.g., see Koellner (2014), §4.

[30] Cf. Wilson (2017a), p. 21: 'I sometimes wish that word processors might administer a mild electric shock to authors whenever they type the phrase "in principle"'.

[31] [Added later … Wilson has confirmed this point: 'I agree with the sentiments you expressed in your reply and don't see any utility whatsoever in making murky demands about "being useful in application". There are a lot of reasonable motives for engaging in math-like investigations' (private communication, quoted with permission).].

arguments from Quinean 'indispensability', Wilson believes there's still something in this vicinity that I've missed:

> in articulating her case for methodological liberation, she occasionally acquiesces in rhetorical motifs that can leave an undischarged residue of Quinean presumption in place with respect to the alleged 'needs' of applied mathematics. (p. 2)

What's intriguing about this is that I've sometimes been criticized as being more dependent on Quinean indispensability than I admit: after all, doesn't my second-philosophical case for treating mathematics differently from astrology and theology rest on the role of mathematics in science? Indeed it does, as noted above, but I take that role to support the actual, often independent methods of mathematics, not to dictate revision or determine its ontology. What Wilson has done here and in 'A Second Pilgrim's Progress' is to demonstrate that the gritty reality of applied mathematics reaches further into abstract and 'starry-eyed' realms than many are willing to admit, but he gets no argument from me on that score: we share the same Quinean residue.

So in the end, I think we can see the criticisms quoted a moment ago in a new light:

> Maddy's ... travels convey her ... to an unfettered Republic of Pure Mathematics whose governing tenets are entirely determined by the 'practices' of its democratically nominated inhabitants – the mathematicians. Like most would-be empiricists, I'd rather not allow any portion of language to escape so flagrantly from the tribunal of experience. (Wilson, 2017b, p. 363)

By now it should be obvious that despite my focus on contemporary set theory's most esoteric questions, I emphatically do not see the entire Republic, or even its Set-Theoretic Branch, as entirely detached from experience—if it were, it would rank with astrology and theology! I never meant to suggest that mathematics as a whole has left behind all worldly ambition; I only meant to suggest that it now has other quite proper and rational ambitions of its own. Likewise, I doubt Wilson would wish to deny contemporary set theorists their explorations into new territories of determinacy and large cardinals any more than he would have wished to deny early set theorists their adoption of Choice. Perhaps he understands his quotation from von Neumann, not as disparaging set theory—as detractors often imagine—but simply as reminding us of the Republic's ultimate grounding in worldly application.[32]

---

[32] [Wilson confirms this interpretive point. I don't pretend to any expertise on von Neumann's thinking; I was only reporting how opponents of set theory have tended to read the passage. E.g., Kline (1980), Chap.ter XIII, 'The Isolation of Mathematics', p. 278, quotes it at even greater length in his forceful critique of mathematicians who 'have withdrawn from the world to concentrate on problems generated within mathematics ... [who] have abandoned science'. Interestingly, the chapter's conclusion strikes me as an eloquent endorsement of what I would call 'deep mathematics': 'In the final analysis, sound judgment must decide what research is worth pursuing. The distinction with which the mathematical world should be concerned is not between pure and applied mathematics but between mathematics that is undertaken with sound objectives and mathematics that satisfies personal goals and whims, between pointed mathematics and pointless mathematics, between significant and insignificant mathematics, and between vital and bloodless mathematics' (ibid., p. 306).].

If this is right, then Wilson's discomfort is that my focus on methodological questions in the starry-eyed portions of set theory has sometimes led me into insufficiently qualified 'rhetorical motifs' of mathematical autonomy or that I might be interpreted as taking the methods and goals of higher set theory to be the methods and goals of all of mathematics. In that case, I happily pledge to be more careful in the future—and Wilson and I remain, without significant reservation, 'wayfarers of a common cloth' (Wilson, 2017b, p. 362).[33]

References

Ernst, M., Heis, J., Maddy, P., McNulty, M. B., & Weatherall, J.O. (eds.) (2015). Mathematical depth. *Philosophia Mathematica, 23*(2).
Kline, M. (1980). *Mathematics: The loss of certainty.* Oxford University Press.
Koellner, P. (2014). Large cardinals and determinacy. In: E. N. Zalta (Ed.), *The Stanford encyclopedia of philosophy* (Spring 2014 Edition). https://plato.stanford.edu/archives/spr2014/entries/large-cardinals-determinacy/
Maddy, P. (1990). *Realism in mathematics.* Oxford University Press.
Maddy, P. (1997). *Naturalism in mathematics.* Oxford University Press.
Maddy, P. (2007). *Second philosophy.* Oxford University Press.
Maddy, P. (2008). How applied mathematics became pure. *Review of Symbolic Logic, 1,* 16–41.
Maddy, P. (2011). *Defending the axioms.* Oxford University Press.
Maddy, P. (2017b). Set-theoretic foundations. In A. Caicedo, J. Cummings, P. Koellner, & P. Larson (Eds.), *Foundations of mathematics: Essays in honor of W. Hugh Woodin's 60th birthday* (pp. 289–322). American Mathematical Society.
Maddy, P. (2019a). What do we want a foundation to do? In S. Centrone, D. Kant, & D. Sarikaya (Eds.), *Reflections on the foundations of mathematics* (pp. 293–311). Springer.
Maddy, P. (2022f). On the question of realism. In [2022h] (pp. 49–90).
Maddy, P. (2022g). A note on truth and reference. In [2022h] (pp. 173–187).
Maddy, P. (2022h). *A plea for natural philosophy and other essays.* Oxford University Press.
Moore, G. H. (1982). *Zermelo's axiom of choice.* Springer.
Van Heijenoort, J. (Ed.). (1967). *From Frege to Gödel.* Harvard University Press.
Wilson, M. (2006). *Wandering significance.* Oxford University Press.
Wilson, M. (2017a). Pragmatics' place at the table, in his [2017c] (pp. 1–50).
Wilson, M. (2017b). A second pilgrim's progress, in his (2017c) (pp. 362–422).
Wilson, M. (2017c). *Physics avoidance.* Oxford University Press.
Wittgenstein, L. (1933/44). *Remarks on the foundations of mathematics*, revised edition G.H. von Wright, R. Rhees, & G.E.M. Anscombe, eds., G.E.M. Anscombe, trans. MIT Press, 1978.
Wittgenstein, L. (1937). *Philosophical grammar*, R. Rhees (ed.), A. Kenny, trans. University of California Press, 1974.
Zermelo, E. (1908). A new proof of the possibility of a well-ordering. Reprinted in van Heijenoort (1967) (pp. 183–198).

[33] I'm grateful to Professor Wilson for several illuminating exchanges on earlier drafts of this reply. Thanks also to David Malament for helpful comments.

Part V
Conversations on Set Theory

Chapter 34
A Conversation Between Penelope Maddy and Menachem Magidor

Jeffrey R. Schatz

Abstract The following conversation took place on March 18, 2022 between Juliette Kennedy (University of Helsinki), Penelope Maddy (University of California, at Irvine), and Menachem Magidor (The Hebrew University of Jerusalem). Henceforth, they will be referred to as JK, PM, and MM, respectively.

JK: PM, you had a question to get started?

PM: I thought we might just start with me asking you whether MM^{++} is still your pick? It sounded like in the EFI paper[1] that if the big theorem got proved, which it did,[2] you would be happy with that as an axiom?
MM: I'm not sure if I would say that it's my 'pick', but it's definitely a very attractive option, that I think should be considered. In the process of trying to get (what I might call) the canonical axioms for set theory—or the accepted axioms, or whatever you might want to call it—I think it should have very serious consideration.

I do have my problems, by the way. For example, a problem is that, while I think MM^{++} gives you a very appealing, concrete theory for $H(\omega_2)$—while for $H(\omega_1)$, Woodin cardinals give us such a theory with AD—with an appealing aspect being that, in the presence of a proper class of Woodin cardinals, this theory can not be changed by forcing, because of the result, by the way, of Asperó and Schindler, that MM^{++} implies $(*)$.[3] So here is a very stable theory—not in a model-theoretic sense, but in the sense that it cannot be changed by forcing—it has many very appealing aspects, it settles a large class of problems, and so it should be very seriously considered.

My problem is that I still don't see how to continue, in the sense that the natural process—I mean you do have appealing axioms for $H(\omega_1)$ in relation to countable

[1] Magidor (2012).

[2] Asperó and Schindler (2021).

[3] Asperó and Schindler (2021).

J. R. Schatz (✉)
Yale University, New Haven, CT, USA

S. Arbeiter and J. Kennedy (eds.), *The Philosophy of Penelope Maddy*,
Outstanding Contributions to Logic 31, https://doi.org/10.1007/978-3-031-58425-1_34

sets, and what I claim is an appealing theory for $H(\omega_2)$ which extends the theory of $H(\omega_1)$—is there a natural way to extend it to find a theory for $H(\omega_3)$ and beyond. And there are problems there, because in some sense the fact that MM, and any appealing theory for me, should violate CH, but anything of that kind should imply that the continuum is $\aleph_2$. So now for $H(\omega_3)$ there must be something different, since you can't have the continuum being ω_3, if you want to have a theory of $H(\omega_3)$ which extends the theory for $H(\omega_2)$. So I think that that's a challenge for accepting [MM^{++}] as an axiom: to what extent it defeats other candidates, to what extent we can continue the process of finding appealing axioms for segments of set theory, if not for the whole thing.

By the way, one thing that I believe that [inaudible][4] did, a significant aspect of this proof—that MM^{++} implies $(*)$—is that it wasn't known until this result whether $(*)$ is consistent with large cardinals. I mean, while you can get it's consistency by forcing over a model of AD, but it wasn't clear that you could have a model where there even exist in that model measurable cardinals or supercompact cardinals or whatever, and that it follows from MM^{++} means that you can force $(*)$ from a supercompact cardinal. If you're just interested in $(*)$, you need much less, but whatever large cardinals you had in the universe before, you can now have a model of $(*)$ with those large cardinals.

I think large cardinals are very appealing. Returning to your first question, I'm not saying that large cardinals are true, or that $(*)$ is true, but this result makes both MM^{++} and $(*)$ more appealing.

PM: So, I take it that the argument for MM^{++} is pretty much that it gives a good theory for $H(\omega_2)$, because it's absolute and it settles lots of questions…

MM: Yes, it's absolute, and it settles lots of questions, and typically—when you ask a question, there are two ways of course that it can be answered, the nice direction and the less-nice direction, or the theorem direction, or the counter-example direction. Of course, dramatically, you usually want the theorem direction, something that will give you some structure, not just an esoteric counter-example. In most cases, MM, or even forcing axioms, give you a theorem instead of a counter-example. There is one exception, unfortunately: for the Whitehead conjecture, for instance, while $V = L$ and $\diamondsuit$ implies (maybe later I'll have to say something about these forcing axioms vs $\diamondsuit$ and $V = Ult(L)$) the nice direction, that the Whitehead Conjecture is true, but all the forcing axioms, even $MA_{\aleph_1}$, implies that it's false. That's a counterexample to what I'm saying. But in most cases it settles a large class of problems in usually the nice direction, and it gives for many, many things a very nice structure theory for a lot of problems.

PM: You say that you're happy with CH being false. Do you have any sympathy for the large continuum kind of picture?

[4] But likely referring to Asperó and Schindler (2021).

MM: Initially, if you asked me 20 or 30 years ago, maybe I would have...But now I think that, I must admit it depends on some conjectures, and they are conjectures, so I'm not sure if everybody believes them, but let me say that the natural value for the continuum should be either $\aleph_1$ or $\aleph_2$. I can at least restrict the options. Let me tell you why I think they're natural. I think that the thing about universally Baire sets, being a nice collection of sets, and what you would like that if the continuum is large then there should be a kind of nice structured way of witnessing that the continuum is large. Of course, you could consistently have that the continuum is large. I'm not just talking about what is possible; you could consistently have many crazy things! If you increase the continuum by forcing Cohen reals, of course there is no structured way, no definable clear way of witnessing that the continuum is large.

Ok, if you want to accept that the continuum is large, then you want to have evidence, a clear object that is evidence for that. There are two ways to have clear evidence. Either you could have simply definable sets, which are countable or have limited sizes, but I dare to conjecture that any universally Baire pre-well-ordering of the reals—think about a pre-well-ordering of the reals as evidence for the size of the reals, since it's splitting the reals into pieces, and I can accept that there is no definable well-ordering of the reals, but if I'm trying to go by splitting the reals and well-ordering the pieces...

PM: The chunks...

MM: Yes, the chunks. So we know that if you define "nice chunks" as being universally Baire, which encompasses a lot of the things that we accept as nice, for example being in $L(\mathbb{R})$ or definable in higher models, like determinacy models, these are universally Baire (this in the presence of large cardinals). It is not known, there is no single model in which you can do such a pre-well-ordering which is of size more than ω_1. It will always be an ordinal between ω_2 and ω_3. So the conjecture, which I am daring to express, is that you cannot have a nice pre-well-ordering of the reals more than ω_2. If you know how you prove from forcing axioms that the continuum is exactly $\aleph_2$ from MM or PFA or almost any strengthening of MA, if you see how the proof goes, usually it tells you that you want to go past $\aleph_1$—if you want to have $\aleph_1$, fine—but if you want to go past $\aleph_1$, there is something that blocks you from going more than one cardinal.

Of course, you know there was this famous attempt by Gödel to show that the continuum is $\aleph_2$ that nobody was able to decipher.

PM: In your EFI paper, one of the motivations for forcing axioms is this idea that "anything that is reasonably imaginable ought to be". And I can see where that kind of arguments can be applied for MM, but how do you get the ++ part?

MM: The ++ is just a matter of when you say "imagine an object exists" you mean that you have a description of the object such that you can imagine that the object exists. The ++ simply allows you to specify more properties of the object you're talking about. I mean, $(*)$ implies the following thing, that whenever you can imagine an object—lets say a subset of ω_1 or a member of $H(\omega_2)$—suppose you also put a predicate to say whether a set is stationary or not and if you've got the least such subset, then that simply gives you a stronger language to express properties of the object. So you can say "imagine an object that has this and that property including the fact that this set is stationary", which is still the same motivation.

PM: So still the same idea.

JK: Can I ask what Pen makes of Menachem's remark that we shouldn't be talking about truth?

PM: (Laughing) Oh, don't worry, we'll get there. But right now, while we're in this vicinity, if it's ok let's postpone that question, but don't worry; I'll ask it.

So, it's the same kind of motivation.

MM: Yes. By the way, one thing that, and here I'm asking questions that I don't exactly know the answer to, you see $(*)$ is equivalent to the maximality of the Π_2 statements that $H(\omega_2)$ can satisfy. And you look at a Π_2 statement, and say "suppose I can force it". Then $(*)$ implies that $H(\omega_2)$ already satisfies it. Namely if you can force a situation in which for every object there exists another object such that so-and-so is true, then it is already true that for every object there exists another object such that so-and-so is true. So in that sense, it's the sort of maximality character of, sort of flavor of "if you can imagine an object, then it does exist".

Unfortunately, there is a competitor for that, which I believe is $\Diamond$, which is a kind of dual (again, this depends on a conjecture), but it implies that $H(\omega_2)$ is maximal as far as Σ^1_2. And you might also say that is also "if you can imagine objects satisfying something..." So why do we prefer maximizing Π_2 statements over Σ^1_2 statements? To be honest, I don't have a very clear answer.

PM: I always thought "why aren't we maximizing the Σ guys, rather than the Π...?"

Stage Directions: There was a technical breakdown with the Zoom setup at this point. The conversation resumed after a brief bit of IT trouble-shooting and a friendly discussion of opera.

JK: So, are we done with MM^{++} for the time being?

PM: I guess, but Menachem do you want to say anything about this inner model business? We've been talking about the prospects for $Ult(L)$...

PM: Hugh I believe did give up the old argument for $\neg CH$.

MM: I actually accepted this as one of my arguments for $\neg CH$.

PM: You liked that argument?

MM: I like this argument.

PM: I had a student working on this argument.[5] It was an interesting argument, that's for sure.

Well maybe we could talk about then, this Jensen contrast.[6] Does that ring any bells with you, the idea that some people just like things that can be built up in a certain way, and other people like wild complexity, and these are just going to be two forces pulling in different directions?

MM: In a very vague sense, I think it is a good metaphor. You're talking about the Pythagorean vs Newtonian, right? This idea of picking out discrete, very discrete objects, vs the idea that everything is continuous?

PM: And very complicated.

MM: I think that there is some truth to it, or maybe some merit to the analogy that inner models—something that is constructed in a very orderly, step-by-step construction vs something that you say right away encompasses everything that you can imagine (I mean it can't be *everything* that you can imagine, but still)—

PM: (Laughing) And maybe includes some stuff that you can't even imagine! I've always felt this in talking to John Steel, for example, that he—and Jensen of course, that was his point of view—but I've always felt this with John, that that's why he likes the inner model program so much. Because that's where he feels his feet are on the ground, and so-on.

In that picture then, the forcing axioms would be on the Newtonian side, I assume?

MM: Yes.

JK: Dima (Sinapova) uses this analogy of the hoarders vs the people who keep their house clean all the time. (All laughing). The forcing axiom people are the hoarders.

MM: (Laughing) As you well know, from my married life, I would be on the hoarder side, and Sarah would be on the cleaner side.

PM: Yes, yes. The thing is, I do like my house kind of clean, but if anything (in mathematics) I'm a hoarder.

---

[5] Rapalino (2010).

[6] Jensen (1995).

MM: You must know Pen, we have a constant fight about buying more books.

PM: Well books don't count! You can hoard books, that's alright! Actually, I did decide at one point that I had enough bookcases in my house, so if any new books came in, other books were going to have to go out. So maybe that is not hoarding.

Anyway, Menachem you were suggesting that the obstacles for getting an inner model of a supercompact suggest that we just need a different way of getting new inner models, and maybe in fact we're just going to end up counting more complex things as inner models.

MM: I'll tell you a conjecture again. By the way, I don't have a very good record of conjectures, so...

PM: Though you did pretty well with MM^{++} implies $(*)$!

MM: But overall, not a very good...But anyway, it has to do with...During the technical interruption, I was talking about inner model theory before Woodin cardinals came out and the proof of determinacy and so-on, that everyone expected the inner models of large cardinals, any large cardinal you could imagine, would be constructed the same way, the same way meaning that—let's talk about the reals, the reals would be constructed by something which is some kind of mouse, and this mouse could be compared by some kind of linear iteration, this comparison would give you a Σ^1_3-ordering of the reals, and so you can't expect any kind of determinacy from it. But then it turned out that life is more complicated, if you go to an inner model for, say, a Woodin cardinal, you need more complicated objects to approximate the large cardinal, and therefore the well-ordering will already become more-and-more complicated, the more large cardinals you put.

And I think that something like that may happen here. In the present, all the inner models that have been constructed, they have a Σ^2_2 well-ordering of the reals: namely, something with a definition of the form "there exists a set of reals such that for every set of reals, blah blah blah", and then only quantifying over reals. So it will have a well-ordering of that type. And I think it's not inconsiderable that simply some large cardinal, maybe not a *very* large cardinal, like a measurable or Woodin or something, would imply—you know the theory already has problems past Woodin limit of Woodins, there is no proper inner model theory for a Woodin limit of Woodins, I mean they do make an assumption that there is an inner model, and we study how this inner model would look like if it existed—but usually you would expect a canonical inner model for a large cardinal to be that if the large cardinal exists, then you can define this inner model. We are not in that situation, unfortunately.

But anyway, I suspect that—and this has to do with the maximization problem that we talked about, namely remember the useful result of Woodin that CH gives you the maximal family of Σ^2_1 statements about the reals, namely that if you (again, assuming large cardinals, what he needs is a class of measurable Woodins), but if a Σ^2_1 statement about the reals can be forced, then CH implies it. And the thing about the Σ^2_1's is that it's an example of an object that exists, so CH gets you the counter-example.

I'm ready to conjecture (and again, take it with a grain of salt because of my record with conjectures) that the same thing for Σ^2_2 would follow from $\diamondsuit$, namely that in the presence of large cardinals, that every Σ^2_2 statement that can be forced, $\diamondsuit$ would imply it.

PM: Huh.

JK: Any particular kind of forcing, or just...

MM: Any kind of forcing. Any set forcing. You know that class forcings are a completely different story, you can create some crazy, crazy things. You can destroy all the large cardinals you have.

So anyways, if that's true, then that would imply that there is no Σ^2_2 well ordering of the reals, that means that inner model theory cannot, at the level of a large cardinal for which such absoluteness result or maximal principle would hold, you cannot have an inner model of the same type, which the inner model theorists are trying to get. So if that conjecture is true, we are to think of other ways of defining canonical inner models for large cardinals, if you can do it at all.

PM: Ok. Before we leave MM^{++} I just want to ask one more question. So we have two kind of arguments on the table. One is this "whatever you can imagine ought to exist", which is, well I'm not quite sure what to call that. It's a methodological maxim. "Try this". And the other one is "look at all these problems it solves in a good kind of way, and it's stable, and so on".

Which of those makes you more inclined to think it's a good axiom to add?

MM: That's difficult, because maybe it's the combination. You know, in a very naive way, even in the EFI paper, I think I get a criteria for adopting axioms. One of them is "whichever can settle a large class of problems which are otherwise independent", [one is] whether it settles them in a nice direction, and [another one is] whether it gets some immunity to forcing extensions. And these criteria can conflict with each other. I think that you would say that it's a matter of finding the right balance.

PM: Yes.

MM: I think it's a matter of time before some other balance will strike. Another example, and I don't know whether you agree with me, but a historical example. Think about the axiom of choice. I think that most of the mathematical community would accept that the axiom of choice is true, and so we use it, we don't hesitate to use it. This was of course not true when you go back to the beginning of the 20th century. And the axiom of choice does settle a lot of questions, does give you a lot of results, it gives you the existence of many sorts of maximal objects. Just consider the fact that the algebraic closure of a field is unique, that requires choice. And many others, Hahn-Banach theorem and so on. But the axiom of choice also gives us many things that are probably less desirable, lets you construct all kinds of pathological sets.

So it's a matter of balance, and somehow it seems that, sociologically, the commmunity settled on accepting it. In a standard course in analysis you'll automatically, or in algebra, if you have to use the axiom of choice Zorn's Lemma, you'll use it. That's all. But you don't try to evade it by making some assumptions, many times you could evade it by saying that the object you're talking about is constructive in some sense, but you don't do that.

PM: Right, right. What I guess I'm trying to get at is…

MM: The important thing about the axiom of choice is that there is a balancing process, you know, the pros vs the cons.

PM: Well there I can see a sort of balancing between a lot of consequences that you like, and some you don't like so much, and the former outweigh the latter. But in the case of the two kinds of arguments for MM^{++}, all of the good consequences stuff is on one side, but on the other side there is this methodological maxim "anything you can imagine, try to make it exist". I guess what I'm really trying to figure out is whether that maxim has any sort of justificatory force…Are you more likely to think it's a good axiom if it comes that way? Does the fact that MM^{++} came to you, was inspired by, was motivated by these considerations, does that make it a better axiom candidate than if it didn't have that?

Let me put it this way: supposed you had an axiom candidate that had all the good consequences that MM^{++} does, but you didn't have any methodological maxim behind it. Would you be less inclined to like it?

MM: Here you can recall a famous passage of Gödel, that there may be an axiom so fruitful in its consequences, that you are likely to adopt it because of its consequences, and not because of some intrinsic value. Let me make a remark, which I'm embarrassed, because it's a little self-serving, which came up in work with Matt Foreman and Shelah, this came through, the idea of Martin's Maximum. Ok, and it was a maximizing principle, rather than the consequences, because I was not aware of any consequences! The problem that was posed was how to generalize Martin's Axiom. And Martin's Axiom, thought of as a maximizing principle, tells you that a lot of objects that you could imagine, if you can force them by ccc forcings, they already exist. That's a way you can figure out Martin's Axiom. So now I thought, ok, why limit ourselves to—and I wasn't the only one to have this thought—just ccc forcings, what about other forcings? And obviously you could imagine an object with the collapse function for ω_1. But it doesn't exist, for the specific ω_1 in your universe.

Ok, so the next attempt is to say "what about axioms that do not collapse cardinals", could you have some kind of forcing axiom for something like that. And again you could find a counter-example exactly by dealing with stationary sets. So the next step was the ask "what about forcings that do not kill stationary sets", and then the issue is potentially consistency, and we showed consistency. And only then came the consequences.

PM: Right, right. But the thing is, that doesn't mean, necessarily, that the motivations there were justification. It could be "this is a way of generating hypotheses, and we generated this hypothesis, and it turned out to be consistent, but now, is it any good?" And then you look at the consequences.

MM: You could put it like that.

PM: I hope that isn't offensive in any way!

MM: You know, I was very happy when we realized, and this happened almost immediately, that Martin's Maximum implies, gives a specific value to the continuum. Because the first observation was that it was $\aleph_2$.

PM: That's a big deal!

MM: Later, later it was derived from weaker axioms.

PM: Yes, from PFA.

MM: Right, PFA.

PM: Ah, I didn't realize that was the order of the thing. That's very interesting. So you've got this tool, and you try this, try that, and find one that it has this specific consequence, and say "Oh wow, I'm onto something now", right?

MM: Yes.

PM: Ok, yes, that's good. Shall we try this thing about truth then, since we are in the vicinity here? So, I think, that we're in agreement, that you can use these tools to generate hypotheses, and then you look to the consequences, and you like them if they're stable, and if they generally settle things in the theorem way rather than the counter-example way.

But is it then important to be able to say that the axiom is true, or, to make it more pointed, is it important to be able to say that CH is either true or false, but that we just don't know it yet?

MM: In some ways, I'm probably going to agree with you, that it doesn't make sense. Every working mathematician, psychologically, is a platonist, about what he is trying to do. There's an expression of Erdös which I call upon all the time, that whenever a problem he was working on turned out to be independent, he'd say that the ugly monster of independence rears its head again. So people hate that, they believe that there is a definite answer.

When we're talking about Erdös, another bit of, I don't know if you'd call it gossip, Erdös was speculating about what would happen, we used to say, when he'd meet God, which he used to call the SF, the "Supreme Fascist". He would say he would ask him if the continuum hypothesis was true. He was speculating about three different answers. One, he would just give an answer, true or false, or whatever. Second answer, it is just a meaningless question. And the third one, there is an answer, but you're not able to understand it.

But anyways, while I would love to somehow pretend that every question has a definitive answer, I could not spend time, you know, as a philosopher, trying to explicate how there could be a way that there is a meaningful, true or false meaning to an independent problem. I could not come up with anything coherent. So I would agree with you out of desparation, in a way.

PM: Well, you know when I started out, I thought there was an answer. For that matter, I thought there was an answer about, say, regions of spacetime, whether there were regions of spacetime with this property or that, and that that was a real question. But, then when I pushed on that, and then came to realize that physicists don't have any confidence about what spacetime is like in the small, so it's not even right to say that our best understanding of the physical world says theres a real continuum somehow, that's when I first began to entertain the possibility that CH just does not have an answer, and that was a big surprise to me. So I guess you went through a similar period.

MM: Yes. Actually, there is a quote of Gowers, that says I was a platonist until I discovered that the continuum hypothesis is independent.

PM: Right. Good, good. That's very interesting.

MM: The quote says that "when I discovered a concrete problem like the continuum hypothesis can be independent, then I lost my naive platonism", or something like that.

JK: But doesn't that imply that the list of axioms that we have, the ZFC axioms, are kind of final, and that there will never be another axiom that is as self-evident as those, and that will decide it? I was in conversation with Tony Martin, and maybe he was just agreeing with me, but this idea that, you know, maybe it's premature, maybe in some time…

PM: I don't think that that's the right way to think of it. It's perfectly open to Menachem and me and Gowers and people like us to say "yeah, we could one day come up with an axiom that we all adopt for all these wide-ranging reasons we have for adopting axioms, in which case then CH will be true or false in the context of this axioms". But, as of now, I don't think we can project that we are sure that we will some day come up with such an axiom.

MM: I agree. But I think that we should try…

PM: We should definitely try!

MM: I dislike very much the attitude, which I call the supreme pluralist attitude, that says that there is only ZFC and that we should just accept that. We have all this jungle of possibilities of what mathematics could look like, and we should leave it at that. I think that there would be a process by which somehow we could converge, obviously I cannot be sure, but we could converge on an axiom that would be widely accepted.

PM: I hope so too.

MM: You know, even the axioms of ZFC, there was a process of adjustment, and they did converge.

PM: Let's take that extreme pluralism to be a sort of surrender, trying to make defeat into a success or something. It papers over the fact that we don't know which way to extend ZFC by saying "oh, well I've got a philosophical view that says we don't do that, because there are all these worlds or whatever it is". But that seems silly. But, well, I think Juliette had suggested to me that she had heard from you that you think—or maybe this was in the EFI paper—there could be a small group of preferred theories, rather than a single one. Is that based on some inkling, or "smell" as you said, of what those kind of theories might be like?

MM: I think it's almost a necessity that we try, if not one, than possibly several, a small number of options. Ok, first, I'm thinking about set theory as a framework for doing mathematics. As uniformizing mathematics and making a common language in which you can express everything. That doesn't mean that when doing mathematics, we actually do that. I mean, the function analyst, when he talks about Banach spaces, he doesn't think about sets of pairs or whatever, he doesn't think about it that way. But the fact that there is a sort of common language, I think, is really important. Especially for independence, for showing that something cannot be solved. What are the rules of the game by which we cannot solve it?

Now, suppose that you don't accept Feferman's claim that the independent problems are—well, he says they are meaningless, I don't know, it means there is some kind of criteria for which problems are meaningful...

PM: I'm always, my gripe about this is that it's perfectly meaningful in any ordinary sense of meaning. We know what it means, it's fine. So you have to have some other idea, some other kind...Give me your account of "meaningful" that is different from any normal account. You have to have your own, tell me what it is, and then we'll have a conversation about this, but until then, I don't get what you're saying.

MM: I think in terms of the pragmatic, the sociological fact about motivations, the Whitehead problem, for example, there were quite a lot of mathematicians working on it who thought it was a very natural problem, trying to solve it, who got all kinds of partial results. Does this mean that they were pursuing a problem that was not meaningful?

PM: Haha, no.

MM: Were they foolish enough to pursue such a problem? Ok, so assuming that the independent problems are meaningful, and you would like to get some way of deciding them. And the pluralist would say, "ok, anything goes, make up your favorite". Now the thing that is not so friendly is that there are combinations, for example you could have the Whitehead conjecture being true and the continuum hypothesis being false, or the opposite, you could have the continuum hypothesis with all sorts of different values, and you can combine it with all sorts of other problems—and I want to get back to it later because it has my favorite possible connection to natural sciences, but the problem of the automorphism of the Calkin algebras, so I hope we will get back to it—so the poor mathematician will have to say "if I cannot have one set theory that will solve all my problems, I would at least like a few choices, so that I can state a theorem assuming CH, let's say, this is correct, assuming $\neg CH$, this is correct".

By the way, CH and $\neg CH$ are not a good problem, in the sense that, for instance, $\neg CH$ does not settle many things.

PM: Right, you have to have some concrete "not CH".

Well, maybe the options would turn out to be the Pythagoreans and the Newtonians?

MM: Or maybe it's exactly the inner models—Ultimate L, if it materializes one way or another—or some kind of super forcing axiom. Which, again, the problem is that we are stuck still at the level of ω_2, and the issue is if there is a way of getting this slogan, of "everything that can imagine does exist"...and by the way, when you say that something you can imagine does exist, that doesn't mean necessarily by forcing, for instance it could be something that follows from large cardinals, for instance $0^\sharp$. $0^\sharp$ is a very concrete object, you know exactly what it will satisfy, it's a set of integers with certain properties etc. Believing that the large cardinals are consistent, you could imagine that such an object could exist, so why assume $V = L$? It looks like, I think this is a term you use somewhere, but artificial gerrymandering. "This object could exist, and I know exactly what it would satisfy..."

PM: But you don't get to have it!

MM: Exactly, you don't get to have it.

PM: I always try to describe it as, you have this theory that is really wonderful, it does all the math things you would ever want it to do, and it has all the virtues you could ever want it to have, but unfortunately it's just not true. The sets aren't like that. Too bad. That just seems absurd to me.

MM: This is analagous to another problem, whether Gorthendieck universes can be eliminated from Wiles' proof. I mean, formally the only proof that has been published so far uses them, and cannot even be done in ZFC.

PM: You need inaccessibles.

MM: Yes, you need inaccessibles. A proper class of inaccessibles.

PM: That's nothing! Go ahead, take it!

MM: There are some attempts though. Some are going around saying it must be done in Peano arithmetic. But it's a lot of work, and it's complicated having a proof. And so it was very natural for Wiles to use Grothendieck universes, and that's the way it was, you know...

PM: Discovered.

MM: Yes.

PM: It might not be surprising, then, that the person I know who is most involved in this is Colin McLarty of philosophy. He's an old pal of mine. But he would want to do this kind of thing.

Hm, let's see...I actually don't think that the question of getting physical applications for set theory is terribly important, but some people do. And I wonder to what extent you're one of those people?

MM: I don't think it's very important, although I definitely think that it could serve to, if selecting axioms is a kind of balancing of pros and cons, it could be an important pro for a particular axiom. It doesn't mean that, you know, that you rule out an axiom because it doesn't have an application, or you don't like—well, though, if you don't like it, if there is a physical application, and you don't like the consequences of that theorem, then maybe it's a little bit of an argument, well it's, if you put the thing on the scale, then it's something that's got some weight.

PM: I see. Well, here's the thing. Going back, one of the arguments that was once given against choice, right, was Tarski-Banach. And to me that just seemed like a misguided argument, because what you have in modeling our space as $\mathbb{R}^3$ is just a mathematical model. And what you find with Tarski-Banach is that you shouldn't press it too far, because you can't assume that all of the mathematical features of your model will be mirrored in the physical thing that you're trying to model. No one expects that. You wouldn't have fluid dynamics if you expected every feature of the model to be there in the world, because the world is not a continuous substance. So, it seems like all you learn is don't press the math too far, because you're going to get some weird results.

I mean, you could insist that you find a theory, not of the reals, but of some related structure, some constructivist thing that didn't have choice, which would be awkward, and wouldn't work as well, but there's no reason to do that. Use the reals! It's the simplest thing, it works the best, just be careful about it. So, if you had this kind of situation where your axiom gave an outright false physical application, *if* you were to regard it as a model in a certain kind way it would come out that way, so don't use it as your physical model, right? I don't see that that counts against the theory. Maybe that wasn't too articulate, but…

MM: Here there is a certain assymetry between, remember the distinction in a mathematical problem between the theorem side/nice side, and the counter-example side. And I think that of course you are completely right that you adopt a mathematical model for a physical model up to the extent that it fit's the physics. And if there is some obscure, esoteric aspects of the mathematical model, then in some sense here you would be completely justified in saying "well, it doesn't apply to the physical model".

PM: Or "it works to a certain extent, but it doesn't work if you do this with it".

MM: But, on the other hand, if you can give an argument that the application of the model would be nicer, or more, elegant if you assume this axiom rather than that axiom…Let me give you an example of what I have in mind. Again, this is pure speculation.

Consider the problem of the Calkin algebra. The Calkin algebra is the algebra of bounded operators of Hilbert space modded by the compact operators. And the question is, what are the automorphisms of the Calkin algebra. You have automorphisms if you take the original Hilbert space, and you take an automorphism of the original Hilbert space, and that of course induces an automorphism on the algebra of bounded operators and the compact operators, so it definitely induces an automorphism of the Calkin algebra. Again, it doesn't matter the technical details themselves. But now, the question is: is that all? Are all the automorphisms of the Calkin algebra of exactly this form, come from automorphisms on the original Hilbert space. And the answer to that is independent of set theory, exactly by the way, CH gets you a counterexample, unsurprisingly I think, while on the other hand forcing axioms get you the nice thing, that the automorphisms are exactly of this form.

It's not quite a physical application that I have in mind, but you get in this situation where operators on Hilbert space are really measurable as far as physics and quantum mechanics are concerned. And suppose that somehow this automorphism of the Calkin algebra would have some physical meaning, again complete, very wild speculation. But for some physical reason you would rather have it simple, that all these automorphisms have symmetry, and this comes from the original symmetry of the Hilbert space. Let's say that's physically desirable.

Now suppose there is a counterexample. Let's say it doesn't effect my physical model, because in my physical model the cases I'm talking about, that I'm interested in, it's a nice automorphism. I'm speculating, going far away...But suppose physicists now picking a theory have an option saying in the cases I'm using this phenomenon doesn't occur or saying it's a theorem of set theory that such a bad thing does not occur. That's what we should prefer.

PM: Sure. Well, what one wants, of course, is a theory that's so rich—and here I sound like a Newtonian again—that it can offer all these different kinds of models within itself. So if the physicist would like a model that didn't have the bad things happening, "ok, work over here!" If you want some other feature to happen, "ok, use this sort of structure."

MM: But now we come to the problem that we said before. It's ok if we have a relatively small number of theories that you can pick from. If you've got a small number of options, then ok.

PM: Yeah, that's why I like the small number of options idea, but—and I think John Steel said this to me one time—as soon you've got a small number of options, what you start looking for is a theory that can encompass them all.

MM: Eh, what do you mean?

PM: Have a nice model of each one.

MM: Ok, ok.

PM: Because then you can just say, "ok physicist, work in this model". I mean, to a certain extent, we give you that now. You can say to a physicist, "if you don't want any non-measurable sets, work inside the projective hierarchy in this situation".

MM: This reminds me of a very interesting story. In our department, Aumann is a Nobelist in economics, and many, many years ago he talked to me since he has a problem in mathematical economics. Again, it's a model, a very Newtonian model. You're talking about continuum-many participants in the economy, and each one has his different preferences and so on that satisfy some axioms, and you'd like to find a utility function for each one that is nice, is continuous, is measurable, over the space of all the members of the economy. And it turns out that the problem, in the generality in which he framed it, is exactly equivalent to every Σ^1_2 set being Lesbegue measurable. So it follows from large cardinals, from measurable cardinals, that every Σ^1_2 set is Lesbegue measurable, but, yeah, ok…I think that they actually solved this problem by putting up a border, and making sure that not every Σ^1_2 set is a counterexample that comes from that is not something that they consider. But the worry is, that whenever a country has economic problems, maybe it's because they don't use the right set theoretic axioms.

PM: Hahaha. Right, it's hard to find examples in real life where they go that far up in the hierarchy.

MM: Well, that's just real life. Where do you see the continuum hypothesis? Of course, this is a process that you know very well, in science, that you abstract and abstract, and you want a model of the economy where each individual has got very different influences and yet still the whole outcome is…

PM: Right. The whole debate about whether economics has gotten too mathematical and lost its connection with reality.

Well, it's getting a little late for you guys. Well then, let me just ask this one other question about the history of forcing, because I sort of remember back in the day when forcing was new, these forcing models were these crazy, wild things and we're just using them to get independence results and so on. And now we don't think of it that way, right, we think of it as a way of expanding the universe, dual to taking inner models. So do you have a sort of sense of when and where that transition happened?

MM: This probably has to do with forcing axioms, maybe with Martin's Axiom.

PM: Really, with Martin's Axiom? I mean, that was just a technical…

MM: Well, it's not a very sharp transition. It could happen gradually. But I think it comes from the fact that in trying to see which sets you can, I don't know what the right term to use is, you can create, define, imagine, then what are the truths that you have for getting descriptions of more sets. You can either have things using something that looks like L, or some way of construction approach, or you could have somehow sets that you get by forcing. Or, actually, the third way is large cardinals, assuming that you want more large cardinals, therefore it now generates things that are not a large cardinal itself, things like $0^\sharp$ and other similar objects. So it becomes—and we don't have any other way of imagining more sets, either how to expand the universe, if you mentally have a certain universe in mind, and say I want to expand it, I want to have other objects, where do you find these objects? Either by adding more cardinals, which have an impact on the objects you have in the universe, or by forcing. And of course there is the dual thing of going to inner models, but there is no...

So at some point, people realized that these are the only tools you have for adding more sets. Again, I don't know what term to use, because it depends whether you are talking about generating new sets, forcing new sets, I don't know.

PM: Maybe it was just as people got more comfortable with forcing and were able to think about it that way? I think I learned forcing from Shoenfeld's paper or something, and it was not really intuitive, but then you got Kunen's book, where it started to get a little more ituitive. And it just gradually started to look like you were adding stuff to the universe...

MM: By the way, Shoenfeld's paper is infinitely more intuitive than Cohen's paper.

PM: Yes! When I was a student, and learning it, people gave me Shoenfeld's paper and said "you're so lucky to be able to learn it from this", right? But now, we realize that it could have been even easier, and so maybe it's just that, that the community assimilated the method in a way that made it seem more natural after a while.

JK: Well I think that this is a perfect place to end our conversation, with the naturality of forcing. Thank you both for a wonderful exchange!

MM: This was very interesting, thank you very much.

PM: Yes, thank you so much.

References

Asperó, D., & Schindler, R. (2021). Martin's maximum $^{++}$ implies Woodin's axiom $(*)$. *Annals of Mathematics, 193*(3), 793–835.

Jensen, R. (1995). Inner models and large cardinals. *Bull. Symbolic Logic, 1*(4), 393–407.

Magidor, M. (2012). Some set theories are more equal. Talk given at the Harvard EFI Project in the Spring of 2012

Rapalino, J. A. (2010). *Generic absoluteness and the continuum hypothesis*. Irvine: University of California.

Chapter 35
A Conversation with David Asperó

Corey Bacal Switzer

Abstract This is the transcript of a conversation between Corey Switzer and David Asperó that took place on February 25, 2023 in Kittilä, Finland.

Switzer: Alright great so, thank you for agreeing to do this. Let's just jump right in. Essentially there are two things I want to talk about in this conversation relating to your work. The first one being your result on MM^{++} and the second one being on higher generalizations of forcing axioms. Let's just start with the MM^{++} result. What do you think is the significance of the result for you? How do you think about it?

Asperó: Right so let's see. So, forcing is the best method we have in set theory to prove independence. By far the most powerful tool we have right? In this sense forcing axioms are very nice axioms. They tell you that many forcing constructions that you could do over V–you have that the intended effect–all of this holds simultaneously over V. So this is a uniform principle–that the universe is very rich in a sort of like fair way. We're not favoring one generic extension over another. You have this uniform theory that will give you interesting outcomes. In this sense, and maybe for other reasons forcing axioms are extremely natural, attractive principles. Okay so we have this family, the notion of properness, Shelah's notion of properness. It's extremely useful because it refers to a class of forcings that preserve ω_1 and moreover the notion is iterable and there are lots of forcings that fall in this family.

Then we have Martin's Maximum which is a provably maximal generalization, strengthening of the proper forcing axiom. It's really a very strong form of this idea–using properness where you are preserving ω_1 and then you will do lots of stuff, you're going to typically reflect a lot of the large cardinals that you have in V. A lot of the large cardinal strength of the ground model is going to reflect to objects of size $\aleph_1$. So in this sense, right. So we have this picture, and we have this, a techincal strengthing of Martin's Maximum, which is Martin's Maximum double plus (MM^{++}).

C. B. Switzer (✉)
Institut für Mathematik, Kurt Gödel Research Center, Universität Wien, Kolingasse 14–16, 1090 Wien, Austria
e-mail: corey.bacal.switzer@univie.ac.at

S. Arbeiter and J. Kennedy (eds.), *The Philosophy of Penelope Maddy*,
Outstanding Contributions to Logic 31, https://doi.org/10.1007/978-3-031-58425-1_35

And then on the other hand we have this–Woodin defined this $\mathbb{P}_{\max}$ axiom $(*)$ in the 90's, which had a completely different flavor but also could be seen as a very attractive maximality principle because it implied that any Π_2 sentences of H_{ω_2} that are forceable, they all hold simultaneously. Also it provided a complete theory for a natural inner model which contains H_{ω_2} where things like the continuum hypothesis or Souslin's hypothesis would live. So it provided a complete theory modulo forcing. So that is a really attractive freature. So you know for example that if there are large cardinals then the theory of $L(\mathbb{R})$ is mostly known. It's frozen under for forcing changes so $L(\mathbb{R})$ feels like a very real object. Very cannonical. So in a similar way you would have such a theory of $L(P(\omega_1))$ from such an axiom that I'm referring to. And you know it was not known if $(*)$ consistent with large cardinals. We have our favorite family of strengthenings of ZFC which is by large cardinals because the idea is that you are in practice able to interpret any theory in some of these large cardinals so you would really like a natural axiom to be compatible with all of them, maybe consistent and this was not known. That is, until with Ralf Schindler we proved that $(*)$ follows from MM^{++} so we unified these two axioms which shows that $(*)$ in particular is consistent with all large cardinals because if you have a supercompact cardinal then you can force and you have $(*)$. So I think that the significance, the foundation signficance of this result follows from what I said.

<u>**Switzer**</u>: So I mean to unpack this a little further–would you say that your result implies, or strengthens $(*)$ in that sense that–how to say?–it builds up importance for $(*)$ in some sense because now we know it follows from something that is perhaps more well founded in some philosophical sense because it's compatible with large cardinals etc ?

<u>**Asperó**</u>: Exactly that's what it says. And also of course it provides further support in favor of MM^{++}. It's really nice in that it's now known that in particular that many–I mean not many but some–things that were known to follow from $(*)$ they can be forced over V. This was not known.

<u>**Switzer**</u>: Maybe this tells you something about forcing also, in some sense. I mean since $(*)$ is about $L(\mathbb{R})$ in some sense and these things that are trying to be away from forcing but MM^{++} says, you know, "all the forcing is good" does it somehow give credence to the idea that forcing is somehow playing a role thematically and not just in our toolbox?

<u>**Asperó**</u>: I mean well for one thing it shows that the class of forcings that Martin's Maximum talks about which preserve stationary subsets of ω_1, that forcing axiom enables to capture Π_2 maximality for all set forcing by $(*)$ axiom. In this sense it shows–well just what I said.

<u>**Switzer**</u>: And I mean, this is a little further–Shelah has this dream that he talks about in one of this papers which is to somehow show that forcing is the only real way to do independence results. And obviously Hugh was sort of questioning that in his

talk.[1] Do you think that the fact that forcing axioms can imply these strong structural things can have anything to say about this goal?

Asperó: I would have to think about it. I would think that this implication tells in itself you much about the forcing methods.

Switzer: Alright. So switching gears a little bit. Obviously there is this big debate in set theory between sort of pluralism and the multiverse view in some sense and then what Hugh pushes this kind of view of "there is a V". What you've done in your result in many ways, one of the many things you've already said, is you've taken two sort of seemingly totally different positions and put them together. Do you think this suggests that there is kind of a right universe or a right axiom or a right way of thinking about things?

Asperó: I think it's too early to make a prediction but it might be an indication. So I would also say–it's interesting you could say well you could try to get a strong forcing axiom that would do something like what MM does but for higher up like H_{ω_3}, some higher fragment of the universe and because, why not? It would be a natural thing to do, try to see if there are higher analogues. Of course not compatible with MM and $(*)$ but give you a strong theory up to that level. Interestingly it looks completely hopeless. You could ask for example for reflection principles that imply the continuum is $\aleph_3$ and we don't know even one example. We have all these reflection principles, there's a number of them like MRP, SRP and others, and they all imply there's exactly $\aleph_2$ many reals. They all give you a nicely definable well order of the reals in length ω_2. But there's nothing like that working for H_{ω_3} or any other H_κ. This is quite interesting. So there seem to be very serious limitations to doing anything strong at these levels, which okay, provides further support that maybe the continuum is just $\aleph_2$. There are good reasons to think that that is the case.

Switzer: And what do you think?

Asperó: Well just what I said. I'm still exploring. If things–I think what we know–the evidence we have so far seems to indicate that but of course it's too early to tell. There's a bunch of very nice, open questions in this area.

Switzer: What would convince you that the continuum was $\aleph_2$? What kind of result would some one have to prove for you to say "yeah that's it. That's the end of the story"?

Asperó: I think it's going to be a cumulative thing. I don't how to answer this question now because I don't think–uh–I wouldn't know how to answer the question.

Switzer: Okay. But somehow you feel like what you're doing is you're gathering evidence and all of this evidence is sort of pointing in one direction and this is sort of the way you view the practice of all of these things?

[1] Hugh Woodin's talk at the 6th Arctic Set Theory Workshop, "The Limits of Forcing".

Asperó: Yes.

Switzer: That's very interesting. So when you are looking for–for example higher analogues of forcing axioms where the continuum would be $\aleph_3$–do you have in the back of your head that maybe they're not there and you're trying to rule things out?

Asperó: Well I'm trying to see. Of course I'm ambivalent about this. I would love to find something interesting which is consistent and work with it. On the other hand it's also very interesting to find that there are limitations and that maybe it's impossible.

Switzer: And what about $(*)$? You talked about higher analogues of $(*)$ and I think that people don't really know how to formulate what that would be.

Asperó: Right. To me it looks to be that this looks hopeless. I mean I cannot think of anything that would play the role of $(*)$. There is no scenario for that. Yes.

Switzer: Okay. I guess just to push this a little bit further, a lot of what you sort of talked about with $(*)$ and MM is these sort of absoluteness principles sort of saying that "look we're going to get rid of some of the forcing by showing that we have a certain amount of elementarity and forcing invariance". Is this a goal of strong axioms? That they should be robust against forcing in this way. Obviously by contrast Ultimate L, if it exists, presumably, you could add a Cohen real and you would be done, right?

Asperó: Well I mean Ultimate L would also be robust. Of course you could force it to fail but the point is that any forcing extensions of the axiom would agree on the theory. In this sense I think it's a very attractive feature that you get this from strong forcing axioms. They tell you that have this inner model that's cannonical - or let's say that its theory is cannonical rather than the model, which is ultimately what we have control of, the theory.

Switzer: To shift gears a little bit–I'm curious–and maybe this is less interesting to some people but I'm very curious about this part: an important feature in your proof that MM^{++} implies $(*)$ is the double plus.

Asperó: Yes.

Switzer: We know you can't get it just from MM. You can't get it from $\mathsf{MM}^{+\omega}$.

Asperó: That's right.

Switzer: You have to go to the double plus. I think that you can sort of say "well okay that's just a technical strengthening and the real result here is that $(*)$ is compatible with large cardinals" and that's all true and that's great but somehow you had to do the stronger one in order to get the implication and I think that there's something structural going on there. You had to strengthen on one side in order to get the implication. What do you think this says? Do you think it says anything?

Asperó: Yes I mean that tells you that the notion of stationary set–which is the extra ε you get from the two pluses is central. I mean it's a central notion when it comes to understanding H_{ω_2} which is what I can say that tells you. You have to be able to talk about the notion of stationary sets in order to be able to prove these things.

Switzer: Why are stationary sets important? Do you have any idea about this? Of course they come up in the proofs but do you think there is something structural going on there?

Asperó: Uhh–I would need to think about it. *laughs*

Switzer: I mean–why am I asking this?–I think...

Asperó: I mean I was trying to say something meaningful.

Switzer: No worries. Stationary subsets of ω_1 are different than stationary subsets of any other cardinal.

Asperó: Yes.

Switzer: Stationary subsets play a very important role in your proof, in forcing axioms, in all this stuff. We've hit exactly the barrier of going above H_{ω_2} which is exactly this barrier that stationary sets stop looking so friendly.

Asperó: Yes.

Switzer: Do you think this is a coincidence? Or do you think that structurally the universe suddenly becomes much more complicated and there's suddenly something that's not going to work there.

Asperó: Yeah I mean I wouldn't know which way to draw a moral here. There are also limitations coming from other areas of combinatorics. When you look at other uniformization principles you have things which are consistent at the level of ω_1 for example Shelah's proof that the Whitehead problem is independent which for one of the possibilities used a unifomization principle that follows from Martin's Maximum. On the other hand you have higher analogues of these things that are simply false. This gives you– you can use these results to show that certain natural forcing axioms are just false. That's one source of obstacles. There are others, there's club guessing where something similar happens, and still some others. And these are not really talking–saying much about stationary sets. For instance these unifomization is really independent of the notion of stationary sets. I'm not going to define it but it does not involve talking about something happening a stationary number of times but happening everywhere.

Switzer: I see. To go back to this issue of getting higher analogues and relating to what you just said, first you run into these ZFC theorems saying you can't do this, you can't do that and you also run into these forcing challenges. Could it be that forcing runs out but that we need some other model construction method that you wouldn't have a *forcing* axiom you would have a ... this other model construction method axiom somehow? Is this something that you think these obstacles could point to? Something like this?

Asperó: I mean that sounds like an interesting idea but I wouldn't see where to start. In the sense that the best that we have really is forcing. I wouldn't know where to start.

Switzer: Yeah that's fair. I think–sorry I'm just thinking now. There is this place where it's just not clear how to continue. I like this idea you talked about that you're sort of building evidence in different areas. Did you expect that MM^{++} was going to imply $(*)$? Were you surprised by this result? Did it change your opinion about... [the evidence you were gathering]?

Asperó: I mean no. That's what we, with Ralf, were conjecturing. That's what I think people were expecting because you had all the–we made the realization that MM^{++} in practice decides everything you would like to ask about H_{ω_2} so that was an indication that it might give you a complete theory for H_{ω_2}. So there was clear evidence that it would be yes.

Switzer: Is this how you think about doing set theory as you move forward? If you have clear evidence for something you follow this line and...

Asperó: ...I mean that's a safe path. Right? Yes. It's been observed since Woodin's work on $\mathbb{P}_{max}$ that that must be the case. We knew that $\mathsf{MM}^{+\omega}$ that we were referring to before does not imply $(*)$. Paul Larson proved that. So that one is consistent with a well order of the reals definable without parameters and this is something we cannot have under $(*)$.

Switzer: Do you think of $(*)$ in some sense as a reflection princple too? Does it somehow change our thinking?

Asperó: I mean now we know–now that we have this implication that MM^{++} implies $(*)$ –actually that gives more information because it gives that $(*)$ is itself a certain bounded forcing axiom, a certain fragment of MM^{++}. So in this sense it is a reflection princple. I mean I like to see forcing axioms as reflection principles because of what they say and they're natural reflection principles for reflecting the low levels of the universe.

Switzer: What do you mean by reflection principles? How would you define this?

Asperó: So okay so large cardinal axioms, at least those that–let's say strong large cardinal axioms and ZFC axioms are either reflection principles or have reflection principles as consequences. It's always the idea that you know you want to have a global phenomenon reflecting that to hold many times below the cardinal or something like this. So now these strong forcing axioms which are obtained by actually obtained by carefully collapsing some large cardinal to something small like ω_2 or maybe ω_3 and then by doing this you guarantee that a lot of these reflections you had from the large cardinal is inherited at the level of the cardinal that it became. Moreover you can often prove that you often have these reflection principles following from forcing axioms.

Switzer: Just to go back to that then also–part of all of these proofs involve in some sense you're showing that ω_2 is in some sense a large cardinal.

Asperó: Yes.

Switzer: And so do you think that–you know if we were to go beyond the forcing axioms that we have–go beyond this $\aleph_2$ sized thing do we need to show that ω_2 and ω_3 are both large cardinals or is this that we just want to look at ω_3 now. What is the direction that you think we should go in?

Asperó: That's interesting. We could try–it could go both ways. We don't know. We could be looking at forcing axioms involving forcings that preserve ω_1 and ω_2 so ω_2 would be small like ω_1 is in the classical forcing axioms or maybe we need to collapse large cardinals and make some large cardinal ω_2 and make some other large cardinal ω_3. We don't really know.

Switzer: Okay. I think this is all the questions I had for now. Thank you so much for your time and for sharing your thoughts.

Asperó: Okay thank you.

Acknowledgements The author would like to thank the Austrian Science Fund (FWF) for the generous support through grant number Y1012-N35.

Chapter 36
A Conversation with Hugh Woodin

Juliette Kennedy and Beau Madison Mount

Abstract This is the transcript of a conversation between the participants of the Arctic Set Theory Workshop VI and Hugh Woodin. The conversation took place on February 24, 2023 at the Biological Research Station of the University of Helsinki in Kilpisjärvi, Finland, and was moderated by Juliette Kennedy. The transcript was created by Beau Madison Mount and is based on detailed notes taken by him during the evening. Questions were asked by David Asperó (DA), Douglas Blue (DB), Juliette Kennedy (JK), Rahman Mohammadpour (RM), Jeffrey Schatz (JS), Corey Switzer (CS), Jouko Väänänen (JV) and Bartosz Wcisło (BW). Unidentified audience members are denoted AM. * * * indicates recording unintelligible or missing.

Keywords Ultimate-*L* conjecture · Generic multiverse · Martin's Maximum · Large cardinals · (*)-axiom

JK: If I may start by asking you about the role of philosophical reflection in your work. How does your philosophical view shape what you're doing in set theory?

HW: I began as a mathematician; I was always in mathematics departments until Harvard. Twenty years ago I would have been sceptical about the role of philosophical reflection. Now I've changed, and I think a philosophical perspective can alter your perspective on which choices you make. I started out as an analyst, with Kaplansky at Cal Tech. The summer year of my undergrad, I did research on Kaplansky's conjecture, and I reduced the conjecture to a combinatorial problem in set theory. Then Dales and Esterle proved that CH implies the negation of Kaplansky's conjecture. Solovay suggested it might be consistent, and my senior year we showed the

The first author would like to thank the Academy of Finland, grant no: 322488. This project has also received funding from the European Research Council (ERC) under the European Union's Horizon 2020 research and innovation programme (grant agreement No 101020762).

J. Kennedy
Department of Mathematics and Statistics, University of Helsinki, Helsinki, Finland

B. M. Mount (✉)
Faculty of Philosophy, Oxford University, Oxford, UK
e-mail: beau.mount@philosophy.ox.ac.uk

S. Arbeiter and J. Kennedy (eds.), *The Philosophy of Penelope Maddy*,
Outstanding Contributions to Logic 31, https://doi.org/10.1007/978-3-031-58425-1_36

consistency of the conjecture with Martin's Axiom. That's why I ended up in set theory. As an undergraduate, I was interested in both physics and mathematics. So that's what I'm doing here.

JK: * * *

HW: In physics, you know there's a universe, but you're not sure what the fundamental building blocks are. In set theory, we know what the fundamental notions are; we're not sure there's a universe. The main motivation was always that I wanted to get to the root of things.

JK: You speak of a change in view. But the Ω-conjecture, the work on the continuum's being $\aleph_2$—that feels as philosophical as the stuff now.

HW: Well, twenty years ago, after the emergence of the $(*)$ axiom, it seemed like not-CH was the case. If the Áspero–Schindler result had been proved then, I'd have viewed it as confirmation. I worked for ten years trying to extend $(*)$ to a global axiom of V: it didn't work. But everything in that programme was based on a misconception: that whatever version of an L-like structure you had, it would be under an anti-large cardinal hypothesis. On that view, how could you have any kind of ultimate axiom? I thought that road was hopeless: the right road was generalized $(*)$. But that was based on a fundamental misconception. If you abstract what inner model theory was doing, large cardinals in a (generalized) L were authenticated from large cardinals in V. And this led to the weak extender model of supercompacts. In this methodology, a supercompact κ in the model M would be authenticated by measures witnessing supercompactness in V. But what you see is: that inner model M inherits all large cardinals from V above supercompact. So you take a feature of known inner model constructions, elevate it to supercompact, and something miraculous happens. But then something else happens. Hamkins was the first to abstract out the notion that you could ask whether V is a generic extension of an inner model (a ground).

AM: Bukovský.

JK: Bukovský was the first.

HW: Well, no, but I'm saying that Hamkins said: let's abstract it out. Bukovský gave a critical reinterpretation of that.

AM: Okay.

HW: What Hamkins did was he isolated two features—these approximation and cover properties—and focussed on those, motivated by being a generic extension. And out of that came a theorem of mine and Laver's: this is a theorem that should have been proved in the '60s.

JK: Maybe it was?

HW: No, because Bukovský was missing the definability.

AM: Yeah.

HW: The theorem that should have been proved in the '60s but wasn't is that, if V is a generic extension of an inner model, then the inner model is uniformly definable from parameters in V. So that becomes a first-order part of the theory of V. And so you can, in a first-order way, examine the inner models of V that V is a generic extension of. Bukovský didn't do that—he missed the definability.

AM: Yeah.

HW: So now you can talk about whether V is a generic extension, you can talk about the so-called generic multiverse: it becomes first-order in V. But then there's a remarkable convergence: approximation and cover notions are motivated by V being a generic extension. And the only known examples were V being a generic extension. So Hamkins proposed the ground axiom: V is not a generic extension. By the uniform definability, that's a first-order axiom. But he didn't propose—and I've never asked him why—the *superground* axiom, which would be that there are no nontrivial inner models of V with δ-cover and -approximation. Remember, δ-approximation is *motivated* by forcing. So something interesting happens: there's this idea of an inner model of a supercompact whose supercompactness is authenticated by measures in V—what I was talking about. So if κ is in an inner model of V and κ is supercompact in N, authenticated by measures in V, it has a κ-approximation and cover property. It's a completely different way to get to these models, and, as a consequence, you get that the superground axiom is falsified by a supercompact. Once you have large enough cardinals—and supercompact is probably the minimum—you get a whole new exotic family of inner models with approximation and cover. For example, if V is a generic extension of N—actually by a theorem I proved as a graduate student—there can be no elementary embedding from N to N.

JK: N being?

HW: The inner model that V is a generic extension of.

AM: From N to where?

HW: To N. So if you have an inner model of V that V is a generic extension of, there can be no elementary embedding in V as an amenable class from that inner model to itself. Now, that's fundamental in grounds. But wait a minute—if you assume κ supercompact, then there's an inner model with κ-approximation and cover. So—let me back up. If you have an inner model N with δ-cover and -approximation—an inner model of V, so it's a class with a closeness property to V above δ. So if you have an elementary embedding from N to N, its critical point would have to be less than δ. That's what the analysis of these models shows. Now let's move over to the context of a supercompact. It gives you these new exotic inner models with approximation and cover, and it gives you examples where there is an elementary embedding from N to N—so these are not grounds. So the existence of a supercompact cardinal in V falsifies the superground axiom; the conjecture is that you can't do it with anything less than a supercompact.

AM: How do you know that there exists an inner model—a weak extender model? Maybe V is the minimal weak extender model.

HW: So the theorem is this: if κ is supercompact, take the least measurable cardinal. In V there's a measurable cardinal below. Pick your favourite normal measure. Take the ωth iterate of V by that normal measure. Call that inner model N. It's a weak extender model for κ supercompact.

AM: Okay.

HW: So it is what I mean. But there's an elementary embedding from N to N. And in fact if you throw the Prikry sequence in, N is closed under ω-sequences. So you can have an inner model with κ-cover and -approximation, closed under ω-sequences, from which there's an elementary embedding from N to N. So a supercompact gives you, from a completely different origin, these exotic models that are somehow close to V. So that changes everything: it suggests that if you could build the inner model for just one supercompact, and realize it as a so-called weak extender model—in other words, you can (for example) imagine the inner model for one measurable and we can realize it as an inner model of V where the measurable is authenticated by a measurable in V; so imagine you could do that at the level of a supercompact: you'd provably build this inner model of V in which there appears to be a supercompact, but its supercompactness is authenticated by measures in V—then that inner model is very close to V and inherits all large cardinals from V. So the idea that Ultimate-L has to be pegged to the ultimate large cardinal axiom looks like it's wrong. You just have to reach one supercompact. Well, that changes everything. That suggests the possibility that there could be an Ultimate-L compatible with all large cardinal axioms. Okay, then, playing into this story is the fact that there's a generic multiverse, which I think it's fair to say that I introduced in that paper I wrote. When I was a graduate student, if I had known about the generic multiverse, I would have bought into that. The generic multiverse is: you're looking at V, and then you're looking at all the models you can reach by generic extensions, and then you follow it in a zigzag way. So you start with V—think of it as a countable model. You start with a countable transitive model and you look at the smallest collection of transitive models where you transitize the relation 'I'm a generic extension or I'm an inner model of a generic extension'.

JK: So what would have been so persuasive to you?

HW: It would have been very persuasive ...

JK: That the truths ...?

HW: Yeah. So if you think about the generic multiverse, think of a countable transitive model and do that, the first thing you think is: the model has no access to the truths of the generic multiverse. And because of the uniform definability of grounds, that's not true. So just very concisely: you have a countable transitive model; we just define the generic multiverse; it's all models you can reach in two steps. You take a generic extension of your current model—that's the outward. And then the inward is that you can drop to an inner model that that model is a generic extension of. So you just close under that. That's the generic multiverse. So from the point of view of a root universe, the root universe would be interested in all the sentences that are true in all the models of the generic multiverse. So you have a model M and you generate its generic multiverse.

AM: You mentioned truths in the multiverse ...

HW: Yeah, so we'll define a multiverse truth as a sentence that's true in every model of the multiverse.

JK: That paper. Were you thinking of the Beatles' song?

HW: Pardon?

JK: The Beatles' song: 'Across the Universe'?
[Laughter.]

HW: If you have a multiverse conception, then the truths are the sentences that are true in every model of the multiverse—those are the intrinsic truths, because your multiverse conception doesn't distinguish one universe from another in the collection of universes. So if you're a formalist, your multiverse is all models, and your notion of multiverse truth is the provable sentences. But if you're doing the generic multiverse, you're looking at sentences true in the generic multiverse. The problem is: it doesn't look like any model has access to that truth.

JK: Because ...?

HW: You're transitizing. But the theorem is ... that's not true. For every sentence φ, there is a recursive transformation of that sentence so that φ is true in the generic multiverse if and only if φ^* is true in M. So every model has access to the multiverse truth. Every model can figure out what sentences are true. The amazing thing about this proof—for forcing, it's not so hard because the transitizing of the relation collapses, because every forcing is absorbed into a collapse. But imagine the c.c.c. multiverse. So the steps are: the outward is c.c.c. forcing, and then you're allowed to drop to an inner model, but you have to be a c.c.c. forcing extension of that inner model. Now it's not so obvious. Or you can restrict to proper, or to semiproper, or to cardinal-preserving.

JK: So . . .

HW: So the point is that all of these notions, they're not productive. But for all of these notions, the theorem applies. The c.c.c. multiverse, the proper multiverse, the semiproper multiverse, the cardinal-preserving multiverse, there's always a transformation of sentences $\varphi \mapsto \varphi^*$, where φ holds in every universe of the multiverse if and only if φ^* holds in any of them.

JK: Are you worried that this privileges forcing?

HW: Pardon?

JK: Your notion of truth privileges forcing.

HW: No . . .

JK: If there's another model construction technique that comes along someday . . .

HW: So one way to frame the debate, to me, about truth in set theory is: let's just be model theorists. Here's our V. What's your conception of truth? Now, the V person would say: the sentence is true in that model. But someone who's not willing to commit to a V is going to have some kind of envelope of universes generated by that universe that defines their notion of truth. So the generic multiverse person might say: I'm not going to distinguish any universe of the multiverse from any other universe. That's the generic multiverse view. The nice thing about that is, CH—you're putting it on the side. Because CH is not uniformly true. That's a lens through which you can just try to frame the debate. In that paper where I introduced this, I argued that, assuming the Ω Conjecture, this was an incoherent view. Now, we can debate that till the end of time, but to me—the Usuba result just ends the story. I showed that if you have the generic multiverse, there's a preferred point of the generic multiverse if and only if there's a root. A generic multiverse generated by a model that thinks $V = L$, that's the root, because L's not a generic extension. Now remember—every model of the universe can figure out whether there's a root by the definability theorems. So the end of that analysis is that there's a preferred point of the generic multiverse if and only if there's a root. That's the only possible preferred point.

AM: Why is it preferred?

HW: It has to do with the definability of truth in the generic multiverse. So every model can figure out whether or not it's the root.

AM: But Usuba showed there is a root.

HW: Well, assuming large cardinals.

AM: Okay, yeah.

HW: So my point is that, before Usuba, if you take any model M and look at its generic multiverse, it can figure out whether the root and identify the root.

AM: But there is no reason to assume the root is the preferable…it depends what you mean by 'preferable'.

HW: Yeah, but then, all models of the generic multiverse are just generic extension of the root.

AM: Okay.

HW: So all truth is reducible to the root.

AM: Yeah, but maybe out of the generic extension of the root there are things which are preferable to the root.

HW: But it can't be first-order, because it's a generic extension of the root. So if there's one preferred point, there's a whole bunch of them. There can be no *preferred* generic extension in the generic multiverse …

AM: There can be a preferred theory …

HW: It's reducible to the theory of the root. So by reductionism—I mean, I don't want to get into an argument, but it's kind of like…you're right, I could do a homogeneous extension of the root, but if you're going to advocate for MM, that's a dead end. There's no homogeneous extension of MM from the root unless MM already holds.

AM: Let me just say: I disagree with you.

HW: So, let me just bring this thread to a conclusion. So I had this whole thing—if you take the inner model for one Woodin, the generic multiverse has no root. It's a theorem. So it's not always that there's a root. And it's easy to get other examples.

JK: So the MM model for one Woodin is not a generic extension of …

HW: No, the envelope of one Woodin has no root. No model can really …

AM: This Usuba result doesn't help, because there are no large cardinals.

HW: But my point is: you bring in the Ω Conjecture and you argue about whether the generic multiverse is a formalism—that's what I did.

AM: But strictly speaking, just the fact that there exists one Woodin doesn't mean that there's no root, because we could have—you need a class of Woodins, because with just one Woodin, suppose there's a real and that real is somehow coded ...

HW: If the inner model of one Woodin is in the generic multiverse, if you're doing the generic multiverse of the inner model for one Woodin, there is no root—that's all I'm saying.

AM: Okay.

HW: So the Ω Conjecture was brought in so you could argue against it, against the generic multiverse, but all that evaporated with Usuba's result. If you have an extendible cardinal in V, the generic multiverse has a root. So this is a theme where large cardinals are sharpening our conception of V. Below an extendible, that theorem is probably not true; once you have an extendible, there's a root.

JV: Is it like—in the generic multiverse, the ideology is that all models in the multiverse are equal ...

HW: It's falsified by an extendible.

JV: But Usuba shows that there is one which is more equal than others.

HW: So there is a theme where large cardinals might be trying to reveal to us more about the truth of V. And if you look at the progression: think about it. If you're in the L view, GCH would be a feature of V, large cardinals can't prove GCH, but wait a minute: by Solovay, a supercompact implies as much of GCH as it can. And now you move to an extendible, which is much stronger—and large cardinals are able to do more. They can say: look, I can't tell you that GCH is true, but I can tell you that the generic multiverse is rooted, which is what you would expect from any generalization of L. L is the root of its generic multiverse. So we're getting more and more features that we would expect of an L-like conception of V as we ramp up the large cardinal hypotheses. It's like they're our friends—they're trying to tell us more and more. And there are many examples of this now ...

JS: This is just a silly question, but I wanted to ask: you mentioned that the discovery that one supercompact was sufficient for Ultimate-L was very surprising. Did you think that Usuba's result was very surprising, or was it more something that you had expected all along?

HW: I found Usuba's result very surprising. I mean, the key to Usuba's result was Bukovský's theorem, which I was unaware of. So for all of you—don't ignore the literature!

[Laughter.]

HW: Bukovský's theorem is an amazing characterization of when V is a generic extension. Once you have the conception of grounds and once you realize they're uniformly first-order, you can start asking: what is the structure of the inner models of V that V is a generic extension of. And a very natural question to ask, if you have set-many inner models of V that V is the generic extension of, is: is there one below them all? Well, imagine trying to prove that—you have set-many models, you're going to work inside the intersection, but if you're going to show that there's an inner model there that V is a generic extension of, there's got to be a partial order and a generic HOD, it looks hopeless. So Bukovský characterized when V is a generic extension in terms of a covering property.

AM: It's not just Bukovský. There are two different theorems. It's Bukovský and it's Vopěnka and . . .

HW: No, I'm talking about Bukovský now.

AM: There is a similar . . .

HW: Well, the difference with Bukovský is that he characterized it in terms of a covering property.

AM: Another one is in terms of some kind of characterization. The Vopěnka characterization is, for every set of ordinals in the extension, there exists a fixed cardinal such that you can find the ground model. * * *

HW: I don't know if that feeds into answering it. I don't know that. So the question is whether you can use that in place of Bukovský to prove Usuba's theorem. So you have set-many inner models of which V is the generic extension, and you have to prove one inside them all. So there Bukovský characterization just works seamlessly. And I'm not sure that the other characterization . . . I don't know.

AM: Bukovský's characterization follows from the other one.

HW: Okay. So I don't know. Anyway, the bottom line is, Bukovský's characterization, which is a generalization of covering, was just the perfect way of viewing it, and you can prove that, if you have set-many models that V is the generic extension of, there is a model below them all that V is the generic extension of, so if you think of it in terms of directedness, the grounds of V are set-directed downward. And then there's another argument, that's not so hard—well, I don't want to say that—that shows that the intersection of all the grounds of V (which you can define) is also a ground of V. So that means the generic multiverse is rooted.

JV: And the root satisfies CH?

HW: No, you can't say anything about the root. Because, if you have a V with an extendible, you can force and preserve the extendible, by class forcing, and make the new model a root of the new generic multiverse.

JV: But you said the root is like a reminiscence of L.

HW: Well it's like L in the sense, to me, that if you want to access the truth of V, it seems that you want to know the truth of the generic multiverse, because V is either the root or a generic extension. So if V is somehow canonical, it's got to be the root, otherwise there's V and a clone of V.

AM: What's the use of the extendible in Usuba's theorem? It's that, in some sense, you can get class-many operations—that's what the extendible buys for you, in the proof.

HW: I mean—are you asking me, or telling me?

AM: No, I'm telling you. You can do things locally, and the extendible—

HW: But … for the point I'm trying to make, how the extendible works its magic isn't relevant. Whatever your conception of V is, it's got to have an extendible, or you're falsifying extendibles. If you have an extendible, you are accepting there's a root. Now you have to defend why you're not that root. And I don't see any coherent way—if you're not the root, then why aren't you the root?

AM: You can change the theory of the root any way you want.

JV: I think there are results that you can have anything in the root.

HW: All right, so …

JV: The root can be anything.

HW: It can be anything, but that just says that … now let's move to the Ultimate-L scenario. Now you sharpen your conception of V. One test question for that is: the axioms should imply that you're the root. Well, we've already argued that you're the root. So if you have an axiom that you're going to propose for V, one test consequence of that axiom should be that you are not a generic extension. Otherwise you haven't sharpened the conception. Now something very interesting happens. If you take the formal axiom $V =$ Ultimate-L, it implies that it's the root of its generic multiverse.

AM: What if Usuba's theorem is true from a supercompact?

HW: Doesn't matter. That root could be V.

AM: No, no, I mean—what if Usuba's theorem is true from a supercompact? Why don't you concentrate on that?

HW: So, I'm taking the view that, whatever conception of truth you have in V, it should have all large cardinals, or some version of large cardinals.

DB: But I could argue—I don't want to, but I could follow your arguments all the way until the last step (and again, I don't believe this), but I could say: Usuba's theorem tells me that extendible cardinals are no good. I could take the contrapositive.

HW: Okay, now I would offer you a challenge: why are extendibles consistent?

DB: Maybe they're not. Maybe this is what I'm learning from Usuba's theorem.

HW: But if that's a defence, then your scepticism is based on the inconsistency of extendibles.

DB: Yeah. I mean, like I said, I don't want to argue that, but ...

HW: But that's an interesting reduction, right? Because extendibles are either consistent or not—that's an arithmetic statement.

DB: I agree.

HW: So you have reduced the debate to an arithmetic statement.

BW: I have a terribly simplistic question, trying to gauge your platonism. Given a pure second-order formula—a formula of pure second-order logic—do you think it has a truth value? Or does the question make sense? What's your view?

HW: My view of truth is V. So I don't ...

JK: Meaning ...?

HW: I'm just saying: can I sharpen the conception of V to answer—is there a coherent sharpening of the conception of V that answers questions like CH, GCH, and all of that. Parallel to this is: whatever one's notion of truth is, you can reduce things down to arithmetic. You'd better come up with an explanation of why large cardinals are consistent.

DB: What if they're not? What if this is the moral of this whole story you're telling.

HW: But, if you're going to take the resolution that large cardinals are inconsistent, then how do you know ZFC is consistent? Where are you going to draw the line?

DB: The extendible, because of Usuba's result.

HW: But how do you know there isn't some follow-on that causes you to doubt supercompacts, and then measurables, and then ZFC? Once you go down that route, you're drawing a line, and it's a moving threshold. Where are you drawing it? Why do you think arithmetic is consistent? …

JV: You were going to say 'why is $10^{10^{10}}$ consistent?', but then you stopped …

HW: I'm a conditional fanatic about platonism.

JK: Sorry, say that again?

HW: I'm a conditional fanatic about platonism, in the following sense. I spent a lot of time with Silver, who thought ZFC was inconsistent. So I'm going to make a very provocative claim. Suppose ZFC + PD is inconsistent. Then I have no confidence in ZFC. My point is: all our understanding in set theory is beyond ZFC. And if we have so misunderstood everything, and PD is inconsistent (or infinitely many Woodins, if you want)—what's the basis for ZFC? Well, wait a minute—what's even the basis for PA?

JK: Well, you can go axiom by axiom and say, this one we like, this one … throw out replacement, for example …

HW: I'll make a very provocative statement now; Juliette has heard this. I've had the opportunity of knowing many of the great mathematicians of our time, many of whom were very close friends. And I've noticed a thread in their views—I'm not talking about the set theorists. Well, Menachem's a great mathematician of our time, but I'm talking about the non-set theorists. The view was: the spectre of inconsistency is not a meaningful spectre.

JK: Because?

HW: Because … if there's a contradiction, we'll define it away. Because what we do has intrinsic truth. And all right, if we have the wrong axioms, we don't go to that dark place where we had a contradiction, we shrink the axioms and so forth …

AM: * * *

HW: Yeah.

AM: But there is a historical example. When Russell came along … the mathematical community didn't care at all.

HW: All right, so … I pointed out to these mathematicians that they're completely wrong. That there is a spectre of inconsistency that they *cannot* rule out, that would wipe out all of mathematics—and there's no way to avoid it. It's a slippery slope. We're all in the same boat. I'm up high. I believe in the consistency of ZFC and Woodin cardinals, not because I'm wedded to them ...

[Laughter]

HW: But we're all in the same boat. We're all assuming a consistency which we cannot verify, which we cannot look at any of our experience to argue against, and it would wipe out everything. So I'm going to take a little bit of a detour. This was a thread which started a few years ago. Fermat's last theorem—there's a big debate: the original published proof is not in ZFC. * * * I e-mailed Richard Taylor, and I asked: well, do you think it's a theorem of second-order, third-order number theory? And he e-mailed back and said: I have no idea. And I asked Solovay, who was really studying the proof, and he said: I think it might be doable in third-order number theory, but I have no idea about second-order number theory. This is Solovay, really studying the proof. And then, you know ... now there are claims that it can be done in PA or whatever.

AM: I think it's worthwhile to point out what Andrew Wiles said to me. He said: I don't care.

HW: Well, that's interesting, because what if it had been—what if they had used a measurable cardinal, or something more? This will get to another pet peeve of mine, but let's stick with Fermat. If you're going to focus on the Fermat and you want to prove whether or not infinity is needed to prove. How can we make precise whether or not infinity is needed? Let's just be logicians. Here's a scenario—instead of working in PA, because if you work in PA or fragments of PA, the non-logicians just glaze over ... so you want to transform the question into something that everyone can engage with. All right, so let's change PA. So PA is Peano arithmetic, FPA is finite Peano arithmetic. So what is this? You start with 1, or 0 depending on your preference, and there's a largest but unspecified number. All right, so now you sit down and you write down what the formal axioms for this are. This is not arithmetic mod n, because you have the order. Plus and times are partial functions. So you just write down the obvious axioms: if $n + k$ doesn't exist, there's a j below $n + k$ such that j is the largest number, and so on. And there's a question what to do about induction. So you could take induction in the usual sense: if you can define a set of integers that's nonempty it has a least element, but in this finite setting, there's a potentially stronger principle, and it's open whether they're equivalent—and that's the pigeonhole property that defines a function from numbers to numbers: it's one-to-one if and only if it's onto. So that's a potentially stronger principle. So I take that as FPA. So the universe starting with 1 is a model of this. An amusing theorem is that FPA proves its own consistency. All right, so we now have FPA, which you could argue is a natural domain for mathematics if you believe our physical universe is finite. We don't know how big it is, but there's a largest number. So now you do some work: we have plus and times; you don't have all the usual coding, but you can still show that exponentiation can be defined as a partial function. (There's a really amusing exercise where you break down the exponents.) The relation $x^y = z$ is a uniformly definable relation in FPA. But wait a minute: that means that Fermat's last theorem is a statement of FPA. Is it provable in FPA? If it is, you're not using infinitary methods at all. So maybe that's the right way to ask that question. So in FPA you can define Gödel sentences which can't be falsified by our experience which would wipe out all of mathematics. So what you can do in FPA is, say: pick your favourite large number. In FPA, the existence of numbers becomes a large cardinal axiom. Because you can't even prove that 2 exists. So you could try to do the Ackermann function—more elegant, I think, are the Paris–Harrington numbers, because they grow much faster. And to a general mathematical audience, you can motivate the Paris–Harrington numbers. So (this is not the normal definition) the nth Paris–Harrington number, Paris–Harrington$_n$, is the least m such that, if you colour the n-size subsets of the numbers from 1 to m with n colours, you can find a subset of those numbers, all of whose n-size sets are coloured the same, and it has at least as many members as its least element (and more than it, to avoid triviality). So it's a theorem of ZFC that for every n, the nth Paris–Harrington number, Paris–Harrington$_n$ exists. These are unbelievably large numbers. The fourth Paris–Harrington number is beyond anything you can write down. So if you look at Paris–Harrington$_{100}$, you can write down a proof in PA that it exists. But take the nth Paris–Harrington number, where n is Paris–Harrington$_{100}$: you can't write down a proof in PA that that exists. It's got to be at least as long as Paris–Harrington$_{100}$, which is beyond beyond .... So you can formalize the existence of Paris–Harrington$_{100}$ in FPA: you just give yourself a couple of exponentials. Now, in FPA everything's about the existence of the largest number: you say that there is a k, and 2^{2^k} exists so that you have enough coding, and I see that k is Paris–Harrington$_n$. So consider the following Gödel sentence. My sentence φ says "there is a proof of my negation of length less than Paris–Harrington$_{50}$ (or some big number), but Paris–Harrington$_{100}$ exists". So if you were in a universe where this sentence is true and you had Paris–Harrington$_{75}$—which is unimaginably bigger than Paris–Harrington$_{50}$—you would have evidence, compatible with all the axioms of FPA, that Paris–Harrington$_{100}$ is inconsistent. So maybe we build a machine from physics that flips quantum coins and we find a proof of length 10^{24} of this sentence. We've just falsified Paris–Harrington$_{100}$. You don't define that away. That wipes out everything.

JK: So what do the number theorists mean by 'we define away the problem'?

HW: They mean: you logicians, you have these axioms, and you're doing what you do, and what we do is uncovering true truth, and what we're doing is intrinsically valid, and—all right, your framework, which we don't need, is contradictory. They're not distinguishing PA from ZFC, or ZFC + a Reinhardt. If ZFC + a Reinhardt is inconsistent, which we know it is, that's not the right foundational framework. What I'm saying is: why is FPA + Paris–Harrington$_{100}$ consistent? Why is that not a contradictory notion? The only argument I see is rooted in the infinite—especially if you do Paris–Harrington$_{\text{Paris–Harrington}_{100}}$. That's explicitly definable, and in PA there's no proof you can write down that that exists. So the only way I see that that's a consistent notion, in any verifiable sense, is to invoke infinity.

DB: I want to bring this discussion way back to the beginning, when you said that twenty years ago you would have been surprised that philosophy would give you anything. You're also in the Cabal, and you invested a lot of time in AD^+. Some would say that that group had a working philosophy, interests in particular things.

HW: I didn't see the philosophy of the group as choosing the directions to investigate. It was just: all right, we're doing AD, we have this idea that all definable sets should be determined, and that was the extent of the philosophy. It wasn't a dynamic, where you kept sharpening the direction you were going to investigate based on updating that philosophy. And the original philosophy of the Cabal, as I understood it, was: definability is orthogonal to the axiom of choice, so definable sets should be determined. Because: how do you produce a non-determined set? The axiom of choice. But if you think about it, that same philosophy would say that definable ordinal games are determined. How are you going to distinguish definable integer games from definable ordinal games? But determinacy for definable ordinal games might well be inconsistent. So the Cabal, as I saw it, had a philosophical perspective, but it was static: it just said, this is our direction, and then the technical development happened. It wasn't that the technical development then updated the philosophical view to suggest a new technical direction. So I think what's happening now is that we have this feedback loop, where you start with a philosophical perspective and a technical line, and then the technical developments feed back to the philosophical perspective, sharpen the tree of possibilities, and then you go back and explore. I really think there's a feedback.

JK: Have you ever entertained any—I don't like to use this jargon, but my sense of you is that you've never entertained any kind of pluralistic view.

HW: If I'd known about the definability of the grounds when I was a graduate student, I would have embraced that pluralism. It would have been just so attractive. But then I was just a graduate student.

[Laughter.]

HW: I would have embraced that, but that realization wasn't present then. Look, when $(*)$ happened, if the theorem of the convergence of MM^{++} and $(*)$ had happened twenty-five years ago, I would have taken that as the smoking gun for $(*)$. So...we have deeply held beliefs that change. And you can think that you have the right view at some stage and it's upended by future developments.

JK: But it seems to me that you've always been committed to the notion that set theory will just become like the rest of mathematics—a yes or no affair, things will be decided. But you're saying there was a time when pluralism would have been on the table.

HW: Well, I'll even go one step further. Set theory is the area of mathematics where we have the opportunity to discover truth which is not formal. There is no other area of mathematics that is doing that.

JK: Can you unpack that? What do you mean by 'not formal'?

HW: Okay—imagine you had access to the oracle for arithmetic truth. You'd be the best mathematician of all time.

JK: For sure.

HW: You won't have an answer to the continuum hypothesis.

JK: So you're saying: arithmetic truth is formal and CH is not?

HW: No, I'm saying—most of mathematics is formal. They're not looking for new axioms; they're working in an environment, usually it's PA, maybe it's some extension of PA, but they always expect that the arithmetic theorems they're proving will be provable in PA. So they're not worried about whether Fermat is provable in PA or not—it's fine if you use a superstructure to prove it. It's interesting: if you look at the famous conjectures of number theory, they can be multi-quantifer statements, like the twin prime conjecture or the ABC conjecture, but many of them—as far as I know, all of them—have sharpened forms that are Π^0_1. In other words, they conceive of a conjecture that's not one-quantifier, but then underneath that conjecture is a deeper conjecture that reduces it to Π^0_1. So for twin primes, there are estimates of how far you have to go to see the next twin prime.

BW: So again I would like to engage with your epistemological approach. So let say—this is a very strange scenario, but maybe imaginable; let's imagine that we've developed, let's say, AI to the point that it answers all the open questions in set theory. And at some point you ask it: 'is the continuum $\aleph_2$'? And it just says 'yes; I have developed strong intuitions'. Is that a good epistemological basis for accepting this axiom?

HW: So…I guess my view is, we have the continuum problem, but we also have the GCH problem. Are we never going to have the solution to the GCH? Is it always going to be an incremental climb? And if you take the $\mathbb{P}_{\max}$ solution, you've solved CH, but now you have the GCH problem. And it doesn't really give you a lot of insight into the GCH. It tells you the SCH, but it doesn't really give you a lot of insight. And we already know that with MM^{++}, there are questions: so where does the ambiguity of MM^{++} start to kick in? We know by the $(*)$ theorem that H_{ω_2} is crystallized out by MM. Go to H_{ω_3}: well, once you're in H_{ω_3} you have access to the universally Baire sets and all these questions, so there are fundamental first-order questions about H_{ω_3} that MM^{++}, unless there's an anti-inner model theorem, is absolutely powerless to answer. All right—so, is there a global generalization of

(∗)? It seems to be where you're led, because otherwise that methodology is seeming a bit hopeless for unravelling GCH. So, you could take the view that GCH will never be settled, because it's always going to incremental. Or you could take the view that we will sharpen our conception of the universe of sets at some stage and we'll get the answer to GCH. My view now is that I don't see the MM^{++} path as leading to the ultimate version of MM^{++} that's going to settle the GCH for us. There's no candidate for an axiom ...

AM: So you're hoping for a global solution.

HW: Yeah. As far as I know, there's no even invented global axiom for MM^{++} that would settle the GCH. On the other side, on the L side, we have an axiom, Ultimate-L, that settles everything. But we don't know whether it's falsified by large cardinals. So on the Ultimate-L side, there is a specific formal axiom which sharpens the universe of sets beyond the reach of forcing.

JK: You were talking to us the other day about how ZFC + large cardinals is much weaker than ZFC + large cardinals + $V =$ Ultimate-L.

HW: Well, we don't know that—that's an emerging possibility. Just think of it abstractly: naïvely there should be one axiom which you add to ZFC which sharpens the conception of V to settle many of these questions. * * * So it would be nice to have potential examples. There's only one. It's like this—people debate about whether all axioms are comparable in consistency strength. And you can do ad hoc counterexamples, so you might say: can you come up with an axiom which you would adjoin to ZFC which would settle all the questions we're interested in. Now, you might say $V = L$, but I want an example which is consistent with all large cardinals. There's only one axiom on the table right now, and that's the Ultimate-L side.

JV: But what is the other reason to adopt that axiom aside from its being consistent with all large cardinals?

JK: That it gives you fine structure, I guess?

HW: Well, I'm just saying there are two lines of inquiry. If you're going to the MM^{++} route—think of it as a game, a debate. So you're debating someone, and they have a different view, and you say: 'give me a candidate for an axiom that is immune to forcing and settles everything and implies MM^{++}'.

AM: Do you like the consequences of this axiom?

HW: I mean—I don't care, just give me an example first.

AM: So, take physics. Suppose that Newtonian mechanics tells you everything you want to know about the universe, and it's a very elegant theory and everything. But you don't want it: there are experimental things that tell you it's the wrong theory.

Now, can I give you a theory of everything that replaces Newtonian mechanics? Right now there is none.

HW: All right—I'll take this bait. V is V. Whether we like it or not—it's not an issue. To me anyway.

AM: * * * I'm issuing a challenge to an alternative axiom. The fact that you are not able does not mean—

HW: No. I'm not taking the nonexistence of the axiom as evidence. I'm just pointing out that, in a snapshot of the current state of affairs, there's an imbalance. There's no axiom on the MM^{++} side; there is an axiom on the Ultimate-L side. And there is a programme to either falsify that that axiom is consistent with large cardinals, in which case it's off the table and we're back, or verify it. That's all I'm saying.

RM: So I have two comments on the question. I think we don't even know any nontrivial axiom that implies that the continuum is $\aleph_3$. But also—this generalization you mentioned of the forcing axiom: there are many results by Shelah that suggest that the analogous situation is * * *. So you cannot expect such a thing. Maybe we need to change our paradigm. But my question is: back to twenty years ago, you mentioned that you thought of generalizations of this axiom. What made you prefer $(*)$ over forcing axioms?

HW: No, no, I didn't *prefer* $(*)$. What happened is: I had $(*)$ and it seemed clear that if the forcing axiom thread was correct, it would be $(*)$: unless $(*)$ was falsified by large cardinals, it's going to win over forcing axioms, because it gives you a canonical theory for H_{ω_2}. How could you not have that theory? So then the question was: how could that be integrated into the global theory of V? And I spent five years trying to pursue * * * how that might happen, and it kept just not working. And the reason was—it was based on a fundamental misconception at the time. So, if you look at inner models from the current perspective—I'm talking about real inner models—there are two varieties of inner models. There are the so-called extender models, and their ancestry is in $L[\mu]$, and then there's the unexpected family, which was only discovered many years later: the so-called strategic extender models. So if you look through the lens of inner model theory, the extender models somehow seem simpler than the strategic ones, because they have extra information. So the disturbing idea is that there might be two incarnations of Ultimate-L: one was the extender model, and one was the strategic extender model. Because the idea was if you can build the strategic extender model, that's harder, so surely you could have built the extender model. That intuition was false. There's now a theorem that says that extender models cannot reach the infinite levels of supercompact. There's only one—it's the strategic extender model. We don't know if it reaches the infinite levels of supercompact, but the extender models cannot. That's a theorem. There's only one Ultimate-L: there aren't two. And we can sharpen this into a very specific question, if you assume supercompact and a proper class of Woodins. In this whole subject, there's AD^+, that's a generalization of AD, and we look at inner models of AD. There's a technical theorem that if you have a supercompact and a proper class of Woodins and you collapse 2^κ, you have a *canonical* AD^+ model, handed to you by the so-called universally Baire sets, and you cannot change the theory of that model by forcing. So it's like a virtual—you don't see it without the collapse, but it's there. So one way to articulate the question is: look at the HOD of that AD^+ model, because we always look at HOD, and look at the rank-initial segments of that HOD—what kinds of large cardinals can exist?

JK: So do we know?

HW: No, that's the whole question. The question of whether Ultimate-L is falsified by a large cardinal hypothesis is a question whether…suppose Ultimate-L were falsified by a huge cardinal. Well, the way you would show that is that in this canonical AD^+ model there are no huge cardinals. So that's the whole question. The axiom $V =$ Ultimate-L is basically saying 'I resemble the HOD of that canonical model'. So we know where to look, and the rank-initial segments—that's the whole question. And that's the origin of the axiom, in this context, of $V =$ Ultimate-L.

DB: I had a question about the forcing axiom discussion. So we have someone in the audience who formulated Martin's Maximum, or helped formulate Martin's Maximum, and someone who showed that Martin's Maximum^{++} implies $(*)$. Can you say what, on your view, the value of research into forcing axioms is?

HW: I think the value is the exploration of the landscape of possibilities. Because if the Ultimate-L picture falls down, it's got to lie there.

DB: But even if the Ultimate-L picture works, what would you say is the value of this?

HW: I think, if Ultimate-L is somehow convincingly shown to be consistent with all large cardinals, V is Ultimate-L. I just don't see how one would argue against it. Now, however one is going to argue about a conception of V, you have to answer the following question: why are large cardinals consistent? Now, if you have a multiverse view, that's a challenge. If you have a single-universe view, you're going to assume they exist cofinally in V. But to me, that's the question—why are huge cardinals consistent?

CS: * * * Are you believing that forcing is the only way to …

HW: But whatever you do, you have to answer the question: are huge cardinals consistent?

CS: I don't have a problem with that …

HW: But why would you say huge cardinals are consistent? What would the basis be? Existence in V seems to me to be the only coherent basis. Why else would you say they are consistent?

CS: We can add it and we can do whatever …

HW: That's the utility hypothesis. But let me tell you a theorem from long before you were born. The first proof of PD was due to me and I used infinitely many Reinhardt cardinals. Now eventually we've come to accept PD. So is that evidence that Reinhardt cardinals are consistent? I had utility—I proved a theorem that was eventually verified from weaker hypotheses. So: what would your framework be? Are you ever going to declare huge cardinals consistent, and if so, what would the basis for that claim be?

CS: Uh … I think huge cardinals are consistent.

HW: Why? Why do you think they are consistent?

CS: For the same reason we, or other mathematicians, think that ZFC is consistent.

HW: What's their real basis for it?

AM: I think their real basis is we can prove theorems from it.

HW: I proved theorems from Reinhardt cardinals. Why is that not an equally sound basis for saying Reinhardt cardinals are consistent?

CS: Well, mathematics develops … we got that Reinhardt cardinals are inconsistent.

HW: Maybe we are going to get that ZFC is inconsistent as mathematics develops. So I don't understand the dynamic here.

CS: My question was not about large cardinals; it was about the multiverse.

HW: So in the multiverse conception, you have to account for the consistency of large cardinals. Now, if you restrict to the generic multiverse, you have a proper class of large cardinals in one * * * . If you expand the generic multiverse, you lose that. So you seem to be locked in—if you allow class forcing, you don't have measurable cardinals. You can do Jensen coding and expand to $V = L[r]$ for a real r.

CS: But maybe there is another way to prove consistency results.

HW: No one has come up with one. That's the 'other way' view. But we know of no example. So someone could say: set theory is misguided; we should be studying widgets. And here are the axioms of widgets. And, lo and behold, in the axioms of widgets, Con(there exists a huge cardinal) is just there. And we've been misguided, because we've been studying sets. And that's always a possibility ...

JK: I think you're very diplomatic in talking about the multiverse view, and the single-universe view, but I think what's dominating now, in the philosophy of set theory conversation, is the pluralistic or the multiverse view, and there hasn't been, apart from yourself, somebody planting their flag on the side of the single-universe view.

HW: So my response is—set theory is great: because of independence, you can change your mind. And I've changed my mind, and, based on my experience, I don't see any coherent view for truth in set theory except a single-universe view. And I can distil that down into a series of specific conjectures: that one formal axiom, $V =$ Ultimate-L, is consistent with all large cardinals. And there are various ways that you can verify that.

JK: So you haven't that through an experience of 'classical mathematics', you've come to that through the developments in set theory.

HW: Yeah. I just don't see another option right now. Now, maybe it's like twenty years—if we'd been having this conversation twenty years ago, $(*)$! But I spent a long time trying to get that to work. And at least my approach wasn't working. And the generic multiverse seemed like an appealing conception, but I think that's ruled out by the Usuba result. So everything seems to be kind of forcing us into a single-universe view.

JK: But, wouldn't it have been—if there's a ZFC line of provability, and then there's all this indeterminacy, wouldn't that have made set theory into something very exotic and not very plausible mathematically? That really the action is below the ZFC line of provability? So I think it's very interesting that you come to a single-universe view through the set theory rather than through 'let's try to have a single theory'.

HW: Well, I think you're compelled to deal with a single-universe view or not, in order to account for the truth of certain arithmetic statements. I know of no argument that large cardinals are consistent other than their existence in V. I mean, maybe there will be—maybe there will be a methodology that just explicates this and you don't have to get into V: the widget scenario.

JK: So it all begins with the large cardinals?

HW: Well, it begins with ...

AM: Where is V?

HW: Well...all right, I'll be more provocative. I get asked this all the time: what is the use of set theory? It's not relevant to core mathematics, this or that, et cetera—and my response is that it's a win–win situation. Look, either set theory is like non-Riemannian geometry, and eventually it will be relevant to whatever, physics or whatever—that would be an extraordinary win. No one would argue with that. It's very dangerous to say that some esoteric area of mathematics is not relevant to the future understanding of physics. Who would have said that infinite-dimensional Hilbert spaces would have been relevant to physics? And that's the basis for quantum mechanics, or at least one version of it. What's the other win? It's never relevant. It's not relevant to core mathematics as currently defined, not relevant to physics as currently conceived, it's just *outside*. That's a better win to me. Because that means set theory is uncovering truths which can't be accounted for any other way. Set theory says large cardinals are consistent: that's a prediction about the physical universe. Design any physical system you want, contact any alien civilization you want, ask them whether ZFC + PD is consistent, they're going to say 'yes, it is'. That's a prediction about our universe. No physical process you implement, I'm predicting, will uncover for you an inconsistency in PD. So I'm claiming that in a billion years, if this planet is still existent, with some descendants of us on it, ZFC + PD will not have been falsified.

AM: * * *

HW: Because I'm so convinced of our understanding of the universe of sets that that kernel of truth, to me, is absolutely true.

CS: Why is PD diffferent from 'there exists a Reinhardt cardinal'?

HW: Because we've understood—there's a moving threshold of our understanding. For example, in the '60s, no one would have made this claim. Or take ZFC + a measurable. In the early '60s, no one would have said 'I bet my career that measurable cardinals are consistent'. It's one of the Banach problems—no one would have said that. But then our understanding evolved. If you'd proved measurable cardinals are inconsistent in '65, all right, I would have been a footnote.

AM: Let me just add a historical twist. If you look at the paper of Keisler and Tarski on large cardinals, they phrased them as what we would now call the negative of large cardinals. They talk about the class of cardinals we would now call *not* weakly compact, the class of cardinals which we would call *not* measurable, the class of cardinals we would call *not* strongly compact—and I think the reason they did it is exactly because of the issue of consistency.

HW: Well, Vopěnka was formulating an axiom he thought was inconsistent, because it was so ridiculous.

AM: What about Berkeley cardinals?

HW: Well, I'll make a prediction: if any of you show measurable cardinals are inconsistent, you're going to win a breakthrough prize, and you'll be on the front page of the *New York Times*.

DB: I don't think that's true.

HW: I absolutely think it's true.

AM: You'll at least be on the front cover of the *Atlantic*.
[Laughs.]

AM: Remember Jensen's 'Modest Remark'.

HW: Well, times have changed. But I think that now 'no measurables' would be front-page news.

JK: Has the work in philosophy of set theory—has that been part of your intellectual or personal experience?

HW: Well * * * I would have never considered the generic multiverse without a foundational philosophical perspective, and that led to some surprising results. My entire research direction over the last five to ten years is driven by a philosophical perspective. What keeps me up at night is I'm trying to understand and sharpen V.

JK: Can you put 'sharpen your conception' in terms of … you're trying to investigate what is true?

HW: So I choose my problems carefully. And I bang my head against the wall a lot. Because that's what I'm interested in.

AM: What drives me—it's not understanding V. I don't know what it means, understanding V. Is there one canonical V? But what drives me is creating a collection of theories that, at the end of the way, the mathematical community or the set-theoretical community will agree that these are theories we would like to adopt. So what drives me is offering a spectrum of possibilities from which the ultimate theory will be chosen. I don't understand what it means to sharpen V.

HW: So you're not in the single-V view?

AM: I don't know what V is—so what is the single-V view. I'm in the business of theories.

HW: But how do you deal with the consistency question?

AM: Simply by the fact that we sharpen our understanding of the concept and we don't run into inconsistencies.

HW: So … at any point …

AM: The people who came up with non-Euclidean geometries, Gauss and Lobachevsky and Bolyai, what was their great invention? It was just this realization: for a long time, there was a lot of work done trying to prove the parallel axiom. What was the great revolution? The great revolution said that, based on all the evidence we have—they didn't have any models, it was forty years later that you were able to prove the relative consistency of non-Euclidean geometry—maybe the negation of the parallel postulate is consistent. It would give us a different geometry, and let's see: maybe it will be useful to study this geometry. And then it turned out that this geometry is very useful.

HW: So it sounds like, in your view, at any point it could happen that ZFC is inconsistent.

AM: That's correct.

HW: And you're never in a position to just declare ZFC is consistent?

AM: Based on the best of my knowledge and the best of my understanding, it is consistent.

HW: But that's a time-dependent thing.

AM: Yes ...

HW: So you would not declare ...

AM: To the same extent that it's the best of my understanding right now that, say, the Big Bang was twelve billion years ago. And maybe that will change if there is new evidence.

HW: Well, I like to think that the question of ZFC is not dependent on the Big Bang ...

AM: I think there's a definite answer as to whether ZFC is consistent or not.

HW: But you're saying you could never know the answer.

AM: Never know *for sure*.

HW: I find that unsettling.

AM: Everything in life is unsettling.

HW: Well, I think we should embrace our intuitions and evolve our confidence level—that's all.

AM: Okay.

HW: Why do you think—PA could be inconsistent. Paris–Harrington$_{100}$ could be inconsistent in FPA.

AM: But it's less likely, because we understand better ...

HW: It's less likely. So is it one chance in a billion? Assign a probability; you say it's less likely.

AM: I don't know what a probability means in this case. I don't understand. You know, when they make a forecast on television and say there's a sixty per cent chance of rain—what does that mean?

HW: So I'm going to push back on this. If one's position is always 'it's unlikely that ZFC is inconsistent', and you can never transition beyond that, that's somehow unsatisfactory. Because I don't know ... I mean, if it's always possible that ZFC is inconsistent, it seems to me that eventually ZFC is inconsistent.

AM: * * * But that almost sounds like a religion.

HW: Well, maybe it's religion.

AM: How do I know that $2 + 2 = 5$?

JK: That's a different conversation ...

HW: That's a different question.

AM: Many years ago I gave a talk at the University of Washington in Seattle, there was a topologist—I don't remember his name. And I explained the consistency—I was talking about something that followed from large cardinals, and the issue of consistency came up, and I explained that we are not sure about the consistency of Peano arithmetic. And he said ... I'm not sure that you'll get there. I keep counting ...

[Laughter.]

JK: I've learned a lot in this conversation, but you know—if you look back at Hilbert, in the 'es gibt kein *ignorabimus*', Hilbert says that human dignity depends on our being able to decide mathematical questions that are easy to ...

HW: Articulate.

JK: So there's one thing that we haven't talked about that I wanted to talk about in this conversation—the unrestricted multiverse. Not the generic multiverse, but anything consistent.

<u>**HW:**</u> Well, the trouble is that no one, as far as I know, has articulated a coherent conception of a multiverse—just think of it abstractly as one model, a multiverse is a collection of models generated by that model. So if we start with a transitive model, we have the generic multiverse and then you might widen that.

<u>**JK:**</u> Well, let's take all consistent.

<u>**HW:**</u> That's formalism: that's just arithmetic truth. So the question is: is there an intermediate?

<u>**JK:**</u> Do you think there is an intermediate?

<u>**HW:**</u> No one has come up with one, as far as I know. An intermediate multiverse view between the generic multiverse and all models. * * * That verifies large cardinals.

<u>**JS:**</u> This is not quite on this particular matter, but I just wanted to ask, since I think this is a similar experience to Juliette's question. What are the ways in which philosophers of set theory—young philosophers trying to follow in Pen Maddy's footsteps—could assist set theorists like you? What is the kind work that philosophers of set theory could do that would be useful?

<u>**HW:**</u> Well, they could assess my philosophical views and say 'you know you're really naïve philosophically'—they can sharpen the philosophical methodology. They can point out 'well, look, these are some issues …

<u>**JK:**</u> Has that happened?

<u>**HW:**</u> To me? Not yet, but I think …

<u>**JK:**</u> So that would be the job.

<u>**HW:**</u> I think the trouble is that the technical results * * * have been accelerating over the last thirty years, you know—there are technical results in the last five years which keep changing the picture. So I think that, for a philosopher, there's a challenge: you don't want to base a philosophical discussion…how do you balance a philosophical critique or analysis with the changing landscape?

<u>**JK:**</u> So why do we even draw this distinction between the philosophical and the mathematical in set theory? Why isn't what you're doing, and Steel and Magidor, just the philosophical exploration of the infinite? Why do we need to draw the distinction?

<u>**HW:**</u> All I can say is, in my own experience, in my own career to this stage, I strongly feel that I have pursued technical lines that I believe are philosophically motivated. I could have gone in many directions, but the directions I've chosen are, I believe, motivated by a philosophical perspective. Now maybe I'm completely …

JK: But is there a way to do philosophy of set theory from the outside?

HW: Well, I mean—if you're going to do philosophy of set theory, it seems to me that it has to be informed by what's going on in set theory.

JK: So do you think, being in a position of proving a theorem and contributing—so you were saying the other night that there's a difference between philosophers and mathematicians, the people who want to advance the field of set theory and the people who have a philosophical view and want to defend it.

HW: I guess—what does it mean to advance the field of set theory?

JK: Well, that's what I'm asking you.

HW: I spoke here on a technical model theory of set theory question, this ω-categoricity result. Getting the answer to that, to me, does not shed light on what V is. I was trying to make the point that in set theory we have two subjects, which it's important to draw a distinction between: we have the model theory of set theory, in which there are really hard problems, and then we have the question, which is harder to articulate: what is V? That question about categoricity that I raised—it has nothing to do with whether CH holds in V, but it's a really hard question, and I think it's a fascinating question, and it's not a question about V.

JK: But when you look at, for instance, the Polish school studying taking away power set, taking away replacement, and so on. It's a whole industry, but one has the feeling that the action is really on the side of what's going on in V. So you're saying it's possible to have a deep question on the side of the model theory of set theory.

HW: In my experience, in set theory, since I was a graduate student, there are lines of set theory which in the end have tremendous impact, and they in the beginning they look like: 'why are you going in that direction'? When I was a graduate student, I remember Solovay telling me: Π^1_2 determinacy is stronger than all large cardinals. That was a deeply held 'truth'. And there were theorems from the perspective of determinacy that supported that view. And then, out of left field, completely unexpected, were the results of Martin's Maximum. Now what did they do? Let's strip away Martin's Axiom: there's the notion of a large cardinal that says that there's an embedding from V into M. And then there's the generic version: you have an elementary embedding from V into M but M is included in a generic extension. Now when you allow the flexibility of the generic extension, then that generic embedding could have critical point ω_1. And when I first heard that floated, I thought: that's going to lift to large cardinal embeddings—it's not relevant. And I was completely wrong. Because if you have a supercompact or a Woodin cardinal, then in V there are generic embeddings with critical point ω_1. That fact, which was so surprising, was the fact from Martin's Maximum that led to Woodin cardinals and all of that. It was completely orthogonal to the deeply held intuition. So the idea of distinguishing between generic embeddings with a given critical point versus elementary embeddings was absolutely essential, and when I first saw it I thought 'this is kind of silly, because there's no way you're going to prove from large cardinals that there's a generic embedding with critical point ω_1', because I had subscribed to the view—and that was completely wrong. And that was the fact from Martin's Maximum—it had nothing to do with Martin's Axiom—it was the fact that the consistency proof for Martin's Maximum just proved from a supercompact there was a generic elementary embedding of V with critical point ω_1. And it was that abstract statement that I used to show, before I knew anything about the proof of Martin's Maximum, that it implies that projective sets are Lebesgue measurable. So it was out of left field. So that kind of thing happens: lines of investigation can lead to very unexpected consequences that feed back and change a line of investigation in a seemingly unrelated realm.

JK: I feel like you are an example of reflecting on the theorems you're proving and reflecting on the broader implications ...

HW: Thirty years are a long time. You'll be here at Arctic 40 or whatver, and you can reflect on how wrong I was.

JK: Well, I think if—if you really want to solve this problem, this technical thing, you have always been directing people toward reflecting on the larger implications of the technical theorem, regardless of whether ultimately you're driven into a single universe. I mean—do you think we could have this kind of conversation if we were complex analysts or something? I don't think we could.

HW: No.

JK: This is the beautiful thing about this field: it's an investigation of the infinite but it's also an investigation of mathematical truth at the same time. I don't think any other field has that.

HW: I strongly believe there are going to be many surprises left to come. This is just a snapshot. There's a process going on in set theory—a kind of synergism with philosophy that seems to me to be unique, that is not happening in the so-called 'other areas' in mathematics, because they're not axiomatically driven. This is not a criticism of the core areas; I'm just saying, we in set theory are dealing with a phenomenon that seems unique to set theory. We shouldn't be embarrassed by this; we should embrace it. And it intertwines with philosophical investigation: where else in mathematics do you get a constructive resonance with philosophy? I don't have any examples.

JK: And the nice thing is—people like Hilbert, they knew their Kant, they knew their history of philosophy: they had an education .... We have David Áspero here, who's proved a great theorem about convergence. Are you a single-universe person or a many-universe person? And what do you think MM^{++} implies $(*)$ has to do with it?

DA: Right. So—yes—I think there's the issue of reference …

JK: That's a fancy philosophical word, 'reference'.

DA: Well, reference, talking about things.

JK: Is mathematics descriptive or not?

DA: Yeah. So, in this sense …

JK: Where do you stand: is your work descriptive, or is it made-up, conventionalist?

DA: I mean, I think there's one universe, but I don't think it's easy to know what the truth about the universe is. I think it's difficult to find natural axioms that give the continuum's greater than $\aleph_2$. So I think there's something there that shows that the continuum is at most $\aleph_2$.

JK: We had this conversation in Mexico in which David was talking about these tremendous obstacles to trying to lift the techniques get any kind of interesting theory with the continuum $= \aleph_3$. So it's bounded by $\aleph_2$. But does that mean that we're too in love with our methods or does that mean that the situation is that that's actually the case—it can't be more. So are we looking for the keys under the lamppost because that's where the light is? We had this very interesting conversation …

HW: * * *

JK: You have a very considered answer to that …

HW: Well, there's a singular universe V which doesn't undermine deep interesting technical questions in the model theory of set theory. So, to me, possibly the extent of canonical theories where the continuum is larger than ω_2, that's in the province of the model theory of set theory. So, for example, it's easy to state problems in the model theory of set theory that are incredibly difficult. So, for example, suppose you have a countable model of ZFC. Is it consistent that every width extension of that model which is cardinal-preserving is a set-generic extension?

JK: The answer is …

HW: Well, that looks just hard beyond hard. We have no large cardinals, global violations of GCH, but just that model. Is it consistent to have a model, a width-extension, where you preserve large cardinals and it's not a set-generic extension? That's an incredibly hard problem, I think. David has some of the best partial results on that. But that problem looks really hard. And it's orthogonal, to me, to the question about whether GCH holds in V or not. Why is it that every problem in the model theory of set theory has to be bundled in, and its value assessed, based on its implications about V? We don't judge theorems in group theory by their implications for the theory of V.

JK: Well, let me play devil's advocate. It's because the action is in what is actually the case. As opposed to what could be the case.

HW: All right, I'll go back to the talk I gave. That question—whether there's a sentence φ such that $\mathsf{ZFC} + \varphi$ is β-categorical and satisfiable and doesn't think that $V =$ Ultimate-L—that's a model theory of ZFC question.

JK: It's a beautiful question.

HW: It has nothing to do with whether CH holds in V. It looks like an incredibly hard problem. I would love to know the answer. If someone solved it and they submitted it to the top journal of mathematics in the world, I would advocate that it should be accepted.

JK: But here's the problem. If you think that depth of a question is truth-conducive, then that's something new that you're saying—that the depth of a question, the hardness of the question—that there are deep questions in the model theory of set theory that have nothing to do with what is actually the case. People often think that things like depth and something's being a hard problem are actually impinging on truth.

HW: No. The difference in set theory is: in number theory, you have a deep, hard problem because there's a conception of the unambiguity of truth in number theory: it's contributing to the theory of the universe of the integers. But we know that's not the case in set theory. You can have a hard question in the model theory of set theory that's not telling you what's true in V, beyond Σ^1_2.

JK: I guess I'm going by—if you look at all this work, on ZF$-$Powerset, ZF$-$Replacement, it's kind of nugatory .... So we have a new situation, that the model theory of set theory can be interesting.

HW: So I guess my cautionary note there is my reaction as a graduate student to generic embeddings of critical point ω_1 was of that ilk. It was just 'why are you even looking at this?' And it led to something unexpected. So it could be that these investigative lines, in unexpected ways, inform not just the model theory of set theory, but our conception of V. I'm just saying: be very careful about what lines of investigation are very relevant to understanding V versus not, because sometimes unexpected things happen. I'll give a good example: so Hamkins's theory of grounds, approximation and cover, turned out, for completely unexpected reasons, to be a key notion.

And the reason is that there are these examples of inner models of approximation and cover that have nothing to with the origin of the concept, namely grounds. So what looked like 'you're just abstracting out when V is a generic extension and running with that, it looks like that's not going to inform our understanding of V'—but then it resonates with an unexpected development, namely that there's another source of inner models of approximation and cover.

JK: So you think that vocabulary was helpful?

HW: Incredibly helpful. It's an introduction of a concept of obvious relevance to the model theory of set, but it doesn't actually seem like it's relevant to a V-type thing, but then connects with an unexpected development. So my reaction is: be very careful about an initial assessment about a line of investigation that one thinks is purely . . .

JK: Technical.

HW: Like the question I raised about β-categoricity. Maybe the answer to that question could tell us something about V in an unexpected way.

DB: On that note, what if—there's forcing, there's Martin's Maximum, there's iterated forcing. What if there's a technique for solving the β-categoricity problem …

HW: Could be.

DB: And there's an axiom, and it's global.

HW: Could be.

JK: So … I noticed you didn't bite when I asked this earlier when I brought this up, but you remember this dinner * * * and everybody asked their question—this was in 2000. 'What is the most pressing question?' And Shelah's question was: is forcing the only model construction technique, or are we too wedded to forcing?

HW: Well, I think I answered that question here. You can't use forcing to solve the β-categoricity problem.

JK: Yeah, so you gave us a nice question … but at a lot of the work in set theory and the philosophy of set theory is aimed at generic absoluteness.

HW: No, but … let's go to forcing. When Shelah's programme of iterated forcing and everything was underway, one might have reasonably conjectured then that the ultimate development of the theory of iterated forcing is not going to change or inform our ocnception of V. And that was completely wrong, because it led to

Martin's Maximum, which drove the prevailing picture. So you've got to be very careful about saying 'that direction is not relevant'.

JK: Yeah.

HW: And I think, if nothing else, what I've learned in my forty years is: never make that claim. The idea that you try to assess a line of investigation in terms of whether it's relevant to, maybe, V if that's what you're interested in: that's a losing proposition. That's like saying some area of mathematics is not relevant ... No. The investigation of forcing axioms beyond MM^{++} or higher values of the continuum could reveal subtleties about the combinatorics at $\aleph_3$ and $\aleph_4$ which could really inform. So I'll give you a problem in the model theory of set theory which is open: assume large cardinals, and—I'll just use the simplest instance—you have prewellordering of the reals that is projective, so definable in second-order number theory: how long can it be? So, with Martin's Maximum, it can be of length ω_2. Can it be of length ω_3 under any hypothesis? That is open. So it could be a theorem from large cardinals that any projective prewellording has a length less than ω_3.

CS: You mean, consistency from large cardinals?

HW: No, an implication. It's in fact a form of the continuum hypothesis. So the continuum hypothesis asks: is every prewellordering of the reals of length less than ω_2? Now you change the question, and you restrict the prewellorderings. So one way to restrict them is to make them projective, or projective only with large cardinals. You can broaden it to universally Baire. So that's really the question. We have a proper class of Woodins and a universally Baire prewellordering. Can it have length ω_3? That's an effective version of the continuum hypothesis, and it's completely open. Now, it could be that it's the subtle combinatorics associated to $\aleph_3$—that kind of collision. Another way to ask the question is: I have an inner model containing the reals of determinacy, so an inner model of AD, you look at the supremum of the ordinals onto which you can map the reals. That's the AD-measure of the size of the continuum. It's not known whether that supremum, in the context, can be bigger than ω_3. So it could be that the ZFC combinatorics on ω_3—you have stationary sets on different cofinalities, all this infinitary combinatorics—collides with the AD view, and that's what prohibits the Θ of the AD model from getting above ω_3. Because whatever ω_3 of V is, from the point of view of the model of determinacy, it's some regular cardinal. And determinacy says: I have all this combinatorics there. And the ZFC world says: this is ω_3; I've got all this combinatorics associated to ω_3. And maybe they're incompatible. So we know that large cardinals can't settle the continuum hypothesis. But we don't know if large cardinals don't prove that the *effective* size of the continuum is less than ω_3.

CS: It is known if it ...

HW: Nothing's known. The only thing that is known is that the effective size can be ω_2, assuming large cardinals. It's not known if it can be bigger than ω_2. It could be a theorem of ZFC + a proper class of Woodins—or it could be a theorem of ZFC—that if you have an inner model of AD, and you look at the supremum of the ordinals onto which you can map the reals in that inner model, that that supremum is less ω_3. That could be a ZFC theorem. That would be remarkable. That would say that: maybe ZFC proves an effective version of the continuum hypothesis. It says: 'Look, I can't tell you how big the continuum is, but I can tell you that if you have any inner model of AD containing the reals, you cannot map the reals onto ω_3 by a function in that inner model'. We know that you can map it onto ω_2—Martin's Maximum implies that. So the next cardinal's ω_3.

DB: So what's your conclusion, if David comes to you tomorrow on the bus and says 'here's a proof that in ZFC there's no projective prewellordering', what's your conclusion there?

HW: That it's a great theorem in the model theory of set theory.

DB: But it doesn't change your perspective.

HW: No, not at all. It just says that there's no $(*)$ generalization for the continuum bigger than ω_3. That would be my takeaway. So $(*)$ tells you that you can be a determinacy model and force ZFC and make the effective continuum bigger than ω_2. One way to test for a generalization of $(*)$ is to ask: can you be an AD-model, like $\mathrm{AD}^{L(\mathbb{R})}$, and force choice, without adding reals, but make the continuum greater than ω_3 in the extension? So my interpretation of that theorem is: there's no generalization of $(*)$ along those lines. If you look at $L(\mathbb{R})$, you can force over $L(\mathbb{R})$ without adding reals and preserve ω_1 and ω_2. If you're willing to sacrifice ω_2, there are many larger regular cardinals, and you can preserve ω_1 and make the larger regular cardinal ω_2. So you can, for example, in the $L(\mathbb{R})$ world, look at the successor of $\aleph_\omega$: in descriptive set theory, it's called $\boldsymbol{\delta}^1_3$. You can actually force over $L(\mathbb{R})$, get choice, ω_1 is preserved—you're not adding reals—but you can make $\boldsymbol{\delta}^1_3$ ω_2?

DB: It's a variation of * * * ?

HW: Yeah, it's a technical variation. Neeman has a paper: there are various regular cardinals you can do. But no one has been able to conceive of a generic extension where you preserve three cardinals—in other words, make the continuum bigger than ω_3. The problem is: you've got ω_1, you're going to preserve that; what's the next cardinal you're going to preserve? $\aleph_2$ and all those are singular: you're not going to preserve those. There's a theorem of Jackson: suppose you have projective determinacy (I'm assuming ZFC). Look at the length of any projective prewellordering: it's less than $\aleph_\omega$. In other words, there are only ω-many cardinals that can survive. But it's not known if more than two can survive. So it could be that it's just a ZFC theorem that not only can there be no projective prewellordering of length ω_3, but there can't be any prewellordering of length ω_3 that exists in any inner model of AD. It has nothing to do with the complexity of the prewellordering; it has to do with existing in that inner model of AD. So this is the effective continuum problem. So some people might argue...let's imagine that someone proved that in any AD model there can be no surjection of the reals onto true ω_3. You might argue that that's additional evidence that the continuum is $\leq \omega_2$. Because you've found that the effective size of the continuum is ω_2, and you just have to sort between ω_1 and ω_2. But maybe the attempt to do forcing axioms beyond will shed light on this. The AD world has very rich combinatorics at its cardinals; ZFC has very rich combinatorics at regular cardinals; and there's a real potential of collision here. That's a great problem; Martin had the conjecture that the right theory would be the supremum of the lengths of the projective prewellorderings would be $\aleph_\omega$; but the funny thing is that it's a really deep theorem of Jackson that, in the AD world, everything is crystallized out. So questions that are trivialized in ZFC are no longer trivial. So questions like cofinalities of successor cardinals, in ZFC we know the answer to that, but in the AD world, that's entirely open. You have two different ways of analysing ordinals: you can take the lengths of prewellorderings calibrated by their definability, and then you try to calibrate them in the $\aleph_\alpha$s. It's a really hard problem. The theorem of Jackson is that in the AD world of $L(\mathbb{R})$, if you take the supremum of the lengths of the projective prewellorderings, it's $\aleph_{\epsilon_0}$. So there's incredible structure there. And he's calculated which of those cardinals are regular. And only ω-many of them are regular. So that's why, if you move out to a model of Choice, the most that the supremum of the projective prewellorderings can be is $\aleph_\omega$. There are just not enough regular cardinals from the perspective of the AD world. But we don't know if that can be realized. We only know you can get two. So in the AD world, because it's a ZF world, questions that are trivialized in ZFC are not trivialized. And when you try to match them, just the problem ... Look at the cardinals in the AD world, the ordinal cardinals (forget about the non-ordinal cardinals, which are the cardinalities of quotients by equivalence classes, which is a whole other aspect of cardinality). Look at the ordinal cardinalities. Well in ZFC, what do we do with ordinal cardinals? We ask about what's the cofinality of a cardinal and so forth; these all become really nontrivial questions in the AD world. So just to give you an illustration. Here is a natural conjecture that I believe is open. Take the simplest setting: $\text{AD} + V = L(\mathbb{R})$. We're supposed to know all about that, because we've lined it up with large cardinals, we have two machines for analysing it. So let's step back and just be set theorists. We know there are successor cardinals that are singular: $\omega_3, \omega_4, \ldots$, those are all singular with cofinality ω_2. All right—well, let's look at the cofinality function on successor cardinals. Is it nondecreasing?

DB: You mean...?

HW: If $\kappa^+ < \lambda^+$, is $\text{cf}(\kappa^+) \leq \text{cf}(\lambda^+)$? It's open.

JK: That's a pretty basic question.

DB: How far up do you have to get...?

HW: Well, I think that up to $\aleph_{\epsilon_0}$, it's true.

DB: I see.

HW: Well, because that whole machinery starts to degrade. So the kind of target problem for that analysis in the AD world was to prove that regular cardinals are measurable below Θ. So that was the holy grail problem for that kind of analysis from below. But then Steel brought in a completely different body of techniques, the fine structure theory, and proved that regular cardinals are measurable, by a completely different method. But that doesn't answer the cofinality question. The Jackson analysis would get the answer by an exhaustive analysis of the combinatorics associated with a cardinal. But it turned out that there was another perspective that answered the holy grail problem and didn't force you to answer that question. So it's a theorem from AD in $L(\mathbb{R})$ that regular cardinals less than Θ that are uncountable are measurable. And then when you bring in a more sophisticated body of techniques ... Steel's analysis leveraged the fact that in $L(\mathbb{R})$ you have the convergence of two machines: you have the determinacy machine and the fine-structure machine. But the fine-structure machine exploits the smallness of $L(\mathbb{R})$. $L(\mathbb{R})$ is a very small model, so there's not an opportunity in that setting for really exotic behaviour. When you shrink down to a really simple setting, things have to be nice. But I introduced another technique, through AD^+, and it's now an AD^+ theorem that regular cardinals less than Θ are measurable. And you might say: what's the interest of that? Well, $\mathrm{AD}_{\mathbb{R}}$ implies AD^+. So if you want to strip away AD^+, assuming DC, then $\mathrm{AD}^+ + \mathrm{DC}$ implies regular cardinals are measurable; and it's a fusion of two different methodologies. But in $L(\mathbb{R})$ we still don't know whether the cofinality function is nondecreasing. So we have all this rich machinery, but it's wide open. There are no independence results here; $L(\mathbb{R})$ is canonical once you have large cardinals. And yet there are fundamental structure properties of $L(\mathbb{R})$ that we just don't know the answer to. And then these connect with questions in ZFC about the effective continuum hypothesis. We don't know the answers. These look like hard questions.

AM: Following this discussion, I have some questions. We have this idea for forcing axioms. We have MM, and now we want higher forcing axioms, so we tinker, consider one of the things you get from $\mathbb{P}_{\max}$. Is there any axiom to generalize that?

HW: Well, one way to ask that question—and that's very much in play now—$\mathbb{P}_{\max}$ is a methodology by which you can start with an AD^+ model, an AD_{R} and Θ regular model, and you can get $\mathrm{MM}^{++}(\mathfrak{c})$. So if you look at the usual consistency proof of $\mathrm{MM}^{++}(\mathfrak{c})$, you need 2^{κ}-supercompact, and AD_{R} and Θ regular is below Woodin limit of Woodins, so it's a different kind of model. One question you can ask—and Steel has asked this—is there a generalization of axiom $(*)$ to ZFC? In other words, $\mathbb{P}_{\max}$ is about $L(\mathcal{P}(\omega_1))$, the $(*)^+$ axiom is an axiom about $L(\mathcal{P}(\mathbb{R}))$, that's what you get with AD_{R} and Θ regular—is there a natural generalization of that to ZFC? Somehow, you start with an AD^+ model, you produce the ZF model over which you force ZFC and whatever: it's a question whether there's a way of getting ...

<u>**AM:**</u> No, I mean * * * we know that we want an iteration for a higher forcing axiom …

<u>**HW:**</u> Okay, so that's a different question. I was asking whether there's a generalization of $\mathbb{P}_{\max}$ beyond $L(\mathcal{P}(\omega_1))$. So you're asking whether there's a methodology to get a forcing axiom that gives you insight into $\mathcal{P}(\omega_2)$ or $\mathcal{P}(\omega_3)$ and so forth. And presumably that axiom's going to be incompatible with Martin's Maximum.

<u>**AM:**</u> Right.

<u>**HW:**</u> Because it's got the continuum big.

<u>**AM:**</u> Right.

<u>**HW:**</u> That's a different question.

<u>**AM:**</u> No, it's not different, because for a higher forcing axiom we don't want to say that I have a strong part of PFA or MM … maybe it is inconsistent with what I want. So there is no reason for us to stick to AD. Maybe I need some part of AD plus something else …

<u>**HW:**</u> But isn't that kind of what I was asking, which was: is there a way of producing a canonical model, possibly starting with an AD model, in which I have the $(*)$ axiom, so $\mathfrak{c} = \aleph_2$, but I have rich combinatorics above? It may not be a forcing axiom but maybe something else. I'll give you the example which I spoke on at the VIG, which is this exotic model—possibly a vacuous theorem—about Martin's Maximum. So if we take the standard intuition about Martin's Maximum, ω_2 should be supercompact in some sense. It has all the properties of supercompact. All right, well: remember I said earlier that if δ is supercompact then there are exotic models with δ-cover and -approximation which are not V up to δ. So if you translate that to the MM^{++} problem, the natural conjecture would be that if MM^{++} holds, there should be an inner model with ω_2-cover and -approximation, corresponding to a version of δ, but it should not be V: it shouldn't contain H_{ω_2}. Now, δ supercompact implies that there's an inner model with δ-covering and approximation which is not V_δ. If you take the analogy seriously, then MM^{++} might imply that there's an inner model of V with ω_2-cover and -approximation which does not contain H_{ω_2}. Well, then you have the Viale–Weiß theorem that, assuming MM^{++} or just PFA, if you have an inner model with ω_2-cover and -approximation, well, let's strengthen the assumption: not just that it can't contain H_{ω_2}, but that ω_2 is strongly inaccessible in the model. Well, then ω_2 is strongly compact in the model. That feature holds in all the known ZFC models of MM^{++}. So you could naturally conjecture: MM^{++} just implies that there's an inner model of V with ω_2-cover and approximation in which ω_2 is strongly inaccessible. But now there's an exotic new model—possibly from a vacuous assumption. So from a ZF hypothesis a little bit beyond Reinhardt—well, quite a bit beyond Reinhardt but below super-Reinhardt—I showed that you can force a ZFC model in which

you have all the large cardinals you want, you have MM^{++}, and if you have any inner model with ω_2-cover and -approximation, it contains H_{ω_2}. So that conjecture is false. Now, this could be a vacuous theorem, because the ZF hypothesis I'm using could be inconsistent. So the feeling, I would say, is that if there's a generalization of $(*)$ to MM^{++}, that model is surely going to have this property—that there's no inner model with ω_2-cover and -approximation where ω_2 is strongly inaccessible. So that's something that would have to be achieved, I would say. So maybe there are exotic models. It's really hard to produce a model of ZFC in which there's no inner model with a degree of cover and approximation that's not somehow trivial, because remember if δ is supercompact, there is a nontrivial inner model of V with δ-cover and -approximation.

AM: But if we assume many Woodin cardinals, it might be that we can construct such a model from models of determinacy ...

HW: Yeah, maybe. But if there is an exotic model of MM^{++} whose origin is in AD^+, then that becomes a really interesting question, because how strong is that AD^+ model? If you look at the usual consistency proof for MM^{++}, then you would naturally conjecture that MM^{++} for partial orders the size of the continuum is very strong, because the usual consistency proof would need 2^κ-supercompact. So you would naturally conjecture that it's stronger than superstrong. But then you have the $\mathrm{AD}^{\mathbb{R}}$ and Θ regular proof that does that for you. So if there's a determinacy-based proof that gets the consistency of MM^{++}, that would challenge the idea that MM^{++} is equiconsistent with supercompacts .... I'm not convinced that MM^{++} is as strong as the community currently accepts. And there's a reason. Maybe this is going to be too technical, but remember inner model theory really changes at supercompact. We know the extender models don't work; you have to be in this other hierarchy. But take the intuition that MM^{++} is equiconsistent with supercompacts—that deeply held intuition. All right: let's change the problem: what about MM^{++} and a proper class of I_0 cardinals; is that equiconsistent with supercompact plus a proper class of I_0 cardinals? In other words, it should be that if MM^{++} thinks ω_2 is kind of supercompact, then you would expect that MM^{++} plus a proper class of whatever cardinals should be equiconsistent with a supercompact plus a proper class. But when you look at it from that perspective, the inner model theory approach looks hopeless—unless the ω_2-cover and -approximation theorem holds. That would do it all, because ω_2-cover and -approximation would inherit the cardinals from V. Now, you have the difference between a strong compact and a supercompact, but let's not worry about that. So if the conjecture that MM^{++} implies that there's an inner model of ω_2-cover and -approximation in which 'ω_2 is strongly inaccessible' holds, you could envision a combinatorial, uniform proof that handles all of these conjectures. Because that inner model of ω_2-cover and -approximation has the strong compact, maybe you tweak it and get the supercompact, but then all the other large cardinals flow from V into the model, by cover and approximation. You have a uniform proof. But if MM^{++} doesn't prove that, that whole scenario collapses. And we have evidence that MM^{++}

doesn't prove that. So either this ZF hypothesis, super-Reinhardt, is inconsistent or this scenario doesn't work.

DB: So, we were talking about the ZF hypotheses possibly being inconsistent. How do you view Farmer Schllutzenberg's results on I_0?

HW: So I see it as opening a door that these hypotheses are consistent. But beyond that . . .

DB: Because they're calibrated by ZFC large cardinals?

HW: Well, I just don't—I see the possibility. Now, I'm a big fan of dichotomies. So up to a year ago, the dichotomy to me was: the Ultimate-L conjecture was syntactic. If the Ultimate-L conjecture, the formal Ultimate-L conjecture, is true, it would render inconsistent these large cardinal hypotheses in ZF, versus the alternative. Now, because of technical developments over the last year, I would predict a different dichotomy: that is, either the Ultimate-L conjecture is true as originally stated, or it's not true and...What Farmer's theorem opens the door for was the following: if Reinhardts and super-Reinhardts and whatever are consistent, how does that fit with the ZFC world? So if Reinhardts and super-Reinhardts and all of that are consistent, one option could be: choice fails high up in the universe, and these are exotic features high up. But let's not go there. Let's assume it's a ZFC world. So how is the consistency of AD reconciled in the ZFC world? Well, ZFC plus large cardinals gives you canonical models of determinacy: $L(\mathbb{R})$ and so forth. So what Farmer's result opens the door for, and actually shows at the level of $V_{\lambda+2}$, is that the ZFC world can provide a canonical inner model where you do have a j from $V_{\lambda+2}$ to $V_{\lambda+2}$, which is inconsistent with choice. So maybe there are generalizations of $L(\mathbb{R})$ inside V where the super-Reinhardt cardinals and so forth exist. Before Farmer's result, I would have thought—and it was the basis of some conjectures I made in early papers—because I thought that if you had an elementary embedding from $V_{\lambda+2}$ to $V_{\lambda+2}$, it would prove that $L(V_{\lambda+1})^\sharp$ exists. This is a generalization of Kunen. So I really thought that I_0 would imply that $L(V_{\lambda+1})^\sharp$ exists. You know, j from L to L implies $0^\sharp$. It seemed a plausible conjecture that if you had the embedding from $V_{\lambda+1}$ to $V_{\lambda+1}$ it would give the sharp, and that would suggest that $V_{\lambda+2}$ to $V_{\lambda+2}$ is inconsistent in ZF. So Farmer proved that not only was that conjecture wrong, it was completely wrong. So now that opens the door for natural inner models inside V, in the ZFC world, we have these ZF large cardinals. Okay, but that runs against the Ultimate-L conjecture. So maybe this is evidence that the *syntactic* form of the Ultimate-L conjecture is actually false, and that in the ZFC world—maybe it's an Ultimate-L endowed with large cardinals where you see these generalizations of $L(\mathbf{R})$, where you have super-Reinhardts and whatever. It feels like a richer picture to me. I tend to describe it thus: in the Ultimate-L perspective, there's a sledgehammer approach. If the syntactic form of the Ultimate-L conjecture is true, end of story. You have no choice, right? Because the syntactic form of the Ultimate-L conjecture says that if there are large cardinals in V, then there's an inner model with δ-cover and

-approximation that satisfies V = Ultimate-L. So you have an ultimate generalization of L that provably exists in close proximity to V. Well, why isn't V that model? It would be a little bit like knowing $0^\sharp$ is inconsistent yet trying to argue that V is not L. So the ultimate-L conjecture is kind of like saying that the analogue of $0^\sharp$ is not consistent. And then you're left with this generalization of L that exists in close proximity to V. It seems to me that you have no choice. But now there's a slightly different picture emerging—the sledgehammer is not true. There's a more nuanced picture. And that comes with a whole separate set of issues. Without getting into technical things, it's hard to say. There's something called an iteration hypothesis that everyone accepted. It's the hypothesis that grounds all inner model constructions—a particular form of UBH. And everyone thought this would be proven. And last summer I showed it was false from I_0. So that kind of, though it didn't have to force me to change my opinion, it was like a wake-up call. So I've kind of pivoted away from the Ultimate-L conjecture. It turns out that, once you see that UBH is false, there's a revision of that hypothesis which that proof can't rule out. And, with generous assumptions about how you build the inner models, the scenario is that this refined UBH hypothesis is equivalent to the existence of Ultimate-L in close proximity to V, under appropriate large cardinal hypotheses. So then everything changes to this refined iteration hypothesis. It's a closure property on V, and if you have this closure property, then Ultimate-L exists in close proximity to V, and therefore arguably V is Ultimate-L, but there are models that aren't closed enough, and it's in these envelopes or fragments where super-Reinhardts and things exist. To me, conceptually, the whole problem with super-Reinhardts and Berkeleys and all of that was: where do they exist? How are you ever going to authenticate their consistency? Now, they might exist in V, but then you're scrapping choice; that seems like an unfortunate option. But there was no candidate, I thought, for generalizations of $L(\mathbb{R})$. It would be like AD—or imagine $AD_{\mathbb{R}}$. You conceive of $AD_{\mathbb{R}}$ and you're trying to debate whether $AD_{\mathbb{R}}$ is consistent. Now, if AD holds, it holds in $L(\mathbb{R})$, so it's reasonable to expect that large cardinals will imply that AD holds in $L(\mathbb{R})$. But that doesn't work for $AD_{\mathbb{R}}$, because in $AD_{\mathbb{R}}$ the strategies are sets of reals; $AD_{\mathbb{R}}$ does not hold in $L(\mathbb{R})$. So how would you, in a ZFC perspective, ever authenticate $AD_{\mathbb{R}}$? Well, it turns out that there is a canonical inner model: there's a generalization of $L(\mathbb{R})$ which is combinatorially defined. You just can't prove that the construction always converges; it does with large cardinals. And if the construction does converge ... well, let me state a theorem and make it more unambiguous. In V you look at inner models of AD, containing the reals. Now they may all line up: if you have two models of of AD containing the reals, maybe the $\mathcal{P}(\mathbb{R})$ of one is just an initial segment of the other, in which case you get a maximal model. But we know from large cardinals that's not going to be the case. So the question is: you try to align models of AD to the point where you see divergent models—that's past the least possible model of $AD_{\mathbb{R}}$, by a theorem of mine. So, if there is an inner model of AD containing $AD_{\mathbb{R}}$, there's a least one, so there's a canonical one, and you can combinatorially define it as a generalization of $L(\mathbb{R})$. And you come to embracing the consistency of $AD_{\mathbb{R}}$ by a deeper analysis. It's not like you point to the model where you're going to check whether $AD_{\mathbb{R}}$ holds. You just know that if there is an inner model of $AD_{\mathbb{R}}$, here's the construction of the model and it's unique. So now we look at Reinhardt and super-Reinhardt, and where do you look? And it doesn't look like there was any candidate for it, if you initialize with the idea that if you have an elementary embedding from $L(V_{\lambda+1})$ to $L(V_{\lambda+1})$, that implies that $L(V_{\lambda+1})^{\sharp}$ exists, that just seems to rule out any way of getting up to $V_{\lambda+2}$ to $V_{\lambda+2}$. But that intuition was just completely false. That's my takeaway—well, that's Farmer's theorem, but now it opens the door for canonical models inside V, inside a ZFC model, where you have these exotic ZF large cardinals.

AM: So you think there might be a natural model for $V_{\lambda+3}$ to $V_{\lambda+3}$?

HW: Yeah, there could be. And there are natural candidates. So it could be—maybe you have to wean Ultimate-L or something, these ZF things that one might think are inconsistent are actually consistent, and they're authenticated not by their occurrence in V but by their occurrence in natural fragments of V, just as we authenticate AD not by asserting that AD is true in V, but that it's true in $L(\mathbb{R})$—and large cardinals tell us that it's true in $L(\mathbb{R})$. So that's a way that you can embrace these ZF large cardinals without abandoning the axiom of choice. And that always seemed really unlikely to me, but that intuition was based on a false assumption. Just as I said, forty years ago, we thought determinacy out of large cardinals—that was based on a false assumption, and MM showed us that. And then there was the assumption that there can't be an Ultimate-L because how could you have it? And that was based on a false intuition. So it's just an evolving series of upending what one's thinks is true. It's a small community; we convince ourselves that certain things are true, and we're wrong. It's like the centre of mass just drifts off the true path.

JK: So, I think we're going to ...

HW: You've all been very polite to sit here this long.

JK: Well, the one thing—having known you, Hugh, for a long time, you make us feel that we're part of something really great, and that's something that working away at the technicalities doesn't. But you give us that sense, and it's the greatest thing, isn't it, for us, to think that we're part of one of the great intellectual projects that's ever been formulated.

RM: I think I've found a new description of you. You're an artist.

JK: Okay. [Laughs.]
[Applause.]

Chapter 37
On Multiversism

Penelope Maddy

Abstract This paper isolates a surprising range of multiverse positions, revealing their sometimes-dubious metaphysical underpinnings and demonstrating that the distinction between multiversism and universism is often muddier than it might appear. Setting metaphysics aside, the operative questions cluster around Axiomatic Universism versus Multiversism—should our fundamental theory be couched in terms of sets or in terms of sets and classes?—and Heuristic Universism versus Multiversism—should set-theoretic practice be guided by an intuitive picture of an iterative hierarchy or of a collection of such hierarchies? It concludes that the mathematical benefits of a switch from (extensions of) ZFC to (extensions of) a multiverse theory like Steel's MV have yet to be demonstrated and that a truly metaphysics-free perspective should happily welcome the heuristic benefits of both intuitive pictures.

Faced with an intractable problem, some philosophers employ a singular strategy: their idea is to dismiss or dissolve the problem in some way, as opposed to meeting it head on with a proposed solution. Multiversism in many of its varieties has recently emerged as a popular application of this approach to the continuum problem: CH is true in some worlds, false in others; the effort to settle it one way or the other is misguided, a pseudo-problem. My goal here is to examine a few actual and possible implementations of this strategy, but first, in the interest of transparency, I should acknowledge a tendency toward the opposing view of CH. At least for now, I believe that one of the most pressing questions in the contemporary foundations of set theory is how to extend ZFC (or ZFC + LCs) in mathematically defensible ways so as to settle CH (and other independent questions) and to produce a more fruitful theory. It seems best to begin by sketching in my own peculiar take on this opposing view. Then, with this as backdrop, I'll turn to multiversism.

P. Maddy (✉)
Department of Logic and Philosophy of Science, University of California, Irvine, CA, USA
e-mail: pjmaddy@uci.edu

S. Arbeiter and J. Kennedy (eds.), *The Philosophy of Penelope Maddy*,
Outstanding Contributions to Logic 31, https://doi.org/10.1007/978-3-031-58425-1_37

37.1 Background

Many defenders of the project of settling CH and other independent questions—I'm going to stop saying that, but please take the broader project as understood—many defenders of the project do so in familiar philosophical terms. Perhaps most familiar is a straightforward embrace of an objectively existing realm of abstracta, of sets, in which CH is determinately true or false (though we don't currently know which). This approach is reasonably attributed to Gödel in his famous paper 'What is Cantor's continuum problem?' (Gödel, 1947/64), but others have proposed alternative versions (including me in my youth (Maddy, 1990)). Years ago, Paul Benacerraf presented a particularly compelling formulation of an epistemological challenge to Gödel—how can we ordinary physical humans come to know anything at all about this non-spatiotemporal, acausal abstract realm? (Benacerraf, 1973)—and since then, that challenge has evolved through various changes in thinking on general epistemology without losing any of its force (see, e.g., Field, 1989, pp. 25–30).

While I think this challenge is serious—my youthful self once tried to meet it—I now see a more fundamental worry about views like Gödel's as logically prior to Benacerraf's. To see this, consider the various defenses that set theorists, past or present, have offered for new assumptions or methods: Cantor was out to extend a theorem on representing functions by trigonometric series; Dedekind wanted characterizations of algebraic ideals or real numbers that don't depend on particular algorithms, series, sequences, or representations, that yield more general theories in abstract algebra and topology; Zermelo hoped to codify the power and reach of the nascent theory of sets, including its key foundational role,[1] while avoiding the threat of contradictions; defenders of large cardinal axioms from inaccessibles to extendables appeal, among other things, to the iterative conception of set, citing their axioms as attempts to maximize the height of V, and to the role of large cardinals as a tool for comparing consistency strengths; contemporary supporters of determinacy hypotheses mount a compelling case so varied and complex that I won't try to summarize it here, but one striking fact is that the determinacy axiom $AD^{L(R)}$ is implied by many, perhaps all, 'natural' theories of sufficiently high consistency strength.[2] These mathematical gains are seen as good reason to adopt and to axiomatize the theory of sets.

At this point, Benacerraf challenges Gödel to explain how these methods manage to track his seemingly inaccessible realm of abstracta, given what we take to be our limited human abilities. This is where I think something has gone wrong. In addition to the mathematical considerations that support a given hypothesis, the Gödelean metaphysics is imposing an additional extra-mathematical condition, the need to check that the hypothesis squares, somehow, with a viable epistemological account that tracks the demanding metaphysics. Starkly put, it seems to me that the mathematical considerations, strong as they are, should be enough by themselves.

---

[1] See (2017), (2019), for what's intended by here by 'foundational role'.

[2] This is what Koellner calls 'overlapping consensus'. See Koellner (2011), §4.5.

Benacerraf wants Gödel to supply a non-trivial epistemology; I think he shouldn't be positing a metaphysics that requires one.[3]

Of course, Gödel takes on (his version of) the Benacerrafian challenge, not my worry. He begins with 'intuitive' or 'intrinsically necessary' claims—for example, axioms that 'force themselves upon us as being true' (Gödel, 1964, p. 268). These are so-called 'intrinsic' considerations. In addition, as is well-known, he recognizes a second mode of justification in terms of an axiom candidate's 'success'—'extrinsic' considerations. The nature of this success depends on the analogy with natural science. First, there are 'verifiable consequences', that is:

> consequences demonstrable without the new axiom, whose proofs with the help of the new axiom, however, are considerably simpler and easier to discover, and make it possible to contract into one proof many different proofs. (Gödel, 1947/64, p. 182/261).

Presumably, the analogy here is with observations and experimental results. More ambitiously,

> there might exist axioms so abundant in their verifiable consequences, shedding so much light upon a whole field, and yielding such powerful methods for solving problems … that, no matter whether or not they are intrinsically necessary, they would have to be accepted at least in the same sense as any well-established physical theory. (Ibid.)

Here we have the counterpart to theory formation and testing.

This approach introduces a range of well-honed tools from the philosophy of science: confirmed predictions, theoretical unification, explanatory power and scope, and so on. In his paper 'Mathematical evidence' (Martin, 1998), Tony Martin gives an impressive example from his defense of the consistency of ZF + AD[4]: he proves a general theorem about sets of recursive degrees from AD (the Cone Lemma), then checks that various implications of this theorem for particular sets of degrees were all provable from ZFC. Martin summarizes: 'I take it to be intuitively clear that we have here an example of prediction and confirmation' (ibid., p. 224).

By itself, this science/mathematics analogy solves neither the neo-Benacerafian problem nor my prior methodological problem: the mechanism of intuition still needs to be spelled out and an external requirement is still being imposed. But even if these could be set aside, there'd be other points of tension in its treatment of extrinsic considerations. Some concern the analogy itself: for example, in what sense is proving a theorem from ZF comparable to observing an experimental outcome? Others concern the corresponding scientific principles that purportedly validate their mathematical cousins: for example, is inference to the best explanation really a reliable rule? Are simple theories more likely to be true? But, in addition to these quite legitimate concerns, I think there's another that's particularly salient from a set-theoretic point of view.

---

[3] A quick note on terminology. I use 'metaphysics' for the study of what there is, what the world includes, what's real and what it's like, as opposed to 'epistemology', theory of knowledge, the study of how we come to know these facts. 'Ontology' is the branch of metaphysics focused on the question of what entities there are.

[4] That is, ZF plus the Axiom of Determinacy.

Consider again the history of set theory. In the aforementioned algebraic work, Dedekind introduced sets simply because he thought representation-free definitions and a more abstract approach would generate good mathematics. (He was right about that.) Zermelo, who pioneered something like the intrinsic/extrinsic distinction well before Gödel (in Zermelo, 1908), was happy to lump all his extrinsic considerations together, appealing simply to their promotion of 'productive' mathematics, with no thought to fitting them into an analogy with natural science. Similarly for contemporary defenses of large cardinals as a natural measure of consistency strength or $AD^{L(R)}$ as an apparent consequence of any natural increase along that hierarchy of those strengths.

Now perhaps there's a way to fit all these into a natural-scientific model, but the more one explores the development of set theory, the more efforts to retroactively impose broad categorizations from the philosophy of science begin to recall Wittgenstein's remark about efforts to translate mathematics into the notation of *Principia Mathematica*:

> ... if tables, chairs, cupboards, etc. are swathed in enough paper, certainly they will all look spherical in the end. (*RFM* III §53, p. 185).

Once again, as a general point, it seems to me that an axiom or concept or method with clear mathematical benefits that blocks no significant mathematical avenues shouldn't be held to any additional philosophical standard. But the specific challenge for those reliant on a science/mathematics analogy is a choice between two unappealing options: reject some seemingly legitimate mathematical justifications because they resist inclusion under the 'science-like' umbrella or take 'science-like' to be so amorphous as to have no actual bite.

Of course, few people these days would explicitly endorse a Gödelian position on the subject matter of set theory. Still, I would argue that many contemporary positions face analogous challenges. This is easy to see for accounts that identify an alternative abstract subject matter for set theory: de-re structures, second-order logical truths, modal facts. Each such account owes an explanation of how our actual set-theoretic methods—including the extrinsic ones—manage to track the relevant metaphysics. They may not face the immediate roadblock of the Benacerraf objection to human knowledge of abstract mathematical *objects*, but this doesn't absolve them of their epistemic duties,[5] nor does it justify the imposition of inappropriate burdens.

[5] Sometimes this obligation is well masked, as for example, Leng's apparently straightforward fictionalism ends up turning on contentful modal facts akin to those of Hellman. See Leng's essay in this volume and my response.

Views that appeal to concepts or meanings may sound less loaded at first blush, but when it becomes clear that these aren't the sorts of concepts or meanings studied by cognitive science or linguistics, questions of precisely what they are and how they're known tend to arise anew.[6] Exploring Gödel's conceptualism, Martin writes:

> this concept of set seems perfectly objective ... We did not create it, though we have singled it out as something to study. (Martin, 2005, p. 362).

Recognizing the epistemological question immediately raised by this objectivity, he continues:

> the objectivity of the concept does not, however, imply that we lack epistemic access to it ... We *understand* the concept and we can *explain* it ... When we work in the mathematical subject of set theory, we can think about what we are doing as finding out what is implied by the concept. (Ibid.)

No one versed in the Benacerraf challenge is likely to be satisfied by this appeal to 'understanding'. It seems more a label for the problem than a solution: how does our human understanding manage to track the truth about this objectively existing concept? Peter Koellner appeals to Gödel's endorsement of 'the belief that for clear questions posed by reason, reason can also find clear answers'.[7] Koellner intends to place the epistemological onus on the human faculty of Reason and rational intuition, as understood by Charles Parsons (Parsons, 2008, Chap. 9), but I doubt that Parsons himself would claim to have given a fully satisfactory account of how this faculty functions.

By now, I think it's clear that each of these proposed subject matters—abstract ontology, truth value realism, second-order validity, modal facts, conceptualism, meanings—requires a non-trivial epistemology of one sort or another. Robust Realism is the label I've used for any such reading of set-theoretic practice. Robust Realisms are subject both to a neo-Benacerrafian epistemological challenge and to my own challenge that the additional burdens imposed by any such epistemology would be inappropriate, that considerations of mathematical advantage alone should be enough. Still, despite the fatal shortcomings of Robust Realism in all its forms, I've suggested that a form of realism is still available (Maddy 2011).

What I call Thin Realism avoids the need for an extra-mathematical epistemology by positing a metaphysically minimal realm of sets that simply *are* the sort of thing accessible by ordinary set-theoretic methods. Though its sets are still objective, nonspatiotemporal, and acausal, this de-natured style of realism is both controversial and unlikely to satisfy anyone with Robust Realistic inclinations—not least of all because it is, I argue, a mere linguistic variant of an alternative that bypasses all matters of truth or reference or ontology or metaphysics more generally.

---

[6] Steel presents a special case: he does appeal to 'meaning'—as in 'the meaning we assign to the word "set"', and to 'synonymy', as in 'translations' from L_{MV} to $L_{\in}$—and not in any ordinary or scientific senses of these terms, but Meadows and I have argued that these appeals are actually inessential (Maddy & Meadows, 2020, §5).

[7] The quotation comes from Gödel (1961), p. 381, cited in Koellner (2006), p. 198.

I call this equivalent position Arealism rather than anti-realism or nominalism because it isn't based on a principled rejection of abstracta for one philosophical reason or another. Instead, I imagine an idealized inquirer beginning from ordinary observations, then generalizing, proposing and testing theories, etc., and soon, in the course of these empirical investigations of the world, coming to develop mathematics, eventually even pure mathematics. How should this inquirer understand this new practice? Its methods often differ from her previous empirical approaches—experiments versus proofs, for example—still, the two practices do share logic and means-ends reasoning. Is mathematics just more of what she's been doing all long, seeking truths about an objective subject matter like the rest of science, or is it something different, immensely useful, but of a different kind? The Thin Realist takes the former stance, the Arealist takes the latter—not out of a prior distaste for abstracta, like an anti-realist or a nominalist, but as a sober-minded assessment of what her methods do and do not support. For her, set theory isn't in the business of describing a subject matter or discovering truths; it's a practice of devising and developing a list of axioms to meet a range of mathematical goals.[8] My claim is that neither Thin Realism nor Arealism is either required or forbidden, that either way of speaking is acceptable. To keep things simple, I speak as an Arealist from now on.[9]

The most compelling objection to views like Arealism is based on the role of mathematics in natural science: crudely put, if you believe what science says, and science says there's a function that does so-and-so, then you must believe in functions.[10] I've argued that this depends on an over-simplified understanding of how mathematized

---

[8] Obviously, one goal of contemporary mathematics is to provide tools for natural science, but this is just one among many.

[9] Candidly, of the two ways of speaking, I've always found Arealism more congenial; Thin Realism was my attempt to spell out what seemed to me equally legitimate (if to me foreign) ideas of John Burgess and John Steel (see Maddy (2011), pp. 60–62). For a more principled reason to prefer the Arealist idiom in the present context, Schatz [ta] argues persuasively that Thin Realism is a convenient way of speaking in ordinary mathematics, but that Arealism is best for foundational discussions. Finally, I've indicated elsewhere (Maddy 2015, pp. 247–248) that Thin Realism (as I intend it) stands or falls with the objectivity of mathematical depth, and while I tend to believe that depth is objective, I have no strong argument to that effect (other than assuming we all agree that 'group' is an objectively better concept than various logically consistent others in the same vicinity and pointing to this as an example). Arealism has the advantage that it would survive the demise of objective depth.

[10] This argument rests on the assumption that 'science' takes the same attitude toward all the 'objects' it employs when even the most cursory glance reveals that mathematics often appears in contexts of explicit idealization, not intended to be understood as literal truth. (Think, e.g., of fluid dynamics.) With an eye toward the determinateness of CH, our best hope for a literal use is in applications of continuum mathematics to the theory of space–time, but, alas, that hope is soon dashed by physicists' clear acknowledgment that no one yet understands the structure of space–time in the small. In fact, I think this picture of mathematized science as a homogeneous expanse of physical entities and mathematical entities (and perhaps entities hovering somewhere in between) is misleading (see, e.g., 'Reply to Burgess', 'Reply to Leng', this volume). A picture closer to the truth sees pure mathematics as an abundant store of abstracta, some of which—when we're lucky, often with a course of non-trivial tinkering (see the exchange with Wilson in this volume)—can serve to model a worldly situation well enough for a particular purpose. To function in this capacity, the mathematical structures needn't exist in some abstract sense, they need only be described in a

science works, but getting into this here would distract from our main quarry. For now, let me just suggest that the Arealist has the means to turn this objection away.

Finally, then, returning to our central theme, though both the Robust Realist and the Arealist take CH seriously, they do so on starkly different terms. For the Robust Realist, there's a determinate truth value waiting to be discovered; for the Arealist, in contrast, set theory isn't a practice that concerns itself with truth values. Given that CH is independent of the current axioms, the remaining question for the Arealist, what's yet to be discovered, is simply this: is there or is there not an additional axiom candidate with both sufficient mathematical advantages to merit adoption and the specific ability to prove or disprove CH? This presents at least as real a question as the Robust Realist's determinate truth value, but it's a question about 'productive' mathematics, to use Zermelo's term, not about the features of an independent subject matter.

37.2 A Partial Taxonomy of Multiversisms

Given this particular way of legitimizing the pursuit of a decision on CH, let's now consider the opposing idea, that this pursuit is somehow or other based on an illusion. One traditional family of such views includes if-thenism: the only real question is which conclusions follow from which assumptions; as far as ZFC and CH are concerned, that question has been answered; analogous questions can be asked and perhaps answered about the relations of other assumptions to CH. To this, the traditional version adds—these implication relations are all on a par, mathematics is just a matter of what follows from what—which is often intended to deflect tough questions from intrusive philosophers. Of course, everyone knows that this isn't strictly correct, that mathematicians prefer, for example, Peano arithmetic, the group axioms, and ZFC over hosts of random collections of assumptions. Any viable form of if-thenism would need to be enhanced with an account of how and why some assumptions are preferred to others. I bring this up only to note: if this 'how and why' is spelled out in terms of the kinds of mathematical advantages we've touched on here, the resulting Enhanced If-thenism more or less coincides with Arealism.[11] But now, multiversism.

To focus discussion, let's limit consideration to multiverses in which every world thinks ZFC + LCs, but worlds differ on CH. An advocate of such a view might consider CH to have been settled, though not by a yae or nay:

> On the multiverse view … the continuum hypothesis is a settled question; it is incorrect to describe the CH as an open problem. The answer to CH consists of the expansive, detailed knowledge set theorists have gained about the extent to which it holds and fails in the multiverse, about how to achieve it or its negation in combination with other diverse set-theoretic properties. Of course, there are and will always remain questions about whether

useful way by our mathematical theory—the Arealist's stock-in-trade. (See Maddy (1997), §§II.6–7, (2008), (2011), Chap. 1.).

[11] See (2022). Also the 'proofism' of Maddy and Väänänen (2023), pp. 48–49.

> one can achieve CH or its negation with this or that hypothesis, but the point is that the most important and essential facts about CH are deeply understood, and these facts constitute the answer to the CH question. (Hamkins, 2012, p. 429)

To the Arealist and her compatriots on CH, this may sound like giving up and declaring victory, but for the multiversist, perhaps discretion is the greater part of valor. Either way, it seems to me there are (at least) three overlapping strains of multiversism, the first of which is ontological, or metaphysical more generally.

37.2.1 Metaphysical Multiversism

The most straightforward Metaphysical Multiversism is a simple ontological variety. Hamkins puts it this way:

> The multiverse view is one of higher-order realism – Platonism about universes ... a realist position asserting actual existence of the alternative set-theoretic universes into which our mathematical tools have allowed us to glimpse. (Hamkins, 2012, p. 417)

Hugh Woodin describes his target, generic multiversism, in less dramatic, but still ontological-sounding terms[12]:

> Let the *multiverse* (of sets) refer to the collection of possible universes of sets. ... The multiverse is the *generic-multiverse* if it is generated from each universe of the collection by closing under generic extensions ... and generic refinements. (Woodin, 2011, p. 14)

Such views are often paired with an analogous universist position to set up the contrast, for example, in Neil Barton's taxomony:

> Universism: There is (up to isomorphism) just one maximal unique universe of set theory, and it contains all the sets. Every set-theoretic sentence has a definite truth value in this universe. (Barton, 2021, p. 22)

> Multiversism: There are multiple equally legitimate universes of set theory, and no-one universe is especially privileged. (Ibid., p. 49)

This is Ontological Multiversism, with Ontological Universism posited as its foil, perhaps the most common approach in the literature.

Koellner complicates the discussion by introducing a second pair of contrasting terms: 'pluralism' and 'non-pluralism'. Joan Bagaria and Claudio Ternullo apparently take these as more or less equivalent to 'multiversism' and 'universism' (Bagaria & Ternullo 2023, p. 3), but Koellner (2019) distinguishes them.[13]

Recounting the history of CH, he first characterizes pluralists as those who.

[12] This isn't to say that Woodin and Hamkins are in agreement. See §IV, below.

[13] Interestingly, the earlier Koellner (2011), p. 1, seems to collapse this distinction. Non-pluralism 'either implies or is a consequence of the fact that there is an objective mathematical realm', thus linking non-pluralism directly to ontology. Similarly, pluralism 'either implies or is a consequence of the fact ... that there is no objective mathematical realm'. (In this context, Koellner distinguishes pluralism and non-pluralism at various levels.).

> ... maintained that the independence results effectively settle the question by showing it *had no answer*. On this view, one could adopt a system in which, say CH was an axiom and one could adopt a system in which ~CH was an axiom and that was the end of the matter – there was no question as to which of two incompatible extensions was the 'correct' one. (Koellner, 2019, p. 2)

'Correct' here apparently means 'true':

> The *non-pluralists* (like Gödel) held that the independence results merely indicated the paucity of our means for circumscribing mathematical truth. (Ibid.)

For the pluralist, while one choice or another for extending ZFC 'has its advantages with respect to certain aims there is no "fact of the matter" as to which one is correct' (Koellner, 2011, p. 49). The pluralist/non-pluralist disagreement over whether or not CH has a truth value then links up with ontology and multiversism like this:

> One way of providing a foundational framework for [pluralism] is in terms of the multiverse. On this view there is not a single *universe* of set theory but rather a *multiverse* of legitimate candidates ... none of which can be said to be the 'true' universe. (Koellner, 2011, p. 25)

So Koellner's pluralism might be called Truth-value Multiversism with the option of Ontological Multiversism as an add-on.

(Let me pause for a moment to note that Barton employs a surprisingly austere version of pluralism: 'we should tolerate multiple competing *theories* of sets, and not necessarily identify one as privileged' (Barton, 2021, p. 75). Presumably, the corresponding non-pluralism would hold that there is one privileged theory of sets, and this theoretical distinction would be expected to map directly to the truth-value or ontological distinction. As it happens, though, Barton draws no such easy correspondence. I come back to this in §III.)

Parallel to the varieties of Robust Realism in §I, there's also the possibility of a multiverse, not of worlds or truth values, but of concepts of set. Hamkins combines this sort of view with his Ontological Multiversism—'the *multiverse view* ... holds that there are diverse distinct concepts of set, each instantiated in a corresponding set-theoretic universe' (Hamkins, 2012, p. 416)—but this could stand alone as Conceptual Multiversism. My point is that there are metaphysical multiversisms of ontological, truth-value, and conceptual varieties, and probably more, as well as many varieties of each—but all of them qualify as versions of Robust Realism because they posit an objective, abstract subject matter that demands a non-trivial epistemological account of how our methods manage to track it. So these views suffer from the same shortcomings that afflict universe versions of Robust Realism: both Benacerraf's challenge and my own methodological concern, as well as the Extrinsicness challenge[14] for views that call on the science/mathematics analogy (as they often do).

All this metaphysical talk is heady enough, but it's natural to wonder how these philosophical maneuvers affect the practice of set theory. Given that these positions

[14] That is, might the requirement that all extrinsic justifications be 'science-like' end up either undercutting some legitimate mathematical considerations or being toothless?

ascribe a distinctive subject matter to the discipline, it's perhaps also natural to expect its fundamental theory to consist of assertions about that subject matter. Just as the universist takes ZFC (ZFC + LCs) to describe her unique universe (or truths or concept or ...), one might expect the multiversist to propose an alternative collection of axioms that describe, instead, the multiverse. Which brings us to the second of the three routes I hope to trace into multiverse thought.

37.2.2 Axiomatic Multiversism

Fitting as it might seem for a Metaphysical Multiversist to opt for a foundational theory of universes and sets to replace a theory, like ZFC + LCs, focused exclusively on sets, the only available list of first-order axioms in a multiverse language is due to John Steel, whose intent is not at all metaphysical. Steel's aim is to collect together all 'natural' candidates for extensions to ZFC + LCs, where 'natural theories' are those 'considered by set theorists, because they [have] some set-theoretic idea behind them' (Steel, 2014, p. 157). In this way, he hopes to compare these theories on 'a neutral common ground', and ultimately perhaps 'to decide whether some such theory is preferable to the others' (ibid., p. 165).[15]

Steel's idea is to locate this common ground by devising a theory of universes strong enough to prove the existence of a realization for each such candidate but weak enough not to include realizations for theories that would 'lose information' (ibid., p. 167). ('Natural' theories like ZFC + V = L or ZF + AD are represented by inner models of worlds, not by worlds.) Observing in practice that each known 'natural' theory is equiconsistent with a theory of the form ZFC + LCA, for some large cardinal axiom LCA, and that the proofs of these equiconsistencies work via inner models and forcing extensions, Steel's axioms end up codifying something very like Woodin's multiverse—closed under forcing extensions and refinements—except that he adds the simplifying assumption of Amalgamation: any two worlds share a forcing extension. This guarantees that Steel's system MV is sound and complete for its intended set of countable models. (There can be no such axiomatization of Woodin's multiverse.)[16]

[15] As indicated, this language of 'comparing' and 'deciding' if one or another theory is 'preferable' comes from Steel (2014). More recently, Steel has called his earlier talk of comparison 'misleading': 'L_{MV} is not a device for comparing the merits of one of those theories with another, it is a way to make more visible their underlying unity' (p. 28 of Steel's essay in this volume). See footnote 18 for more.

[16] See Maddy and Meadows (2022) for discussion and proofs. Here let me make one correction. Our paper calls Amalgamation into question on the grounds that it limits the array of worlds and hence of theories—how do we know we haven't ruled out some good candidates for extending ZFC + LCs? Though I wasn't able to see this during our exchanges while the paper was being drafted, Steel subsequently got through to me that this is only a problem if you focus on the worlds and wonder whether important ones might be missing. What matters for Steel isn't the worlds but the theories: as long as the axioms of Extension and Refinement are satisfied, all the live candidates will be represented. So, as long as Amalgamation is consistent with Extension and Refinement,

The system MV offers a second route to multiversism—it could be offered as an alternative to ZFC—but Steel himself is not an Axiomatic Multiversist. His central concern is the status of CH: his axioms are designed to represent the range of viable theoretical options; his leading question is whether one (or some) of those options can be seen to be preferable to others. In other words, he isn't tempted by the typical multiversist move of declaring the question of CH to have no answer, true in some worlds, false in others; he wonders whether one (or some) of those worlds—for him, theories represented by those worlds—stands out from the rest.[17] So, he reasons: what if there were 'a distinguished reference world ... an individual world that is definable in the multiverse language' (Steel, 2014, p. 168)? Woodin observed that if there were such a definable world, it would be unique, and it would be contained in all the other worlds; it would be what's now called the 'core' of the multiverse, defined as the intersection of all worlds, assuming that intersection is itself a world. Steel argues that this core, if it exists, would represent the sought-after preferred theory,[18] that CH would emerge as the legitimate question whether it is true or false in the core.

Of course, since Steel's (2014) paper, Toshimichi Usuba (2017) proved that the multiverse does have a core, ultimately assuming an extendible (Usuba, 2019). For Steel, this means that his foray into multiverse thinking, his willingness to entertain his multiverse of possible theories, has returned the answer that CH is a legitimate question, after all.[19] The focus, then, settles on determining the theory of the core, a question on which MV + LCs is largely uninformative.[20] At that point, the problem devolves into one familiar to the Arealist: how to find good mathematical reasons to adopt axioms beyond ZFC + LCs (whether for the theory of the core or for the theory of V makes no difference). Steel regards V = Ult-L as a promising candidate, but however that may be, multiverse thinking has done its job and drops out of the discussion.

---

which it is, there's no danger of significant omissions. The axiom is then amply justified by its mathematical benefits (e.g., the completeness theorem and the translation function from L_{MV} to $L_{\in}$). (Steel would say that Amalgamation is justified because it's provably true on the interpretation of L_{MV} determined by the translation function t. See footnote 53 of Maddy and Meadows (2020) for discussion of this approach.).

[17] It's worth noting Hamkins's remark that 'there is no reason to consider all universes in the multiverse equally, and we may simply be more interested in the parts of the multiverse consisting of universes satisfying very strong theories' (Hamkins, 2012, p. 436). It's possible in principle that the search for more interesting universes in his multiverse could end up answering to the same mathematical considerations as the analogous search in Steel's multiverse, but this seems unlikely in practice.

[18] This is a non-trivial argument, sketched in the paragraph overlapping pp. 508–509 below. On Steel's more recent understanding of his project (see footnote 15), we might say that the unification of the candidate theories by L_{MV} allows us to see the theory of the core as preferable because it 'includes' all the other theories (in the sense of that same passage on pp. 508–509).

[19] That is, if there's good reason to include an extendable among the large cardinals posited in our fundamental theory. I leave this important question aside.

[20] MV + LCs is MV plus the assumption that each world has large cardinals.

Still, a Metaphysical Multiversist could help themselves to MV + LCs or some variation as a mathematically explicit expression of their multiversism, though for Hamkins, in particular, MV + LCs wouldn't suit, because his multiverse is much more generous than Woodin's or Steel's generic versions.[21] In fact, though Hamkins does offer axioms of his own, he doesn't appear to be an Axiomatic multiversist: his axioms don't take the form of a live first-order alternative to ZFC + LCs or MV + LCs; he doesn't propose a multiverse replacement for our universist foundational theory. He sees himself, rather, as offering a fruitful way of thinking about the set-theoretic project:

> The mathematician's measure of a philosophical position may be the value of the mathematics to which it leads. Thus, the philosophical debate here [over the subject matter of set theory] may be a proxy for: where should set theory go? Which mathematical questions should it consider? (Hamkins, 2012, p. 440)

He goes on to describe two lines of mathematical thought inspired by the multiverse perspective, namely, the modal logic of forcing and set-theoretic geology. Neither of these is stated in a multiverse language, but in ordinary $L_\in$, and proved in ZFC or an extension thereof. This practice-oriented approach brings us to the third and final route to multiversism.

37.2.3 Heuristic Multiversism

Consider the role of the iterative hierarchy. For the Metaphysical Universist, it's the abstract subject matter of set theory, but the intuitive picture it provides—of sets arranged in a succession of ranks, taking all possible subsets at each step, extending as far as possible into the transfinite—this picture has also served, for roughly a century now, as a prolific inspiration for all manner of set-theoretic progress: reflection principles for large cardinals, for example, or closure principles for forcing axioms. The intuitive picture provides an effective heuristic for set-theoretic practice, quite independently of any metaphysical status. In an analogous way, a Heuristic Multiversist might hold that the intuitive picture of an array of such hierarchies plays the same sort of role and does a better job of suggesting new mathematical developments—like those Hamkins cites—and promising more. Instead of replacing ZFC, this version of multiversism simply recommends a different attitude toward it, as a theory shared by many distinct universes. This substitution, the Heuristic Multiversist claims, is a more effective inspiration, leading to new developments of ZFC, new theorems of ZFC, and perhaps even new axioms candidates!

For an early version of this approach, consider Kenneth Kunen's highly influential textbook treatment of forcing (Kunen, 1980). In an appendix to the chapter introducing the method, he lays out several ways of understanding what's going on in applications of forcing. His own preference, the method of countable transitive

[21] See §IV.

models, is familiar and effective for most purposes, but there are alternatives like the Boolean-valued models deployed in Jech (2003). Another is what Kunen calls 'forcing over V', the purely syntactic approach with the definable forcing relation. Consider Kunen's intuitive description of how it works. With 'M' understood as the countable ground model in the c.t.m. approach, he writes of 'forcing over V':

> We may think of this approach as putting ourselves (in V) in the place of the M-people of the c.t.m. approach; so we are making up names for, and talking about, objects in some generic extension of V which does not exist at all (to us). (Kunen, 1980, p. 234)

In a later rewriting of the book, he puts this even more colorfully:

> We, living in V, are in the position of the M-people ... We can dream about some ideal universe $V[G] \supset V$ and make use of these dreams to motivate our discussion of [the forcing relation], but these dreams cannot be mentioned in rigorous proofs of the theorem about [the forcing relation]. (Kunen, 2013, p. 281).

With his 'dreams', Kunen perfectly and delightfully illustrates how a multiverse intuitive picture can serve as an effective heuristic in a context that remains squarely within $L_{\in}$ and ZFC, in other words, without sacrificing Axiomatic Universism. So, Heuristic Multiversism isn't entirely new!

Now recall Steel's multiverse project. We've seen that his foray into multiverse thinking leads him to focus on the core, and in particular, to propose, at least tentatively, the axiom $V = C$.[22,23] We now recognize this development as a particularly dramatic application of Heuristic Multiversism, but to probe a bit deeper, consider a question left hanging in our earlier discussion: what argument is there for identifying V with the core? Answer: C is contained in every world, every world is a generic extension of C.[24] Here Kunen's dreamers have reappeared: the resident of C has imaginative access to every world of the multiverse, and that's enough. Steel's Heuristic Multiversism leads to the core and returns us to Axiomatic Universism with a new axiom in $L_{\in}$, , a candidate for addition to ZFC + LCs. (Would that it were more informative!).

---

[22] As mentioned earlier, Steel goes on to suggest that $V = \text{Ult-L}$ is true in C, but I leave that further step aside here.

[23] As it happens, this axiom candidate also emerged in the context of set-theoretic geology under the name 'Ground Axiom'. See Reitz (2007).

[24] See Maddy and Meadows (2020), pp. 146–147. [Added later: In his interview in this volume, Woodin appears to echo this general idea in the context of his own version of the generic multiverse: pressed on why the core is 'preferable' to other worlds, he replies, 'All models of the generic universe are just generic extensions of the root ... so all truth is reducible to the root' (Woodin, p. 468).].

37.3 Multiversism and the Arealist

With this taxonomy in place, imagine the Arealist presented with the possibility of multiversism. Keeping the focus on CH, given the Arealist's belief—tentatively, of course—that it remains a real question, she would be expected to resist the characteristic multiversist dissolution of the problem, and for this reason, expected to side with the universist. Surprisingly, though, the various characterizations of the universist/multiversist dichotomy we've reviewed don't deliver this outcome so straightforwardly. Consider, then, her attempts to locate herself on one side or the other of this dichotomy.

Starting from the ontological Hamkins/Barton version from §II, she sees herself nowhere; she's no more inclined to assert the existence of 'just one maximal unique universe' than of 'multiple equally legitimate universes'. Koellner's choice, again from §II, between pluralism and non-pluralism is even more befuddling. On the one hand, she agrees with his pluralist that there is 'no fact of the matter' about CH; on the other, she holds out hope that it isn't a matter of indifference which of CH and ~CH we adopt, siding this time with the non-pluralist. Another of Koellner's characterizations focuses the point more precisely: for the pluralist,

> although there are *practical* reasons that one might give in favour of one set of axioms over another—say, that it is more useful for a given task—, there are no *theoretical* reasons that can be given. (Koellner, 2011, p. 1)

In contrast, for the non-pluralist, 'theoretical reasons *can* be given for new axioms' (ibid.). For the Arealist, the distinction between practical and theoretical reasons is obscure; being 'more useful for a given task' is just the sort of thing that counts as a good reason, full stop. So, as perhaps would be expected, she sees herself in neither Truth-value Multiversism nor Truth-value Universism. To fully inhabit the Arealist's perspective, we must eschew all metaphysics—ontology, truth values, concepts, meanings—and attend exclusively to mathematical needs and benefits.

In this respect, the prospects are better for Axiomatic Multiversism. One version of this idea would be the austere pluralism of Barton, mentioned in passing above, to repeat: 'we should tolerate multiple competing *theories*, and not necessarily identify one as privileged' (Barton, 2021, p. 75). The corresponding Axiomatic Universist would presumably hold that one theory is properly preferred, for now, presumably ZFC + LCs. Here the Arealist would seem a clear fit for team universist, but Barton imposes unexpected obstacles. One is predictable, if onerous: the universist …

> … thinks that every statement of first-order set theory has a determinate answer, [so] there is a unique privileged set theory; the true one. (Barton, 2021, p. 75)

Obviously, the Arealist won't endorse the idea of a 'true' set theory; only slightly less obviously, she sees the question—can the first-order statement CH be satisfactorily settled one way or the other?—and thus the question of determinateness, as open. Further sealing the case, Barton imposes a second, purely methodological condition on his universist: she cannot 'tolerate the use of different theories' (ibid.). As

Barton himself points out, no reasonable set-theoretic practitioner could accept this restriction:

> In virtue of the kinds of ways in which she might be ignorant, [she] should tolerate many different theories of sets (at least for now) (ibid.).

If this Axiomatic Universist can't endorse, for example, the development of both ZFC + LCs + V = Ult-L and ZFC + LCs + MM, then the Arealist once again falls between stools.

Barton aside, there's a more straightforward version of the Axiomatic Universism/Multiversism distinction, namely, the choice between ZFC + LCs and a multiverse theory like MV + LCs. A less specific contrast would be between those who hold that our fundamental theory should be written in the language of sets, $L_{\in}$, and those who hold that it should be written in the language of worlds and sets, something like L_{MV}. Here, at last, the Arealist's position is straightforward, siding with the Axiomatic Universist in both cases (tentatively, as always). The trouble this time is that no multiversist known to me takes the other side, the Axiomatic Multiversist side, in either of these forms. So the Arealist once again finds herself outside any live universist/multiversist debate.

This leaves the final option, Heuristic Universism versus Heuristic Multiversism. Here, perhaps, the distinction will gain some real traction, recommending that the practice be guided by contrasting intuitive pictures. There can be little doubt of the effectiveness of the iterative picture, and Kunen's, Steel's, and Hamkins' examples demonstrate the initial promise of a multiverse picture. Still, any hope of pinning the Arealist down to one argumentative position over the other is once again dashed; from her point of view, the question is, why choose? Metaphysics aside, nothing precludes exploiting now one, now the other of these intuitive guides to maximal effect. One needn't forgo large cardinals to pursue set-theoretic geology or vice versa! So, the opportunistic Arealist remains stubbornly unclassified.

37.4 Digression: Two Contrasting Multiversisms

That's pretty much the story I want to tell here, but my tight focus on the supposed conflict between universism and multiversism has obscured the contrast within multiversism between the broad Hamkins multiversism and the narrower Woodin-Steel generic multiversism. It seems irresponsible to leave the subject without at least a nod in this direction, besides which I think we're now in a position to make a few potentially fresh observations from the Arealist point of view. The story begins with their respective motivations.

For Hamkins,

> The most prominent phenomena in set theory has been the discovery of a shocking diversity of set-theoretic possibilities. Our most powerful set-theoretic tools, such as forcing, ultrapowers, and canonical inner models, are most naturally and directly understood as methods of constructing alternative set-theoretic universes.

> This abundance of set-theoretic possibilities poses a serious difficulty for the universe view … We have a robust experience in these worlds, and they appear fully set theoretic to us. The multiverse view … explains this experience by embracing them as real, filling out the vision hinted at in our mathematical experience, that there is an abundance of set-theoretic worlds into which our mathematical tools have allowed us to glimpse. (Hamkins, 2012, p. 418)

Hamkins motivation is clear: he's out to explain mathematical experience, the phenomenology of set-theoretic practice. He does so by positing a metaphysics that takes that experience at face value, 'placing no undue limitations on what universes might exist' (Hamkins, 2012, p. 437). Indeed, he suggests that 'there seems no reason to restrict attention only to ZFC models' and goes so far as to include second-order arithmetic as 'set-theoretic in a sense' (ibid., p. 436). So my focus here on multiverses where the worlds agree on ZFC + LCs, though differing on CH, is already a distortion, a taming, of Hamkins' position.

Compare now what Woodin describes as the source of his generic multiverse:

> The refinements of Cohen's method of *forcing* in the decades since his initial discovery of the method and the resulting plethora of problems shown to be unsolvable, have in a practical sense almost compelled one to adopt the generic-multiverse position … The argument that Cohen's method of forcing establishes that the Continuum Hypothesis has no answer, is implicitly assuming the generic-multiverse conception of truth. (Woodin, 2011, pp. 16–17)

Here Woodin sees forcing in particular as 'almost compelling' us to generic multiversism and thus as a threat to the determinacy of independent statements like CH. If forcing extensions were all obviously unintended, not viable, they wouldn't present such a problem, but they aren't. As Steel points out, they often represent 'natural' theories, legitimate contenders for extensions of ZFC + LCs; this line of thought is what inspires Steel's sharper generic multiverse with Amalgamation. For Woodin and Steel, the goal isn't to explain mathematical experience but to explore the possibility that forcing has undermined the universist's picture of the world of sets.

Enter the Arealist. How does the contrast between Hamkins generous multiverse and the Steel/Woodin generic multiverse look from this perspective? On the motivational contrast just noted, the Arealist's sympathies lie with the generic version: her potential interest in multiversism, like theirs, arises from a concern with CH, not from an analysis of set-theoretic experience. Motivations aside, the core disagreement here is perhaps most often seen as metaphysical—how many set-theoretic universes are there?—but these intramural debates between forms of Metaphysical Multiversism strike the Arealist as ill-posed.[25] She does care about axiomatics, but Hamkins hasn't offered a clear alternative to ZFC and its extensions, and as far as heuristics are concerned, those cited by Hamkins seem equally available to Woodin and Steel,[26] with Steel's case for V = C as an added attraction. In sum, then, once the metaphysical questions have been set aside, the generic multiverse appears to provide a more promising heuristic, at least for now. Still, just as the Arealist remains open to

[25] In particular, the Arealist sees no basis for Hamkins's inference from what happens in model theory, as a branch of mathematics conducted within ZFC and its extensions, to a conclusion about an ambient, pre-theoretic metaphysics.

[26] E.g., the modal logic of forcing and set-theoretic geology are both studies of forcing models.

the heuristic benefits of both Universism and Multiversism, she's perfectly happy to explore both versions of Multiverism. Once again, without the illusory metaphysical contrast, there's no need to rule out any helpful heuristic.

37.5 Conclusions

I hope to have shown that multiversism is no one doctrine, indeed no one type of doctrine, that many versions are uncomfortably metaphysical, and that the contrast between universism and multiversism is considerably muddier than the either-or presentations suggest. Looking past the metaphysics, different questions take center stage: should our fundamental assumptions involve just sets or sets and worlds; should our axioms be formulated in $L_\in$ or something like L_{MV}; should our fundamental theory be ZFC and its extensions or something like MV and its extensions; which ways of thinking, which heuristics can deliver the best mathematics? Some of these are questions of Axiomatic Universism versus Axiomatic Multiversism. Multiverse axioms have indeed been offered, most prominently by Steel and Hamkins, but no one appears to be arguing that replacing ZFC + LCs with such a theory would leave us better off mathematically: Steel's line of multiverse thinking ends up endorsing the extension of ZFC + LCs with V = C (back to Axiomatic Universism), and Hamkins employs his multiverse thinking to suggest new theorems in ZFC + LCs (Heuristic not Axiomatic Multiversism). For now, Axiomatic Multiversism is unmotivated; Axiomatic Universism remains dominant.

On the question of heuristics, Heuristic Universism, in the form of the iterative picture, has proved itself dramatically since Zermelo's time, but we've seen that Heuristic Multiversism has made a recent showing, as in Kunen's 'forcing over V', Steel's case for V = C, and Hamkins' set-theoretic geology. For the Arealist, these new developments are unalloyed good news, the rise of a new guide to good mathematics. In the end, it's only the warring metaphysical camps that make us think we need to choose between two effective heuristics; this impediment removed, the Arealist can reap the rewards of both. Shouldn't anyone be pleased with a shiny new tool added to the toolbox?

So, what morals should we draw? Obviously, I think we can and should approach matters of set-theoretic method as Arealists, unencumbered by metaphysical delicacies, helping ourselves to a full range of heuristics. From this perspective, metaphysics-free multiversism has its heuristic successes: (1) developments like set-theoretic geology, (2) the possibility of an Axiomatic Multiversism promoting MV + LCs or something like it as our fundamental theory, and especially, (3) Steel's employment of Heuristic Multiversism to mount a case for V = C. Returning at last to CH, as long as our language remains $L_\in$ and our theory ZFC and its extensions, the Arealist's understanding of CH survives: is there or is there not an axiom candidate with both sufficient mathematical advantages to merit adoption and the specific ability to prove or disprove it? In this form, CH remains a legitimate question. And finally—until someone seriously proposes a multiverse alternative to ZFC + LCs as our fundamental theory—until that happens—my considered opinion on universism versus multiversism is that we'd be better off retiring the whole debate.[27]

References

Bagaria, J., & Ternullo, C. (2023). Steel's programme: Evidential framework, the core and ultimate-L. *Review of Symbolic Logic, 16*, 788–812.

Barton, N. (2021). Indeterminateness and 'the' universe of sets: multiversism, potentialism, and pluralism. In M. Fitting (ed.), *Selected topics from contemporary logics*. College Publications, pp. 105–182.

Benacerraf, P., & Putnam, H. (Eds.). (1983). *Philosophy of mathematics* (2nd ed.). Cambridge University Press.

Benacerraf, P. (1973). Mathematical truth. Reprinted in Benacerraf and Putnam [1983], pp. 403–420.

Feferman, S., Parsons, C., & Simpson, S. (Eds.). (2010). *Kurt Gödel: Essays for his Centennial*. Cambridge University Press.

Field, H. (1989). Introduction. In: *Realism, mathematics and modality*. Basil Blackwell, pp. 1–52.

Gödel, K. (1947/64). What is Cantor's continuum problem?. Reprinted in Benacerraf and Putnam [1983], pp. 470–485, and Gödel [1990], pp. 176–187, 254–270.

Gödel, K. (1961). The modern development of the foundations of mathematics in the light of philosophy. In Feferman et al. (1995), pp. 375–387.

Gödel, K. (1990). *Collected papers*, vol. II. In: S. Feferman et al. (eds.). Oxford University Press.

Gödel, K. (1995). *Collected works*, vol. III. In: S. Feferman et al. (eds.). Oxford University Press.

Hamkins, J. (2012). The set-theoretic multiverse. *Review of Symbolic Logic, 5*, 416–449.

Jech, T. (2003). *Set theory*. Springer.

Koellner, P. (2006). On the question of absolute decidabilty. *Philosophia Mathematica*. Reprinted in Feferman et al. (2010) (pp. 189–225).

Koellner, P. (2011). Independence and large cardinals. In: E.N. Zalta (ed.), *The Stanford encyclopedia of philosophy* (Summer 2011 Edition). https://plato.stanford.edu/archives/sum2011/entries/independence-large-cardinals/

Koellner, P. (2019). The continuum hypothesis. In: E.N. Zalta (ed.), *The Stanford encyclopedia of philosophy* (Spring 2019 Edition). https://plato.stanford.edu/archives/spr2019/entries/continuum-hypothesis/

Kunen, K. (1980). *Set theory: An introduction to independence proofs*. Elsevier.

Kunen, K. (2013). *Set theory*. College Publications.

Leng, M. [ta]. Enhanced "if-thenism", fictionalism, and realist anti-platonism. This volume.

Maddy, P. (1990). *Realism in mathematics*. Oxford University Press.

Maddy, P. (1997). *Naturalism in mathematics*. Oxford University Press.

Maddy, P. (2008). How applied mathematics became pure. *Review of Symbolic Logic, 1*, 16–41.

Maddy, P. (2011). *Defending the axioms*. Oxford University Press.

Maddy, P. (2015). 'Afterward', special issue on Mathematical Depth. *Philosophia Mathematica, 23*, 245–248.

Maddy, P. (2019). What do we want a foundation to do? In S. Centrone, D. Kant, & D. Sarikaya (Eds.), *Reflections on the foundations of mathematics: Univalent foundations, set theory and general thoughts* (pp. 293–311). Springer.

[27] Thanks to Joan Bagaria, Juliette Kennedy, Claudio Ternullo, and Georgio Venturi for their kind invitation to address their conference in Barcelona and to the audience there for a stimulating discussion. I'm also grateful to John Steel for his thoughts on §II.2, to Toby Meadows and Jeffrey Schatz for illuminating discussions related to the addendum to footnote 24, and to Adam Chin, Chris Misch, and Evan Sommers for helpful comments on an earlier draft.

Maddy, P., & Meadows, T. (2020). A reconstruction of Steel's multiverse project. *Bulletin of Symbolic Logic, 26*, 118–169.
Maddy, P., & Väänänen, J. (2023). *Philosophical uses of categoricity arguments.* Cambridge University Press.
Maddy, P. [ta]. Reply to Leng. This volume.
Maddy, P. (2017). Set-theoretic foundations. In: A. Caicedo, J. Cummings, P. Koellner, & P. Larson (eds.), *Foundations of mathematics: Essays in honor of W. Hugh Woodin's 60th birthday.* American Mathematical Society, pp. 298–322.
Maddy, P. (2022). Enhanced if-thenism. In: *A plea for natural philosophy* (pp. 253–293). Oxford University Press.
Martin, D. A. (1998). Mathematical evidence. In H. G. Dales & G. Oliveri (Eds.), *Truth in mathematics* (pp. 215–231). Oxford University Press.
Martin, D. A. (2005). Gödel's conceptual realism. *Bulletin of Symbolic Logic* 11, 207–224. Reprinted in Feferman et al. (2010), pp. 356–373.
Parsons, C. (2008). *Mathematical thought and its objects.* Cambridge University Press.
Reitz, J. (2007). The ground axiom. *Journal of Symbolic Logic, 72*, 1299–1317.
Schatz, J. [ta]. Arealism, Thin realism, and the problem of objective depth. In: C. Antos-Kuby, N. Barton, & G. Venturi (eds.), *Palgrave companion of philosophy of set theory.* Palgrave.
Steel, J. (2014). Gödel's program. In J. Kennedy (Ed.), *Interpreting Gödel* (pp. 153–179). Cambridge University Press.
Steel, J. Generically invariant set theory. This volume.
Usuba, T. (2019). Extendible cardinals and the mantle. *Archive for Mathematical Logic, 58*, 71–75.
Usuba, T. (2017). The downward directed grounds hypothesis and very large cardinals. *Journal of Mathematical Logic, 17.*
Van Heijenoort, J. (ed.) (1967). *From Frege to Gödel.* Harvard University Press.
Wilson, M. The flight from application. This volume.
Wittgenstein, L. (1939/40) (1978). Part III, *Remarks on the Philosophy of Mathematics.* Revised edition, G.E.M. Anscombe, trans., G.H. von Wright, R. Rhees, G.E.M. Anscombe (eds.) (pp. 143–221). MIT Press.
Woodin, H. (2011). The continuum hypothesis, the generic-multiverse of sets, and the Ω-conjecture. In J. Kennedy & R. Kossak (Eds.), *Set theory, arithmetic, and foundations of mathematics* (pp. 13–42). Cambridge University Press.
Woodin, H. A conversation with Hugh Woodin. This volume.
Zermelo, E. (2010). *Collected works*, vol. I. In: H.-D. Ebbinghaus & A. Kanamori (eds.). Springer.
Zermelo, E. (1908). 'A new proof of the possibility of a well-ordering. S. Bauer-Mengelberg, trans., reprinted in van Heijenoort (1967), pp. 183–198, and Zermelo (2010), pp. 121–159.

Name Index

S. Arbeiter and J. Kennedy (eds.), *The Philosophy of Penelope Maddy*, Outstanding Contributions to Logic 31, https://doi.org/10.1007/978-3-031-58425-1

Subject Index

S. Arbeiter and J. Kennedy (eds.), *The Philosophy of Penelope Maddy*, Outstanding Contributions to Logic 31, https://doi.org/10.1007/978-3-031-58425-1

The manufacturer's authorised representative in the EU is Springer Nature Customer Service Centre GmbH, Europaplatz 3, 69115 Heidelberg, Germany. If you have any concerns regarding our products, please contact ProductSafety@springernature.com

Printed and bound by CPI Group (UK) Ltd, Croydon, CR0 4YY
01/10/2026
02225909-0017